Herbal Plants
and their
Applications in
Cosmeceuticals

Herbal Plants
and their
Applications in Cosmeceuticals

Kuntal Das M Pharm, PhD

Assistant Professor
Department of Pharmacognosy and Phytochemistry
St. John's Pharmacy College
Vijayanagar, Bangalore

Affiliated to
Rajiv Gandhi University of Health Sciences

CBS Publishers & Distributors Pvt Ltd

New Delhi • Bengaluru • Chennai • Kochi • Kolkata • Mumbai
Hyderabad • Jharkhand • Nagpur • Patna • Pune • Uttarakhand

Herbal Plants and their **Applications** in **Cosmeceuticals**

ISBN: 978-81-239-2296-6

Copyright © Author and Publisher

First Edition: 2014

Reprint: 2018, 2021

Published by Satish Kumar Jain and produced by Varun Jain for

CBS Publishers & Distributors Pvt Ltd

4819/XI Prahlad Street, 24 Ansari Road, Daryaganj, New Delhi 110 002, India
Ph: 011-23289259, 23266861, 23266867 Website: www.cbspd.com
Fax: 011-23243014   e-mail: delhi@cbspd.com; cbspubs@airtelmail.in
Corporate Office: 204 FIE, Industrial Area, Patparganj, Delhi 110 092

Ph: 011-4934 4934   Fax: 011-4934 4935   e-mail: publishing@cbspd.com;
                                               publicity@cbspd.com

Branches

- **Bengaluru:** Seema House 2975, 17th Cross, K.R. Road, Banasankari 2nd Stage, Bengaluru 560 070, Karnataka
  Ph: +91-80-26771678/79   Fax: +91-80-26771680   e-mail: bangalore@cbspd.com
- **Chennai:** 7, Subbaraya Street, Shenoy Nagar, Chennai 600 030, Tamil Nadu
  Ph: +91-44-26680620, 26681266   Fax: +91-44-42032115   e-mail: chennai@cbspd.com
- **Kochi:** 42/1325, 1326, Power House Road, Opp. KSEB, Power House, Ernakulam 682018, Kochi, Kerala
  Ph: +91-484-4059061-65   Fax: +91-484-4059065   e-mail: kochi@cbspd.com
- **Kolkata:** No. 6/B, Ground Floor, Rameswar Shaw Road, Kolkata 700014 (West Bengal), India
  Ph: +91-33-2289-1126, 2289-1127, 2289-1128   e-mail: kolkata@cbspd.com
- **Mumbai:** PWD Shed, Gala No. 25/26, Ramchandra Bhatt Marg, Next to JJ Hospital, Gate No. 2 Opp. Union Bank of India, Noorbaug, Mumbai 400009, Maharashtra, India
  Ph: +91-22-66661880/89   e-mail: mumbai@cbspd.com

Representatives

- **Hyderabad** 0-9885175004 • **Jharkhand** 0-9811541605 • **Nagpur**   0-9421945513
- **Patna**   0-9334159340 • **Pune**   0-9623451994 • **Uttarakhand** 0-9716462459

Printed at Glorious Printers, Jhilmil Industrial Area, Delhi, India

to

my beloved
Parents, wife and son

Preface

The global scenario totally changed with the plant derived medicinally useful formulations, drugs and healthcare products, because of National and International market demand. It is estimated that about 70% modern medicines in India are derived from natural products. Medicinal and aromatic plants have played an important role in many ancient traditional system of medication worldwide. Not only that the rich diversity in herbals but also focuses towards the application in pharmaceuticals, agricultural, food industries and even now in cosmeceuticals.

Commercialization of plant base products gaining importance due to the effective therapeutic values and minimal side effects rather than chemical products. But most of the advancements related to herbals and their products are made in this regards either confined to research organizations or revealed scattered way in the research journals. Not much information is available in bound and readily applicable at one place. Focus on that the book entitled *"Herbal Plants and their Applications in Cosmeceuticals"* has made an ultimate effort to fulfill all the requirements about related plants and their products available in market which are used in beautification. Ultimately, the aim of this book is to provide a solid platform for eminent researchers, academicians, students, physicians, traditional medical practitioners, farmers, manufacturers and exporters of herbal products and marketing consultants from all over the world. All kinds of information were compiled together and represented in this book in a simple manner to serve the basic concept to the readers.

This book compiled with five chapters, viz. 1. Introduction, 2. External Parts of Human Body and their Different Complications, 3. Antioxidants, 4. Introduction to Topical Cosmetics and their Evaluations, and 5. Individual Herbs with their Details in Cosmetics Formulations.

The first chapter, "Introduction" covers a few words about herbal plants and their systemic utilization from traditional to modern medicines. Further, a few descriptions have given about cosmetics and their demand in Indian market. Role of herbs in cosmetic preparations, modern formulations and focus of Indian market globally also revealed in this chapter. The second chapter, "External parts of human body and their different complications" covers general description about external parts of human body for easy to understand by all readers. Different complicated diseases related to various parts of body have also described in this section. The third chapter, "Antioxidants" covers general introduction about the same and their role in human body. Various evaluation methods for determination of antioxidant, antiseptic and preservatives has also described in this section. The herbs that are useful for the above same applications have enlisted with their parts used. Chapter four, "Introduction of topical cosmetics and their evaluations" focuses on the various herbal formulations for hairs in terms of conditioner, dyes, shampoos, skins as creams, lotions, eye make-up products, dentifrices, various mouthwash products, toiletries preparations and their various evaluation techniques. Chapter five, topic entitled "Individual herbs with their details in cosmetics formulations", has enlisted about forty-one individual plants with all their details. Likewise this chapter discussed, plant profiles, vernacular names (state wise names), introduction to herbs, distribution, main chemical constituents, medicinal uses, cosmetic uses, pharmacological action, toxicity studies, clinical investigations, market formulations, mechanism of action in particular formulation, method of preparation of formulations, industries concerned with particular products, storage, dosages, patent related information and references. This chapter is the major focus in which all the latest information have gathered from various sources and recent formulations related to plants with their manufacturers in India.

It is expecting that this book will provide up-to-date knowledge with all types of information to the readers and manufacturers who are relating to this field.

This is my first attempt in this new direction, any criticism and suggestions from the readers are always welcome. So that further editions can be rectified.

It is my great pleasure to acknowledge the help from Dr Raman Dang and Dr Sobha Rani RH, from Al-Ameen College of Pharmacy, Dr Lalitha BR, from Government Ayurvedic College, Bangalore, for their valuable suggestions. I feel elysian pleasure to express my deepest sense of gratitude and heartiest gramercy to Dr TN Shivananda from Indian Institute of Horticultural and Research, Bangalore. I express my sincere thanks and respect to Principal, Vice-Principal (Mrs Sobha Rani G), all my colleagues, especially Manjunath U Machale, M pharm students especially Mr Nilesh Kumar Gupta, Anil Kumar Kathiriya, Manan Joshipura, Janak, Syed Ali, Sanjay, Manoj, Nishit Mandal, non-teaching staffs especially Jagadish B, Director General and Deputy Director of my college for their active co-operation and encouragement.

I would like to give special and sincere thanks to my wife Smt Sangita Das and our beloved son master Niladri Das for their heartfelt inspiration to pursue the work in all the situations.

I felt no word to express sense of indebtedness to my parents, whose silent blessings, encouragement, and helping me put my best foot forward in all my endeavors of chasing my dreams in life.

I am most thankful to my in-laws for their constant moral support, love, encouragement, affectionate care as a source of energy.

Lastly I am giving thanks to the CBS Publisher & Distributors, New Delhi, for giving me an opportunity to write this book.

Kuntal Das M Pharm, PhD

Contents

1 | Introduction to Herbal Plants and Cosmetics in Indian Market

INTRODUCTION

Herbs are the plants that are used for culinary, medicinal or fragrant properties since the ancient times. In spite of the recent advances seen in modern medicines, the plants are having important contribution to health care. Among worldwide-distributed medicinal plants, most of all are abundantly distributed in tropical countries. It has estimated that about 25–30% of all modern medicines are directly or indirectly derived from natural herbs. Of late, cosmetic is the branch of science in which natural herbs have been use ever since the beginning of civilization but over the past few decades, focus towards herbal drugs used in the same field has increased expressively. In herbal, especially in cosmetics sector, the current focus not only in Asian countries like India but also in countries like USA, UK, Australia and others.

As per Drug and Cosmetic Act (1940) and Rules (1945), cosmetics can be defined as any article intended to be rubbed, poured, sprinkled or sprayed on, or introduced into or otherwise applied to the human body or any part thereof for cleaning, beautifying, promoting attractiveness, or altering the appearance and includes any article intended for use as a component of cosmetic. Cosmetics include skin care creams, lotions, powders, perfumes, nail polish, eye and facial make-up, hair colors, hair sprays, gel, deodorants, baby products, bath oils and many other type of products. The term Cosmeceutical has created in 1990 and mainly used as skin care products. Even though FDA, and D&C Act does not recognize the term Cosmeceutical, but the cosmetic industry uses this word to refer to cosmetic products that have drug like benefits and marketed as cosmetics. The cosmetic product relies on various vegetable materials, minerals and animal origin having medicinal properties.

The first archaeological evidence of cosmetics usage was in Egypt around 3500 BC. In the Western world, the usage of cosmetics was in the middle ages. In India, Ayurveda is the concept of beauty has an age-old origin. Especially for females, beautification is the main attraction. Several ways are known by the world since centuries. Tilak, Kajal, Alita, Agaru, Chandan, Mehendi that were use as body decorative and to create beauty spots on the chin and cheeks from Vedic period of India. In modern days herbal beauty treatments were carried out throughout India to sensual appeal and to maintain well hygiene.

Fruits, leaves roots, bark, oil extracted from seeds and other various parts of plant have been used directly or in derived form for the preparation of herbal cosmetics that are usually free from side effects. Some of the well known herbs like *Aloe vera*, Amla, Turmeric, Sandal wood, Tulsi, Neem, Ginger, Cucumber, Peach, Apricot and many others which are used traditionally for various herbal preparations in the form of creams, lotions, face packs, soaps and other forms to possesses remarkable antiseptic, antibacterial, antioxidant, protective, soothing, nourishing and beautifying effect on skin. Not only the crude plant extracts but also essential oils obtained from plants like Ajowan oil, Black

1

Pepper oil, Clove oil, Eucalyptus oil, Rose oil and other essential oils are also playing an important role in both medical and cosmetic purpose.

The demand for cosmetic and skin care products consumption gradually increasing due to increasing number of middle class population. Focus on that, many companies have jumped into this area and have launched their exclusive range of herbal beauty care and skin care products. Many foreign companies with their popular cosmetics brands entered to the Indian market and Indian market gaining steady momentum in this sector. According to figures given by the Confederation of Indian Industries (CII), the current size of Indian cosmetic and toiletries, market is approximately US$ 950 million and showing growth between 15 and 20% per annum. Of this, the color cosmetic is the fastest growing sector that captured around US$ 60 million of the total market. The skin care market and the hair care market in India are about US$ 180 million and US$ 200 million respectively, whereas herbal cosmetic market in India captured only US$ 100 million of total market. From the beginning, most consumers are using face creams, antiwrinkle creams, moisturizing lotions, fairness creams, face cleansers and the usages of the same prepared from herbal plants seems to have grown steadily in recent years due to very minimal side effects. People have now become aware of the benefits and positive points of usage of herbal cosmetics rather synthetic cosmetics that causes some skin allergies and even cost aspect.

Today, Indian companies are responsible for the production of a wide range of herbal raw material products and side-by-side manufacturers are also provide their products to big cosmetic brands. So many Indian companies are capable of complying with the global grading system and Indian herbal medicines are produce nowadays to meet International quality standards. Indian herbal product manufacturers are making a wide range of products that include herbal toiletries and herbal cosmetics brushes and cosmetic products on one-end and herbal powder treatments on the other. There are several industries in India involved with herbal cosmetic preparations; some of them are having international focus with respect to import and export, viz. Dabur India, Dew Herb Cosmetics, Himalaya Drug Company, Sri Baidyanath Ayurvedic Bhawan Ltd, The Emami groups, Vicco Laboratories, etc. There are so many recently grown herbal cosmetic industries that all are growing fastly to take up respective position in world market, some of them are Agarwal Herbals, Ayur Herbals, Bajaj Herbals, Daxal Cosmetics Pvt Ltd, Gayatri Herbals, Pearl Naturals, Nature Plus Pvt Ltd, Radico, Reshlon Cosmetics Pvt Ltd, Zigma Herbal Remedies and so on.

The Indian manufacturers have to stress on various standardization processes to establish their products in the global markets. Standardization starting from production of quality materials, analysis of raw materials for authentication, foreign matter, organoleptic evaluation, microscopic examination, extractive values, chromatographic profiles, pesticides residue, heavy metal detection, etc. are necessary for standardization of drugs. Similarly, the standardization methods of medicinal plants and its extracts have great importance in the fields of cosmetics which is emerging as one of the most important segment in the global markets. In order to withstand competition in the global market, it is necessary to create a brand image, especially in cosmeceuticals and natural products. Out of the Rs.12,000 crores industry, Rs.700 crores belong to skincare products and Rs.100 crores for general cosmetics. Our country has a lot of medicinal plants, plants with essential oils and the demand in the overseas markets for its concentrates is growing fast to meet the present and future consumer demand, it is also necessary to set up world standard research and development facilities in this cosmeceuticals sector.

Recent novel approach in herbal cosmetics is nutracosmetics which is an emerging class of health and beauty aid products. Herbal products are mild, biodegradable and have low toxicity profile and to enhance these properties, research is being done in the development of newer approaches, which could improve both the aesthetic appeal and performance of a cosmetic product.

External Parts of Human Body and their Different Complications

The human body has the most complex organization, which is mainly build by trillions of cells and they are the structural and functional unit of the body. Further the aggregation of cells having similar origin, structure and functions and forms tissue. There are mainly four types of tissues, viz. epithelial, connective, muscular and nervous. Different tissues are combine together and perform various particular functions is called as organ and various organs further working in coordination for the particular functions form an organ system. In simple form:

Cell   →   Tissue   →   Organ   →   System
(unit of   (collection   (collection   (collection
life)   of cells)   of tissues)   of organs)

(Fundamental of human body)

As the interest, the chapter has described the external parts of human body, which have the corelation with the herbal cosmetics. It is necessary to know all the people about some information related skin, hair, nails, ear and mouth. They are known as special sense organs in body. People should aware about the application of herbal cosmetics by knowing their anatomy and various disorders.

SKIN

The skin is the membranes lining the body orifices, which contain sensory nerve ending of pain, temperature and touch, protect from injury and from invasion by microbes and mostly it regulate the body temperature. For the average adult human, the skin has a surface area of between 1.5 and 2.0 square meters. The skin holds about 650 sweat glands, 20 blood vessels, 60,000 melanocytes and more than a thousand nerve endings. Skin has three layers: Epidermis, dermis and hypodermis. Between the skin and underlying structures a subcutaneous fatty layer is situated.

The epidermis provides protection from environmental pathogens and serves as a barrier to infections. Epidermis do not contains any blood vessels, and its functions are nourished by diffusion from the dermis. The main type of cells that make-up the epidermis are keratinocytes, melanocytes, Langerhans cells and Merkels cells. Epidermis is divided into several layers where cells are formed through mitosis at the innermost layers. It has five sublayers, viz. *Stratum corneum, Stratum lucidum, Stratum granulosum, Stratum spinosum* and *Stratum germinativum.*

Dermis serves as a location for the appendages of skin. Beneath the epidermis the dermis layer is located which are consists of connective tissue and cushions the body from stress and strain. The layer is tightly connected to the epidermis by a basement membrane. It contains the hair follicles, sweat glands, sebaceous glands, apocrine glands, lymphatic vessels and blood vessels. The dermis is structurally divided into two areas, i.e. a superficial area adjacent to the epidermis is known as papillary region that composed of loose areolar connective tissue and a deep thicker area known as the reticular region that is composed of dense irregular connective tissue.

Finally, the hypodermis which is a subcutaneous adipose layer. Practically, it is not a part of the skin but it lies below the dermis.

The main function of this layer is to attach the skin to underlying bone and muscle as well as supplying it with blood vessels and nerves. It consists of loose connective tissue and elastin. The main cell types are fibroblast, macrophages and adipocytes.

posed laws. Many of these are models derived from fitting of uniaxial experimental data. These models have relatively simple mathematical expressions and can successfully solve some simple mechanical problems.

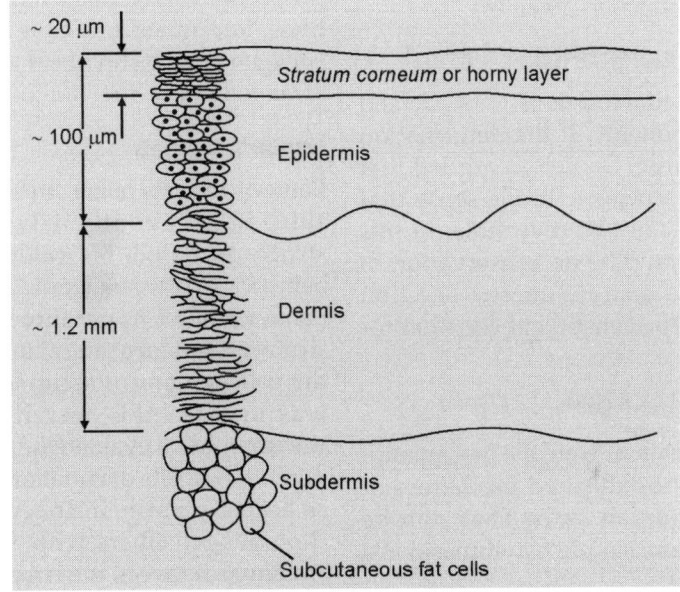

Fig. 2.1: Layers of skin

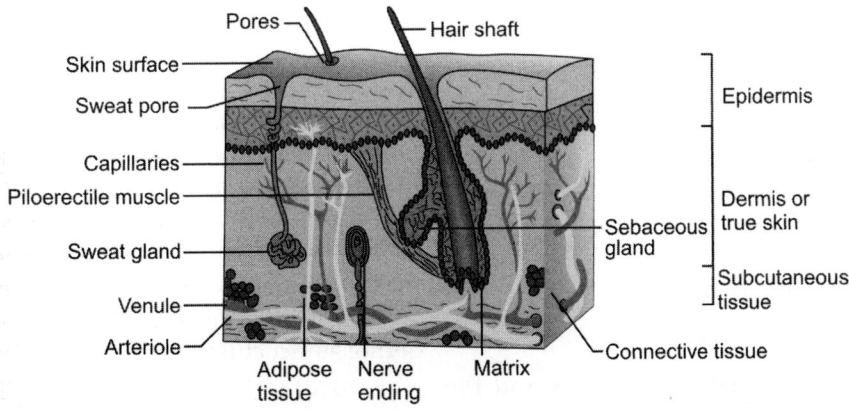

Fig. 2.2: Cross-section of skin

Few of the Mathematical Methods of Human Skin

The stress-strain relationship in human skin is mainly based on the variety of pro-

Danielson's Model

Danielson's model describes the human skin as an elastic membrane. He modeled the mechanical behaviour under two different scenarios. For relatively small deformations, he stated that the skin would be elastic and the strain energy preserved.

The constitutive equation in this is:

$$N^{\alpha\beta} = \delta W/\delta\gamma_{\alpha\beta}$$

Where $N^{\alpha\beta}$ are the stress tensors and W is the strain energy function given by:

$$W = A^{\alpha\beta\lambda\mu}\,\gamma_{\alpha\beta}\,\gamma_{\lambda\mu}\,exp\,(B^{K\delta}\gamma_{K\delta} + C\gamma^{\rho\varphi}\,\gamma_{\rho\varphi})$$

A and B are material constants, C a scalar and γ the strain components. If the deformation were relatively large, Danielson argued that skin would be viscoelastic in the sense that deformations may not be reversible. In this situation there would be no conservation of strain energy and a strain energy function would not exist. The constituent equation for this is:

$$N^{\alpha\beta} = A^{\alpha\beta\lambda\mu}\,\gamma_{\lambda\mu}\,exp\,(B^{K\delta}\gamma_{K\delta} + C\gamma^{\rho\varphi}\,\gamma_{\rho\varphi})$$

Danielson's equations were the first attempt to apply the well-established mechanics of membranes to human skin. They can be used to solve several simple two-dimensional problems.

Tong and Fung's Model

Tong and Fung's model is based on an extensive collection of data from experiments on rabbit skin, they derived a model that gave the best fit to the observed behavior. Their model was also based on the preservation of strain energy, but their strain energy function different to that of Danielson. So this model can be said as the modified method of Danielson's model. The stress-strain relationship is given by:

$$S_{ij} = \delta W/\delta e_{ij}$$

Where S_{ij} are the stress tensors and the strain energy function is given by:

$$W = \tfrac{1}{2}\,(\alpha_1 e_1^2 + \alpha_2 e_2^2 + \alpha_3 e_{12}^2 + 2\,\alpha_4 e_1 e_2) + \tfrac{1}{2}\,c\,exp\,(\alpha_1 e_1^2 + \alpha_2 e_2^2 + \alpha_3 e_{12}^2 + 2a_4 e_1 e_2 + \gamma_1 e_1^3 + \gamma_2 e_2^3 + \gamma_4 e_1^2 e_2 + \gamma_5 e_1 e_2^2)$$

α, a, γ and c are material constants and e the strain components. Both of the above models

need to be aligned to symmetry lines in the human skin. In practice, this means that the coordinate system has to be chosen to coincide with the Langer lines (Langer derived a map of the human body consisting of all the lines/directions in the skin with the least cross-sectional tension. Mechanical properties can be determined by this Langer lines. Recent studies have proved that both collagen and elastin fibers are predominantly arranged in the direction of the Langer lines).

Lanier's Models

Lanier's models relate stress and strain in the skin, based on an analysis of its micro-structure. As per his model the mechanical behavior in a tissue would be the sum of the behavior of all its constituents. Hence, if the structure and mechanical interaction between the collagen and elastin fibers in the skin was known, the overall response of the tissue could be evaluated. Lanier accordingly derived two sets of equations. The first model assumes elastin-induced undulation of the collagen fibers with a high density of crosslinks between the two.

Bischoff's Model

Bischoff, Arruda and Grosh have developed a model for the stress-strain relationship where the constituent equation expressed the skin as an orthotropic, hyperelastic material. This was based on the microstructure of the skin and modeled the non-linearity as stretching of long-chain molecules, these being the collagen fibres. This model accounted for anisotropy in the way that the initially isotropic constitutive law is transformed when the tissue is subjected to an anisotropic stress state.

Few of the important functions of skin

1. **Protection action:** Skin is an anatomical barrier from pathogens.

2. **Sensation:** Skin contains a variety of nerve endings that react to environmental stresses like heat, cold, touch, pressure, vibrations, and other tissue injury.

3. *Heat regulation:* The skin contains a blood supply far greater than its requirements that allows precise control of energy loss by radiation, convection and conduction.

4. *Control of evaporation:* The skin provides a relatively dry and semi-impermeable barrier to fluid loss.

5. *Water resistance:* The skin acts as a water resistant barrier so essential nutrients are not washed out of the body.

6. *Storage and synthesis:* Skin acts as a storage center for lipids and water, as well as a means of synthesis of vitamin D by action of UV on certain parts of the skin.

7. *Excretion:* Excretion by sweating is a most important secondary function to temperature regulation.

8. *Absorption:* Oxygen, nitrogen and carbon dioxide can diffuse into the epidermis in small amount, in addition, medicine can be administered through the skin. The skin is an important site of transport in many other organisms.

Table 2.1 depicted the components of the skin and their functions:

Different skin diseases/disorder

Skin diseases are most common form of infections occurs in all ages of the people. Skin disorders occur mainly due to ugliness. Most of the skin infections, treatment take long time to show their effects. In general, of about 10–30 percent of patients seeking medical advice suffer from skin diseases. The skin conditions are prevalent across all parts of the world. There are various skin diseases are reported which are broadly divided in to infective and non-infective categories. In infective category, viral, bacterial and fungal related and in case of non-infective category, Eczema, dermatitis, Psoriasis and *Acne vulgaris* like skin diseases are commonly observe.

Bacterial skin infections are very common, and they are caused by two bacteria viz. *Staphylococcus aureus* and a form of *Streptococcus*. These bacteria are the causes of Impetigo and Cellulitis.

(a) *Impetigo:* It is one of the bacterial infections. It results in a honey-colored, crusty rash. It is mainly located on the face near the mouth and nose.

Table 2.1

Components of the skin	Functions
Stratum corneum	Prevent skin desiccation, protects against external chemical and antigen damage, microbial, fungal and parasitic attacks
Viable keratinocytes	Produce keratin and *Stratum corneum*
Basement membrane zone	Attaches epidermis to dermis
Dermal collagen, elastin and glycosaminoglycans	Protects against mechanical shearing effects, provide strength and stretchability
Melanin	Protects against UV lights
Dermal vasculature and eccrine sweat glands	Thermoregulation, vasculature provides nutrients to epidermis
Dermal nerves and neuroreceptor network in upper dermis and hair follicles	Monitor environment
Langerhans cells	Process antigens and function as macrophages
Fibroblast	Produce collagen and elastin and glycosaminoglycans
Mast cells	Synthesize substances that mediate inflammatory responses

Contd.

Table 2.1 *Contd.*	
Components of the skin	*Functions*
Sebaceous glands	Produce sebum, which maintain the soft texture of the skin surface
Apocrine glands	Secrets sweat
Hairs over scalp, body, eyelashes and nasal hairs	Provide aesthetic adornment and protect eyes from small particles and filter air breathed into nose
Fingernails	Assist in grasping objects and in self defense

(b) Cellulitis: Cellulitis is an infection of the skin and subcutaneous tissue that typically occurs when bacteria are penetrates through a puncture, bite, or other break in the skin. The area with cellulitis is usually warm and has some redness.

Fungal skin infections: Few of the important fungal infections are likely candidal dermatitis, tinea infection, tinea pedis, etc.

(a) Candidal dermatitis: It is mainly finds in the folds of the skin in the diaper area of infants. This infection is due to perfect for growth of the yeast *Candida*.

(b) Tinea infection: It is also finds as ringworm infection but is not a worm at all, is a fungus infection that can affect the skin, nails, or scalp. *Tinea* fungi can infect the skin and related tissues of the body. Medically this ringworm of the scalp is called as *tinea capitis*, ringworm of the body is called *tinea corporis* and ringworm of the nails is called *tinea unguium*. *Tinea corporis* can cause scaly, ring-like lesions on the body.

(c) Tinea pedis: It is the type fungal infection occurs on the feet and caused by same types of fungi that cause ringworm. It is also known as athlete's foot. This disease is commonly find in adolescents and is more likely to occur during warm weather.

(d) Tinea versicolor: It is a fungal infection caused by a yeast that normally inhibits the skin. It is also known as *Pityriasis versicolor* that occurs infection of the skin. Certain factors can cause these yeasts can convert to a pathogenic form known as *Malassezia furfur*, which causes the rash of *tinea versicolor*.

Viral skin infections: Many viruses cause characteristic rashes on the skin. They are mainly includes chicken pox, shingles, herpes simplex, warts.

(a) Chickenpox: Chickenpox is a common, usually self-limited, infection caused by the varicella virus. This infection is mainly occurs to the human in the age of 14 years and younger. Chickenpox is transmits by direct contact with infected blister fluid or by inhalation of respiratory droplets. When a person with chicken-pox coughs or sneezes, they expel tiny droplets that carry the varicella virus. A person without exposed to chicken-pox, inhales these droplets and the virus enters the lungs and then is carried through the bloodstream to the skin where it causes a rash.

(b) Shingles: Shingles is a painful rash caused by reactivation of the chicken-pox virus, *Varicella zoster*. This painful rash is more common after the age of 60. After a person has infected with the varicella virus, the virus travels back into the body and waits. For various reasons, viz. illness, trauma, suppressed immune system, chemotherapy, etc. the virus can reactivate and travel down the nerve to the skin that finally causes the shingles rash. The preliminary symptoms of this rash are usually itching, tingling or significant pain with just a light touch.

(c) Herpes simplex: Herpes simplex infection of mucosal cells results in cold sores and occasionally encephalitis and genital herpes. This virus comes in contact with the broken skin or the lining of the mouth, vagina or anus. Then goes to the nuclei of the cells and tries to reproduce itself. Sometimes the virus's replication process destroys the cells it has invaded causing blisters or ulcers to form on the skin.

(d) Warts: Warts are actually benign tumors of the epidermis caused by a virus. The virus is called as papilloma viruses that cause skin cells to proliferate and produce a benign growth. That is known as a wart or papilloma. This disease transmitted through direct contact.

Non-infective category diseases: The other types of skin disorders are like eczema and dermatitis, psoriasis, acne, skin cancer, etc. These types of diseases are easily curable by using various modern herbal formulations.

Dermatitis and eczema: Dermatitis is used to describe a specific type of rash with many causes. It is refers to any inflammation (swelling, itching, and redness) possibly associated with the skin. There are many types of dermatitis, including:

(a) Atopic dermatitis (eczema): This is a very common hereditary dermatitis, which causes an itchy rash primarily on the face, trunk, arms, and legs. It generally develops in infancy, but can also appear in early childhood. It may be associated with allergic diseases such as asthma and seasonal, hay fever, environmental and food allergies.

(b) Contact dermatitis: This is developed when the skin comes into contact with an irritating substance. The best-known cause of contact dermatitis is poison ivy, but there are many others, including chemicals found in laundry detergent (irritant contact dermatitis), cosmetics, perfumes and metals like nickel plating on jewelry, belt buckles, and the back of a snap.

(c) Seborrheic dermatitis: This oily rash, which appears on the scalp, face, chest, and back, is related to an overproduction of sebum from the sebaceous glands. This condition is common in infants and adolescents.

(d) Allergic contact dermatitis: It is a delayed hypersensitivity reaction involving allergens and antibodies.

(e) Stasis dermatitis: It occurs on the ankles and lower legs of people with venous insufficiency.

(f) Dyshidrotic dermatitis: It is also known as pompholyx. It is generally located on the hands or feet. It is characterized by redness, scaling and deep blisters.

(g) Nummular dermatitis: Coin shaped patches that occur anywhere on the body in relation to dry skin.

(h) Lichen simplex chronicus: It is a rash caused by long-term scratching of an area producing thickened skin.

Psoriasis: Psoriasis is a skin disease that causes itchy or sore patches of thick, red skin with silvery scales. It is a chronic, non-contagious autoimmune disease. It affects the skin and joint and is generally located on elbows, knees, scalp, back, face, palms and feet. The scaly patches caused by psoriasis, called psoriatic plaques, are areas of inflammation and excessive skin production. Skin rapidly accumulates at these sites and takes on a silvery-white appearance. Fingernails and toenails are frequently affected (psoriatic nail dystrophy) and can be seen as an isolated finding. Psoriasis arthritis is also a kind of disease due to inflammation of the joints caused by psoriasis. Generally, ten to fifteen percent of people with psoriasis are suffering by this psoriatic arthritis pain.

Acne: Acne is the most common cutaneous disorder. Generally, it is cause by *Acne vulgaris*. Nearly 80% of people aged between 11 and 30 years are affected. It is a disease of pilosebaceous units, which consist of sebaceous glands and hair follicle. Acne lesions are commonly referred to as pimples, blemishes, spots, zits, or acne. *Acne vulgaris* is defined as a chronic inflammatory disorder of pilosebaceous follicle. It develops as a result of blockages in follicles. It is almost a universal disease occurring in all races and affecting 95% of 16-year-old boys and 83% of 16-year-old girls to some degree. The incidence and severity of acne peak at 40% in fourteen to seventeen-year-old girls and 35% in boys aged 16–19 years. Acne is seen in 85% of adolescents, and nearly all teens have the occasional pimple, blackhead, or whitehead.

BIBLIOGRAPHY

1. Bischoff J.E., E.M. Arruda, and K. Grosh. *A microstructurally based orthotropic hyperelastic constitutive law.* Transactions of the ASME, 2002; 69: p. 570–79.

2. Bischoff J.E., E.M. Arruda, and K. Grosh. *Finite element simulations of the ortho-tropic hyperelasticity.* Finite Elements in Analysis and Design, 2002; 38: p. 983–98.

3. Danielson, D.A., *Human skin as an elastic membrane.* Journal of Biomechanics, 1973; 6(5): p. 539–46.

4. Lanir Y. *A structural theory for the homogeneous biaxial stress-strain relationships in flat collagenous tissues.* Journal of Biomechanics, 1979; 12(6): p. 423–36.

5. Lanir Y. *Constitutive equations for fibrous connective tissues.* Journal of Biomechanics, 1983; 16(1): p. 1–12.

6. Tong P. and Y.C. Fung. *The stress-strain relationship for the skin.* Journal of Biomechanics, 1976; 9(10): p. 649–57.

HAIR

Hair is mainly a protein filament that grows through the epidermis from follicles deep within the dermis. Hair is responsible for beauty and attractiveness of the people. The main component of hair fiber is keratin. Lengthwise, the hair can be divided into three parts, viz. the bulb, which is a swelling at the base which originates from the dermis, the root, which is the hair lying beneath the skin surface, and lastly the shaft, which is the hair above the skin surface. Also in cross-section of hair, there are three parts: (1) the medulla, an area in the core that contains loose cells and airspaces, (2) the cortex, which contains densely packed keratin and (3) the cuticle, which is a single layer of cells arranged like roof shingles. Growth of hairs mainly depends on cycle of three phases, viz. an anagen, catagen and telogen phases. Different hair colors and follicle shapes affects the timings of three phases. Anagen is the active growth phase of hair follicles that grows about 1 cm every 28 days because of the cells in the root of the hair are dividing rapidly. Normally, up to 90% of the hair follicles are in anagen, 10–14% are in telogen and 1–2% in catagen phase. The catagen phase is a short transition stage that occurs at the end of the anagen phase. It signals the end of the active growth of a hair. This phase lasts for about 2–3 weeks. Lastly, the telogen phase indicates the resting or quiescent phase of the hair follicle. It lasts for about 3 months. Hair color is the pigmentation of hair follicles due

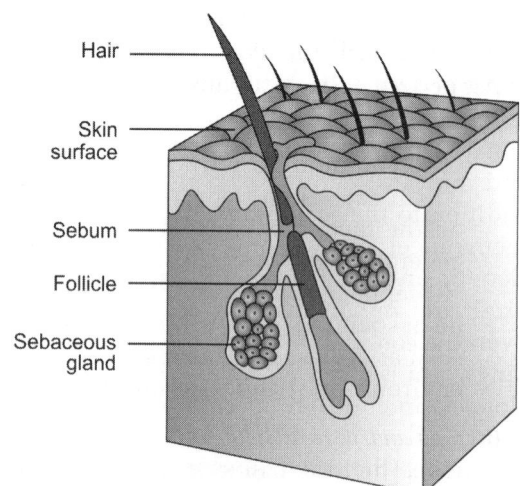

Fig. 2.3: Anatomy of hair

to two types of melanin, viz. eumelanin (red) and pheomelanin (black or brown). Generally, if more melanin is present, the color of the hair is darker and if less melanin is present, the hair is lighter.

Hair Disorders/Diseases

Common problems affecting the hair and scalp include hair loss, infections and disorders causing itching and scaling. Various types of lice, bacteria and fungi can invade the scalp and cause numerous problems of hair. These infectious agents of the scalp and other regions of hair are collectively called as "dermatophytes". Knowing that, the major hair diseases are broadly divided into two different categories in accordance to their occurrence in the scalp. They are mainly hair shaft abnormalities and infectious hair diseases. Hair damage may occur with different reasons like lack of proper diet, sleep or frequent use of chemical based shampoos. That results falling hair, thinning, dandruff and baldness, etc. Hair loss is a common disease for all. Medically it is known as alopecia. Certain diseases (e.g. thyroid problems, diabetes, and lupus), adverse effects of medicines or poor nutritional condition may cause hair loss. *Alopecia areata* is a causative organism that causes hair loss in small and form round patches. *Traction alopecia* may cause hair loss at the hairline.

Alopecia areata is an autoimmune condition resulting in hair loss for both males and females. Hair loss may be patchy or sparse and may involve the rest of the body in addition to the scalp. Hair in most people, recurrences of the condition are typical. Genetic and environmental factors play a vital role in this condition of hair loss. Even the condition may be seasonal as well. Ringworm is a common cause of temporary alopecia in children. In case of males, it is known as *androgenetic alopecia* (male pattern baldness). The hair loss is non-scarring and has a genetic basis. Sex steroids (androgens) specifically, dihydrotestosterone plays a significant role in this form of balding. In other case, females, female pattern baldness (alopecia) is a form of hair loss affecting due to an inherited susceptibility. Generally, this type of hair loss noticed after menopause. *Traction alopecia* is an another type of serious hair loss caused by the mechanical pulling of hair. It is more often seen in women of African-American origin. Prolonged this condition can lead to scarring, resulting in permanent hair loss and the formation of bald patches.

Scalp infections are one of the major causes of hair damage and loss through the promotion of hair disease. Scalp infections can treat and further it helps to hair growth. The main causes of scalp infections are various pathogens like bacteria, fungus, virus or parasites. Among these, the most common pathogens involved in hair infections are bacteria and fungus. Infections of the scalp include bacterial infection of hair follicles (folliculitis), infestation of head lice (*Pediculosis capitis*), and fungal infection of scalp ringworm (*Tinea capitis*) and Piedra. Itching and excessive flaking of the scalp is seen with dandruff (seborrheic dermatitis).

Folliculitis: It is a skin condition caused by an inflammation of one or more hair follicles in a limited area. It is a type of bacterial infection caused by *Staphylococcus* that results scalp infections. It is typically occurs in areas of irritation, such as sites of shaving, skin friction or rubbing from clothes. The most common causes of damage to hair follicles/infection includes, irritation from shaving, friction from tight clothing, a pre-existing skin condition, viz. eczema, acne, or another dermatitis, injuries to the skin (abrasions), etc. This is usually appears as small, white-headed pimples around one or more hair follicles. Most infections are superficial and hence it is also known as superficial folliculitis. There are several forms of folliculitis like **Staphylococcal folliculitis**, which is marked by itchy, white, pus-filled bumps that can occur anywhere on the body. **Pseudomonas folliculitis**, this is also known as hot tub folliculitis. This is marked by a rash of red, round, itchy bumps will appear that later may develop into small pus-filled blisters. *Tinea barbae,* it is a type of fungal infection. It develops in the beard area in male, causing itchy, white bumps but the surrounding skin forms red. Apart from that there are several infections are reported like *Pseudofolliculitis barbae, Pityrosporum folliculitis, Herpetic folliculitis, etc.*

Pediculosis capitis: It is a Head lice. It is common for all males and females and mainly occurs in nurseries, day care centers, and schools. It can be very itchy.

Tinea capitis: It is a contagious fungal infection of the scalp and term for ringworm infection. Generally, it is seen in children of aged 4 to 14 years. It is contagious and is acquired by contact with infected people, animals or objects like towels, combs, and pillows. The affected area can develop drastic hair loss.

Piedra: It is a very common fungal infection located in scalp. It affects both curly and straight hair and leads to the formation of nodules of different hardness on the infected hair. It can co-exist with bacterial infection.

Seborrheic dermatitis: It is known as dandruff. It is a common non-contagious condition of skin areas marked by flaking. A seborrheic dermatitis associated fungal scalp infection and the causative species of fungus is *Pityrosporon ovale.* It occurs sometimes redness and itching (inflammation) of the scalp, varying in severity from mild flaking of the scalp to scaly and red patches.

NAILS

Nails are a type of modified skin. Nails are epidermal cells converted to hard keratin. The bed on which the nail situated, is highly vascular, making the nail appear pink and whitish. Nails protect the sensitive tips of fingers and toes. Human nails are provides support for the tips of the fingers and toes, protect them from injury and aid in picking up small objects. Nails grow at an average rate of 3 to 6 millimeters in a month but the growth rate is dependent upon age, gender, season, exercise level, diet and hereditary factors.

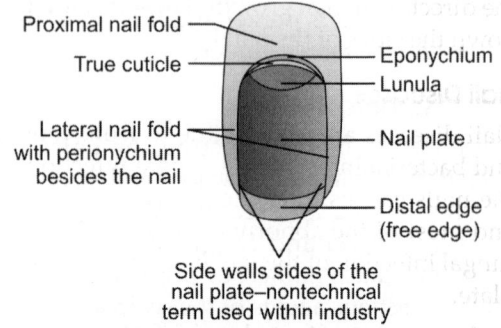

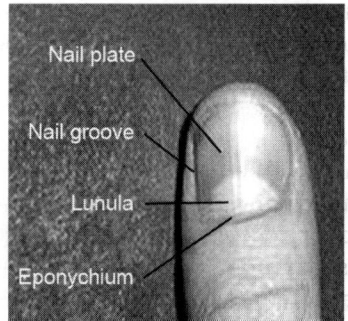

Fig. 2.4: Anatomy of nails

Nails are formed by keratinization process that means when the nail cells accumulate, the nail is pushed forward. The nail root move up to the surface of the skin then increase in number. Those closest to the nail root become flattened and pressed tightly together. Each cell is transformed into a thin plate. These plates are piled in layers to form the nail. The skin below the nail is called the matrix which is only living part of the nail. It is situated behind and underneath the nail fold and produces protein keratin which makes up the nail plate. The larger part of the nail is known as the nail plate. The plate looks pink because of the network of tiny blood vessels in the underlying dermis, is nail bed. The area of attachment between the nail plate and nail bed that lies underneath the free edge is known as *hyponychium.* The dead skin that forms around the cuticle area is known as *eponychium* and the live skin that folds around the cuticle, which provides protection to the matrix, is known as *peronychium.* The whitish crescent-shaped area at the base of the nail is called the lunula. Nail groove gives the direction of nail growth. They are situated down the sides of the nail fold.

Nail Diseases

Nail diseases are mainly due to from fungal and bacterial infections. A fungal infection of the nails causes thickened, discolored nails and most of the abnormal nails are due to a fungal infection of the nail bed, matrix or nail plate.

Appearance of fungal nail infections: There are four different types of fungal nail infections classified by the part of the nail involved. The most common infection involves the end of the nail when the fungi invade the hyponychium. Initially, the nail plate splits from the nail bed (onycholysis). The end of the nail then turns yellow or white and keratin debris develops under the nail causing further separation. Finally, the nail becomes fragile and crumble. Commonly, the fungal organism responsible for most fungal nail infections is *Trichophyton rubrum*.

Onychomycosis: This is one of the most important fungal infection in the proximal and lateral nail folds as well as the eponychium. This type of infection is characterized by onycholysis, i.e. nail plate separation, with evident debris under the nail plate. It normally appears white or yellowish in color. Some-times due to this infection the texture and shape of the nail also may change. This type of fungal infections are also known as *tinea unguium*.

Tinea unguis: This is also known as ring-worm of the nails. This is characterized by nail thickening, deformity, and eventually results in nail plate loss.

Nail psoriasis: It is characterized by raw, scaly skin of the nails. When it attacks to the nail plate it will leave it pitted, dry and it will often crumble. The plate may separate from the nail bed and may also appear red, orange or brown, with red spots in the lunula.

Onychotrophia: It is an atrophy or was-ting away of the nail plate. This infection mainly causes it to lose its luster and become smaller. Injury or disease may account for this irregularity.

Onychogryposis: These are claw-type nails that are characterized by a thickened nail plate and are often the result of trauma. This type of nail plates are characterized by curve inward or pinching the nail bed.

Few of the important bacterial nail infections are like:

Paronychia: This infection located at the area of the nail fold and can be caused by bacteria, fungi and some viruses. The proximal and lateral nail folds act as a barrier between the nail plate and the surrounding tissue. If a break occurs in this barrier, the bacterium can easily enter. This type of infection is characterized by pain, redness and swelling of the nail folds. Paronychia is divided into acute and chronic paronychia depending on the amount of time the infection has been present. An acute infection is associated with trauma to the skin such as a ingrown nail or nail-biting. It appears as a red, warm, painful swelling of the skin around the nail. The most common bacteria responsible is *Staphylococcus aureus*. The other one, chronic infection is associated with repeated irritation such as exposure to detergents and water. The main difference from the former one is the redness and tenderness are less noticeable. The nail surround skin can get boggy. Most chronic infections are caused by *Candida albicans*.

Pseudomonas bacterial infection can occur between the natural nail plate and the nail bed. The nail may have a green discoloration due to *Pseudomonas* infection. *Pseudomonas* feeds off the dead tissue and bacteria in the nail plate, while the moisture levels allow it to grow. The after effects of this infection will cause the nail plate to darken and soften underneath an artificial coating.

EAR

The ears are another paired sensory organs comprising the auditory system, nothing but involved in the detection of sound. The visible

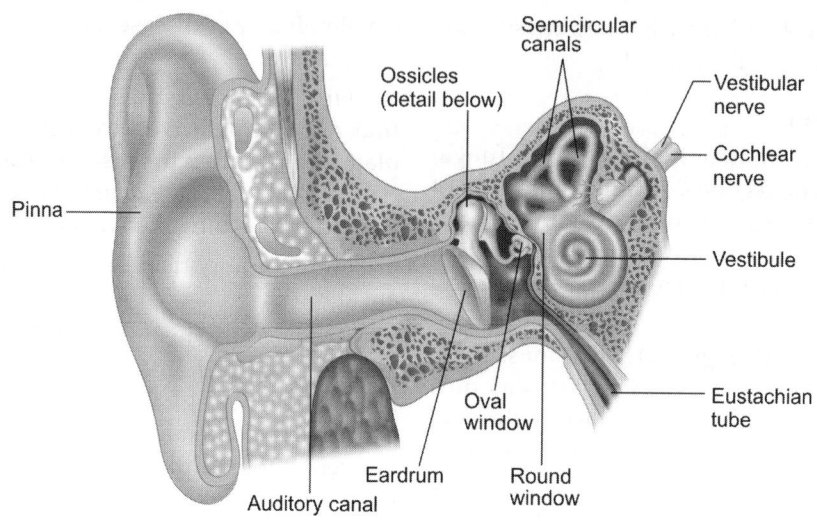

Fig. 2.5: Anatomy of the ear

ear is a flap of tissue that is also called the pinna. The pinna may be all that shows of the ear, but it serves only the first of many steps in hearing. The ear divides anatomically and functionally into three regions: The **external ear**, the **middle ear** and the **inner ear**. All three regions are involved in hearing. Only the inner ear functions in the vestibular system that plays a major role in the sense of balance and body position.

The main three regions of the ear are external, middle and inner regions. The external ear is also called as pinna which serves to protect the tympanic membrane or eardrum. This region collects and directs sound waves through the ear canal to the eardrum. The middle ear, separated from the external ear by the eardrum, is an air-filled cavity (tympanic cavity) carved out of the temporal bone. It connects to the nasopharynx via the eustachian tube, which equalize air pressure on both sides of the eardrum. Adjoining the eardrum are three linked, movable bones called "ossicles (smallest bones in the human body)," which convert the sound waves striking the eardrum into mechanical vibrations. The hammer (malleus) joins the inside of the eardrum. The middle bone, connects to the hammer and to the stirrup (stapes) which leads to the inner ear.

The last region is inner ear that consists of fluid filled tubes, connected through the temporal bone of the skull. The inner ear has two membrane—covered outlets into the air-filled middle ear—the oval window and the round window. The oval window situated behind the stapes, the third middle ear bone, and begins vibrating when "struck" by the stapes. This sets the fluid of the inner ear sloshing back and forth. The round window serves as a pressure valve, bulging outward as fluid pressure rises in the inner ear. Nerve impulses generated in the inner ear travel along the vestibulocochlear nerve, which leads to the brain.

Few of the Ear Related Diseases/Infections

Secretory otitis: It is the most common disease process where somewhat thicker fluid fills the middle ear. This is common in small children and is often "outgrown" by the time they reach their teens. This thicker fluid has components that are actually "secreted" by the mucous glands of the middle ear. There are actually tissue breakdown enzymes in this fluid; that, if left untreated, can gradually eat away bone and cause chronic hearing loss/damage.

Serous otitis: A sudden descent of an airplane with poor pressurization or a bad cold are two of the most common causes of acute serous otitis media. Usually, decongestants will clear the fluid or even blood that can be sucked from the mucosa into the middle ear with wither of these processes. If the fluid does not clear within a few weeks, it is considered chronic serous otitis which causes sudden obstruction of the eustachian tube. Generally, to the older people with poorly functioning eustachian tubes commonly have recurrent serous otitis.

Acute otitis media: It occurs when pus fills the middle ear. It is associated with sudden obstruction of the eustachian tube at the same time infections bacteria are present to cause the acute otitis, i.e. sudden perforation of the eardrum, with profuse drainage from the ear. The eardrum may be bright red or the creamy color of the fluid can sometimes be seen through the eardrum. It sometimes looks "soggy." Pain and fever mainly accompany an ear infection.

Chronic otitis media: It occurs when chronic infection fills the middle ear space and mastoid cavity. True chronic otitis media is almost always a form of chronic mastoiditis, where the bone of the mastoid cavity (the honey-combed bone behind the ear) is chronically infected along with the tissues of the middle ear space.

Cholesteatoma: It is a common disease along with chronic otitis and mastoiditis. It is a skin sac that grows back into the middle ear or mastoid from the eardrum, creating a mass of skin and debris in the middle ear. With spreading this infection gradually the ear bones of the inner ear and also the facial nerve can all be damaged or destroyed.

Otosclerosis: Otosclerosis is a kind of ear disease in which the otic capsule which is present only around the inner ear, is replaced in patches by soft bone at random locations around the inner ear. When this soft bone starts growing at the edge of the stapes, the stapes bone cannot move like it should and sound is not passed properly from the middle ear to the inner ear, creating hearing loss.

Menière's disease: Menière's disease is caused by an imbalance in the fluid in the sacs in the inner ear. This results sudden hearing loss, vertigo, ringing (tinnitus) in the ear.

MOUTH

The mouth is also known as buccal cavity, or oral cavity. It is the first portion of the alimentary canal that receives food and begins digestion by mechanically breaking up the solid food particles into smaller pieces and mixing them with saliva. It also plays important role in speech. It is divided into

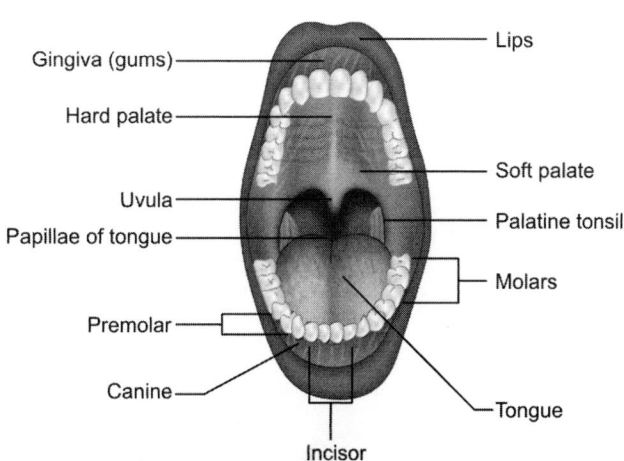

Fig. 2.6: Mouth cavity

two sections, viz. the vestibule, the area between the cheeks and the teeth, and the oral cavity. The oral mucosa is the mucous membrane epithelium lining the inside of the mouth.

The chief structures of the mouth are the teeth, which grind ingested food into small pieces for digestion. The tongue make positions and mixes food. It also carries sensory receptors for taste. The palate make separates the mouth from the nasal cavity. allowing separate passages for air and for food. At the surface of the tongue many small papillae are situated that hold the taste buds.

Few of the Mouth Diseases

There are a number of common mouth diseases that can affect human bacterias, fungus, etc. Diseases like hand-foot-mouth disease, thrush and gingivitis are most common to the human.

Hand-foot-mouth disease: This kind of viral desease is usually seen in babies and children. General symptoms include for this disease like blistering in the mouth, sore throat, fever, and lack of appetite. This common mouth disease also presents as a rash on the palms of the hands and soles of the feet.

Thrush: This is one more common disease in mouth. It is also called as candidiasis. This is an oral fungal mediated infection that affects children and adults. Thrush looks like a white lesions on the tongue and in the mouth. The tissue underneath the lesions will appear red in color.

Gingivitis: This is an another common mouth diseases. This condition causes the swelling of the gums. If gingivitis is left untreated, it can lead to periodontitis, which is a more serious mouth disease.

Herpes simplex: This is one more type of viral infection which causes oral herpes. This disease mainly affects to the HIV patients. It is a chronic infection which finally appears as whitish red painful ulcers. Infants can acquire this infection from their mothers during pregnancy time.

Aphthous ulcer: This is also known as canker sore or mouth ulcer. This is a painful ulcer which appears solely or in groups. The ulcers are less than 5 mm in diameter appears as a grey/white centre, with a red, inflamed border.

3

Antioxidants

The term antioxidant originally was used to refer specifically to a chemical that slowing or preventing the consumption of oxygen by oxidation of other molecules. Oxidation is a chemical reaction that transfers electrons from a substance to an oxidizing agent by produced free radicals, which start chain reactions that damage cells. Finally, antioxidants terminate these chain reactions by removing free radical intermediates, and inhibit other oxidation reactions by being oxidized themselves. Earlier they were used in prevention of oxidation of unsaturated fats. Recently, its versatile applications are led to the realization of the importance in daily life. The uses of antioxidants are in medical science as treatments for stroke and neurodegenerative diseases. Antioxidants are also widely used as ingredients in dietary supplements for maintaining health and preventing diseases such as cancer and coronary heart disease. In addition to these uses, recent natural antioxidants are also has many industrial uses, such as preservatives in food and cosmetics and preventing the degradation of rubber, gasoline and prevention of metal corrosion.

Antioxidants are classified into two broad divisions, depending on the solubility in water (hydrophilic) or in lipids (hydrophobic). In general, water-soluble antioxidants react with oxidants in the cell cytosol and the blood plasma, while lipid-soluble antioxidants protect cell membranes from lipid pero-xidation. The action of an antioxidant may depend on the proper function of other members of the antioxidant system.

The amount of protection provided by an antioxidant will also depend on its con-centration, its reactivity towards the particular reactive oxygen species and the status of the antioxidants with which it interacts. Some compounds acts as antioxidant defense by chelating transition metals. Selenium and zinc are commonly referred to as *antioxidant nutrients*, but they are not having antioxidant action themselves. Some known antioxidants are vitamins A, C, E, beta carotene, idebenone, coenzyme Q10. They assist in skin repair and the strengthening of blood vessels.

a. *Ascorbic acid:* This is also known as "vitamin C", which is found in both animals and plants. It cannot be synthesized in humans and must be obtained from the diet. In cells, it is maintained in its reduced form by reaction with glutathione. Ascorbic acid is a reducing agent and acts as pro-oxidants. It can reduce and thereby neutralize reactive oxygen species such as hydrogen peroxide. It will also reduce metal ions that generate free radicals through the Fenton reaction.

$$2\,Fe^{3+} + Ascorbate \rightarrow$$
$$2\,Fe^{2+} + Dehydroascorbate$$
$$2\,Fe^{2+} + 2\,H_2O_2 \rightarrow 2\,Fe^{3+} + 2\,OH\cdot + 2\,OH^-$$

b. *Glutathione:* It is a cysteine-containing peptide found in most forms of aerobic life. It is not required in the diet and is instead synthesized in cells from its constituent amino acids. Glutathione has antioxidant properties due to presence of thiol group in its cysteine molecule. In cells, glutathione is maintained in the reduced form by the enzyme glutathione reductase and in turn reduces other metabolites and enzyme systems like glutathione peroxidases.

c. *Melatonin:* It is a powerful antioxidant that can easily cross cell membranes and the blood brain barrier. Melatonin does not undergo redox cycling, i.e. once oxidized, cannot be reduced to its former state because it forms several stable end-products upon reacting with free radicals. Therefore, it has been referred to as a terminal antioxidant.

d. *Tocopherols and tocotrienols (vitamin E):* Vitamin E are the fat soluble vitamin with antioxidant properties. Among them, the α-tocopherol form is the most important lipid-soluble antioxidant. It can protect membranes from oxidation by reacting with lipid radicals produced in the lipid peroxidation chain reaction. This removes the free radical intermediates and prevents the propagation reaction from continuing. This reaction produces oxidised α-tocopheroxyl radicals that can be recycled back to the active reduced form through reduction by other antioxidants, such as ascorbate. The most important function of α-tocopherol is as a signaling molecule.

Role of Antioxidant on Skin

Free radicals are byproducts that are formed when oxygen is used by the body and hence antioxidants are required because they combat free radicals. Free radicals start a chain reaction under the surface of the skin. The process of destruction as:

- Antioxidant agents have an unpaired electron in their outer orbital and hence molecule acts as unstable.

- Free radicals attempt to gain an electron from natural proteins in the skin to acquire stability as a result the structure of the skin is damaged and its cellular structure is weakened. Some time free radicals also alter DNA, which results in aging and illness.

Antioxidants can help to protect skin from the damaging effects of the sun. They can protect the skin from the inside out by guarding the cells from damage. Vitamins A, C and E and the mineral selenium are thought to be particularly helpful in skin care. In addition to helping fortify cells against free radicals, vitamins A and C also encourage cell and tissue growth, helping the body to repair itself. Antioxidant agents protect cells and encourage cell growth which could be helpful in an anti-aging regimen and also help to fight fine lines and wrinkles. Skin comes in many different colors. These different colors, referred to as skin tones, are determined by our outer layer's supply of a pigment called melanin. Melanin serves as body's defense against ultraviolet rays. When UV rays start penetrating our bodies, melanocytes gets highly activates (skin cells) and start producing melanin, which results in a tan. The ingredients contained in the best antioxidant skin care products are natural substances that have been tested in clinical studies. They will work at reducing the lines and wrinkles on the skin. Coenzyme Q10 in the nanoemulsion form enhances skin cell activity and has a dramatic antiwrinkle effect because it gobbles up free radicals. This coenzyme Q10 is also provides significant protection against harmful UV rays.

Enzyme Systems

Cells are protected against oxidative stress by an interacting network of antioxidant enzymes with the help of chemical antioxidants. This system is an alternative pathway for detoxification of reactive oxygen species.

a. *Superoxide dismutase:* Superoxide dismutases (SODs) are related enzymes that catalyse the breakdown of the superoxide anion into oxygen and hydrogen peroxide. Superoxide dismutase enzymes contain metal ion cofactors that, depending on the isozyme, viz. copper, zinc, manganese and iron. In humans, the copper/zinc SOD is present in the cytosol, while manganese SOD is present in the mitochondria. SOD also present in extracellular fluids that have copper and zinc in its active sites. In plants, SOD isozymes are present in the cytosol and mitochondria, with an iron SOD found in chloroplasts.

b. *Catalases:* Catalases are enzymes that catalyse the conversion of hydrogen peroxide to water and oxygen, using cofactors such as iron. They are located in peroxisomes of eukaryotic cells.

c. *Peroxiredoxins:* They are peroxidases that catalyze the reduction of hydrogen peroxide, hydroperoxides and peroxynitrite. They are divided into three categories, viz. typical 2-cysteine peroxiredoxins, atypical 2-cysteine peroxiredoxins and 1-cysteine peroxiredoxins. These enzymes are having basic catalytic mechanism, in which a redox-active cysteine (the peroxidatic cysteine) in the active site is oxidized to a sulfenic acid by the peroxide substrate.

d. *Glutathione systems:* This system includes glutathione, glutathione reductase, glutathione peroxidases and glutathione *S*-transferase. These enzymes are generally situated in liver and acts for detoxification of metabolites. Not only animals but these enzymes are also located in plants and microorganisms. Glutathione peroxidase is an enzyme which is abundantly available, containing four selenium cofactors that catalyze the breakdown of hydrogen peroxide and organic hydroperoxides. Glutathione peroxidase is most active with lipid hydroperoxides where as the glutathione *S*-transferases show high activity with lipid peroxides.

Considering the importance of this area, the text has listed some important enzymatic and non enzymatic models for evaluating antioxidant activity.

Enzymatic Antioxidant Models
Superoxide Dismutase (SOD)

SOD activity was performed by various methods. Misra and Fridovich (1972) were described the method as epinephrine rapidly oxizides at pH 10.5 and produced adrenochrome which is a pink colored product that can be detected at 480 nm. The addition of samples containing SOD inhibits epinephrine auto-oxidation. The inhibition rate was monitored during 150s at intervals of 10s. The amount of enzyme required to produce 50% inhibition at 25°C was defined as one unit of enzyme activity. The SOD activity was expressed as μ/ml erythrocytes.

Further the model was modified by Kakkar et al. (1984) and as per them, the assay is based on the inhibition of NADH-phenazine methosulfate nitroblue tetrazolium formazan formation. The reaction was initiated by the addition of NADH. After 90 sec incubation, the reaction is stopped by addition of glacial acetic acid. The color developed at the end of the reaction was measured at 520 nm. Required reagents for the study are sodium pyrophosphate buffer pH 8.3, phenazine methosulphate 186 μM, nitroblue tetrazolium 300 μM and reduced nicotinamide adenine dinucleotide 780 μM. The procedure described by them as follows 0.5 ml of tissue homogenate in 0.052 M sodium pyrophosphate buffer pH 8.3 or 0.5 ml of erythrocyte lysate was diluted to 1 ml of ice-cold water followed by the addition of 2.5 ml ethanol and 1.5 ml chloroform. Shaken well for 90 sec at 4°C and then was centrifuged. The enzyme activity in the supernatant was determined. The assay mixture contained 1.2 ml sodium pyrophosphate buffer, 0.1 ml phenazine methosulphate, 0.3 ml nitroblue tetrazolium and exact diluted enzyme preparation in a total volume of 3.0 ml. The reaction was started by addition of 0.2 ml reduced nicotinamide adenine dinucleotide. After incubation of 90 second at 30°C, the reaction was stopped by addition of glacial acetic acid. The reaction mixture was stirred and shaken with 4 ml of n-butanol and was allowed to stand for few minutes. After centrifugation, the color intensity of the chromogen in butanol layer was measured by colorimeter at 520 nm against control. The specific activity of the enzyme is expressed as enzyme required for 50% inhibition of nitroblue tetrazolium reduction/min/mg Hb for erythrocyte lysate and enzyme required for 50% inhibition of nitroblue tetrazolium reduction/min/mg protein for tissues.

Further the same method was modified by Guzman et al. (2001). The superoxide radical generated by hypoxanthine and xanthine oxidase system was determined. Basic principle involved the reaction was performed in the absence and presence of plant extracts or allopurinol or superoxide dismutase (reference compounds), containing 1 mM EDTA, 100 μM hypoxanthine, 100 μM NBT with final volume adjusted to 1.2 ml by 50 mM phosphate buffer (pH 7.4). The $O_2^{\bullet-}$ generation was initiated by the addition of 0.066 U xanthine oxidase and detected by the NBT reduction/min spectrophotometrically at 560 nm.

A reaction mixture with a final volume of 1 ml/tube was prepared with 50 mM $KH_2PO_4^-$ KOH pH 7.4 containing 1 mM EDTA, 100 μl hypoxanthine, 100 μM nitroblue tetrazolium and 0.666 U per tube of xanthine oxidase in 100 μl of phosphate buffer. The reaction mixtures were incubated at 25°C for 5 minutes and the absorbance was measured at 560 nm against standard ascorbic acid. The results are expressed as the percentage inhibition of nitroblue tetrazolium reduction rate with respect to the reaction mixture without test compound.

Catalase Enzyme Activity

The catalase activity was determined in erythrocyte lysate and tissue homogenate by the method described by Sinha (1972). The assay was based on the principle of dichromate in acetic acid was reduced to

chromic acetate, when heated in the presence of hydrogen peroxide with the formation of perchromic acid as an unstable intermediate. The catalase preparation was allowed to split hydrogen peroxide for various period of time. The reaction was stopped at different time intervals by the addition of dichromate acetic acid mixture in hot condition. The remaining hydrogen peroxide forms hydrogen peroxide-chromic acetate which was determined colorimetrically at 590 nm. The used reagents are phosphate buffer 0.01 M, pH 7.0, hydrogen peroxide 0.2 M, 5% potassium dichromate, dichromate-acetic acid reagent (potassium dichromate and glacial acetic acid were mixed in the ratio of 1:3. From this mixture 1.0 ml was diluted with 4.0 ml acetic acid) and standard hydrogen peroxide 0.2 mM. The procedure as follows, tissue homogenate was prepared in phosphate buffer. To 0.9 ml of phosphate buffer, 0.1 ml of erythocyte lysate or tissue homogenate and 0.4 ml hydrogen peroxide were added. The reaction was arrested after 15, 30, 45 and 60 sec by adding 2.0 ml dichromate-acetic acid mixture. The tubes were kept in a boiling water bath for 10 minutes, cooled and color developed was measured at 590 nm. The specific enzyme activity is expressed as micro mole of hydrogen peroxide utilized/min/mg of Hb for erythrocyte lysate and micro moles of hydrogen peroxide utilized/min/mg of protein for tissues.

Further the method was modified by the method described by Aebi (1984). As per the method, packed erythrocytes were hemolyzed by adding 100 volumes of distilled water, then 20 μl of this hemolyzed sample was added to a cuvette and the reaction was started by the addition of 100 μl of freshly prepared 300 mM hydrogen peroxide in phosphate buffer (50 mM, pH 7.0) to give a final volume of 1 ml. The rate of hydrogen peroxide decomposition was measured spectrophotometrically at 240 nm during 120s. The catalase activity was expressed as μM H_2O_2/ml erythrocytes/min.

Glutathione Peroxidase Activity

This activity was described by Rotruck et al. (1973). The assay was based on the principle of a known amount of enzyme preparation was allowed to react with hydrogen peroxide and with reduced glutathione for a specific time period. The reduced glutathione content remaining after the reaction was measured at 340 nm. The reagents requires for the assay are, Tris-HCl buffer 0.4 M, pH 7.0, sodium azide solution 10 mM, 10% trichloroacetic acid, EDTA 0.4 mM, hydrogen peroxide 1.0 mM and reduced glutathione 2.0 mM. The described method is as follows, 0.2 ml of Tris-HCl buffer, 0.2 ml of EDTA, 0.1 ml of sodium azide and 0.2 ml of enzyme preparation were added and mixed well. To this mixture, 0.2 ml reduced glutathione followed by 0.1 ml hydrogen peroxide were added. The mixed solution was incubated at 37°C for 10 minutes. Further the reaction was stopped by adding of 0.5 ml trichloroacetic acid. Then centrifuged and the remaining reduced glutathione was determined colorimetrically at 340 nm.

Glutathione Reductase

This method was described by Carlberg and Mannervik (1985) and the method was based on the principle of the oxidation of NADPH was measured spectrophotometrically at 340 nm. The required reagents are potassium phosphate buffer 0.2 M, pH 7.0 containing 2 nm EDTA, NADPH 2 nm in 10 nm Tris-HCl, pH 7.0 and GSSG 20 mM. The basic procedure involves that 0.5 ml phosphate buffer (30°C), 50 µl NADPH, 50 µl oxidized glutathione (GSSG) and a volume of deionized water were mixed in a test tube to make volume up to 1.0 ml. The reaction was initiated by added erythrocyte lysate or tissue homogenate to the tube. The decrease in absorbance at 340 nm was measured at 30°C for every 5 minutes and the change in absorbance was measured. The specific enzyme activity is expressed as micro mole of NADPH oxidized/min/mg of Hb for erythrocyte lysate and micro moles of NADPH oxidized/min/mg of protein for tissues.

δ-Aminolevulinate Dehydratase (δ-ALA-D) Activity

This enzymatic activity method was described by Berlin and Schaller (1974), assayed in whole blood by measured the rate of porphobilinogen formation in 1 hour at 37°C. The enzyme reaction was initiated after 10 minutes of preincubation of blood with 1 mM zinc chloride. The reaction was started by added ALA to a final concentration of 4 mM in a phosphate buffer solution and incubation was carried out for 1 hour at 37°C and the reaction product was measured at 555 nm and expressed as nmol porphobilinogen/ml blood/hour.

Xanthine Oxidase Method

This is one of the recent methods for evaluation of antioxidant activity. The method was described by Opoku et al. (2002). The percentage inhibition in the xanthine oxidase activity in presence of antioxidants was measured. Xanthine oxidase enzyme produces uric acid together with superoxide radicals from xanthine and the amount of uric acid is measured at 292 nm.

Nonenzymatic Antioxidant Models

Thiobarbituric Acid Reactive Substances (TBARS)

This assay method described by Lappena et al. (2001) by measures the peroxidative damage to lipids that produced by excessive reactive oxygen species generation. Lipid peroxidation was estimated in plasma using 1% phosphoric acid and 0.6% thiobarbituric acid (TBA). The pink chromogen produced by reaction of TBA with malondialdehyde was measured spectrophotometrically at 532 nm. The results were expressed as nmol TBARS/ml plasma, using malondialdehyde as standard.

β-Carotene Linoleate Model

This method was described by Hidalgo et al. (1994) and was based on the principle that linoleic acid (an unsaturated fatty acid) gets oxidized by reactive oxygen species (ROS) produced by oxygenated water. The products formed would initiate the β-carotene oxidation which leads to discoloration of the β-carotene. This was prevented by antioxidants hence decrease in extent of discoloration indicates the activity, i.e. more the discoloration, less will be the antioxidant activity and *vice versa*. Due to insolubility of β-carotene in water, it was first made into emulsion. β-carotene would dissolve in chloroform and using tween-40 (polyoxyethylene sorbitan mononitrate) as surfactant, it was mixed with linoleic acid to form emulsion. The resulting mixture was diluted with distilled water and further made up with oxygenated water which indicates the process of oxidation immediately. It would transfer to tubes containing samples. Initial reading was read, followed by measured the absorbance at 15 minutes intervals at 470 nm. All tubes were placed in water bath at 50°C throughout the experiment.

Materials required for the models were β-carotene linoleic acid emulsion mixture, 100 ml. The emulsion were prepared by 0.4 mg of β-carotene, 40 mg of linoleic acid and 400 mg of tween-40 added in 1 ml of chloroform, which was further mixed with 20 ml water and 80 ml of oxygenated water. Butylated hydroxyanisole (BHA) solution was prepared using 4.5 mg of BHA dissolved in 1 ml of ethanol.

The method was described as 50, 125, 250 and 500 μl of extract solution (prepared in ethanol) and 50 and 125 μl of BHA solution were added to separate tubes according to concentration (100, 250, 500 and 1000 ppm) and volume was made up to 0.5 ml with ethanol. Four milliliter ml of β-carotene linoleic acid emulsion was added to each tube. Absorbance was taken at zero time and tubes were placed at 50°C in water bath. Measurement of absorbance was continued at an interval of 15 minutes, till the color of β-carotene disappeared in the control reaction (t =180 minutes). A

mixture prepared as above without β-carotene emulsion served as blank and mixture without extract served as control. Dose response of antioxidant activity for various extracts was determined at different concentrations. Finally, the antioxidant activity (AA) of extracts was evaluated in terms of bleaching of β-carotene using the formula:

% AA = 100 [1-(A° − A^t)/ (A°0 − A°t)],

where A° = Initial absorbance of sample, A^t = Absorbance of sample after incubation for 180 minutes, A°0 = Initial absorbance of control, A°t = Absorbance of control after incubation for 180 minutes.

The method was further modified by Tepe et al. (2005). A stock solution of β-carotene and linoleic acid was prepared by dissolved 0.5 mg of β-carotene in 1 ml of chloroform and added 25 μl of linoleic acid together with 200 mg of tween-40. The chloroform was evaporated. Hundred milliliter ml of aerated water were added to the residue. To 2.5 ml of this mixture 300 μl of each extract were added. The test tubes were incubated in boiling water for 2 hours together with two blanks, one containing the antioxidant butylated hydroxytoluene (BHT) and the other one without antioxidant. With this the latter tube a complete oxidation is obtained with the initial yellow color disappeared, while in the test tube with BHT the yellow color would maintain during the incubation period. The absorbance was measured at 470 nm.

α-Tocopherol (Vitamin E)

This method was described by Desai (1971). The method was described as a modified version of Emmerie-Engle method in which ferric ions were reduced to ferrous ions in the presence of tocopherols and a pink colored complex was formed with bathophenanthroline. Orthophosphoric acid was added as a chelating agent to reduce carotene interference by preventing its oxidation and standardization of color by binding excess ferric ions and thus preventing their photochemical reduction. Absorbance of the stable chromophore was measured spectro-photometrically at 536 nm.

The reagents for that method were batho-phenanthroline reagent which is 0.2% solution of 4,7-diphenyl-1,10-phenanthroline in absolute ethanol, ferric chloride reagent, orthophosphoric acid reagent and standard α-tocopherol reagent (prepared in ethanol).

Three milliliter of extract was evaporated to dryness. To the residue 1.0 ml of ethanol, 0.2 ml of bathophenanthroline reagent and 0.2 ml of ferric chloride reagent were added and mixed properly. After one minute 0.2 ml of orthophosphoric acid reagent was added and measured at 536 nm. The amount of α-tocopherol is expressed as mg/dl plasma and µg/mg tissue.

Ascorbic Acid (Vitamin C)

Omaye et al. (1979) were described the method of measurement of Ascorbic acid as antioxidant. According to the method, the basic principle involved as the ascorbic acid was oxidized by copper to form dehydroascorbic acid and diketogulonic acid. These products were treated with 2,4-dinitrophenylhydrazine to form the derivative bis-2,4-dinitrophenylhydrazone. This compound reacts with strong sulphuric acid and form a rearranged product which was measured at 520 nm. The reaction was continued in presence of thiourea that helps to prevent interference from nonascorbic acid chromogens.

The reagents were used for the methods are 10% thiazolidine-4-carboxylic acid (TCA), 65% sulphuric acid, DNPH-thiourea-copper sulphate reagent (DTC) which was prepared by dissolved 0.4 g thiourea, 0.05 g copper sulphate and 3.0 g of 2,4-dinitrophenylhydrazine in 100 ml of 9 N sulphuric acid.

The method was as follows: 1.5 ml of ice cold TCA was added to the 0.5 ml of plasma or tissue homogenate and mixed well, then centrifuge for 10 minutes. To 0.5 ml of the supernatant, 0.1 ml of DTC reagent was added and mixed well. Then the tubes were

incubated at 37°C for 3 hours and 0.75 ml of ice-cold 65% sulphuric acid was added and the tubes were allowed to stand at room temperature. A set of standards containing 10–50 µg ascorbic acid was made up to 0.5 ml and were processed in similar way along with a blank containing 0.5 ml TCA. The color developed was measured at 520 nm.

The amount of ascorbic acid is determined as mg/dl plasma and µg/mg tissue.

The method was modified by Jacques-Silva et al. (2001). Plasma was precipitated with 1 volume of a cold 5% trichloroacetic acid solution followed by centrifugation. An aliquot of 300 µl of the supernatants were mixed with 2,4-dinitrophenylhydrazine (4.5 mg/ml) and 13.3% trichloroacetic acid and incubated for 3 hour at 37°C. Then 1 ml of 65% sulphuric acid was added to the medium and the orange red compound was measured at 520 nm.

The content of ascorbic acid was calculated using a standard curve (1.5–4.5 µmol/L ascorbic acid freshly prepared in sulphuric acid) and expressed as µg ascorbic acid/ml plasma.

Reduced Glutathione (GSH)

Boyne and Ellman (1972) was described the method which was based on the principle on the developement of yellow color in the presence of 5, 5'-dithio(bis-nitrobenzoic) acid (DTNB) reagent containing sulfhydryl groups.

The reagents for this method were phosphate buffer 0.2 M (pH 8.0), 10% thiazolidine-4-carboxylic acid (TCA), Ellman's reagent (40 mg DTNB in 10 ml 0.1 M phosphate buffer), 100 mg reduced glutathione (standard) and working standard as stock standard was diluted with water to get concentration of 100 µg/ml.

1.0 ml of erythrocyte lysate or tissue homo-genate was precipitated with 2.0 ml of TCA and centrifuged. To 1.0 ml of the supernatant, 3.0 ml of phosphated buffer and 0.5 ml of Ellman's reagent were added. The yellow color developed was measured in a colorimeter at 412 nm. A series of standards (20–100 µg) were treated in a similar manner along with a blank containing 1.0 ml of buffer. Finally, the amount of GSH was calculated as µg/dl erythrocyte lysate and mM/mg of tissue.

The method was modified by Jacques-Silva et al. (2001), which is also known as protein thiol groups (P-SH) assay. The method was described as the reduction of 5, 5'-dithio(bis-nitrobenzoic) acid (DTNB) in 0.3 M phosphate buffer (pH 7.0) and measured at 412 nm. Quantification of total protein-SH groups indicated thiol status and the general state of thiol containing proteins that indirectly indicated the redox state of the blood cells. The P-SH group was calculated from a standard glutathione curve. The results were expressed as nmol P-SH/ml plasma.

Further the non-protein thiol group (NP-SH) assay method was described by the same scientist in 2001. Red blood cell pellets (300 µl) was collected after centrifugation then hemolyzed with 10% triton solution (100 µl) for 10 minutes. Then the protein fraction was precipitated with 200 µl of 20% TCA followed by centrifugation. Finally the NP-SH groups were calculated from a standard glutathione curve by colorimetric determination at 412 nm. The result was expressed as nmol NP-SH/ml erythrocytes.

Conjugated Diene Assay

The method was described by Ashok (2001). The principle of this assay is that during linoleic acid oxidation, the double bonds are converted into conjugated double bonds, which are characterized by a strong UV absorption at 234 nm. The activity is expressed in terms of inhibitory concentration (IC_{50}). This method allows dynamic quantification of conjugated dienes as a result of initial polyunsaturated fatty acids oxidation by measuring UV absorbance at 234 nm.

DPPH Method (1,1-Diphenyl-2-Picryl-Hydrazyl)

This is most widely used method for screening of antioxidant activity of many plant drugs. Blios in 1958 was described this assay method. The basic principle of this method was free radical scavenging potentials of the extracts were tested against a methanolic solution of DPPH. Antioxidant reacts with DPPH and converts it to α-α-diphenyl-β-picryl hydrazine. The degree of discoloration indicates the scavenging potentials of the antioxidant extract. The change in the absorbance produced at 517 nm has been measured for antioxidant activity. The method was as follows: aliquots of 50, 100, 250 and 500 µl of extracts were taken in different test tubes and to these all 4 ml of methanolic solution of DPPH solution (prepared by 39.4 mg of DPPH was dissolved in one liter of methanol) was added and shaken well. The mixture was allowed to stand at room temperature for 20 minutes. The blank was prepared and measured at 517 nm.

α-α-diphenyl-β-picrylhydrazyl   α-α-diphenyl-β-picrylhydrazine

The scavenging activity was expressed as the inhibition percentage calculated using the formula:

%Antioxidant activity =

$$\frac{\text{Absorbance of control–absorbance of sample}}{\text{Absorbance of control}} \times 100$$

Sanchez-Moreno et al. (1999) were reported the method and the method was based on the reduction of methanolic solution of colored free radical DPPH by free radical scavenger. The procedure involves measurement of decrease in absorbance of DPPH at its absorption maxima of 516 nm, which was proportional to concentration of free radical scavenger added to DPPH reagent solution. The activity is expressed as effective concentration EC_{50}.

The method was further modified by Germano et al. (2002) and was evaluated in terms of hydrogen donating or radical scavenging activity using DPPH radical. The method was described as follows: 200 µl filtrate was taken in test tubes and volume made up to 1 ml with methanol. Three ml of the freshly prepared DPPH solution (200 µM in methanol) was added to the sample tube and mixed thoroughly. The sample tube was kept in a water bath at 37°C for 20 minutes. The absorbance of the sample was measured at 517 nm by UV spectrophotometer. Gallic acid, BHA and trolox were used as standard references. The DPPH radical scavenging effect was calculated as percentage inhibition using the formula:

$$\%\text{Inhibition} = [A_{c\,(0)} - A_{a\,(t)} / A_{c\,(0)}] \times 100$$

where, $A_{c\,(0)}$ = absorbance of the control DPPH solution at zero minute and $A_{a\,(t)}$ is an absorbance of the test sample after 20 minutes.

Hydroxyl Radical Scavenging Activity

The assay was adopted from a method described by Halliwell et al. (1987). In case of nonsites–pecific hydroxyl radical system, reaction mixture contained 28 mM 2-deoxy-2-ribose, 1 mM EDTA plus 200 µM $FeCl_3$ (1:1), 1 mM H_2O_2 and 1 mM ascorbic acid in 10 mM

phosphate buffer solution (pH 7.4). In site-specific hydroxyl radical system, EDTA was replaced by the phosphate buffer. This reaction mixture was incubated at 37°C for 1 hr in the absence and presence of plant extracts and standard. The generated OH radicals reacted with the deoxyribose forming malondialdehyde (MDA) which was measured spectrophotometrically at 532 nm by thiobarbituric acid (TBA) reaction.

Further the method was modified in 1998 and the basic principle involved in the method was the hydroxyl radicals formed by the oxidation react with dimethyl sulfoxide (DMSO) to yield formaldehyde which detect the formation of hydroxyl radicals formed during the oxidation of DMSO by the Fe^{3+} ascorbic acid system that was used to detect hydroxyl radicals.

The materials used for the study were iron-EDTA (1:2) mixture (prepared by 0.13 g of ammonium ferrous sulphate and 0.26 g of EDTA in 100 ml of water), ascorbic acid solution (0.22 g of ascorbic acid in 100 ml distilled water), 1.70 ml of dimethyl sulfoxide in 200 ml of phosphate buffer (pH 7.4), 17.5% trichloroacetic acid in water, EDTA solution, Nash reagent (prepared by added 750 g of ammonium acetate, 3 ml of acetic acid and 2 ml of acetyl acetone in one liter of distilled water) and extract solution, prepared by 7.0 mg/ml of the extract in 2% alcohol.

The method was as follows: 5.0, 10.0, 20.0, 30.0, 40.0 and 50.0 µl of the extracts are prepared in 2% alcohol were taken in different test tubes and evaporated on a water bath. To that 1 ml of iron-EDTA solution, 0.5 ml of EDTA and 1 ml of DMSO were added and the reaction was initiated by added 0.5 ml of ascorbic acid to each of the test tubes. Test tubes were then heated on the water bath at 80–90°C for 15 minutes. The reaction was then terminated by the addition of 1 ml of ice-cold TCA to all the test tubes, kept aside for 2 minutes and the formaldehyde formed was determined by addition of 3 ml of Nash reagent which was form yellow color, after kept for 10–15 minutes. The intensity of the color formed was measured by spectro-photometrically at 412 nm against blank reagent.

%Hydroxyl radical scavenging

$$= \frac{1 - \text{Absorbance of sample}}{\text{Absorbance of blank}} \times 100$$

Further Babu et al. (2001) were described the method which was involved the *in vitro* generation of hydroxyl radicals by Fe^{3+}/ascorbate/EDTA/H_2O_2 system using Fenton reaction. Scavenging of this hydroxyl radical in presence of antioxidant is measured. In one of the methods the hydroxyl radicals formed by the oxidation is made to react with DMSO (dimethyl sulphoxide) to yield formaldehyde. Formaldehyde formed produces intense yellow color with Nash reagent. The intensity of yellow color formed is measured at 412 nm spectrophotometrically against reagent blank.

Superoxide Radical Scavenging Activity

This activity was described by Sabu and Ramadasan in 2002. The basic principle of this method was based on generation of superoxide radical (O_2^-) by auto-oxidation of hydroxylamine hydrochloride in presence of NBT (nitroblue tetrazolium reagent) which gets reduced to nitrite. Nitrite in presence of EDTA gives a color which measured at 560 nm.

Test solutions of extract (20–200 µg/ml) were taken in a test tube. To this, reaction mixture consisting of 1 ml of (50 mM) sodium carbonate, 0.4 ml of (24 mM) NBT and 0.2 ml of 0.1 mM EDTA solutions were added to the test tube and immediate measured at 560 nm. About 0.4 ml of (1 mM) of hydroxylamine hydrochloride was added to initiate the reaction then reaction mixture was incubated at 25°C for 15 minutes nad reduction of NBT was measured at 560 nm. Ascorbic acid was used as the reference compound. Decreased absorbance of the reaction mixture indicates increased superoxide anion scavenging activity.

Percentage of inhibition was calculated according to the following equation:

%Inhibition = $(A_0 - A_t)/A_0 \times 100$

where, A_0 was the absorbance of the control (blank) and A_t was the absorbance in the presence of the samples of the extract.

Nitric Oxide Radical Scavenging Activity

The method was described by Marcocci et al. 1994. The method was based on the generation of nitric oxide from sodium nitro-prusside and measured by Greiss reaction. Sodium nitroprusside in aqueous solution at physiological pH spontaneously generates nitric oxide which interacts with oxygen to produce nitric ions that can be estimated by using Greiss reagent. Scavengers of nitric oxide compete with oxygen leading to reduce production of nitric oxide. Sodium nitroprusside (5 mM) in phosphate buffer saline (PBS) was mixed with 3.0 ml of different concentrations (20–200 µg/ml) of the extract and incubated at 25°C for 2.30 hours. The samples were added to Greiss reagent (1% sulphanilamide, 2% H_3PO_4 and 0.1% naphthylethylenediamine dihydrochloride). The absorbance of the color (formed during the diazotization of nitrite with sulphanilamide) was measured at 546 nm and referred to the absorbance of standard solution of ascorbic acid treated in the same way with Greiss reagent as a positive control. The percentage inhibition was measured by the formula:

%Inhibition = $(A_0 - A_t)/A_0 \times 100$

where A_0 was the absorbance of the control (blank) and A_t was the absorbance in the presence of the samples of the extract.

Ferric Reducing Ability of Plasma (FRAP) Assay Method

This FRAP method was described by Benzie and Strain in 1996. The basic principle of this method is based on the reduction of a ferric-tripyridyl-triazine complex to its ferrous, colored form in the presence of antioxidants.

The method was described as the FRAP reagent contained 2.5 ml of a 10 mmol/L TPTZ (2,4,6-tripyridyl-s-triazine) solution in 40 mmol/L hydrochloric acid and 2.5 ml of 20 mmol/L hydrated ferric chloride solution and 25 ml of 0.3 mol/L acetate buffer (pH 3.6) and was prepared freshly and warmed at 37°C. Aliquots of 40 µL sample filtrate was mixed with 0.2 ml of distilled water and 1.8 ml of FRAP reagent. Then the absorbance was measured at 593 nm using spectrophotometer after incubation at 37°C for 10 minutes. Gallic acid, ascorbic acid, BHA and trolox were used as the standard. The final result was expressed as the concentration of antioxidants having ferric reducing ability equivalent to that of mg of standard used per gram of sample on dry weight basis.

Trolox Equivalent Antioxidant Capacity (TEAC)

The method was based on the ability of antioxidant molecules to quench the ABTS$^{•+}$ [2, 2'-azino-bis (3-ethylbenzthiazoline-6-sulfonic acid)], a blue-green chromophore with characteristic absorption at 734 nm, compared with that of Trolox (water-soluble vitamin E analog). The addition of antioxidants to the preformed radical cation reduced it to ABTS, determining a decolorization. The method was developed by several researchers namely, Rice-Evans et al., 1994; Miller and Rice-Evans, 1996 ; Poungbangpho et al., 2000. The required chemicals were ABTS$^{•+}$, Trolox (6-hydroxy-2,5,7,8-tetramethylchroman- 2-carboxylic acid) and myoglobin. All other chemicals were of analytical grade. Metmyoglobin was purified after adding the stock myoglobin solution (100 µM) in 5 mM isotonic phosphate buffer saline (PBS buffer), pH 7.4 to an equal volume of freshly prepared 740 µM potassium ferricyanide. The solution was dialysed against PBS buffer, pH 7.4 twice at 4°C for 24 hours, and the metmyoglobin was collected and calculated for the concentration as described by Miller and Rice-Evans (1996). All compounds were dissolved in PBS buffer, except Trolox or samples dissolved in 95% ethanol. The standard antioxidative curve of Trolox (concentration 0–2.5 mM) was obtained at the concentration of 108 µM H_2O_2 and 100 µM metmyoglobin at 20 minutes. The percentage inhibition of Trolox was calculated of the blank at an absorbance 734 nm by UV-vis

spectrophotometer (Shimadzu, UV-1201v) and then was plotted as a function Trolox concentration. The total antioxidant capacity of several plant extracts were examined and compared with a standard antioxidant, Trolox. The Trolox equivalent antioxidant capacity was defined as the antioxidant capacity of 1 mg crude extracted to imol of Trolox and represented as TEAC value.

This method was described by Schlesier et al., 2002. The method is based on the reduction of ABTS$^{•+}$ [2,2'-azino-bis (3-ethylbenzthiazoline-6-sulphonic acid)] radical nation by antioxidants. 25 mM of Trolox was prepared in ethanol and use as standard that was prepared by mixing ABTS$^{•+}$ (7 mM) with 2.45 mM of potassium persulphate in water. The mixture was kept for 12–24 hours at room temperature in the dark until reaction to be completed. To prepare ABTS$^{•+}$ working solution, ABTS$^{•+}$ stock solution was diluted with water to an absorbance of 0.700 ± 0.02 at 734 nm. A stock solution of Trolox (1 mM) was prepared with water. Then for photometric assay, 1 ml of ABTS$^{•+}$ working solution and 10 µl of 1 mM Trolox stock or 10 µl of a range of dilutions of plant extracts were mixed for a few seconds and measured immediately after 1 minute at 734 nm. The antioxidant activity of the test substances was calculated by determined the decreased in absorbance using the following formula:

$$\%\text{Antioxidant activity} = \{(E[ABTS^{•+}]/E[standard]/E[ABTS^{•+}]\} \times 100$$

Trolox equivalents were estimated by linear extrapolation of the antioxidant activity from the Trolox standards.

Hydrogen Peroxide Method

This is one of the most simple method for determination of antioxidant activity of the compounds. Jayaprakasha, et al. 2004, were described the method. Hydrogen peroxide solution (20 mM) was prepared in phosphate buffer saline (PBS) at pH 7.4. Various concentration of extracts or standard were prepared in 1 ml of methanol and further 2 ml of hydrogen peroxide solution in PBS were added. After 10 minutes the absorbance was measured at 230 nm.

Oxygen Radical Absorbance Capacity (ORAC)

The methodology of ORAC was originally developed by Dr. Guohua Cao of the National Institute of Aging in 1992. The method was based on the principle by the free radical damage to a fluorescent probe through the change in its fluorescence intensity. The change of fluorescence intensity is an index of the degree of free radical damage. In the presence of antioxidant, the inhibition of free radical damage by an antioxidant, which is reflected in the protection against the change of probe fluorescence in the ORAC assay, is a measure of its antioxidant capacity against the free radical.

A peroxyl radical (ROO$^\bullet$) is formed from the breakdown of AAPH (2,2'-azobis-2-methyl-propanimidamide, dihydrochloride) at 37°C. The peroxyl radical can oxidize fluorescein (3,6'-dihydroxy-spiro [isobenzofuran-1[3H], 9'[9H]-xanthen]-3-one) to generate a product without fluorescence. Antioxidants suppress this reaction by a hydrogen atom transfer mechanism, inhibiting the oxidative degradation of the fluorescein signal. The fluorescence signal is measured over 30 minutes by excitation at 485 nm, emission at 538 nm, and cutoff = 530 nm. The concentration of antioxidant in the test sample is pro-portional to the fluorescence intensity through the course of the assay and is assessed by comparing the net area under the curve to that of a known antioxidant, Trolox.

The general method was describes as below

Prepare fluorescein working solution from the stock solution by transferring 16.8 ml of assay buffer to an empty tube and add 1.2 ml stock fluorescein solution. Mix and protect from light.

Prepare Trolox standards as follows

Briefly spin down the contents of the 1.5 mM Trolox standard tube after thawing. Pipette 580 µl of assay buffer into the 1.5 mM Trolox standard tube provided and mix well by vortexing. This produces a diluted stock Trolox standard of 50 M. Pipette 50 µl of assay buffer into 6 tubes. Mix each new dilution thoroughly before proceeding to the next. The 50 µM stock dilution serves as the highest standard, and the assay buffer serves as the zero standard.

Add 150 µl of the working fluorescein solution (stored at 4°C) to each well of the assay plates. Then add 25 µl of samples or Trolox standards to individual wells of the assay plates, add 25 µl of assay buffer to individual wells as a negative control. Place plate at 37°C for at least 5 minutes. While the assay plate is equilibrating to 37°C, prepare the AAPH working solution (stored at −20°C) by added 2.7 ml assay buffer to the tubes and were gently inverted. Kept the working solution on ice. AAPH solution is good for 8 hours if kept on ice. To start the assay, 25 µl of the AAPH working solution were added to each of the wells containing standards and samples. Then placed the assay plate in the plate reader and begin kinetic fluorescence reading and was set-up plate reader to perform a kinetic read for 30 minutes with 1 minute intervals. Excitation = 485 nm; emission = 538 nm; cutoff = 530 nm.

Flow chart 3.1: Reactive oxygen species

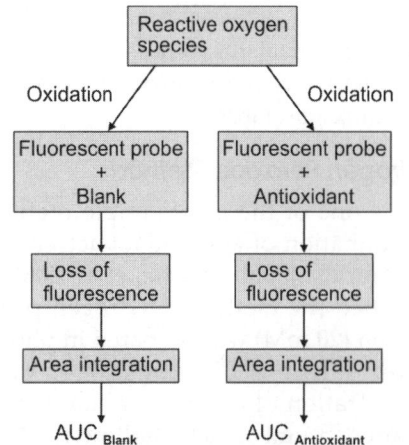

Antioxidant capacity will be determined by following formula

Antioxidant capacity = $AUC_{antioxidant}$ – AUC_{blank}, where AUC stands for area under curve.

The method was further described by Cao and Prior, 1999 and as per their study, the required reagents are:

Phosphate buffered saline: (PBS, 75 mM, pH 7.0) was prepared by first mixing 0.75 M K_2HPO_4 with 0.75 M Na_2HPO_4 in the respective ratio 61.1:38.9 (v/v). This mixture was diluted 1:9 (v/v) with Milli Q water and pH adjusted to 7.0.

Human serum: Blood (20 ml) was taken from a healthy volunteer (20–24 years; nonfasted) into serum clot tubes (Greiner Laboratories). These were centrifuged at 2500 × g for 10 min. Serum was removed and diluted further (1:100 v/v) in PBS before being used in the assay.

Phycoerythrin (R-PE): A stock solution of 0.17 mg/ml was prepared by dissolving 1 mg PE in 5.9 ml PBS. This was diluted further to give a working solution of 3.38 mg/L.

AAPH: (2,2'-azobis(2-aminopropane) dihydrochloride; molecular probes). A working solution of 320 mM was prepared fresh by adding 10 ml PBS to 868 mg AAPH in a universal container. This was stored on ice until used for analyses.

Trolox: (6-hydroxy-2,5,7,8-tetramethylchroman-2-carboxylic acid). A stock solution (100 mM) was prepared by dissolving 5.0 mg Trolox in 200 ml PBS. This was further diluted 1:4 v/v in PBS to give a working solution of 20 mM.

The method was followed as per the method described above.

Calculation

ORAC value (mM) = $20 k (S_{Sample} – S_{Blank}) / (S_{Trolox} – S_{Blank})$, Where k is the sample dilution factor.

Herbs as Antioxidant

Antioxidants are essential to better health. Herbs have been found to be ideal sources for antioxidants. They work to enforce immune system and make us stronger and healthier. It was seen that the intake of herbs contribute significant to the overall daily intake of antioxidants. Intake of 1 g of culinary herbs may contain 1 mM antioxidants. Culinary herbs are even a better source of antioxidants than other foods such as fruit, vegetables and cereals. Fruits and vegetables are the primary dietary sources of antioxidant but herbs have higher antioxidant activity than fruits, vegetables and some spices. As per research revealed, the use of fresh herbs showed greatest amount of antioxidants from culinary herbs. Drying of herbs tends to deplete the concentration of antioxidant compounds in the herbs and therefore, the use of only dried herbs will not be sufficient to give proper antioxidants to fend off free radicals in the body. For example, the antioxidant activity of fresh garlic is 1.5 times higher than dry garlic powder.

The scientists found that the medicinal herbs have strong antioxidant actions as expected. These herbs include periwinkle, ginkgo biloba, garden sage, St. John's wort, valerian and the sweet Annie herbs. The other herbs like rosemary, dill, thyme, peppermint are also showed good antioxidant content. Some popular food materials with their high antioxidant activity are:

Fruits

Berries (cherry, blackberry, strawberry, raspberry, crowberry, blueberry, bilberry/wild blueberry, black currant), pomegranate, grape, orange, plums, pineapple, kiwi fruit, grapefruit. Apricots, prunes, dates, figs, apples are having high antioxidant in dried form.

Vegetables

Kale, chili pepper, red cabbage, peppers, parsley, artichoke, Brussels sprouts, spinach, lemon, ginger, red beets, tomatoes, etc.

Legumes: Broad beans, pinto beans, soybeans, etc.

Nuts and seeds: Almonds, pecans, walnuts, hazelnuts, groundnut, pistachio, etc.

Cereals: Barley, millet, oats, corn, etc.

Spices: Cloves, cinnamon, oregano, etc.

Table 3.1: Antioxidants levels found in dried commercial herbs

Herbs	Antioxidant level (mM/100 g)
Clove	465
Allspice	102
Cinnamon	98
Rosemary	67
Thyme	64
Marjoram	54
Bayberry leaves	24
Ginger	22
Nutmeg	20
Mustard	10
Garlic	02
Coriander	02

Table 3.2: Some of the antioxidant plants with their functions

Herbs	Functions	Dosage
Green tea (Camellia sinensis)	Prevents skin, lung and stomach cancers, lowers blood pressure and LDL cholesterol levels to prevent heart disease and boosts immune function to protect against the flu	Tea: Two to three cups a day Capsules: 250 mg a day
Grape seed extract (Vitis vinifera)	Obtained from seeds of red and blue grapes. Protects capillaries. It also improves circulation, prevents blood clotting and lowers cholesterol levels	Capsules: 30 to 150 mg a day in divided doses
Garlic (Allium sativum)	Enhances immune function, destroys cancer cells, regulates blood pressure and cholesterol levels and inhibits blood clotting to prevent heart disease and stroke	Capsules: Two grams (g) a day of standardized extract with food
Ginkgo (Ginkgo biloba)	Enhances overall circulation, particularly in the brain's small blood vessels. Studies show it slows the development of Alzheimer's disease	Capsules or tincture: 120 mg a day. Look for supplements standardized for ginkgo flavo-glycosides
Billberry (Vaccinium myrtillus)	Protects arteries and capillaries and inhibits blood clotting	Capsules: 160 to 480 mg a day in divided doses
Asian ginseng (Panax ginseng)	Regulates blood pressure, reduces the effects of physical and emotional stress on the body, and promotes homeostasis	Tea: Two to three cups a day. Capsules: 250 to 500 mg a day

Table 3.3: The foods antioxidant power as per their serving sizes

Type	Food item	Serving size	Total antioxidant capacity per serving size
Beans/Legumes	Red beans (dried)	Half cup	13727
Fruit, berry	Wild blueberry	One cup	13427

Contd.

Table 3.3: The foods antioxidant power as per their serving sizes *Contd.*

Type	Food item	Serving size	Total antioxidant capacity per serving size
Beans/Legumes	Red kidney beans (dried)	Half cup	13259
Beans/Legumes	Pinto beans	Half cup	11864
Fruit, berry	Blueberry	One cup	9019
Fruit, berry	Cranberry	One cup (whole)	8983
Vegetable	Artichoke (cooked)	One cup (hearts)	7904
Fruit, berry	Blackberry	One cup	7701
Fruit	Prune	Half cup	7291
Fruit, berry	Raspberry	One cup	6058
Fruit, bery	Strawberry	One cup	5938
Fruit	Red delicious apple	One	5900
Fruit	Granny Smith apple	One	5381
Nut	Pecan	One ounce	5095
Fruit, bery	Sweet cherry	One cup	4873
Fruit	Black plum	One	4844
Vegetable	Russet potato (cooked)	One	4649
Beans/Legumes	Black beans (dried)	Half cup	4181
Fruit	Plum	One	4118

BIBLIOGRAPHY

1. Aebi, H., Catalase *in vitro*. Methods Enzymol. 105, 1984, 121–126.

2. Ashok, K.J., Imbalance in antioxidant defence and human diseases:Multiple approach of natural antioxidant therapy, Current Science. 2001, 1179–1186.

3. Babu, B.H., Shylesh, B.S. and Padikkala, J., Antioxidant and hepatoprotective effect of Alanthus icicifocus, Fitoterapia, 72, 2001, 272–277.

4. Benzie, I.F.F., Strain, J.J., The ferric reducing ability of plasma as a measure of antioxidant power: the FRAP assay. Analytical Biochemistry. 239, 1996, 70–76.

5. Berlin, K., Schaller, H., European standardized method for the deter-mination of δ-aminolevulinic dehydratase activity in blood. Z klin Chem Klin Biochem. 12(8), 1974, 389–390.

6. Blois, M.S., "Antioxidant determinations by the use of a stable free radical", *Nature* 26, 1958, pp. 1199–1200.

7. Boyne, A.F., Ellman, G.L., A methodology for analysis of tissue sulfhydryl components. Anal. Biochem. 46, 1972, 639–53.

8. Cao, G., Prior, R.L., The measurement of oxygen radical absorbance capacity in biological samples. *Methods enzymol.* v. 299 'Antioxidants and Oxidants' Part A; 50–62. 1999. Ed. Lester Packer; Academic Press.

9. Carlberg, I., Mannervik, B., Glutathione reductase. Methods Enzymol. 113, 1985, 484–490.

10. Desai, I. D., Vitamin E analysis methods for animal tissues. Methods Enzvmol. 105, 1971, 108–147.

11. Germano, M.P., Pasquale, R. D., D'Angelo, V., Catania, S., Silvari, V., Costa, C., Evaluation of extracts and isolated fraction from *Capparis spinosa* L buds as an antioxidant source. Journal of Agriculture and Food Chemistry. 50, 2002, 1168–1171.

12. Guzman, S., Gato, A., Calleja, J.M., Antiinflammatory, analgesic and free radical scavenging activities of the marine microalgae *Chlorella stigmatophora* and *Phaeodactylum tricornutum*. Phytotherapy Research. 15, 2001, 224–230.

13. Halliwell, B., Gutteridge, J.M.C., Aruoma, O.L., The deoxyribose method: a simple test tube assay for determination of rate constants for reaction of hydroxyl radicals. Analytical Biochemistry 165, 1987, 215–219.

14. Hidalgo, M.E., Quilhot, F.W., Lissi, E., Antioxidant activity of depsides and depsidones. Phytochem. 37, 1994, 1585–1587.

15. Jacqes-Silva, M.C., Nogueira, C.W., Broch, L.C., Flores, E.M.M., Rocha, J.B.T., Diphenyl disselenides and ascorbic acid changes deposition of selenium and brain of mice. Pharmacol Toxicol, 88, 2001, 119–125.

16. JayaPrakasha, G.K., Jaganmohan Rao, L., Sakariah, K.K., Antioxidant activities of flavidin in different *in vitro* model systems. Bioorg Medical Chem. 12, 2004, 5141-5146.

17. Kakkar, P., Das, B., Viswnathan, P.N., A modified spectrophotometric assay of superoxide dismutase. *Indian J. Biochem.* 21, 1984, 130–132.

18. Lapenna, D., Ciofani, G., Pierdomenico, S.D., Giamberardino, M. A., Cuccurullo, P., Reaction conditions affecting the relationship between thiobarbituric acid reactivity and lipid peroxides in human plasma. Free Radic Biol Med. 31, 2001, 331–335.

19. Marcocci, L., Maguire, J.J., Droy, M.T., The nitric oxide scavenging properties of *Gingo biloba* extract EGb761. Biochem. Biophys. Res. Commun. 15, 1994, 748–755.

20. Miller, N.J., Rice-Evans, C., Spectrophotometric determination of antioxidant activity. Redox Rep. 2(3), 1996, 161–171.

21. Misra, H.P., Fridovich, I., The role of superoxide anion in autoxidation of epinephrine and a simple assay for superoxide dismutase. J Biol Chem, 247, 1972, 3170–3175.

22. Omaye, S.T., Turnbull, T.D., Sauberlich, H.E., Selected methods for the determination of ascorbic acid in animal cells, tissues and fluids. Methods Enzymol. 62, 1979, 3–11.

23. Opoku, A.R., Maseko, N.F. and Terblanche, S.E., The *in vitro* antioxidative activity of some traditional Zulu medicinal plants, Phytother. Res., 16, 2002, S 51–S56.

24. Poungbangpho, S., Sang-Arun, J., Limmongkon, A., Suttajit, M., Optimization of H2O2 and metmyoglobin concentration for measuring antioxidant capacity and application to determine antioxidant capacity in plant extract. Naresuan Univ. J. 8(1), 2000b, 19–25.

25. Poungbangpho, S., Sang-Arun, J., Suttajit, M., The use of ABTS technique to determine total antioxidant capacity of plant extract. pp. 31–34. *In* The Second Bio-Systems Symposium and Work-shop. Faculty of Science, Khon Kaen University. 2000a.

26. Rice-Evans, C., Miller, N.J., Bolwell, P.G., The relative antioxidant activities of plant derived polyphenolic flavonoids. Free Rad Res. 22, 1994, 375–383.

27. Rotruck, J.T, Pope, A.L, Ganther, H. T., Swanson, A. B., Hafeman, D.G., Hoekstra W.G., Selenium: Biochemical role as a component of glutathione peroxidase. Science. 179, 1973, 588–590.

28. Sabu, M.C., Ramadasan, K., Anti-diabetic activity of medicinal plants and its relationship with their antioxidant property. J. Ethanopharmacol. 81, 2002, 155–160.

29. Sanchez-Moreno, C., Larrauri, J., Saura-Calixto, F., Free radical scavenging capacity of selected red and white wine, J. Sci.Food. Agric., 79, 1999, 1301–1304.

30. *Schlesier, K., Harwat, M., Böhm, V., Bitsch, R., Assessment of Antioxidant Activity by Using Different In Vitro Methods. Free radical Research, 36(2), 2002, 177–187.*

31. Sinha, K.A., Colorimetric assay of Catalase. *Anal.Biochem.* 47, 1972, 389–394.

32. Tepe B, Sokmen M, Akpulat HA, Daferera D, Polissiou M, Sokmen A. Antioxidative activity of the essential oils of Thymus sipyleus subsp. sipyleus var. sipyleus and Thymus sipyleus subsp. Sipyleus var. rosulans. J. Food. Eng. 66, 2005, 447–454.

ANTISEPTICS

Antiseptics are the chemical agents that inhibit or resist the growth of microorganisms on external surfaces of the body, i.e. skin and tissue. It helps to prevent infections, sepsis or putrefaction. Hence they are also known as antimicrobial substances. Antiseptics should be distinguished from antibiotics that destroy microorganisms inside the body, and from disinfectants, which destroy microorganisms found on inanimate (non-living) objects. Most chemical agents can be used as both an antiseptic and a disinfectant. The purpose for which it is used is determined by its concentration. For example, hydrogen peroxide 6% solution is used for dressing wounds, while stronger solutions (> 30%) are used as oxidizing and bleaching agent in industries. There is much variation in the ability of antiseptics to destroy microorganisms and in their effect on living tissue. For example, mercuric chloride is a powerful antiseptic, but it irritates delicate tissue whereas, silver nitrate kills fewer germs but can be used on the delicate tissues of the eyes and throat. There is also a great difference in the time required for different antiseptics to work. Iodine, one of the fastest-working antiseptics, kills bacteria within 30 sec. Other antiseptics have slower, more residual action. Some antiseptics are true *germicides*, which are capable of destroying microbes, i.e. bacteriocidal, while others are only prevent or inhibit their growth, i.e. bacteriostatic. Antibacterials are antiseptics that have the proven ability to act against bacteria especially if they target systems which kill only bacteria. Microbicides which kill viruses are called viricides or antivirals.

Types of Antiseptics

Antiseptics can be classified according to their chemical structure. Commonly used antiseptic groups include alcohols, quaternary ammonium compounds, chlorhexidine, antibacterial dyes, chlorine and hypochlorites, inorganic iodine compounds, metals, peroxides and permanganates, halogenated phenol derivatives and quinolone derivatives. Few are described here as:

- *Alcohols:* Commonly used alcohols are ethanol (60–90%), n-propanol (60–70%) and isopropanol (70–80%) or mixtures of these alcohols. They are commonly referred to as "surgical alcohol".
- *Quaternary ammonium compounds:* They are also known as *Quats* or *QACs*. They include benzalkonium chloride (BAC), cetyltrimethylammonium bromide (CTMB), cetylpyridinium chloride (cetrim, CPC) and benzethonium chloride (BZT). Benzalkonium chloride

is abundantly used in some preoperative skin disinfectants (conc. 0.05 – 0.5%), eye drop preservative and antiseptic towels. Cetalkonium chloride, dofanium chloride, domiphen bromide are also used chemical in this group.

- *Boric acid:* It is a known popular antiseptic. They are used in suppositories to treat yeast infections, in eyewashes, and as an antiviral to shorten the duration of cold sore attacks. It also mixed with creams to treat burns. Trace amounts used in eye contact solution.
- *Brilliant green:* It is a triarylmethane dye. It is used for treatment of small wounds and abscesses. It is mostly efficient against gram-positive bacteria.
- *Chlorhexidine gluconate:* It is a biguanidine derivative and used in concentrations of 0.5–4.0% alone or in lower concentrations in combination with other compounds, such as alcohols. It is used as a skin antiseptic and to treat inflammation of the gums.
- *Iodine:* Iodine, one of the fastestworking antiseptics, kills bacteria within 30 second. It is used as tincture of iodine (an alcoholic solution) or as Lugol's iodine solution as a pre- and postoperative antiseptic. Gentle washing with mild soap and water or rinsing a scrape with sterile saline is a recommended application. Novel iodine antiseptics containing povidone iodine (an iodophor, complex of povidone, a water-soluble polymer, with triiodide anions I_3^-, containing about 10% of active iodine). The great advantage of iodine antiseptics is the widest scope of antimicrobial activity, killing all principal pathogens and given enough time even spores, which are considered to be the most difficult form of microorganisms to be inactivated by antiseptics.
- *Hydrogen peroxide:* It is used as a 6% (20 Vols) solution to clean and deodorize wounds and ulcers. More common 3%

solutions of hydrogen peroxide have been used in household first aid for scrapes, etc. Gentle washing with mild soap and water or rinsing a scrape with sterile saline is a recommended practice.

- *Phenol (carbolic acid) compounds:* Strong solution of phenol is act as germicidal agent. It is used as a "scrub" for pre-operative hand cleansing. It is used in the form of a powder as an antiseptic baby powder. It is also used in mouth washes and throat lozenges.

Few Applications of Antiseptics

- *Hand washing:* Chlorhexidine gluconate and povidone iodine solutions are often used in hand scrubs.
- *Preoperative skin disinfection:* Antiseptics applied to the operation site to reduce the resident skin flora.
- *Mucous membrane disinfection:* Antiseptic irrigations may be instilled into the bladder, urethra or vagina to treat infections or cleanse the cavity prior to catheterization.
- *Preventing and treating infected wounds and burns:* Antiseptic preparations are available over-the-counter to treat minor cuts, abrasions and burns.
- *Treating mouth and throat infections:* Dequalinium chloride has both anti-bacterial and antifungal properties and is the active ingredient in antiseptic throat lozenges.

General Mode of Action

Antiseptics bind readily to bacteria, the amount absorbed increasing with an increasing concentration solution. The most important site of absorption is the cytoplasmic membrane.

The extent of killing of the bacteria is governed by three principal factors, viz. (i) concentration of the antiseptic, (ii) bacterial cell density and (iii) time of contact. The absorption of a given amount of the compound per cell leads to the killing of a definite fraction of the bacterial population in a specified time interval. In general, the necessary characteristic of antiseptics is their bactericidal action, but there is often a low and rather narrow concentration range in which their effect is baceriostatic. At these low concentrations, certain biochemical functions associated with the bacterial membrane may be inhibited. At higher concentrations and after prolonged treatment of antiseptics, the compound usually penetrates the cell and brings about extensive ill-defined disruption of normal cellular functions.

Herbs as Antiseptics

Antiseptics are mainly used to reduce levels of microorganisms on the skin and mucous membranes. However, when the skin or mucous membranes are damaged or breached in surgery, antiseptics can be used to disinfect the area and reduce the chances of infection. There are number of herbs available for use of this purpose. Hence, their role in this field cannot be ignored. Natural antiseptic herbs are used to kill or inhibit the growth of microorganisms when applied to living tissue or skin. It kills or inhibits the growth of microorganisms on the external surfaces of the body. They reduce the possibilities of infection, sepsis or putrefaction. Commonly, use strong antiseptic medicinal plants are like acacia, agave, anise oil, barberry, beth root, bilberry, birthroot, bistort, black walnut, buchu, calendula, chaparral, clove, cubeb, echinacea, eucalyptus, feverweed, garlic, golden seal, guggul, juniper berries, iceland moss, lavender, marigold, musturd, myrrh, nasturtium, nettles, olive, onion, oregon grape root, plantain, rosemary, sandalwood, saw palmetto, sage, sassafras, stevia, sweet gum, thyme, tormentil, white oak bark, white pond lily, white willow, wild indigo, wormwood, etc. Among them few of the herbs like lavender is the most popular anti-bacterial herb that can be use as homemade antibacterial skin cleanser. Rosemary has astringent properties. Patchouli helps regenerate skin cells, it has antiseptic, antifungal and antibacterial properties, eucalyptus is also an antibacterial herb. Tea tree is used for relieving skin irritation and skin problems where as Sandalwood is used as strong antiseptic and astringent herb. All members of the mint family have antiseptic, antifungal and antibacterial properties. An antiseptic skin washes are one of the formulations with herbs for treatment of skin problem and wound care.

Comfrey: Scientifically it is known as *Symphytum officinale* belongs to the family Boraginaceae. Its roots are used in all natural bath and beauty products. It has slight antiseptic properties. It was added to baths to promote youthful skin. These days it is mostly used to combat skin irritation.

Carrot: Scientifically, it is known as *Daucus carota* belongs to the family Apiaceae. Mash the boiled carrots and apply them to sores to extract pus and heal the area. It is a very strong antiseptic.

Cinnamon bark: Scientifically it is known as *Cinnamomum zeylanicum* belongs to the family Lauraceae. Cinnamon bark contains an active fragrant, volatile, antigermicidal oil. The bark is use in mouthwashes, gargles, combinations of herbal teas for taste and antiseptic ability.

Clove: Scientifically, it is known as *Eugenia caryophyllus* belongs to the family Myrtaceae. Clove oil is strong germicidal. Bruised cloves are use in teas and externally to deaden skin pain.

Eucalyptus: Scientifically, it is known as *Eucalyptus globulus* belongs to the family Myrtaceae. The leaves and stems of this fragrant plan are widely used in all natural recipes. This plant can be used in topical preparations for its antiseptic properties, but nontopical uses are more common. Creating steam infused the eucalyptus scent is used for clearing sinuses and lungs. It is, after all, in many popular cough drops.

Ginger: Scientifically, it is known as *Zingiber officinale* belongs to the family Zingiberaceae. Wild ginger contains a broad spectrum antibiotic active against bacteria and fungi. Ginger root or ginger powder added to herbal drinks at the onset of a cold causes perspiration and alleviates cold symptoms.

Rosemary: Scientifically, it is known as *Rosmarinus officinalis* belongs to the family Lamiaceae. An infusion of rosemary stems and flower tops is often used to help treat dandruff. It is act as an antiseptic for skin too. The infusion can prepare by soak rosemary in clear vinegar for a period of time. Strain this solution and apply to the scalp or use it as a rinse after shampooing.

Thyme: Scientifically, it is known as *Thymus vulgaris* belongs to the family Labiatae. The pure oil of the thyme plants is used as an antiseptic properties. The stems and leaves have can be crushed and used to infuse natural oil to create a useful mixture. This antiseptic skin wash is very popular in natural tooth-pastes and mouthwashes. Thymol may be used in miniscule amounts (one part to a thousand of water) for the dressing of wounds.

Sage: Scientifically it is known as *Salvia officinalis* belongs to the family Lamiaceae. Sage leaves added to tea or gargles for sore throat and other antiseptic action prepared by boiling in water for seven to ten minutes.

Evaluation of Antiseptics

Antiseptics, disinfectants and antibiotics are used in different ways to combat microbial growth. Antiseptics are used on living tissue to remove pathogens. Disinfectants are similar in use but are used on inanimate objects. Antibiotics are substances produced by living organisms, such as *Penicillium* or *Bacillus*, that kill or inhibit the growth of other organisms, primarily bacteria. Various methods exist for the examination of the antimicrobial activities of antiseptics. These methods include disk diffusion, phenol coefficient (for disinfectant), and dilution assays that determine minimal inhibitory and minimal bactericidal concentrations.

Disk diffusion test is also called the Kirby Bauer test. The disks containing an antimicrobial agent are placed on the surface of an agar plate containing a medium that has been inoculated with the disease agent being tested, which will grow and fill the disk. The antimicrobial agent diffuses into the medium, killing some of the disease agent around where the anitmicrobial agent was innoculated, depending on how susceptible the disease agent is to the antimicrobial agent. A zone of inhibition develops around the disk if the organism is unable to grow in the presence of the antimicrobial agent. The limitations of this assay include the solubility and size of the molecule. This assay is quick and easy to perform. Many conditions can affect a disc diffusion susceptibility test. Conditions that must be constant from test-to-test include the agar used, the amount of organism used, the concentration of chemical used and incubation conditions, viz. time, temperature and atmosphere. The amount of organism used is standardized using a turbidity standard. This may be a visual approximation using a McFarland standard 0.5 or turbidity may be determined by using a spectrophotometer (optical density of 1.0 at 600 nm).

Phenol coefficient test is a method for evaluating water-miscible disinfectants in which a test organism is added to a series of dilutions of the disinfectant. It is a measure of the bactericidal activity of a chemical compound in relation to phenol. To calculate phenol coefficient, the concentration of the test compound at which the compound kills the test organism in 10 minutes, but not in 5 minutes, is divided by the concentration of phenol that kills the organism under the same conditions. In standard phenol coefficient assays the traditional organisms used are *Pseudomonas aeruginosa*, *Staphylococcus aureus*, and *Salmonella choleraesuis*. Following 5-, 10-, and 15-minutes exposure to the chemical and subsequent incubation in nutrient broth, the tubes are examined for the bactericidal properties of the chemical as compared to phenol. For example, the Rideal-Walker method gives a Rideal-Walker coefficient.

Assays to determine minimum inhibitory concentrations (MICs) and minimum bactericidal concentrations (MBCs) of antimicrobial chemicals are more useful. To determine MI, a known amount of organism is incubated with increasing amounts of the antimicrobial agent. Following a 24-hour incubation period, the tube containing the lowest amount of the antimicrobial agent that exhibits no growth is defined as the MICs. Aliquots of each tube are then removed and placed in fresh broth containing no antimicrobial agent. Following an additional incubation period, the tubes can be examined for growth. The tube containing the least amount of the antimicrobial agent that still shows no growth is considered the MBCs.

PRESERVATIVES AND PROTECTIVES

In general the preservatives are defined as a natural or synthetic substances or chemicals that are added to the products such as foods, pharmaceuticals, paints, biological samples, wood, etc. to prevent decomposition or degradation of products by microbial growth or by undesirable chemical changes. The most important thing is that the cosmetic product should and must be safe for use. Just like food, all natural skin care products will eventually deteriorate and become rancid. Hence preservatives can play a vital role in cosmetic products as an ingredient in formulation and prevent the products from any degradation, i.e. to control microbial growth and to stabilize any cosmetic product.

Important properties of ideal preservatives
Preservatives can overcome the broad spectrum of microbes (bacteriae and fungi), should not to be harmful to the skin and deleterious to other ingredients in a cosmetic product. Apart from they should have some other propreties likely:

a. They should be effective over the anticipated shelf life.
b. They should preferably liquid and water soluble.
c. They should effective over a wide pH range.

d. They should not be deactivated by other ingredients.

e. They should odorless, colorless and safe.

f. They should be free of toxic effects and non-irritating on the skin, mucous membranes and in the intestinal system.

Mechanism of Action of Preservatives

A preservative actually is a protective agent to keep cosmetics in safe and stable through killing or inhibiting microorganisms in cosmetic products. It does not show strong instant killing effect until the preservative contacted the microbes' cells directly at much concentration. They can destroy the cleavage of cells to inhibit the microbes grow by acting on the target nucleus of microbes, such as cell membranes, cell wall, enzymes of cell. Actually, it affected the activity of enzymes or the structure of heredity particles in bioplasm, nothing but they interfere cells growth through inhibiting synthesis of enzymes, proteins and nucleic acids. For example, the acting target nucleus of phenoxyethanol or alcohol was cell membrane, bronopol (2-bromo-2-nitropropane-1, 3-diol) acted on the SH enzymes, etc.

Factors of Affecting Activity of Preservatives

An efficacy of the ideal preservative are depends on many factors. The preservative may be highly effective in a special system but it may be of no efficacy in others. Many factors can cause the preservative inactivation in a formula. The following are several major factors that influence the efficacy of preservatives.

a. *Influence of pH:* The pH of the system can affect the activity of preservatives through influencing the dissociation of organic acid. For example, bronopol (2-bromo-2-nitropropane-1, 3-diol) is very stable when its pH is 4; it also could keep active for a year when its pH is 6; but its activity only lasts several months when its pH is 7.

b. *Influence of particles and gels:* Some particles like kaolin or magnesium

Aluminum silicate in cosmetics could reduce the preservative activity because of its adsorption of preservatives but the particles could sometimes strengthen the bacteriostatic effects through adsorbing the microbes as well. The activity of preservatives may be affected by combination with water-soluble high molecular weight polymer, because they can reduce the concentration of free preservatives in the formulation.

c. *Influence of nonionic surfactant:* All kinds of surfactants in cosmetics, especially the nonionic surfactants, can interfere the activity of preservatives by solubilization and complexation methods. The oil soluble nonionic surfactant (HLB value is 3–6) has more deactive effect to preservatives than the water soluble one with higher HLB value. The activity of preservatives decreases with its free concentration reduction. So, the use level of preservative should be increased in the nonionic surfactants system in order to keep the sustainable preservation power.

d. *Influence of preservative decomposition:* Some factors may reduce the preservation efficacy through the preservatives' decomposition. For example, light or heat cause preservative decomposition, chemical reaction may lead to preservation inactivation or radiation sterilization may reduce the action of preservatives. Proper care should be taken during adding the preservatives in the formulations.

e. *Other influencing factors:* In addition, there are many other factors, such as preservative distribution in oil/water phase, packaging, fragrance, chelating agents, may all interfere with preservative activity.

By looking at these properties the preservatives can categorized by synthetic or chemical and natural preservatives. Chemical preservatives are generally used because they are much cheaper and extend the shelf life of the product more than natural alternatives. But synthetic preservatives can cause allergies in susceptible people, including dermatitis and other side effects. Not only that, natural preservatives can also be harmful if they are uses in excess but natural preservative are not as toxic as synthetic preservative. Natural preservative are much safer because they are widely existed in nature and are known to the immune system of the body. Most preservatives derived from plants are safe for humans but the main drawback against natural preservative is that they are no powerful enough. Ideally no preservatives should be used but many products would not stay fresh even for short time without them. Hence preservatives should add to the cosmetic formulations in balance amount. Some of the synthetic preservatives which are commonly use by the cosmetic industries are Imidiazolidinyl urea (Germall 115) and diazolidinyl urea, DMDM hydantoin, methyl paraben, propyl paraben, 2-bromo-2-nitro-propane-1 3-diol, benzalkonium chloride, chloromethylisothiazolinone and isothiazolinone, butylated hydroxytoluene (BHT) and butylated hydroxyanisole (BHA), methylisothiazolinone, etc. All these chemicals are having side effects commonly cause contact dermatitis, allergies and skin rashes. Whereas, commonly used natural preservatives are: Tea tree essential oil, Thyme essential oil, grapefruit seed extract, d-alpha tocopherol acetate (vitamin E), citric acid, neem oil, lemon, honey, bee propolis, rosemary extract, etc.

Tea tree oil: It is also known as Melaleuca oil. It is with pale yellow color fixed oil. The oil is procure from leaves of tea tree, scientifically known as *Melaleuca alternifolia*, belongs to family Myrtaceae.

Thyme oil: The oil is obtained from thymus vulgaris, belongs to the family Labiatae. It is mainly uses as antiseptic agent.

Grapefruit seed extract: It is obtained from *Vitis vinifera* belongs to the family of Vitaceae. It is rich in vitamin E, flavonoids, etc. It is a natural antibiotic, antiseptic, disinfectant and preservative. It is used to promote the healing of almost any atypical skin condition.

Alpha tocopherol acetate: It is also known as vitamin E. It is an antioxidant and nutrient. Vitamin E is abundant in whole wheat, rice germ, and vegetable oils. It is destroyed by the refining and bleaching of flour. Vitamin E is used because it prevents oils from going rancid. Recent studies indicate that large amounts of vitamin E may help reduce the risk of heart disease and cancer.

Citric acid: It is an important metabolite in all living organisms and is especially abundant naturally in citrus fruits and berries. It is an acid, flavoring and chelating agent in ice cream, sherbet, fruit drink, candy, carbonated beverages. It is used as an powerful antioxidant.

Neem oil: Neem oil is one of the most powerful oils obtained from seeds of *Azadirachta indica*, belongs to the family Meliaceae. The oil having great demand on the market today. It is antifungal, antibacterial as well as anti-protozoan and a spermicide.

Lemon: It is obtained from *Citrus lemon* belongs to family of Rutaceae. The acid produced mostly by the citrus and identified as $C_6H_8O_7$ promotes preservation. The lemon is rich in vitamin C and much like salt removes moisture to prevent spoilage and rotting.

Honey: It is known for being highly stable against microbial growth because of it is low moisture content and water activity, low pH and antimicrobial constituents. It is the saccharine liquid prepared from the nectar of the flowers by the hive bee *Apis mellifera* belongs to family Apidae.

Bee propolis: It is a mixture of beeswax and resins collected by the honey bee from plants, particularly flowers and leaf buds, it is used to line and seal the comb. The propolis is effective in protecting the hive offering both antibacterial and antifungal properties.

Table 3.4: Effectiveness of essential oils in killing bacteria

Essential oil	Minimum amount in %	Essential oil	Minimum amount in %
Thyme	0.070	Rosemary	0.430
Origanum	0.100	Cumin	0.450
Sweet orange	0.120	Neroli	0.475
Lemongrass	0.160	Birch	0.480
Chinese cinnamon	0.170	Lavender	0.500
Rose	0.180	Melissa balm	0.520
Clove	0.200	Ylang ylang	0.560
Eucalyptus	0.225	Juniper	0.600
Peppermint	0.250	Sweet fennel	0.640
Rose geranium	0.250	Garlic	0.650
Meadowsweet	0.330	Lemon	0.700
Chinese anise	0.370	Cajeput	0.720
Orris	0.380	Sassafras	0.750
Cinnamon	0.400	Heliotrope	0.800
Wild thyme	0.400	Fir, pine	0.860
Anise	0.420	Parsley	0.880
Mustard	0.420		

Rosemary extract: It is obtained from the plant of *Rosmarinus officinalis*, belongs to the family Lamiaceae. It helps against aging processes, viz. browning, thickening and wrinkling, etc. Melanoma and other skin cancers are thought to be accelerated by the accumulation of peroxides in the skin tissues. These peroxides are produced by environmental factors such as heat and ultraviolet radiation from sunlight, a primary cause of sunburn and melanoma. Carnosic acid which is isolated from rosemary, can protect the skin from UV damage. It is also has antibacterial and antimicrobial properties and found to be effective against HIV-1.

Essential oils have also been shown to be effective in killing the viruses and assist in healing the affected skin. Products in aromatherapeutic preparations, viz. foam baths, soaps, bath oils and massage oils, etc. are prepared with a large content of essential oils because of their own antiseptic properties.

Evaluation Parameters of Preservatives in Cosmetic Preparations

Preservation has drawn increased attention in cosmetic manufacturing and distribution process nowadays due to its important role in cosmetic formulation. Preservatives in a cosmetic product, keeps the product safe and stable. With rapid growth of cosmetic industry, a vast number of antimicrobial agents are widely used. Various approaches to preservative efficacy testing (PET) have been developed over the years by regulatory agencies, standards organizations, industry organizations and individual companies, viz. USP method, BP method, and EP method. Of the various protocols and approaches proposed and developed, the microbial challenge test has evolved as the most commonly used and accepted evaluation criterion.

Microbial challenge test: This test is established by international practice, is a most widely used method to evaluate the efficacy of preservatives. The test is performed for a period of time to simulate clinical scenarios in the process of manufacturing and handling. In this test, samples are inoculated with test microorganisms and then inspected visually and by plate count to determine if the preservative system is effective.

Principle: The test is based on the concept of measuring the survival ability of selected microorganisms that are purposely introduced into a preserved test product system. Conventional preservative efficacy tests or preservative challenge test methods generally require microbial assays at multiple test points over extended periods of time. Test durations typically range from a minimum of 28 days to 12 or more weeks.

Some preservative challenge tests some time added relatively large inocula of various laboratory culture to aliquots of the formulation and determine their rate of inactivation by viable counting methods, known as single challenge test. Some time re-inoculatation repeatedly at set intervals and monitoring the efficacy of inactivation until the system fails is known as multiple challenge test. This technique will give better estimation of the preservative capacity but the method is very time-consuming and expensive.

The CTFA method is the internal method for cosmetic, toiletries and fragrance Association in USA. Besides standard USP method and CTFA method, there are some rapid evaluation techniques, such as linear regression method, presumptive challenge test and accelerated preservation test methods. There were some differences in the time needed for analysis and performance criteria for all the tests and could lead to different results. These methods are given in Table 3.5 with various aspects.

Table 3.5

Items	USP method	CTFA method	Linear regression method
Strains	*Staphylococcus aureus, Escherichia coli, Pseudomonas aeruginosa, Candida albicans* and *Aspergillus niger*	Gram-positive and gram-negative bacteria, mold, yeast and aerobic spore-forming bacillus some strains separated from antipreservative samples	*Staphylococcus aureus, Escherichia coli, spore forming bacillus, Aspergillus niger* and *Aspergillus flavus*
Inoculums level Unit:cfu/ml(g)	1.0×10^5–1.0×10^6	1.0×10^6 cfu/ml (g) the level of yeast and mold in eye-cosmetic is 1.0×10^4 cfu/ml (g)	1.0×10^6
Measuring process of sampling	0, 7, 14, 21 and 28 days after original inoculation	0, 7, 14, 21 and 28 days after original inoculated. Take sample additional after 28 days if it was re-challenged	0, 2, 4, 24, 28 and 168 hours after original inoculated. Or up to the last APC is less than 10 cfu/ml (g).
Performance criteria	A 99.9% reduction of bacteria by the 14th day. The concentration of viable yeasts and molds remain at or below the original inoculated within 14 days. The concentration of test organism remains at or under required levels of the 28-day test period	A more than 99.9% reduction of bacteria within 7 days. The concentration of yeast and mold reduced and continued reduction for the duration of the test. Bacteriostatic activity against spores through the entire test	The D value of pathogenic bacteria is less than 4 hours; the D value of non-pathogenic bacteria is less than 28 hours. [Linear egression method is expressed by 'D' value and determined the decimal-reduction time of microorganism in challenge test]

The term protective is use for versatile meaning. In this section it is mainly deals with protection of skin from any stressed conditions especially from sunlight. There are 3 types of burns, viz. 1st degree, 2nd degree and 3rd degree burns. 1st degree burns are described as having the red color and pain. Second degree burns are accompanied by the red, pain, as well as water bubbles. Third degree burns are all of the above and including patches of white skin. This is the most dangerous form of sunburn.

Drinking water helps in flushing out the toxins from the body. It helps not only skin healthy but also promotes great health for all organs in the body. General cleanliness is another inexpensive way of herbal skin protection. Daily shower, wearing clean clothes and sleeping on a clean mattress/pillow are all part of general cleanliness. Regular exercise also can increase the flow of blood that helps in getting rid of body toxins and keeping you healthy. Healthy food and eating habits are also recommended for herbal skin protection. Oily type of food may cause spots and should be avoided as much as possible. Raw fruits and vegetables are known to provide freshness to the body and help in getting rid of body toxins. Protection of the skin may be using chemical agents or by using herbals. But usage of chemical agents in longer times may cause several skin infections instead of protection. Hence, these days the role of herbals cannot be ignored. Herbs are use in maintaining and enhancing human beauty because herbs have many beneficial properties, such as sunscreen, antiaging, moisturizing, antioxidant, anticellulite and antimicrobial effects. Generally, they are use in combinations to get the effective action against any stresses. Skin damages are mostly related with the sunlight and air pollution. Excessive sun-exposure can lead to loss of moisture from the skin, dark patches and even premature aging. The skin needs physical protection against harmful rays which is provided by sunscreens. They can protect against the UV rays of the sun.

Evaluation Test for Measurements of Sunscreen Protection

Sun Protection Factor (SPF)

The sun protection factor of a sunscreen is a laboratory measure of the effectiveness of sunscreen. The more protection a sunscreen shows against UV-B (the ultraviolet radiation that causes sunburn) with the higher SPF. The SPF is the amount of UV radiation required to cause sunburn on skin with the sunscreen on, relative to the amount required without the sunscreen. It is mainly a measure of UV-B protection and ranges from 1 to 45 or above. In practice, the protection from a particular sunscreen depends on several factors such as:

- The skin type of the user
- The amount applied and frequency of re-application
- Physical activities in which one engages (for example, swimming leads to a loss of sunscreen from the skin)
- Amount of sunscreen absorb by the skin.

The SPF is an imperfect measure of skin damage because invisible damage and skin aging are also caused by ultraviolet type A (UV-A, wavelength 320 to 400 nm), which does not cause reddening or pain but causes DNA damage to cell deep within the skin and resulting the risk of malignant melanomas. Hence, broad spectrum sunscreens are designed to protect against both UV-B and UV-A. The best UV-A protection is provided byproducts that contain zinc oxide, avobenzone and ecamsule.

The SPF can be measured by applying sunscreen to the skin of a volunteer and measuring how long it takes before sunburn occurs when exposed to an artificial sunlight source. It can also be measured *in vitro* with the help of a specially designed spectrometer. In this case, the actual transmittance of the sunscreen is measured, along with the degradation of the product due to being exposed to sunlight. In this case, the transmittance of the sunscreen will be measured over all wavelengths in the UV-B range (290–320 nm).

Mathematically, the SPF is calculated from measured data as:

$$SPF = \int A\ (\lambda)\ E(\lambda)d\lambda / \int A\ (\lambda)\ E(\lambda)/MPF\ (\lambda)\ d\lambda$$

where, $E(\lambda)$ is the solar irradiance spectrum, $A(\lambda)$ the erythemal action spectrum, and $MPF(\lambda)$ the monochromatic protection factor, all functions of the wavelength ë. The MPF is roughly the inverse of the transmittance at a given wavelength.

Persistent Pigment Darkening

The persistent pigment darkening (PPD) method is an another method of measuring UV-A protection. The method was originally developed in Japan. Instead of measuring erythema or reddening of the skin, the PPD method uses UV-A radiation to cause a persistent darkening or tanning of the skin. The PPD method is an *in vitro* and *in vivo* test method.

Star Rating System

The boots star rating system is a proprietary *in vitro* method used to describe the ratio of UV-A to UV-B protection offered by sunscreen creams and sprays. Based on original work by Prof. Brian Diffey at Newcastle University, the boots company in Nottingham, UK, developed a standard method which has described as one-star products provide the least ratio of UV-A protection and five-star products are best. The method still uses a spectrophotometer to measure absorption of UV-A vs UV-B; the difference stems from a requirement to preirradiate samples (where this was not previously required) to give a better indication of UV-A protection, and of photostability when the product is used.

Mechanism of Skin Protection due to Sunlight

The principal ingredients in sunscreens are usually aromatic molecules conjugated with carbonyl groups. This general structure allows the molecule to absorb high-energy ultraviolet rays and release the energy as lower-energy rays, thereby preventing the skin-damaging ultraviolet rays from reaching the skin. Some sunscreens also include enzymes like photolyase, which are claimed to be able to repair UV-damaged DNA.

Role of Herbs as Skin Protectants

Skin care herbal is used not only for the routine nourishing of skin but also for the treatment of skin disorders like eczema and psoriasis. Most body care products based on plants have no side effects (the most important reason to prefer them to synthetic products). Moreover, the products of skin care herbal can be easily at home, making them more attractive. Thus, skin care herbal is the way forward. Plants have long been the focus of studies concerning sun protection, as they are exposed to the sun daily with minimal damage. Oils extracted from plants can be used in almost all aspects of daily care and it is only consider that sun protection be one of these aspects.

Herbs Against Sun Protection and other Skin Problems

Alexandrian laurel: Alexandrian laurel, more commonly known as the Ballnut tree, (*Calophyllum inophyllum*), family of Gutteriferae or Clusiaceae. It contains calophyllolide, fatty acid methyl ester that are believed to regenerate skin tissue. The oil has no toxicity, which means that it can be used on the skin with no irritations or adverse effects. The oil can also be added to eye drops to provide natural sun protection for the eyes. The oil is also known as Pinnai or Dilo oil.

Alfalfa oil: Alfalfa, also known as Lucerne. Scientifically, the plant is known as *Medicago sativa*, family of Fabaceae. The oil can be used as a pretreatment along with other sunscreens. It contains rich in minerals and vitamins. It is known to reduce the damage caused by sunburn and is also a rich source of antioxidants. Traditional Indian medicines used alfalfa leaves for treating skin conditions, such as sores or boils. Alfalfa extracts were also used in Indian medicines for the treatment of water retention.

Almond oil: The oil is obtained from the herb of *Prunus dulcis* (Family: Rosaceae). The oil does reduce the aging effect that UV light causes on the skin, preventing skin damage and aging. The oil contains fat and carbohydrates. The oil can be used as a daily moisturizer instead of commercial moisturizers, hydrating the skin and providing a natural skin protection.

Caution: A high rate of allergic reactions can be attributed to almonds and almond extracts, so the oil should be tested on a small section of skin before used as a moisturizer.

Aloe: This is a very common and easily available ancient natural skin care herb protects the skin's immune cells against the damaging effects of the sun. Researches have been proved its sun protection activity. Scientifically, the herb is known as *Aloe vera*, belongs to the family Liliaceae. The herb is belongs to the C-anthracene glycoside. Fresh *Aloe vera* may be applied after any type of sun exposure. Its anti-inflammatory, moisturizing, emollient, and antimicrobial actions make it especially useful for preventing sunburn damage due to the presences of its chemical constituents like aloin, barbaloin and others.

Calendula: Commonly this plant is known as marigold. Scientifically, known as *Calendula officinalis* belongs to the family Asteraceae. This herb has been used clinically for skin conditions including sunburn. The flowers are used for its healing and tissue regenerating properties for all types of skin. It is use in cell regeneration especially in cases of sunburn and sores where the skin is red and irritated. It is also good as a regular wash for spots and boils.

Carrot-seed: An essential oil extracted from carrot seed (Botanical Name: Roots of *Daucus carota*, family: Apiaceae) which is especially beneficial to sun-damaged skin and is even used to treat precancerous skin

conditions. The β-carotene which is main active constituent has been proven to protect against ultraviolet-induced skin cancer.

Chamomile: This herb is known as *Matricaria recutita,* belongs to the family of Asteraceae. Main useful parts are flowers. Chamomile herbs have well known sedative and relaxing properties. It is use in skin and hair care for its cleansing and soothing benefit. This plant is also use for sensitive and irritated skin and for scalp irritations. The main constituents, viz. farnesene, chamazulene, terpene bisabolol, flavonoids, etc. are responsible for these beneficial actions.

Chichweed: This is the whole plant scientifically known as *Stellaria media* and other species. Family belongs to Caryophyllaceae. This plant is best known in its use for treating itchy skin, to treat skin allergies and other skin problems due to the presence of rich ascorbic acid content and β-carotene.

Cleavers: This plant is scientifically known as *Galium aparine* (arial parts) belongs to the family of Rubiaceae. The variety of tannic acids, rubichloric acid, citric acid, glycoside asperuloside in cleavers have an strong astringent and antiseptic effect and this makes the herb good for healing burns and wounds.

Coconut oil: The oil is extracted from the tree of *Cocos nucifera* (Family: Arecaceae). The oil has been found to absorb a large majority of the sun's rays. Coconut milk can also be extracted directly from a fresh coconut to produce the same results. Coconut oil and milk are also natural skin moisturizers and can be applied to the skin daily instead of commercial moisturizers, providing a natural daily sunscreen. The activity is mainly due to the presence of fatty acids, proteins.

Comfrey: The leaves of this herb are useful and the herb is known as *Symphytum officinale* and other species, belonging to the family Boraginaceae. The herb has rich in carbohydrate content (inulin) and along with that allantoin, steroidal saponin, tannins. It is considered to rejuvenate the skin and to have healing, soothing and moisture retaining properties. They are good for rough, damaged skin and can, with time, alleviate wrinkling and enable skin tissue to regain its youthful elasticity. Allantoin promotes skin cell regeneration, stimulates the growth of new cells and helps sensitive skin to become more resilient, counteracting dryness and cracking.

Cornflower: This herb is known as *Centaurea cyanus* (Family: Asteraceae). Main useful part is flowers. It is used as a face toner for its soothing and astringent effect on the skin. It has antibacterial as well as antioxidant properties. It is also soothes inflamed and irritated skin. It is very effective to use for washing out wounds and for mouth ulcers.

Echinacea: The herb with its root and aerial parts is known as *Echinacea purpurea*, belongs to the family of Asteraceae. The presence of phenol, cichoric acid, caftaric acid it is use to regulate and soothe the skin and to enhance overall immunity. It is use in ointments, creams, compresses and toners for wounds, skin regeneration, skin infections and inflammatory skin conditions.

Green tea: Scientific researchers have proved that green tea is a promising skin protectant against sunburn. Scientifically, it is *Camellia sinensis* belongs to the family of Theaceae. Its may be applied topically or consumed as a beverage or dietary supplement. It protects against direct damage to the cell and moderates inflammation. Its having strong antioxidant power due to presence of catechins.

Hemp seed oil: The oil is obtained from Cannabis genus, like *Hibiscus cannabinus*, belongs to the family of Cannabaceae. It contains high 3:1 ration of omega-6 to omega-3 essential fatty acid. The nutrients that are found in hemp seed oil, closely resembles the naturally occurring molecules in the human body. These molecules are known as lipids and include elements such as bodily waxes, fats and fat-soluble vitamins and hence the nutrients are easily absorbed into human skin, providing added sunscreen.

Hemp seed oil is also stimulate keratin formation in the human body, adding many benefits to the skin due to the presence of essential amino acids and fatty acid.

Lavender: This is the herb of *Lavandula angustifolia*, family: Lamiaceae. The flowers are useful and considered to be beauty product. Lavender has many sedative compounds like lavandin, camphor that can penetrate the skin. It is good for normal and dry skin, used as a tonic for all types of skin and also can be used in deodorant.

Nettle: This plant is the entire herb of *Urtica dioica* and other species of *Urtica*, belongs to the family of Urticanceae. It is used as a compress, in lotions and creams for oily skin. It is supposed to be cleansing, clarifying and emollient and soothes sensitive skins with surface blood capillaries due to the presence of histamine, 3,4'-divanillyl-tetrahydrofuran, etc. It is used extensively in hair care products where is supposed to counteract hair loss, prevent and cure dandruff, reduce oily secretions and improve the quality of dry, lifeless hair and stimulates the scalp. It is also be used in deodorants.

Peppermint: The herb is commonly known as *Mentha piperita L* (family: Lamiaceae). It cools, refreshes, stimulates and revitalizes. It is considered to restore elasticity to the skin, to tone tissue, close pores, reduce swellings, counteract bad smells, reduce redness and irritation and clarify the skin due to the presence of menthone, methyl esters, methyl acetate, etc.

Rose petals: This herb is known as *Rosa damascene* and other species of Rosa, belongs to the family of Rosaceae. Rose oil, rose water and an infusion made from rose petals are used in skin care products for tender, dry, sensitive skin and have a cleansing, astringent, toning, moisture retaining, stimulating and soothing effect due to the presence of volatile oil. When rose is included in a cream or lotion, it stimulates and protects the skin, while moisturising and hydrating it. It gives an effective use to dry, mature and sensitive skin.

Rosemary: The leaves of the herb is *Rosmarinus officinalis,* belongs to the family Lamiaceae. Rosemary is used as a fragrant additive in soaps and other cosmetics. An infusion can also be used as an invigorating toner and astringent. This plant is considered to stimulate blood circulation and to restore elasticity to the skin due to the presence of ursolic acid, rosmanol, rosmarinic acid, caffeic acid. It is also effective for oily skin, blackheads and spots.

Senna: This is well known common plant in India which is mainly use as laxative purpose. The plant is known as *Cassia angustifolia* belongs to the family of Leguminosae. The plant is belongs to O-anthracene glycoside. Its powdered leaves in vinegar are very much useful against wounds and burns and to also remove pimples.

St. John's wort: The herb is scientifically known as *Hypericum perforatum* belongs to the family of Clusiaceae. It soothes irritations of the skin and it is also makes the skin photo-sensitive due to the presence of hyperforin, amentoflavone so the skin should not be exposed to sunlight after St. John's wort has been applied externally. It is use as in a cream, ointment or macerated oil forms.

Tropical ferns: Golden polypody (golden serpent fern or cabbage palm fern) is a tropical fern native to America. This fern has been found to contain beneficial antioxidants, as well as properties that lower the risk of skin damage caused by the sun. The fern can help to maintain the skin's elasticity and most importantly, can reduce the risk of skin cancer. In medicine, extracts from this fern have been used to treat serious skin conditions, such as psoriasis.

Samambaia (scientifically known as *Polypodium decumanum* and other species of Polypodium, family: Polypodiaceae) is a fern that grows in the rainforests of South America as well as drier tropical forests in Latin America. The plant extract for centuries for the treatment of inflammatory disorders and skin diseases. Recently, clinical research has shown that it has antioxidant and photoprotective properties. The extract of the plant is taken orally and provides protection against the harmful effects of ultraviolet (UV) radiation from the sun and other sources. The plant is rich source of lipid and fatty acids. The chemicals are identified include adenosine, alkaloids, arachidonic acid, arabinopyranosides, calagualine, ecdysone, ecdysterone, eicosapentaenoic acid, elaidic acid, juglanin, kaempferols melilotoside, oleic acid, ferulic acid, polypodaureine, ricinoleic acid, rutin, selligueain, and sulfo-quinovosyldiacylglycerols.

Witch hazel: This herbs is known as *Hamamelis virginiana,* of family Hamamelidaceae. It is use in the cosmetic and toiletry industry in face toners, shaving creams, face packs and aftershave creams. The steam distillate, infusion or tincture of fresh leaves is used because of its astringent and anti-inflammatory effects, and has also been recommended for certain skin conditions, insect bites, bruises, etc. Its high tannin content has a strengthening astringent effect on veins. It can be used as a facial toner for oily skin and to control minor pimple formation.

Wild pansy: This herb is commonly known as *Viola tricolor* belongs to the family of Violaceae. It is healing, cleansing and soothing and is used in compresses, baths, steam baths and in creams and ointments for skin care. With combined application of this herb and comfrey, it is reduced pores and rejuvenate the skin due to the presence of rutin, violanthin, violutoside, scoparin, orientin, etc.

4

Introduction to Topical Cosmetics and their Evaluations

Cosmetic is a Greek word which means to adorn, i.e. addition of something decorative to a person or to a thing. In one sentence it is define as a substance which comes in contact with various parts of the human body like skin, hair, nail, lips, teeth, etc. The most popular cosmetics are hair dyes, powders, lotions, lipsticks, creams, nail polishes, face make ups, baby products, deodorants, hair colorants, sprays and toiletries. Hence cosmetics are broadly categorized into four types, viz. skin cosmetics, hair cosmetics, Nail cosmetics and cosmetics for hygiene purpose. Hence, the herbal cosmetics can be classified based on dosage form and part or organ of the body to be applied for.

a. Dosage forms: **(i) Emulsions:** Cold cream, vanishing cream, liquid cream, etc. **(ii) Powders:** Face powder, talcum powder, tooth powder, etc. **(iii) Oils:** Hair oils, **(iv) Mucilages**: Hand lotion, **(v) Solutions:** Aftershave lotion, Hair set solutions and lotions, **(vi) Suspensions:** Cosmetic stockings, **(vii) paste:** Tooth paste, deodorant paste, **(viii)** Soaps: Shampoo soap, shaving soap, Toilet soap, etc. **(ix) Jellies:** Hand jelly, wave set jelly, etc.

b. Part or organ of the body to be applied for: **(i) Herbal cosmetics for skin:** Powders, creams, lotions, deodorants, bath and cleaning products, Make-up preparations, **(ii) Herbal cosmetics for hair:** Shampoos, tonics, hair dressing, shaving media, **(iii) Herbal cosmetics for nails:** Nail polishes and polish removers, manicure preparations,

(iv) Herbal cosmetics for teeth and mouth: Tooth paste, dentifrices, mouth washes, **(v) Kindered preparations:** Eye preparations, foot powders, insect repellants, etc.

HAIR CARE

Hair Conditioner

The term hair conditioners are those which are used for the attractive, healthy looking hairs. They should be capable of giving life, softness, body and silky touch, control of flyaway and ease of styling to the hair. Hair conditioners fall into different groups according to what people want to accomplish with their hair. People with different types of hairs need a specific kind of "conditioner". Broadly, hair conditioner can be divided into six major categories viz. moisturizers, reconstructors, acidifiers, detanglers, thermal protectors, glossers and oils.

Moisturizers: They are concentrated with humectants. Humectants are compounds that attract and hold moisture into the hair. *Recommendations:* Sunset hair elements moisture plus.

Reconstructors: They normally contain protein. Hydrolyzed human hair keratin protein is the best source of it, because it contains all 19 amino acids found in the hair. The main purpose of a reconstructor is to strengthen the hair. *Recommendations:* Sunset hair elements hair rebalancer.

Acidifiers: An acidifier is a product when it carries a pH of 2.5 to 3.5. This pH will close to the cuticle layer of the hair. The result is shiny, bouncy hair. This pH range will adjust the beta bonds to alpha bonds (hydrogen bonds). They do create shine and add elasticity. Finally, results bouncy hair. *Recommendations:* Sunset hair elements hair rebalancer.

Detanglers: Most detanglers are acidifiers. They close the cuticle of the hair which cause tangles. Some detanglers are instant and some take 1–5 minutes to work. *Recommendations:* Sunset hair elements hair rebalancer.

Thermal protectors: Thermal protectors safeguard the hair against heat. They are normally use heat absorbing polymers that distribute the heat, so the hair does not get heat damage (a major cause of hair damage). *Recommendations:* Sunset hair elements thermal spray.

Glossers: Most glossers contain dimethicone or cyclomethicone (very light oils derived from silicone). They are used in small amounts that reflect light. Also, they are one of the best products to control the "frizzies."

Oils: Oils especially essential fatty acids are required for dry hair. The scalp produces a natural oil called sebum. Oils are the closest thing to natural sebum (sebum contain essential fatty acid). It can take very dry and porous hair and transform it into soft pliable hair. *Recommendations:* Safflower oil (used sparingly), sunset hair elements vanilla bean treatment.

Apart from these some of the other ingredients like surfactants, lubricants, sequestrants, antistatic agents and preservatives are also added.

Surfactants: Hair consists of approximately 97% of a protein called keratin. The surface of keratin contains negatively-charged amino acids. Hair conditioners therefore usually contain cationic surfactants, which do not wash out completely, because their hydrophilic ends strongly bind to keratin. The hydrophobic ends of the surfactant molecules then act as the new hair surface. **Lubricants** like fatty alcohols, panthenol dimethicone, etc. are present, **sequestrants**, for better function in hard water.

A hair conditioner should be capable of providing following few characters like, improved wet and dry combing, reduced flyaway, increased shining, increased volume, simple and ease of handling and finally over all lucrative appearance.

Some of the important herbs that produce conditioner effect for hair are listed in Table 4.1.

Table 4.1

Plant name	Biological source	Family	Part used
Aloes	Aloe vera	Liliaceae	Leaves
Almond	Prunus dulcis	Rosaceae	Nuts
Amla	Emblica officinalis	Euphorbiaceae	Fruits
Avocado	Persea americana	Lauraceae	Fruits
Ginger	Zingiber officinale	Zingiberaceae	Rhizomes
Hemp	Cannabis sativa	Cannabaceae	Fruit
Henna	Lawsonia inermis	Lythraceae	Leaves
Lavender oil	Lavandula latifolia	Lamiaceae	Flowers
Lemon grass	Cymbopogon citratus	Poaceae	Leaves
Neem	Azadirachta indica	Meliaceae	Leaves
Rosemary	Rosmarinus officinalis	Lamiaceae	Leaves
Shikakai	Acacia concinna	Fabaceae	Pods
Stinging nettles	Urtica dioica	Urticaceae	Leaves
Soapnut/reetha	Sapindus mukorossi	Sapindaceae	Nuts
Tea tree oil	Melaleuca alternifolia	Myrtaceae	Leaves
Yarrow	Achillea millefolium	Asteraceae	Roots

Ideal Properties of Hair Conditioners

- The primary function of a good hair conditioner is to treat the hair after shampooing has taken place and keep it nourished and tamed.

- Essentially, hair conditioners help replenish the look and feel of the hair after shampooing so they look shiny, healthy and smooth.

- Effective hair conditioners help moisturize, improve gloss and manageability to hair after it has been subjected to a head-bath with shampoo or other detergents that deplete its natural protective coating.

Tips for Application of Hair Conditioners

- It is important to understand that hair conditioners must be applied to the hair shaft at the ends only. Never apply a hair conditioner to the scalp as it will lead to buildup and create a limp look. The ends and the hair shaft are dead protein so are chemically inert and insoluble, which is why they can be coated protein or polymer to simulate smoothness and thickness.

- Select a hair conditioner after considering the hair types (like dry, frizzy, oily, undernourished, limp, colored, curly, etc.) and hair growth (whether thick or thin).

Choosing Hair Conditioners

- A hair conditioner helps to replace natural oils removed by the detergents in shampoos. Therefore, the best product for hair that needs the oil, i.e. dry, curly, often frizzy and coarse hair are conditioners containing cetyl or stearyl alcohol, panthenol and methicones, silicone/dimethicone, or essential oils and botanicals like avocado, jojoba oil or shea butter.

- Increased volume of hair conditioners give body and temporarily fluff up fine, limp or damaged hair to give it a lush appearance. These are labeled reconstructors and coat the hair with a layer of protein that fills in gaps if the outer cuticle is damaged, giving the hair a smoother look.

- Those with color treated hair will benefit from a conditioner meant for chemically processed hair.

Evaluation Tests for Hair Conditioners

A. Physicochemical test

1. Net content
2. Description of the color, odor, physical state of hair conditioner
3. Ash value at 600°C
4. Nonvolatile matters ar 105°C for two hours
5. Water content
6. *pH of the hair conditioner:* It should be between 6 and 7.5
7. *Test for ammonia:* It should be negative.

B. Performance test

1. *Wetting test:* It should be within 4–150 second
2. *Luster and softness:* Up to satisfaction
3. Conditioning effect on greasy and dry hair.

C. Physiological test

1. *Dermal toxicity:* It should be neagtive
2. *Eye toxicity:* It should be negative
3. *Cytotoxicity:* It should be negative
4. *Stability test:* Should give good stability.

Some Formulations

Jojoba conditioner: Ingredients: One cup rose floral water, 1 tablespoon jojoba oil, 10 drops vitamin E oil.

Procedure: In the top of a double boiler, gently warm the rose water. Then jojoba oil is added. Pour the mixture in a blender and add the vitamin E. Blend at high speed for 2 minutes.

To use

1. Wet hair with warm water.
2. Pour the conditioner onto your hair and scalp, massaging in thoroughly.

3. For damaged hair or extraconditioning, leave on for several minutes, perhaps while bathing.

4. Rinse thoroughly with warm water.

5. Shampoo lightly and rinse again with cool water.

Rosemary hair conditioner: It contains birch sap, jojoba, chamomile, green tea, vegetable protein, nettle and rosemary. Birch sap has astringent and antibacterial properties, jojoba has hardening of sebum in the scalp, chamomile has decongestant properties, green tea and rosemary have powerful antioxidant properties.

IHT 9 natural hair conditioner: It contains rose water (*Rosa damascena*), cyclomethicone, vegetable glycerin, cutina shine, genamin KDMP, genapol LA 230, olive oil (*Olea euroea*), henna extract (*Lawsonia inermis*), *Aloe vera* (*Aloe barbadensis*), soya protein, holy basil (*Ocimum basilicum*), rosemary ext. (*Rosmarinus officinalis*), thyme ext. (*Thymus vulgaris*), grape seed oil (*Vitis vinifera*), coco butter (*Theobroma cacao*), coconut oil (*Cocos nucifera*), liquorice ext. (*Glycyrrhiza glabra*), natural vitamin E (tocopherol), potassium sorbate, sodium benzoate.

Flower and plant deep hair conditioner: It contains jojoba oil, avocado oil, pure essential oils of ylang ylang, carrot seed, birch, peppermint and clove. It is used for dry and brittle hair. Even though it both conditions and nourishes the scalp, giving relief to dandruff and eczema while promoting hair growth.

Avocado Deep Conditioner

Ingredients
1 small jar of mayonnaise and 1/2 avocado.

Method: Peel avocado and remove pit. Mix all ingredients in a medium-sized bowl with your hands until it is a consistent green color. Smooth into hair being careful to work it to the ends. Use shower cap or plastic wrap to seal body heat in. Leave on hair for 20 minutes. For deeper conditioning wrap a hot, damp towel around your head over the plastic, or use a hair dryer set to a low to medium heat setting.

Hair Dyes

Hair dye or hair color is a chemical tool that is used to change the color of human hair. Hair dye is used mostly to change gray hair, since gray hair is a sign of an advanced age. Younger people want to change their hair color to make the style or as a fashion. Sometimes hair colors are used to return hair to its original color after chemicals (e.g. tints, relaxers, sun bleaching) have discolored it. A variety of hair colors are available may be due to the combination of pheomelanins and eumelanins, the quantity of the pigment present, the different sizes of granules and their distribution in the pigments. An artificial hair dye used properly and produce useful results. Metallic hair dyes contain lead compounds which gives startling colors, viz. a vivid red, a violent purple, a sober green or a dashing violet. The advantage is that these colors wash of easily as they are deposited only on the surface of the hair fiber. Godrej consumer products (GCPL) said the powder hair dyes, make up 50% of the Rs. 580 crore hair color market in India. In terms of volumes it makes up an even greater share of 75% of the total market. The rest of the market is made up of cream based hair colorants, which are catching up fast among the affluent consumers. GCPL is still the market leader in the overall category with a share of 35.2%. Although its share has gone down in the last six years from 46% cent in 2008 as many companies both global and local have entered the Indian market since then like Schwarzkopf, Cavin Kare, L'Oréal, Garnier, etc. which entered the Indian personal care market in 1994 has already garnered a share of nearly 20% from local hair dye brands with its cream-based colorant.

The four most common classifications of hair dyes are 'temporary', 'semi-permanent', 'demi-permanent' (or 'deposit only') and "permanent".

Some Ideal Properties of Hair Colors

1. The formulation of the hair colorant should be stable
2. They sould be color the hair evenly
3. The shaft of the hair must not be damaged
4. Must be nontoxic
5. Must impart stable color to the hair
6. Must be nonirritant and nonsensitizing
7. The colored hair must be unaffected by air, water, sunlight, sweat, shampoos, gels, lotions, etc.
8. The natural moisture of the hair must not be lost
9. They should maintain the texture and gloss of the hair
10. The pH of hair dye should be so far from neutral (similar as skin pH) and temperature during application should not exceed 40°C.

Permanent hair color: The most popular way to achieve permanent hair coloring is through the use of oxidation dyes. The mechanism of oxidation dyes involves three steps: 1) Oxidation of 1,4′-diaminobenzene derivative to the quinone state. 2) Reaction of this diamine with a coupler (more detail below). 3) Oxidation of the resulting compound to give the final dye. The preparation (dye precursors) is in the leuco (colorless) form. Oxidizing agents are usually hydrogen peroxide, and the alkaline environment is usually provided by ammonia. The combination of hydrogen peroxide and the primary intermediate causes the natural hair to be lightened, which provides a blank canvas for the dye. Ammonia opens the hair shaft so that the dye can actually bond with the hair, and ammonia speeds up the reaction of the dye with the hair.

Semi-permanent hair dye: Semi-permanent hair dye has smaller molecules than temporary dyes, and is therefore, able to partially penetrate the hair shaft. For this reason, the color will survive repeated washing, typically 4–5 shampoos or a few weeks.

Semipermanents contain no, or very low levels of developer, peroxide or ammonia, and are therefore, safer for damaged or fragile hair.

Demi-permanent hair color: Demi-permanent hair color is permanent hair color that contains an alkaline agent other than ammonia (e.g. ethanolamine, sodium carbonate), and while always employed with a developer, the concentration of hydrogen peroxide in that developer may be lower than used with a permanent hair color. These product provide no lightening of hair's color during dying.

Temporary hair color: Temporary hair color is available in various forms including rinses, shampoos, gels, sprays, and foams. Temporary hair color is typically brighter and more vibrant than semi-permanent and permanent hair color. The pigment molecules in temporary hair color are large and cannot penetrate the cuticle layer. The color particles remain adsorbed or adhered to the hair shaft and are easily removed with a single shampooing. Temporary hair color can persist on hair that is excessively dry or damaged in a way that allows for migration of the pigment to the interior of the hair shaft.

Herbal hair color: Natural hair dye is hair dye extracted from plants and vegetables. It is usually used in a pure extract form and does not contain chemicals like ammonia, resorcinol and phenylenediamine which are found in commercial hair dyes. Natural hair dyes do not pose a health or environmental hazard due to its non toxic in nature. They do not harm the hair structure, can even help with conditioning and hair moisturizing and so are generally good for long-term use. Herbal hair dyes are generally less expensive than artificial colors, contain safer ingredients and if manufactured properly are as easy to use.

Dyes are made from respectively boiling potatoes, black coffee and black tea. Other effective rinses include those made from various herbs like rosemary, sage, parsley, catnip, lemon, raspberry leaves, hibiscus flowers, calendula, rosehips, betony, ivy berries, black coffee and black tea. An extract of the flowers of the chamomile plant was long used to lighten hair, and this is still used in many modern hair preparations. The bark, leaves, or nutshells of many trees were used for hair dyes. Wood from the brazil wood tree yielded brown hair dyes, and another hair dye known in antiquity as *fustic* was derived from a tree similar to the mulberry. Some of the important herbs that produce different colors for hair are listed in Table 4.2.

General Method for Herbal Color Preparation

1. Create an infusion that is extremely strong.
2. Add enough clay to create a smooth paste.
3. Apply to hair in small sections, starting at the scalp and working towards the ends.
4. Cover your hair with plastic to retain heat.
5. Leave on hair for 30 minutes.
6. Rinse with tepid water and let air dry.
7. Repeat as needed.

Some Tips

1. For red hair, use alkanet root or hibiscus.
2. For black hair, use elderberry.

3. For grey hair, help cover yellow tones with hollyhock. Use sage to darken grey hair.
4. To lighten hair, use mullein flowers or chamomile flowers
5. To darken hair, try sage, rosemary, red raspberry leaves.

Method for Preparation of some Hair Dyes/Colors

1. ***Blue dyes with woad:*** For dyeing, fresh leaves are put into a jar and covered with almost boiling water. The jar is covered to exclude all air. In a while the liquid becomes colored and produces small bubbles. Alkali is added to the colored liquid, and then the solution is shaken until it becomes greenish. Woad is a tricky dye to get correct, and many do not have success even following the detailed recipes that exist. Some type of green liquid is the form of the solution for dyeing. The fabric is dyed greenish-yellow but turns blue when exposed to the air (air oxidation) and becomes relatively fast when put in an acid and then a soapy rinse.

2. ***Dyes with indigo:*** Dye is obtained from the processing of the plant's leaves. They

Table 4.2

Colors	Plant name	Biological source	Family	Part used
Red/Brown	Henna	Lawsonia inermis	Lythraceae	Leaves
	Walnut	Juglans regia	Juglandaceae	Leaves, nuts
	Catechu	Acacia catechu	Fabaceae	Heartwood
Blondes/Yellows	Cassia	Cassia obovata	Fabaceae	Leaves
	Catechu	Ourouparia gambir	Rubiaceae	Leaves
	Saffron	Crocus sativus	Iridaceae	Flowers
	Chamomile	Anthemis nobilis	Asteraceae	Flowers
	Rhubarb root	Rheum rhapoticum	Polygonaceae	Roots
Black	Vashma	Partially fermented Indigo	Fabaceae	Leaves
	Karchak	Ricinus communis (castor bean)	Euphorbiacea	Beans
	Shoe flower	Hibiscus Rosa cynensis	Malvaceae	Flowers
Blues	Indigo	Indigofera tinctoria	Fabaceae	Leaves
	Woad	Isatis tinctoria	Brassicaceae	Fresh leaves

are soaked in water and fermented in order to convert the glycoside indican naturally present in the plant to the blue dye indigotin. The precipitate from the fermented leaf solution is mixed with a strong base such as lye, pressed into cakes, dried, and powdered. The powder is then mixed with various other substances to produce different shades of blue and purple.

3. *Dye with henna:* The leaves of henna are powdered and make paste form with the hot water. The directly applied on hair and a warm towel is wrapped around the head to enhance the coloring effect. It gives reddish color due to presence of lawsone which is imparting the color.

4. *Dye with chamomile:* The flowers are used to make the dye. The paste form is to be made with hot water and then applied to the hair.

Evaluation Test for Hair Dyes

A. *Performance test*

1. Color uniformity of the dye

2. Compatibility of color with hair

3. Extent of reaction in case if permanent dye

4. Washability of color

5. Color stability

B. *Physiological test*

i. Dermal toxicity

ii. Eye toxicity

iii. Cytotoxicity

iv. Stability test

C. *Physicochemical test*

a. Net content

b. Ash value at 600°C

c. Nonvolatile matters at 105°C for two hours

d. Effect on hard water

e. pH

f. Assay for permanent dyes

g. IR spectroscopy for determination of surfactants.

Skin irritation and allergy: In certain individuals, the use of hair coloring can result in allergic reaction and/or skin irritation. Symptoms of these reactions can include redness, sores, itching, burning sensation and discomfort. To help prevent or limit allergic reactions, the majority of hair color products recommend that the client conduct a patch test before using the product. This involves mixing a small quantity of tint preparation and applying it directly to the skin for a period of 24 hours. If irritation develops, manufacturers recommend that the client not use the product. A skin patch test is advised before the use of every coloring process, since allergies can develop even after years of use with no reaction. In some cases, allergic reactions are caused by the aniline derivative and/or p-Phenylenediamine (PPD) found in permanent hair color.

A strand test is also recommended and helpful for checking how the dye works on the hair, but this is not always accurate and the final hair color application could look different.

The toxic effect test: Toxic effects are studied in animals to know about the long term effects of the preparations.

Skin discoloration: Skin is made of the same type of keratinized protein as hair. That means that drips, slips and extra hair tint around the hairline can result in patches of discolored skin. This discoloration will disappear as the skin naturally renews itself and the top layer of skin is removed (typically takes a few days or at most a week). A good way to prevent dye discoloration is to put a thin layer of vaseline or any oil-based preparation around the hairline. It is recommended that latex gloves be worn to protect the hands.

Some Formulations

Formulation of natural dye with mordant: 30 g aqueous extract of *Cymphomandra betacea* containing 35% flavanoid and 10% tannins, 30 g aqueous extract of *Tagetes erecta* containing 20% carotenoid and 40 g *Aloe vera* gel, as natural mordant, containing 0.3% polysaccharide and 98.5% water were mixed together.

Kali Mehandi: Kali Mehandi is entirely produced by Ayurvedic Process. It contains Mehandi Powder, Amla, Aritha, Shikakai, Bhringraj and other Hurbs.

- It gives relief in common hair problems like hair loss, hair breakage, dandruff and hair dryness.
- Rinse hair after 60–90 minutes with clean water.

Naaz herbal henna powder: It is available in Black, Brown, Burgundy colors. It contains Henna, Amla, Shikakai, Tulsi, Reetha, Kapur, Bhringraj and other herbs. It fights dandruff, gives hair uniform and lasting color leaving them extra silky and feather soft without any side effects. Take Henna into a glass or plastic bowl. Gradually, add water. Henna may be applied on cleaned wet or dry hair. Rinse the hair thoroughly after 60 to 90 minutes.

Super Vasmol-Amla natural black and natural brown—herbal powder hair dyes: Super Vasmol Amla herbal powder hair dye is a special herbal formulation that darkens the hair in natural way to give it a natural black and brown color. The formulation is enriched with the goodness of herbal ingredients like Aamla (*Emblica officianalis*) and Bhringraj (*Eclipta alba*) for revitalising and darkening hair, preventing dandruff and hair fall. Besides Mehandi (*Lawsonia alba*) and Jasund (*Hibiscus Rosa sinensis*) also were added which nourish and condition hair to make it look shiny and lustrous. This formulation is available in three shades in market—natural black, natural brown and chest nut color.

Natural Brunette Hair Dye

Ingredients
Triple strength black coffee.

Method
Shampoo hair: Place a large bowl in sink and rinse hair with cooled coffee. Repeat several times, reusing the coffee. Leave final rinse in hair for at least 15 minutes. Rinse with clear water.

Red Hair Color Enhancer

Ingredients
1/2 cup beet juice

1/2 cup carrot juice

Method
Mix ingredients together, pour over clean, damp hair. Wrap head in plastic and apply hot towel, medium dryer heat, or sit in the sun for one hour, shampoo.

Natural Color Restorer for Gray Hair

Ingredients
1/2 cup organic dried sage

1/4 cup organic dried rosemary

Method
Simmer rosemary and sage in two cups of water for 30 minutes, then steep for several hours. Apply to gray hair and allow to dry, then shampoo. Repeat weekly until desired shade is reached, then once a month for maintenance.

Chamomile Brightener for Blonde Hair

Ingredients
6 organic chamomile tea bags
1/2 cup plain yogurt and oil of lavender

Method
One cup of water should boil and steep tea bags for 15 minutes. Discard tea bags. Combine yogurt and seven drops of lavender oil with chamomile tea, mix thoroughly. Apply the mixture to dry hair, working through to ends. Cover head in plastic wrap and condition for minutes. Shampoo hair.

Hair Shampoos

Hair shampoos are viscous cosmetic preparation. It is a preparation of a surfactant (i.e.

surface active material) in a suitable form—liquid, solid or powder—which when used under the specified conditions will remove surface grease, dirt, and skin debris from the hair shaft and scalp without adversely affecting the user.

Requirements of a Shampoo

1. It should effectively and completely remove dust or soil, excessive sebum or other fatty substances and loose corneal cells from the hair.

2. It should produce a good amount of foam to satisfy the psychological requirements of the user.

3. It should be easily removed on rinsing with water.

4. It should leave the hair non-dry, soft, lustrous with good manageability and minimum fly away.

5. It should impart a pleasant fragrance to the hair.

6. It should not cause any side-effects/irritation to skin or eye.

7. It should not make the hand rough and chapped.

Types of Shampoos

a. Powder shampoo

b. Liquid shampoo

c. Lotion shampoo

d. Cream shampoo

e. Jelly shampoo

f. Aerosol shampoo

g. Specialized shampoo

h. Conditioning shampoo

i. Medicated shampoos like antidandruff shampoo, anti lice shampoo and anti baldness shampoo, etc.

j. Baby shampoo

Ingredients use in Shampoos

Surfactants are the main component of shampoo. Mainly anionic surfactants are used. *Principal surfactants:* Provide detergency and foam.

Secondary surfactants: Improve detergency, foam and hair condition. A surfactant consists of two part—one hydrophilic (water loving) while the other is hydrophobic in nature.

Surfactants

Surfactants: Anionic surfactants are mostly used (good foaming properties). The hydrophilic portion carries a negative charge which results in superior foaming, cleaning and end result attributes. Non-ionic surfactants have good cleansing properties but do not have sufficient foaming power. Cationic surfactants are toxic and are hence not used. However, they may be used in low concentration in hair conditioners.

Additives

Additives conditioning agents: Lanolin, mineral oil, herbal extracts, egg derivatives. Foam builders: Lauroyl monoethanolamide, sarcosinates viscosity modifiers: Electrolytes – NH_4Cl, NaCl, natural gums: Gum karaya, tragacanth, alginates cellulose derivatives: Hydroxyethyl cellulose, methylcellulose carboxy vinyl polymers: Carbopol 934 others: PVP, phosphate esters. Sequestering agents: EDTA opacifying agents: Alkanolamides of higher fatty acids, propylene glycol, Mg, Ca and Zn salts of stearic acid, spermaceti, etc. Clarifying agents: Solubilizing alcohols like ethanol, isopropanol phosphates; non-ionic solubilizers: Polyethoxylated alcohols and esters.

Additives

Additives perfumes: Herbal, fruity or floral fragnances. Preservatives: Methyl and propyl paraben, formaldehyde (most effective). Antidandruff agents: The shampoos contain small amount of these actives, which are in contact with the scalp for only a short time.

Herbal Shampoos

There are thousands of herbal hair shampoos available in the market. There is growing evidence to support the view that some herbal remedies offer real hope to many people. Ideally, a herbal shampoo should have all herbal ingredients and it should be 100% natural.

Herbs used for Hair Loss Treatment

There are lots of natural herbs for hair available or in fact being used since hundreds of years by the different countries all over the world. Table 4.3 is the list of regularly used herbs:

Ayurvedic herbs: Some of the Ayurvedic herbs that are widely use for the formulation of the herbal shampoos. They are namely, *Bacopa monnieri, Indian spikenard, Roots of Aegle marmelos, Gmelina arborea, Oroxylum indicum, Clerodendrum phlomidis, Stereospermum chelonoides, Desmodium gangeticum, Uraria picta, Solanum indicum, Solanum surattense, Tribulus terrestris.*

Some Formulations for Shampoos

Hot oil treatment for damaged hair.

Ingredients

1/2 cup organic soybean oil or sunflower oil
8 drops sandalwood oil
8 drops oil of lavender
8 drops oil of geranium

Method

Mix all ingredients well. Warm oil to a comfortable temperature and apply the mixture to damp hair. Wrap hair in plastic wrap and apply a hot towel for 20 minutes, shampoo.

Lavender Rosemary Hot Oil Treatment

Ingredients

1/2 cup organic soybean oil or sunflower oil
5 drops oil of rosemary
10 drops oil of lavender

Directions

Mix all ingredients well. Warm slightly and apply the mixture to damp hair. Wrap hair in plastic wrap and apply a hot towel for 20 minutes, shampoo.

Horse Tail Hair Rinse

Ingredients

2 1/2 teaspoons dried horsetail

Methods

Steep horsetail in boiled water for 20 minutes. Shampoo hair and rinse thoroughly. Pour horsetail rinses through hair and leave in for ten minutes. Rinse with clear water.

Table 4.3

Plant name	Biological source	Family	Part used
Aloes	Aloe vera	Liliaceae	Leaves
Arnica	Arnica montana	Asteraceae	Flower
Birch	Betula alba	Betulaceae	Leaves
Burdock	Arctium lappa	Asteraceae	Roots
Catmint	Nepeta cataria	Lamiaceae	Leaves and flowers
Chamomile	Anthemis nobilis	Asteraceae	Flowers
Fenugreek	Trigonella foenum graecum	Fabaceae	Seeds
Horsetail	Equisetum fluviatile	Equisetaceae	Leaves
Licorice	Glycyrrhiza glabra	Leguminosae	Roots
Marigold	Calendula officinalis	Calenduleae	Flowers
Nettles	Urtica dioica	Urticaceae	Leaves
Parsley	Petroselinum crispum	Umbelliferae	Leaves
Rosemary	Rosmarinus officinalis	Lamiaceae	Leaves
Sage	Salvia officinalis	Lamiaceae	Seeds and buds
Southern wood	Artemisia abrotanum	Asteraceae	Leaves

Natural Shampoo for Normal Hair

Ingredients

1/4 cup water

1/4 cup liquid castile soap

1/2 teaspoon organic sunflower oil

Method

Mix together all the ingredients. Store in a bottle. Use as you would any shampoo, rinse well.

Dry Shampoo for Oily Hair

Ingredients

Cornstarch

Directions

Dip large soft make-up brush in cornstarch and dust through oily spots in hair. Allow to sit for five minutes then brush out with a natural boar bristle brush. Repeat as necessary.

Chamomile Shampoo

Ingredients

6 organic chamomile tea bags

4 tablespoons pure soap flakes

1 1/2 tablespoons pure vegetable glycerin

Method

Steep the tea bags in 1 1/2 cups of boiled water for 20 minutes. Remove the tea bags and discard. Add the soap flakes to the tea and let stand until the soap softens. Stir in glycerin until well blended. Keep in a dark, cool place in a sealed bottle.

Desert Shampoo for Dry Hair

Ingredients

1/4 cup liquid castile soap

1/4 cup organic *Aloe vera* gel

1 teaspoon pure vegetable glycerin

1/4 teaspoon organic avocado oil

Table 4.4: Some of the hair care products with their evaluation tests

Biophysical methods	*In vivo methods*	*Products*
• Wet and dry combing force	• *In vivo* tolerability tests for hair care, hair cleaning, and hair styling products	• Hair shampoo
• Combing out force		• Hair conditioner
• Flexural strength		• Colorization
• Flexural strength hysteresis	• Consumer test	• Hair styling (gel, foam, spray)
• Hair gloss panel-test	• Consumer test to assess the scalp tolerability	• Hair dryer spray
• Hair volume		• Anti-dandruff shampoo
• Curl retention	• Anti-dandruff	• Anti-hair loss
• Long lasting hold	• Hair loss	• Hair growth
• Assessment of color protection on colored hair tresses applying radiation or elution	• Hair growth	• Hair removal Products (wax, cremes, shaver)
	• Epilation	
	• Shaving studies	
• Friction force		
• Suppleness on hair		
• Tensile strength on single hair fibers		
• Breaking force on single hair fibers		
• High pressure differential scanning calorimetry		
• Assessment of heat damage		
• Atomic force microscopy		

Method

Mix together all the ingredients. Store in a bottle and always shake well before using. Apply to hair and allow to absorb for a few mintues. Rinse well with cool water.

Sunshine Shampoo for Blonde Hair

Ingredients

1/4 cup liquid castile soap

2 tablespoons fresh lemon juice

1/4 cup water

1 teaspoon lemon zest

Method

Mix together all the ingredients and heat in the microwave on high for 1 to 2 minutes until it is hot but not boiling. Cool the mixture completely and strain to remove the lemon zest. Store in a bottle. Shampoo as you normally would. Rinse well with cool water.

Evaluation Tests for Herbal Shampoos

Evaluation of shampoos comprises the quality control tests including visual assessment and physiochemical controls such as pH, density and viscosity. Sodium lauryl sulfate based detergents are the most common as additive but the concentration will vary from brand-to-brand and even within a manufacturer's product range.

1. *Physical appearance/visual inspection:* The formulation of shampoo is evaluated in terms of their clarity, foam producing ability and fluidity.

2. *Net contents:* At the beginning of the experiment, mark the outside of the bottle at the surface level of liquid then at the end of the experiment emptied the bottle and note the volume of water required to fill it up to the mark.

   If the formulated materials are paste or solid forms then place the materials in an open can with the frozen material in a beaker and let it warm to room temperature. Discard the liquefied content and weigh part of container together by taking the weight of the container. Note the net content.

3. *Determination of pH:* The pH of 10% shampoo solution in distilled water is determined at room temperature 25°C.

4. *Determine percent of solids contents:* Weigh a clean dry evaporating dish and record the result. Add approximately four grams of shampoo (not the 1% solution) to the evaporating dish. (four grams is a little less than the weight of a nickel.) Weigh the dish and shampoo and record. Calculate the exact weight of the shampoo only and record. Put the evaporating dish with shampoo on the hot plate until the liquid portion has evaporated. After drying, weigh the dish and shampoo solids and record result. Calculate the weight of the shampoo only (solids) after drying and record result. Calculate the percent of solids in the shampoo and record.

5. *Rheological evaluations:* The viscosity of the shampoos is determined by using Brookfield viscometer (model DV-l plus, LV, USA) set at different spindle speeds from 0.3 to 10 rpm. The temperature and sample container's size is kept constants during the study.

6. *Dirt dispersion:* Two drops of shampoo is added in a large test tube contain 10 ml of distilled water. One drop of India ink is added to the test tube and shakes it ten times. The amount of ink in the foam is estimated as none, light, moderate, or heavy.

7. *Cleaning action:* Five grams of wool yarn is placed in grease, after that it has placed in 200 ml of water containing one gram of shampoo in a flask. Temperature of water is maintained at 35°C. The flask has to shaked for four minutes at the rate of 50 times a minute. The solution has to removed and sample is taken out, dried and weighed. The amount of grease

removed is calculated by using the following equation:

$$DP = 100\,(1 - T/C)$$

where DP is the percentage of detergency power, C is the weight of sebum in the control sample and T is the weight of sebum in the test sample.

8. **Surface tension measurement:** Measurements are carried out with a 10% shampoo dilution in distilled water at room temperature. Thoroughly clean the stalagmometer using chronic acid and purified water. The data calculated by following equation given below:

$$R_2 = \frac{(W_3 - W_1)n_1}{(W_2 - W_1)n_2} \times R_1$$

Where, W_1 is weight of empty beaker. W_2 is weight of beaker with distilled water. W_3 is weight of beaker with shampoo solution. n_1 is no. of drops of distilled water. n_2 is no. of drops of shampoo solution. R_1 is surface tension of distilled water at room temperature. R_2 is surface tension of shampoo solution.

9. **Detergency ability:** The Thompson method is used to evaluate the detergency ability of the samples. Briefly, a crumple of hair is washed with a 5% sodium lauryl sulfate (SLS) solution, then dried and divided into 3 g weight groups. The test sample is suspended in a n-hexane solution containing 10% artificial sebum and the mixture is shaken for 15 minutes at room temperature. Then sample is removed, the solvent is evaporated at room temperature and their sebum content determined. In the next step, sample is divide into two equal parts, one washed with 0.1 ml of the 10% test shampoo and the other considered as the negative control. After drying, the resided sebum on sample is extracted with 20 ml n-hexane and reweighed. Finally, the percentage of detergency

power is calculated using the following equation:

$$DP = 100\,(1 - T/C)$$

where, DP is the percentage of detergency power, C is the weight of sebum in the control sample and T is the weight of sebum in the test sample.

Other two methods are likely

Finger method: This technique is an accurately mimics the "real world" shampoo procedure in that the solution being tested is applied and agitated in a manner consistent with actual consumer use. A 1.5-gram soiled tress is wetted under running water for five seconds per side (about 250 ml of water). Next 0.1 gram of undiluted shampoo (a pproximately 10% surfactant) is applied to the length of the tress. The tress is rubbed 15 times between the fingers as evenly as possible. Then the tress is reversed and rubbed 15 more times. Next the tress is rinsed under warm (40°C) water for 10 seconds per side and dried as before using a handheld dryer. This yields a total rinse volume of approximately 500–600 ml.

Sponge method: This technique is a modification of the "finger method". The procedure is modified to provide a more consistent pressure upon the hair tress during the rubbing portion of the sample treatment. A 1.5-gram soiled tress is held under warm (40°C) running water for 5 seconds per side, and then 0.1 gram of 10% surfactant solution is applied to the length of the tress. The tress is drawn 15 times between two pre-wetted sponges. 100 gram weight is placed on the top sponge to simulate the approximate pressure applied by the fingers in that method. The tress is rinsed under 40°C running water for 10 seconds on each side and dried as previously stated. The sponges are cleaned before each use to prevent any build-up of sebum on them.

10. *Surface characterization:* Surface morphology of the hairs are examined by scanning electron microscopy (Leo 430, Leo Electron Microscopy Ltd., Cambridge, England). The hair samples are mounted directly on the SEM sample stub, using double side stitching tape and coated with gold film (thickness 200 nm) under reduced pressure (0.001 mm of Hg). The photomicrographs of suitable magnification are obtained for surface characterization.

11. *Stability studies:* The thermal stability of shampoo is study by placing in glass tubes and they have placed in a humidity chamber at 45°C and 75% relative humidity. Their appearance and physical stability is inspected for a period of 3 months at interval of one month.

12. *Foam evaluation and foam stability:* In this test there are several methods are available. In each of these methods the temperature of the water and water hardness may be varied.

    Ross miles test: In this method a dilute solution is dropped from a fixed height into a pool of the same dilute solution and the foam volume is measured. This test produces an airy foam but mainly it is used by the suppliers of surfactants. It does not give an accurate reading on foam volume, foam density or foam longevity.

    Cylinder shake: This method was developed in 1941 and it is one of the most widely used method for evaluation of foam. A fixed amount of dilute shampoo is poured into a graduated cylinder. A stopper is placed onto the cylinder and it is inverted for a fixed number of times. The foam volume is then measured.

    Perforated disk: This foam evaluation method was developed in 1958 by Barnetts and Powers. 200 g of shampoo solution is placed into a glass cylinder (6.3 cm in diameter and 30 cm in length).

A perforated disk (6 cm in diameter) is moved up and down in the tube (26.5 cm) at a speed of 30 strokes per minute and then the foam height is measured. This method is having consistency but the foam it produces is loose and airy.

Moldovanyi-Hungerbubler method: By this method 500 ml of shampoo is prepared and poured in to a flask. The flask has an input tube to permit nitrogen gas to flow into the solution (from the bottom) at a rate of 17 liters per minute. The time required to produce 2 liters of foam is measured. The liquid is drained off and the flask is weighed for foam density. If is waited for a fixed period of time and drained off additional liquid then foam stability can measured. This method produces is loose and airy foam.

Hart-deGeorge blender method: This method was the first to incorporate a blender to generate the foam evaluation. The foam produce in this method is creamy and thick in nature. A 200 ml of shampoo is agitated in one liter of vessel for one minute. The foam is poured into a funnel placed on a sieve with a mesh of 0.5 mm. The funnel measures 182 mm from top to 23 mm at bottom. A gauging wire is placed 80 mm from the bottom of the funnel. The time for the level of the foam to reach the wire (seconds) is recorded. The higher the number, the better the foam.

Blender-foam density/stability/lubricity: In this method 10% shampoo solution is prepared. 4 g of this solution are added to 146 g of water at 29°C. The solution is agitated for 10 seconds at medium speed in a blender. The foam is poured into a 100 ml graduated cylinder to overflowing. A rubber stopper is gently dropped into the foam. The time for the rubber stopper to pass between two points (80–40 ml) is measured. A longer time indicates denser and more stable foam. This method is an excellent for determination of foam quality.

13. *Safety study:* It become essential to evaluate the safety of the shampoo. The shampoo must be nontoxic and non-irritative. The safety is usually evaluated by toxicity determination using Draize test which suggests two separate methods for testing skin and eye toxicity respectively.

Skin sensitization test: The guinea pigs or albino rats are selected. On the previous day of the experiment, the hairs on the backside area of animals are removed. Then shampoo is applied onto nude skin of animals. A 0.8% v/v aqueous solution of formalin is applied as a standard irritant on animal. The animals is applied with new patch/formalin solution up to 72 hours and finally the application sites are graded according to a visual scoring scale, always by the same investigator. The erythema scale is as follows: 0) none; 1) slight; 2) well defined; 3) moderate; and 4) scar formation (severe). Based on the results obtained the shampoo is considered as either safe or toxic.

Eye irritation test: Animals (albino rats) are selected. About 1% shampoo solutions is dripped into the eyes of albino rabbits with their eyes held open with clips at the lid. The progressive damage to the rabbit's eyes is recorded at specific intervals over an average period of four seconds. Reactions to the irritants can include swelling of the eyelid, inflammation of the iris, ulceration, hemorrhaging (bleeding) and blindness.

14. *Antimicrobial activity:* Mainly preservatives are added to prevent the microbial growth of the shampoos because shampoos are the liquid or viscous preparations. Hence, antimicrobial study has to perform to evaluate the preservative activity. This is generally done by using challenge antimicrobial study. According to this study, the product is said to be preserved when it does not support microbial growth even after repeated attacks of various microorganisms. As per this examination, initially an appropriate strain of microorganism is selected and is considered as test organism. Usually, the species of *Pseudomonas* are selected for this test and culture is prepared. Then the shampoo is innoculated repeatedly in the culture medium. The studied are carried out for a peroid of 12 weeks. Along with the test, two control samples are prepared, i.e. one with preservative and another one without preservative. Then comparative study is done and the test comes to a conclusion only when it has been proven that the product has not supported the growth of any microorganisms.

BIBLIOGRAPHY

1. Aghel N., Moghimipour B. and Dana R.A., Iranian Journal of Pharmaceutical Research. 2007, 6(3), 167–172.
2. Evaluting shampoo foam. Cosmetics and toiletries magazine. 2004, 119 (10), 32–35.
3. Gaud R.S. and Gupta G.D., Practical Physical Pharmacy, 1st ed., 2001. C.B.S. Publisher and Distributer, New Delhi, 81–105.
4. Hadkar U.B. and Ravindera R.P., ijper. 2009, 43, 187–191.
5. http://beautytips.ygoy.com/haircare/conditioners.php
6. http://consumergoods. indiabizclub. com/
7. http://herbal-hair-shampoo.com/
8. http://herbgardens.about.com/od/herbalcraftsandgifts/ht/Herbalhaircolor. htm
9. http://www.in-cosmetics.com/Exhibitor Library/61/Quick_Facts_Cosmetic_2.pdf
10. http://www.indiamart.com/excelimpex/herbal-hair-color.html
11. http://www.indianchild.com/beauty/hair_dye.htm
12. http://www.lasscosmetics.com/natural-hair-conditioner.php
13. http://www.longlocks.com/hair-care-recipes-cookbook.htm#dye5
14. http://www.welstarorganics.com/index.php/skin-body-care/hair-nail/flower-plant-deep-conditioner-4-oz.html

15. http://www.wisegeek.com/what-is-natural-hair-dye.htm
16. Mainkar A.R., and Jolly C.I. International Journal of Cosmetic Science, 2000, 22(5), 385 – 391.
17. Nilani Packianathan and Saravanan Karumbayaram. Formulation and Evaluation of Herbal Hair Dye: An Ecofriendly Process. J. Pharm. Sci. and Res. Vol.2 (10), 2010,648–656
18. Sharma P.P., Cosmetic Formulation Manufacturing and Quality Control, 3ed ed., Vandana Publication, Delhi, 644–647.
19. Thompson D., Lemaster C., Allen R., and Whittam J., Evaluation of relative shampoo detergency. Journal Of The Society Of Cosmetic Chemists. 1985, 36, 271–286.

SKIN CARE

Creams

A cream is a topical preparation for application to the skin. They are considered pharmaceutical products as even cosmetic creams are based on techniques developed by pharmacy and unmedicated creams are highly used in a variety of skin conditions. They are semi-solid emulsions, that is mixtures of oil and water. They are broadly divided into two types: (a) **Barrier cream**, that locks moisture in the skin over a period of time, such as a "protective cream, a healing cream, a night cream, a cleansing cream, a sunscreen cream" etc. They should be used for short periods of times. (b) **Daily moisturizer cream** which is acts as a daily moisturizer for skin like lotions. This type of cream allows the skin the breathe while drawing moisture to the skin so it does not become dehydrated during the daytime. This cream functions as good foundation under make-up, a daily moisturizer after your bath, a facial day cream, or a spot cream for dry skin areas.

As per the emulsifiers used for the above two types of creams, they are divided into i) **oil-in-water (O/W) creams** which are composed of small droplets of oil dispersed in a continuous phase and **water-in-oil (W/O) creams** which are composed of small droplets of water dispersed in a continuous oily phase. Oil-in-water creams (e.g. Vanishing creams) are more comfortable and cosmetically acceptable as they are less greasy and more easily washed off using water. Water-in-oil creams are more difficult to handle but many drugs which are incorporated into creams are hydrophobic and will be released more readily from a water-in-oil cream than an oil-in-water cream. Water-in-oil creams (e.g. cold creams) are also more moisturising as they provide an oily barrier which reduces water loss from the stratum corneum, the outermost layer of the skin.

Herbal creams are extracted from natural herbs, which keep the skin healthy and protect facial tissues from dust, pollution and direct sunlight. Further, these creams restore natural oils and nourish the skin well.

Uses of Creams

- The provision of a barrier to protect the skin
- This may be a physical barrier or a chemical barrier as with sunscreens
- To aid in the retention of moisture (especially water-in-oil creams)
- Cleansing
- Emollient effects
- As a vehicle for drug substances such as local anesthetics, hormones, antibiotics, antifungals, etc.

Some of the Herbal Cream Preparations

1. Basic cold cream

Ingredients: Mineral oil, wax, IPM, glycerin, olive, marigold, avocado, almond, jojoba, perfume and preservative.

Instructions: Heat oil and wax in double boiler. Add all other ingredients in a thin stream, stirring vigorously in one direction. When mixed remove from heat and check temperature. At 140°, add scent. At 120°, pour into jars. Finally, add the preservative and perfume as required amount.

Action: Olive, marigold, avocado, almond and Jojoba, which provides vital nourishment to the skin. Olive and marigold oil acts as a natural emollient with effective, healing and soothing properties for minor skin problems. Jojoba, avocado and almond oil, high in vitamins and rich in nutrients and proteins, nourish the skin from deep inside, keeping it soft, smooth and beautiful even in the most severe winter conditions.

2. Night cream

Ingredients: Three tablespoons olive oil, 1 tablespoon *Aloe vera* gel, 1 tablespoon shea butter, 2 teaspoons grated beeswax, 1 tablespoon rosewater, 1/4 teaspoon lecithin, oil from two capsules of vitamin E, 3 drops chamomile or lavender essential oil.

Instructions: Melt oil, butter, and beeswax in heatproof glass measuring cup, over boiling water or in microwave (low heat). Remove from heat just before beeswax is completely melted. Finish melting by stirring in warm oil. Add the vitamin E oil and lecithin and stir in between additions. Mix rosewater and aloe together and slowly add to main mixture. Continue stirring with a small metal whisk. Once mixture is cooled, add the essential oil and stir. Keep away from sun and heat.

3. Fairness cream

Ingredients: Aqueous extract of *Berberis aristata* 20 mg, *Glycyrrhizin glabra* 30 mg, *Vetiveria zizanioides* 20 mg, saffron, coconut water as required quantity, oil of sandalwood 0.005 g, almond 0.005 g, wheat germ 0.005 g.

Instruction: Take all the extracts as required quantities and mixed them in previously prepared mixed solution of water phase. Then perfume to be added as required quantity.

Ingredients: Extract of *Aloe vera*, *Juglans regia* (walnut), *Rosa centifolia*, *Citrus reticulata* and other oil and aqueous phase.

4. Vanishing cream

They are when applied on the skin surface, they spread as a thin oil less film and completely miscible on skin which is not visible to the naked eyes. They are used to hold powder on the skin as well as to improve an adhesion.

Ingredients: Stearic acid 24 g, glycerin 10 g, potassium hydroxide 1.34 g, methyl paraben 0.10 g, Propyl paraben 0.05 g, purified water 64 g, rosewater required amount, *Stevia* white extract 2.5 g.

Instruction: Ingredient of oil phase was melted in a beaker by using water bath on constant stirring. Components of aqueous phase were mixed together and warmed to about same temperature of oil phase (70°C). The preservative propyl paraben and methyl paraben was added into aqueous phase after dissolved in little quantity of warm water. Then oil phase was added to water phase little by little on constant stirring and purfume was added to it when the temperature was 35 – 40°C. After some time the mixture appeared as pearl like semi, solid mass. Further-required quantity of the preparation was taken out from total formulation and specified amount of white powdered *Stevia* extract was mixed uniformly by levigation method to get 2.5% *Stevia* vanishing cream.

Aloe Moisturizing Hand Cream

1/4 cup rosewater, 12 drops aloe extract, 1 tbsp aloe gel, 1/8 cup sweet almond oil, 1/2 tbsp peanut oil, 1 tsp cocoa butter, 1/2 tbsp olive oil, 1/4 tsp anhydrous lanolin, 1/2 tbsp liquid lecithin, 1/8 oz. grated beeswax, 400 IU vitamin E, 6 drops tincture of benzoin.

Method: Melt waxes, oils, cocoa butter, lanolin and lecithin and allow to cool a bit. Place aloe gel in blender and whip for a bit. Add rosewater to aloe and blend. Stir tincture of benzoin (or grape seed extract, if desired) and vitamin E into the warm oils. With blender running, slowly drizzle the oils into the aloe-water mix. Put up into suitable containers and store in refrigerator. Shelf life is about one month.

Lotion: A lotion is a liquid preparation, applied externally on the skin to produce or enhance a beautification. Lotions are used for washing the skin and to remove the oily secretions. It increases the blood circulation, emolliency, extended to astringency, skin freshness, bleaching and other medicinal properties.

General formula for a lotion includes alcohol, water and glycerin with some special astringents, gums, honey, antiseptics and preservatives.

Ideal Properties

1. They should give cooling effect on application.
2. They should be free from particles.
3. They should produce emollient effect.
4. They should remove the oily secretion upon application.
5. They should spread uniformly on the skin surface.
6. They should not cause any skin toxicity.
7. They should be compatible with skin pH.

Some Formulations

1. *Skin toning lotion:* Sunflower oil 45 ml, wheat germ oil 8 ml, witch hazel extract 24 ml, sodium benzoate 5 g, lanolin as required quantity to makeup the volume up to 100 ml.

2. *Sunscreen lotion:* Quinine oleate 2.5 g, olive oil 450 g, oil of cassia 0.50 g, perfume oil 0.50 g and refined peanut oil quantity sufficient to make volume up to 100 g.

3. *Skin care lotion:* Henna paste 3.0%, turmeric, tincture of benzoin, perfume, rosewater.

Evaluation Tests of Skin Cosmetics

The study can be perfomed at a dermatological clinic or a skin test institute. Usaully, the toleracne and cosmetic effects of the products are evaluated. Pertly, bioengineering methods are integrated in the test design to contribute objective parameters. Generally, common skin cosmetics are evaluated on the basis of three parameters. Flow chart 4.1 has described brief about the details of evaluation tests.

Bioengineering Methods for Skin Moisturizer Tests

Methods to measure skin finction parameters with the help of special devices.

a. pH-meter for measuring skin surface pH

b. Corneometer, infrarotspektroskopie, or magnet resonnance tomography to measure skin hydration

c. Sebumeter for skin surface lipid determination

d. Tewameter or Evaporimeter for barrier function assessment via water evaporation cutometer for skin elasticity

e. Chromameter for skin color

f. Ultrasound sonography to measure skin thickness

g. Profilometry, visiometry, or skin surfametry to detect the skin surface profile and quantify wrinkles, etc.

Chromameter

Device to measure skin color, suited to monitor skin inflammation via redness. It is also used to quantify skin cleansing effects. For this purpose, a red soiling paste is applied and the fading of the red color by standardized washing is recorded.

Fig. 4.1: Chromameter

The chromameters CR-400 supersedes the internationally recognized and acclaimed series CR-100, CR-200 and CR-300. The data memory now can store up to 100 target colors and up to 2000 measurements.

Corneometer

An adequate moisture content of the stratum corneum is essential for a well functioning

Flow chart 4.1

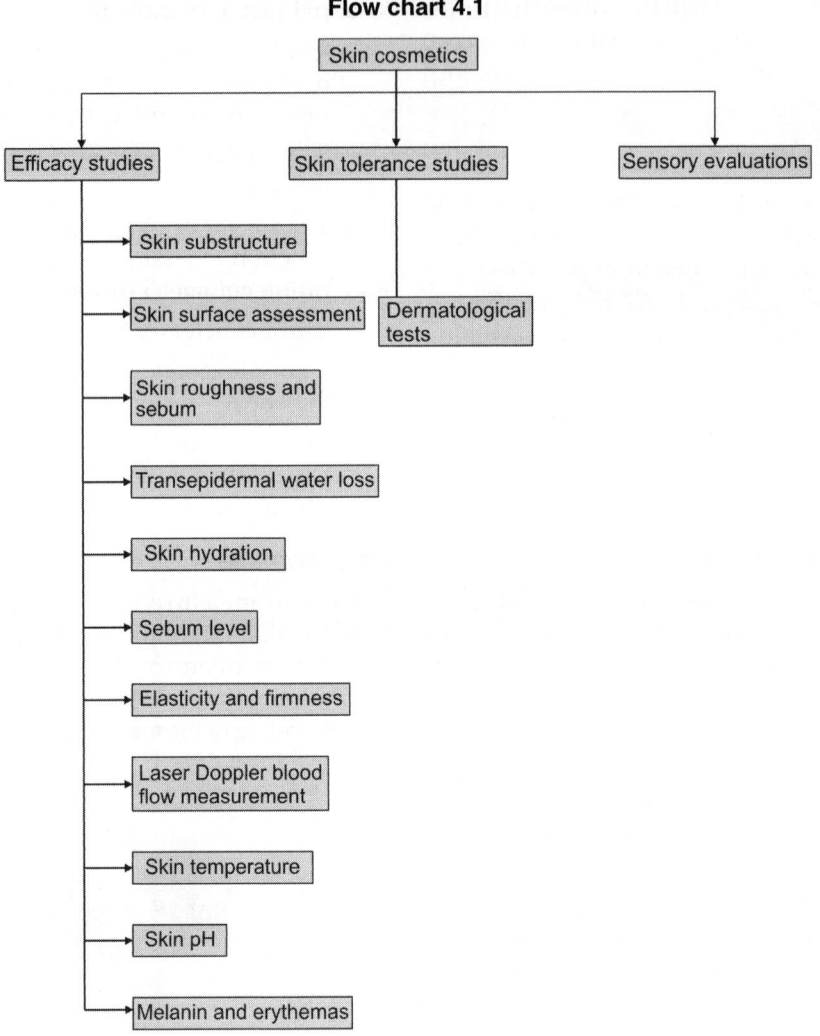

skin barrier. NMF (natural moisturizing factor) provides an effective scavenging function against radicals and nitrogen oxides penetrating from the outside, and consequently protects the skin against premature skin aging. Hence, the moisture content is the most important parameter of skin diagnosis and measured with the Corneometer®. The analysis of the moisture retention capacity of the skin is easy to perform, based on the dielectric constant of the water and measured in the superficial layers of the stratum corneum as deep as 10–20 μm to ensure that the measurement is not influenced by capillary blood vessels. It determines the electrical capacitance of the skin which is proportional to its water content, or hydration.

The measuring principle of the Corneometer® is based on capacitance measurement of a dielectric medium. Any change in the dielectric constant due to skin surface hydration variation alters the capacitance of a precision measuring capacitor. The reproducibility of the measurement is very high and the measurement time is very short (1s). Due to the construction of the measuring head, the measurement depth is very small (in the first 10–20 μm of the stratum corneum).

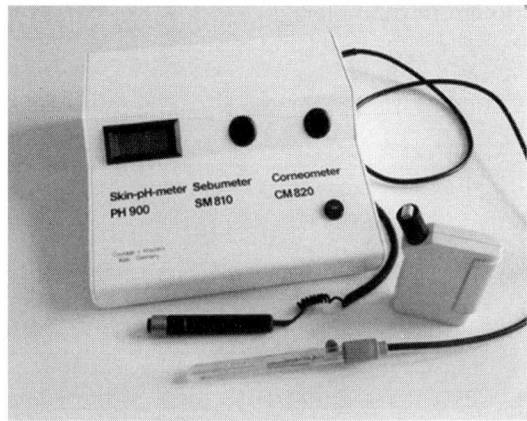

Fig. 4.2: Corneometer

Cutometer

Cutometer used to instrumentally measure the viscoelasticity of the skin under experimental conditions. The method is non-invasive. The measuring principle is based on the suction method. Negative pressure is created in the device and the skin is drawn into the aperture of the probe. Inside the probe, the penetration depth is determined by a non-contact optical measuring system. This optical measuring system consists of a light source and a light receptor, as well as two prisms facing each other, which project the light from transmitter to receptor. The light intensity varies due to the penetration depth of the skin. The resistance of the skin

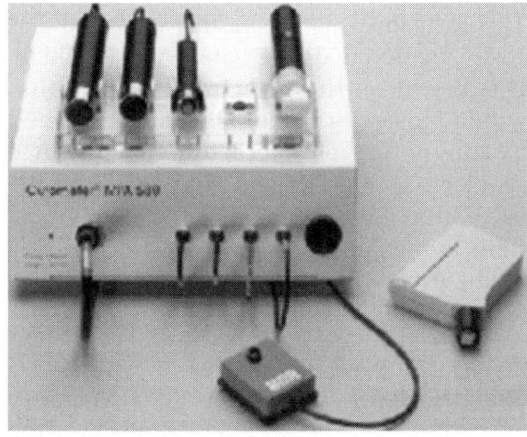

Fig. 4.3: Cutometer

to be sucked up by the negative pressure (firmness) and its ability to return into its original position (elasticity) are displayed as curves at the end of each measurement. From these curves interesting measurement parameters can be calculated.

The signal delay is indicative of the elasticity of the skin: the faster, the higher it is.

Epicutaneous Patch Testing

Skin tolerance test for cosmetic products performed on 50 to 100 persons, similar to dermatological allergy test. The test proudt is applied on a patch to the back of the test persons and left there for 24 hrs. The effect on the skin is read immediately after removal of the patch as well as 1 and 2 days later. The appearance of redness, flaking, blistering, papules or pustules is recorded as a sign of irritation potential of the test product. The test gives no indication for the sensitizing or allergenic potential of a test product. The repetitive patch test is even more sensitive. Here, the patches with the test product is repeated at certain intervals for several times. Thus, it also allows conclusions on the sensitizing potential of the test product.

In the Duhring chamber test (soap chamber test) dilutes test products are applied to the skin in a similar way as described above for 24 hrs. Irritation reactions can also be measured by bioengineering methods, e.g. TEWL, chromametry or Laser-Doppler-Flowmetry.

Evaporimeter

The device measures the evaporation of water from the skin surface and thus the skin's barrier function against water loss. This gives an indication for the risk of skin dryness, the water binding capacity and, indirectly, also for the ease with which substances from the environemnt can penetrate into the skin.

The device consists of a sharp hook suspended from a micrometer cylinder, with the body of the device having arms which rest on the rim of a still well. The still well serves to isolate the device from any ripples

Fig. 4.4: Evaporimeter

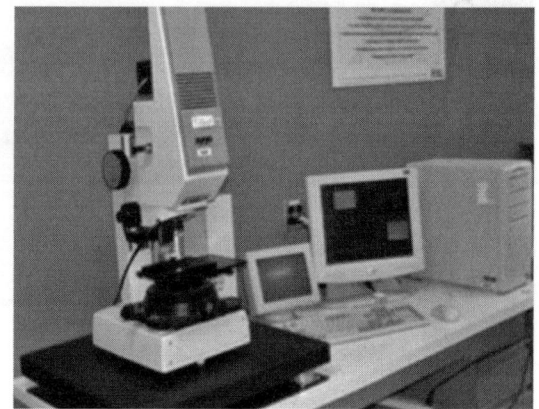

Fig. 4.5: Profilometry

that might be present in the sample being measured, while allowing the water level to equalize. The measurement is taken by turning the knob to lower the hook through the surface of the water until capillary action causes a small depression to form around the tip of the hook. The knob is then turned slowly until the depression "pops," with the measurement showing on the micrometer scale. Evaporation rate is determined by a sequence of measurements over a set time interval.

Skin Surfametry

This instrument is use for recording the skin surface profile. Under standardized conditions with light coming from one side the skin surface is digitally photographed. The data are computerized to analyse the surface profile according to methods adapted from the material sciences, especially parameter for surface roughness like the average depth of roughness. Thus it is possible to quantify the increase in skin roughness as a result of dehydration as well as number, depth and abundance of wrinkles and the effect of anti-aging products.

Profilometry

Profilometry is also use to monitor the skin surface profile. Silicon replicases of the skin surface are taken. There surface is a negative reproduction of the skin. By mechanical or

optical (laser) scanning of the replica surface the roughness parameters from material sciences can be determined.

Histology

Section of skin tissue taken during surgery or biopsies is investigated. Very thin slices are cut and often stained with dyes to make fine structures visible under the microscope.

Infrared Spectroscopy

This instruments use to measure skin hydration by reflection of infrared light in the upper skin layers. Skin tissue with a high water content reflects more infrared light than dry areas.

Magnet Resonance Tomography

This instrument is technically advanced and costly method to detect water within the skin. It is based on the principle of oscillation within molecules in the skin, haut is induced by magnetic fields and thereby produce an electromagnetic field. The fading of the oscillation after switching off the magnetic field is measured. Water slowsdown this fading process. Via computer the time course of the fading process can be analyzed and correlated to a certain water content.

Skin pH-Meter

Skin pH-meter is a measurement device for the skin suface pH. The normal value is 5.5 on

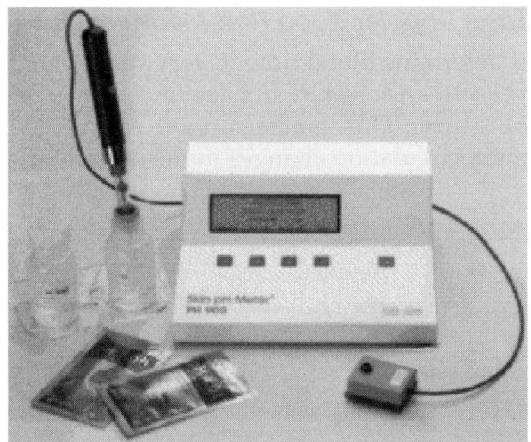

Fig. 4.6: Skin pH-meter

average. Skin care products can alter the skin surface pH.

Recommended measuring areas are the back of the hand, the forearm, the front and the cheeks, however readings are also possible on any other body part.

Sebumeter

This instrument measures the skin surface lipids. A semi-transparent tape is pressed onto the skin surface, absorbing the lipids present there. The lipids increase the transparency of the tape. A photometer (combination of a light source which sends light through the tape to a mirror who reflects it back through the tape again to be detected by a device which measures the light intensity), quantifies the increase in light

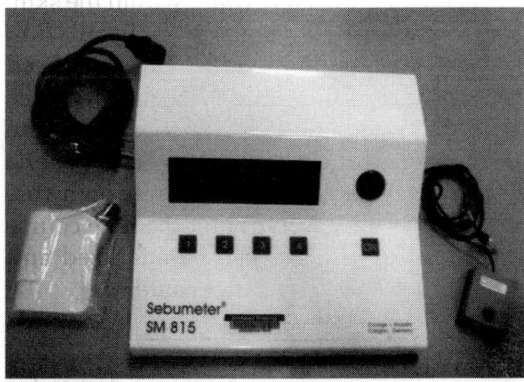

Fig. 4.7: Sebumeter

transmission through the tape after lipid absorption as compared to before. This allows a quantification of the amount of skin surface lipids: The more lipids the higher the transparency.

Sniffing-Test

Test of the efficacy of deodorants. Trained people assess the intensity of body odour developing in the armpits of 10 to 30 persons after cleansing over the course of 24 hrs. In one armpit, no deodorant is used, in the other, the test deodorant is applied. The intensity of body odour is scored from 0 = none to 5 = very intensive is determined every 6 hrs and normal acitivties of the test persons without intermittent washing. The difference of the average score between the two sides is the basis for the evaluation of efficacy. If there is a significant difference between the treated and untreated side after 24 hrs the efficacy is proven. This, however, does not guarantee that no body odour develops at all when a deodorant is used, but confirms only a reduction of the intensity of the malodour.

Ultrsound Sonography

This method is to measure the skin thickness. Ultrasound echoes from the skin are detected. The mode of reflection is indicative for different density and composition of the skin layers. Higher frequencies of ultrasound allow a higher resolution of skin structures but reduce the depth of the measurements.

The ultrasound skin imagery is a non-invasive, quantitative and reproducible technique to measure and to display the skin structures, viz. epidermis, dermis strates and collagen. The Dermascan, is a unique medical ultrasound scanning technology, analyze the skin and obtain readings on the conditions and construction of the skin down to the deepest layers. The Dermascan-C device with its 10–50 MHz frequency, allows to determine and display the substructure of the skin up to a depth of 20 mm.

Fig. 4.8: Dermascan-C

Visiometry

This is another instrumental method to measure the skin surface profile. Digital photographs of the skin surface are mathematically analyzed and translated into surface parameters adapted from the material sciences.

Infrared Pyrometer

Skin temperature can be measured by using this instrument. This instrument allows assessing the cooling effect or the stimulating effect of a cosmetic formulation. Temperature resolution is very high with ± 0.03°C.

Laser Doppler Blood Flow Measurement

It determine blood flow in very small blood vessels under skin surface and skin temperature. This measurement can reveal microcirculation changes in the dermis.

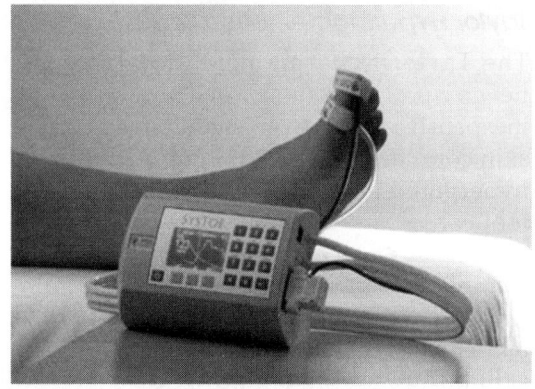

Fig. 4.10: Laser Doppler

Mexameter

Melanin and erythema index can be measured by using this instrument. The measuring principle for the melanin and erythema readings is based on a source of light with three specific wavelengths whose radiation is absorbed by the skin and diffusely reflected. A photodetector analyzes the diffuse reflection from the skin. If the skin is well-supplied with blood also the hemoglobin value is increased. Consequently, it is possible to evaluate the stimulation of the microcirculation before and

Fig. 4.9: Infrared pyrometer

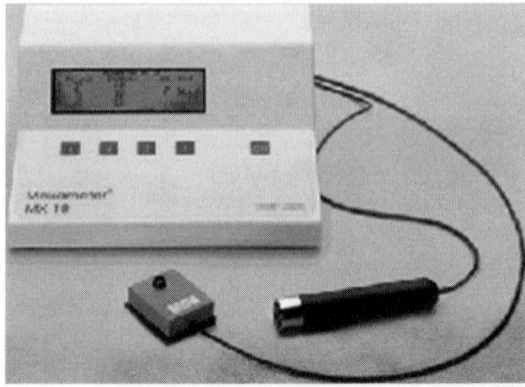

Fig. 4.11: Mexameter

after applications with the measuring of the hemoglobin value.

The same measuring probe is also used to quantify the skin redness (erythema) and determine the degree of skin tanning (melanin).

Taylor Hyperpigmentation Scale

The Taylor hyperpigmentation scale is a new visual scale developed to provide an inexpensive and convenient method to assess skin color and monitor the improvement of hyperpigmentation following therapy. The tool consists of 15 uniquely colored plastic cards spanning the full range of skin hues and is applicable to individuals with Fitzpatrick skin types I to VI. Each card contains 10 bands of increasingly darker gradations of skin hue that represent progressive levels of hyperpigmentation.

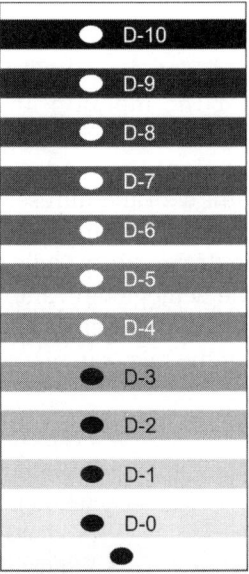

Fig. 4.12: Taylor hyperpigmentation scale

The predefined primary efficacy endpoint was a decrease in the symptom score for dark complexion. The predefined secondary safety endpoints were acute and chronic safety, as assessed by the incidence of adverse events of the volunteers.

Dermatological Test

Various skin irritation tests are performed under this test. The tests are may be occlusive, semi-occlusive or open irritation tests. The various tests includes:

- Single patch tets for 24 hours or 48 hours (open type or close type).

- Repeated patch test (4 applications, 5 days/over 2 weeks)

- HRIPT–human repeat irritation patch test (9 applications, 21 days)

- Duhring chamber test (5 applications, 1 week)

Sensory evaluation: This test can be done on human volunteers upon request. Volunteers are selected as per the sex, origin, skin type, habits, and sensitivity.

The Epiderm *in vitro* three-dimensional human skin tissue model was introduced in response to demand for non-animal model for determination of skin irritation effect. Epiderm is a normal (non-transformed), human cellderived, metabolically active, three-dimensional organotypic *in vitro* skin model. It is also known generically as reconstructed human epidermis (RhE), which closely mimics human epidermis, both structurally, biochemically and does so in a very reproducible manner.

Eye Make-up

Eye make-up is an essential part of facial make-up. Hence it becomes important to choose any eye makeup products very carefully. These products include **eye shadow** or **eye shade** cream, stick or liquid. The cream is emulsion or a liquefying type, **eyebrow pencil** and **mascara. Eye shadow** is a cosmetic that is applied on the eyelids and under the eyebrows. It is commonly used to make the wearer's eyes stand out or look more attractive. Eye shadow comes in many different colors and textures. It is usually made from a powder and mica, but can also be found in liquid, pencil or mousse form.

Table 4.2: Some of the skin care products with their various evaluation studies are listed below

Efficacy studies	Tolerability studies	Products
• Skin moisturization • Short-term kinetics • Deodorant efficacy • Anti-wrinkle test • Skin roughness • Skin care test • Forearm-wash test • Shaving irritation • Antitranspirant-hotroom- and sauna test • Wound healing • Skin moisturization use test • Barrier recovery • Barrier protection • Patch scratch test • Anti-cellulite test • Efficacy test pre-acne • Skin Bleaching, depigmentation • Adhesive forces determination • Skin elasticity • Skin blood flow • Anti-facial erythrosis • Anti-stinging efficacy • Melasma and age spots efficacy	• Consumer test • Soap chamber test • Simple epicutaneous test • Open epicutaneous test • Repetitive epicutaneous test • ROAT–repeated open application test • HRIPT–human repeat irritation patch test • 21 Cummulative irritation patch test • Forearm-wash test • Shaving irritation • Stinging test • Anti-itch test	• Moisturizer • Anti-aging • Anti-wrinkles • Anti-cellulite • Antitranspirants • Acne • After-shave • Tapes and bandages • Decorative cosmetics • Textiles • Cleansing products

Table 4.3: Some of the sun care products and their evaluation studies are listed

Standard studies	Customized studies	Products
• In vivo sun protection Factor determination according to international norm 2006 • In vivo sun protection Factor determination with water resistancy according to international norm 2005 • In vivo UV-A protection (JCIA-guideline) • In vivo UV-A protection (Koreanische method) • In vitro UV-A protection Measurement (Colipa 2007) with pre-radiation • In vitro UV-A protection (Boots 2008) with pre-radiation • In vivo phototoxicity • In vivo photosensibilization	• UV-erythema • Assessment of anti-irritative efficacy (sunburns, burns) • Efficacy assessment of self-tanning products • Efficacy assessment of skin bleaching cosmetics • Sweat resistancy of sun protection products • Efficacy assessment of aftersun products • Skin bleaching, depigmentation in UV-models	• Sun protection products • Water resistant Sun protection products • Cosmetics with UV-protection • Self-tanning products • Whitening products • Aftersun products • Anti-inflammatory

Table 4.4: List of herbs are used for skin care products

Plant Name	Biological source	Family	Part use	Uses
Shikakai	Acacia concinna	Fabaceae	Pods	Various skin diseases
Sweet flag	Acorus calamus	Acoraceae	Rhizome	Aromatic, dusting powders, skin lotions
Garlic	Allium sativum	Amaryllidaceae	Bulb	Promotes skin healing, antibacterial
Aloe	Aloe vera	Liliaceae	Leaf	Moisturizer, sunscreen, emollient
Galanga	Alpinia galanga	Zingiberaceae	Rhizome	Aromatic, dusting powders
Oat	Avena sativa	Poaceae	Fruit	Moisturizer, skin tonic
Neem	Azadirachta indica	Meliaceae	Leaf	Antiseptic, reduce dark spots, antibacterial
Marigold	Calendula officinalis	Asteraceae	Flower	Skin care, anti-inflammatory, antiseptic, creams
Gotu kola	Centella asiatica	Umbelliferae	Plant	Wound healing, reduce stretch marks, creams
Chicory	Cichorium intybus	Asteraceae	Seed	Clears skin of blemishes
Orange	Citrus aurantium	Rutaceae	Peel	Skin creams, anti-acne, anti-bacteria
Turmeric	Curcuma longa	Zingiberaceae	Rhizome	Antibacterial, antimicrobial, skin creams
Nagarmotha	Cyperus rotundus	Cyperaceae	Roots	Suntan, astringent, anti-inflammatory
Carrot	Daucus carota	Apiaceae	Seed	Natural source of vit A, creams
Spurge herb	Euphorbia hirta	Euphorbiaceae	Entire	Skin diseases, cracked lips
Tulsi-basil	Ocimum sanctum	Labiatae	Leaves	It has got properties of removing dark spots and scars from the face
Kalawalla	Polypodium leucotomos	Polypodiaceae	Leaves	It reduce sun burn severity and that it may help prevent skin aging and decrease the risk of cancer from UV radiation.
Sandalwood	Santalum album	Santalaceae	Wood	Glowing of skin by removed dead skin
Honey leaf	Stevia rebaudiana	Asteraceae	Leaves	Skin moisturizer

Eye shadow can be applied in a variety of ways depending upon the desired look and formulation. Typically, application is done using sponges, fingers and/or brushes. The most important aspect of applying eye shadow and make-up in general, is blending well.

Eye liner is a cosmetic used to define the eyes. It is applied around the contours of the eye to create a variety of esthetic illusions.

Depending on its texture, eyeliner can be softly smudged or clearly defined. There are four main formulas available on the market; each produces a different effect:

- *Liquid eyeliner* is an opaque liquid that usually comes in a small bottle and is applied with a tiny brush or felt applicator. It creates a sharp, precise line.

- *Powder-based eye pencil* is eyeliner in a wood pencil. It is generally available in dark matte shades.
- *Wax-based eye pencils* are softer pencils and contain waxes that ease application. They come in a wide variety of intense colors as well as paler shades such as white or beige. Wax-based eyeliners can also come in a cone or a compact with brush applicator.
- *Kohl* eyeliner is a soft powder available in dark matte shades. It is most often used in black to outline the eyes. It comes in pencil, pressed powder, or loose powder form. This type of eyeliner is more likely to smudge.

Formulation

1. Herbal kohl eyeliner (kajal) color: Cleopatra black.

Key ingredients: Sinapis alba (white-mustard), cera alba (beeswax), oil of *prunus dulcis* (almond), *Brassica nigra* (black-mustard kajal), *Camphora officinarum.*

Table 4.5: Some of the herbs that are used in preparation of the eye make-up products

Plant name	Biological source	Family	Uses
Azalea	Azalea arborescens	Ericaceae	It is the deepest coral shade of our coral shimmer mineral make-up color range
Cinnamon	Cinnamomum zeylanicum	Lauraceae	This neutral medium brown color can be used as eyeliner, eye shadow, cheek definition, lip liner or body shimmer
Cocoa	Theobroma cacao	Malvaceae	This neutral brown color can be used as eyeliner, eye shadow, cheek definition, lip liner or body shimmer
Hydrangea	Hydrangea macrophylla	Hydrangeaceae	It can be used for eyes, or to get a little blue sparkle
Indigo	Indigofera tinctoria	Fabaceae	It can be used for eyeliner or eye shadow to get dark blue shade
Larkspur	Delphinium parishii	Ranunculaceae	It can be used for eyeliner, eye shadow or wherever to get medium blue highlights
Mahogany	Swietenia mahagoni	Meliaceae	This neutral deep brown color can be used as eyeliner, eye shadow, cheek definition, lip liner
Maidenhair	Adiantum capillus veneris	Pteridaceae	It can be used for eyeliner, eye shadow or to get dark green shade in eyes
Neem oil	Azadirachta indica	Meliaceae	It can be used for mascaras
Oatstraw	Avena sativa	Poaceae	It is the lightest shade of our taupes mineral make-up color range
Primrose	Primula vulgaris	Primulaceae	It is the deepest shade of the pink shimmer mineral make-up color range
Almond oil	Prunus amygdalus	Rosaceae	It can be used for eyeliner, mascaras
Sage	Salvia officinalis	Lamiaceae	It can be used for eyeliner, eye shadow or to get light green accents
Triphala (Amalaki, bibhitaki and haritaki)	Emblica officinalis Terminalia bellirica Terminalia chebula	Phyllanthaceae Combretaceae Combretaceae	It can be used for eyeliner, mascaras

Mode of action: Sinapis alba (white-mustard): White mustard oil has high levels of omega- 3-fatty acids (6–11%) and helps to improve skin tone and texture; the oil also helps to preserve product freshness. ***Brassica nigra*** (black-mustard): Kohl/Kajal is self-generated carbon collected from the traditional Ayurvedic processing of *Brassica nigra* (mustard oil).

Method of use: Apply as eyeliner (to inner and/or outer eyelid). Can be used everyday. Does not require sharpening, it sharpens itself as it is used on an angle. Easily removed with moist cotton and make-up cleanser.

Mascaras: Mascara as a cosmetic substance for darkening, coloring, and thickening the eyelashes, applied with a brush or rod. Normally, mascara has three forms: Liquid, cake or cream but the modern mascara product has various formulas. Commonly, all mascaras contain the same basic components of pigments, oils, waxes, and preservatives.

Preparation: There are two main methods of production, i.e. anhydrous method (in this method all waxes, oils, and pigments are mixed, heated, and agitated simultaneously in formulated ratios. The result is a semi-solid substance that is ready to be placed in tubes, packaged, shipped and sold) and emulsion form (in this method, water and thickeners are first combined. Separately, waxes and emulsifiers are mixed and heated. Pigment is then added individually to both mixtures. Finally, all is combined in a homogenizer, which acts as a high-speed agitator in order to thoroughly mix the oils, water, waxes, and emulsifiers — ingredients that naturally repulse each other).

Evaluation Test for Eye Care Products

1. Schirmer test: Schirmer's test determines whether the eye produces enough tears to keep it moist. This test is used when a person experiences very dry eyes or excessive watering of the eyes. This test consists of placing a small strip of filter paper inside the lower eyelid (conjunctival sac). The eyes are closed for 5 minutes. The paper is then removed and the amount of moisture is measured. The results can be calculated as:

1. Normal which is ≥15 mm wetting of the paper after 5 minutes.
2. Mild which is 14–9 mm wetting of the paper after 5 minutes.
3. Moderate which is 8–4 mm wetting of the paper after 5 minutes.
4. Severe which is < 4 mm wetting of the paper after 5 minutes.

Table 4.6: Some eye care products with their evaluation method

Methods	Products
• Eye tolerability test no tears	• Shampoos
	• Artificial tear fluid
• Eye instillation study	• Facial care products
	• Sun protection products
• Safety test with ophthalmological control	• Eye shadow
	• Mascara
	• Eye creme
• HET-CAM test	• Kajal/Eyeliner
• Tear break-up time (BUT) test	• Concealer
	• Make-up remover
• Schirmer test	• Facial powder

Lipsticks: In general lipsticks may be defined as dispersion of the coloring matter in a base consisting of a suitable blend of oils, fats and waxes with suitable perfumes and flavors moulded in the form of sticks. When it applied on lips it imparts attractive gloss, color and can provide moist appearance to the lips.

Some Ideal Characters

1. It should efficiently cover lips with color with maintained intensity without any alteration in the degree of its shade.
2. It should be able to adhere firmly to the lips.
3. It should possess good thixotropic property and plasticity.
4. It should be easily dried.

5. It should not be gritty and dry easily on storage.

6. It should be safe and nonirritant to the lips.

7. It should not produce sweating of the lips.

8. It should maintain its strength at varying temperature up to 55–60°C.

Some Formulations

1. **Liquid lipstick:** Chamomile extract: 0.001% to 1.00%, jojoba oil-0.01% to 1.00%, allantoin-0.01% to 1.00%, *Aloe vera* gel: 0.01% to 1.00%; and octyl methoxy cinnamate-0.01% to 1.00%.

2. **Ecosense herbal lipstick:** Neem, aloe, wheat germ with vitamin E.

   Mode of action: Neem helps prevent blackening of lips, repairs damaged lips and *Aloe vera* with vitamin E keeps lips supple, soft and prevents chapping. It cures chapped lips.

   Vitamin E protects against free radicals, it helps soothe and calm skin.

3. **Herbal lipstick:** Two spoons almond oil, half spoon beeswax, 1 spoon beetroot juice and add two spoons little warm oil in beeswax and mix vigorously. Then add beetroot juice and refrigerate for 3 hours. Finally, the lipstick is ready.

Evaluation Test of Lipsticks

1. **Determination of color dispersion:** This test is performed in order to determine the uniform dispersion of the color particles. The size of the particles is determined by the microscopic studies and it should not be more than 50 µ.

2. **Determination of surface characters:** The study of the surface property of the product is carried out to evaluate the formation of the crystals on the surface or any contamination by microorganisms, etc.

3. **Determination of thixotropic nature:** This test is performed to check the uniformity of the viscosity of base. The thixotropic character is determined by penetrometer where the lipstick is placed at 25°C and a standard needle of known diameter is made to penetrate through it for 5 second and the extent of penetration measures.

4. **Microbial tests:** Extent of any contamination either from the raw materials or the mould is determined by microbial tests. The test involves the plating of known mass of samples on two different culture media for the growth of microorganisms and incubating them for a specific period of time. The extent of contamination can be estimated by colony counter instrument by counting the number of colonies.

5. **Test for rancidity:** This test is performed to check the oxidation of oils and many other ingredients which are used as additives in the lipsticks which show stickyness. The test can be done by using hydrogen peroxide and determining its peroxide number.

6. **Breaking load point test:** This test is done in order to determine the strength and hardness of the lipstick. In this method, the lipstick is placed in horizontal position half inch from the base and weights with increasing loads

Table 4.7: Some herbs that are use in lipstick preparations

Plant name	Biological source	Family	Part use	Colour
Annatto tree	*Bixa orellana*	Bixaceae	Seed	Red, orange
Beetroot	*Beta vulgaris*	Chenopodiaceae	Roots	Red
Turmeric	*Curcuma longa*	Zingiberaceae	Roots	Yellow, orange
Patang	*Caesalpinia sappan*	Fabaceae	Wood	Red, violet

are attached to it. The weight at which the lipstick start breaking known as the breaking load point.

7. *Melting point determination test:* This test is performed to determine the storage characteristic of the products. The melting point of lipstick base should be between 60 and 65°C in order to avoid sensation of friction or dryness during application. In this method, about 50 mg of lipstick is taken and is inserted into a glass capillary tube open at both ends. Then the capillary tube is ice cooled for about 2 hours and then places in a beaker containing hot water and a magnetic stirrer. The temperature at which material starts moving out from capillary tube is known as melting point temperature. Another parameter is droop point which determines the temperature at which the product starts oozing out the oil and becomes flattened out. Hence drooping point should be lesser than melting point to make safe handling and storage of finished products.

8. *Storage stability:* This test is done in order to determine the stability of product during storage.

9. *Test for application force:* This test is performed to determine the force to be applied during application. In this method, two lipsticks are cut to obtain flat surfaces which are placed one above the other. A smooth paper is placed between them which is attached to a dynamometer to determine force required to pull the paper.

BIBLIOGRAPHY

1. Bharat Vishwakarma, Sumeet Dwivedi, Kushagra Dubey and Hemant Joshi. Formulation and evaluation of herbal lipstick. International journal of drug discovery and herbal research. 1(1): 18–19, 2011.

2. http://altmedicine.about.com/od/herbsupplementguide/a/Polypodium. htm

3. http://emedicine.medscape.com/article/1067211-overview

4. http://en.wikipedia.org/wiki/Cream_(pharmaceutical)

5. http://en.wikipedia.org/wiki/Eye_shadow

6. http://en.wikipedia.org/wiki/Lipstick

7. http://en.wikipedia.org/wiki/Lotion

8. http://en.wikipedia.org/wiki/Sunscreen

9. http://library.thinkquest.org/04apr/01297/html/cosherb.htm

10. http://www.a1-natural-beauty.com/Sunscreen_Lotion.html

11. http://www.agriinfotech.com/htmls/PDF-Files/HERBS/Cosmetic % 20Herbs. pdf

12. http://www.biotecharticles.com/Health care-Article/Herbal-Cosmetics-Part-1-Cleansing-Cream-Vanishing-Cream- and-Sunscreen-896.html

13. http://www.eris.pl/files/Plakat % 201 % 20EpiDerm_EADV.PDF

14. http://www.glenbrookfarm.com/face_creams.htm

15. http://www.hennausa.com/Natural_Eye_Makeup.php

16. http://www.in-cosmetics.com/Exhibitor Library/813/STI_Livret_final_compr_corr4_2

17. http://www.in-cosmetics.com/Exhibitor Library/813/SKIN_TEST_INSTITUTE_Cosmetic_Dermatological_Skin_Investigations_1.pdf

18. http://www.medicinenet.com/sun_protection_and_sunscreens/article.htm

19. http://www.neemjeevan.com/herbal-cosmetics.php

20. http://www.xomba.com/how_prepare_natural_herbal_lipstick_home

21. Taylor SC, Arsonnaud S, Czernielewski J. The Taylor Hyperpigmentation Scale: a new visual assessment tool for the evaluation of skin color and pigmentation. Cutis. 2005 76(4):270–4.

DENTIFRICES

Dentifrices such as tooth pastes, tooth powders and tooth gels are use for cleaning the surface of the teeth by removing the food debris and plaque adhered to the surface of the teeth which may cause tooth decay or foul smell and other tooth problems.

Some Characters of Dentifrices

1. They should capable of cleaning the teeth adequately by removing food debris, plaque and stains efficiently.
2. They should give pleasant, cool and refreshing sensation in the mouth.
3. They should be harmless and nontoxic.
4. They should not cause any irritation in the mouth or any ulcers in the buccal cavity.
5. They should maintain constant flow properties.
6. They should be easy to pack and easy to use.

The ingredients that are used to determine the quality and efficiency of tooth pastes like, polishing or abrasive agents which provides cleansing action, humectants and gelling agents which are responsible for the formation of toothpastes, sweetening and flavoring agents which improves palatability and other miscellaneous agents like coloring agents, whitening agents, preservatives, etc.

Some Formulations

Herbodent

Ingredients: Choti ilaychi-100 mg, lavang-100 mg, saunf-100 mg, neem-100 mg, babool-100 mg, majupha-200mg, ashok-300 mg, bakul-200 mg, tejowati-200 mg, yasti-madhu-200 mg, khadir-200 mg, kulanjan-200 mg. Along with that various oils are used in the formulation like coriander, ginger, eucalyptus, lemon, spearmint, kapur, sata-jwayan, jyotismati, satpodina.

Action of Herbodent Toothpaste

Choti ilaychi (Elettaria cardamomum) 100 mg: it helps in refreshing mouth, stimulates taste buds and also helps in increasing the salivary secretion. Its antibacterial action helps in fighting against the bacterial invasion occurring in the oral cavity.

Lavang (Syzygium aromaticum) 100 mg: Traditionally being used in suppressing toothache by its strong actions of its volatile oil. It helps in reducing the pain of the tooth that is very severe and unbearable. It promotes circulation in the mouth because of its astringent properties. It also heals up ant kind or wounds and injuries occurring in oral cavity.

Saunf (Foenieulum vulgare) 100 mg: Its cold potency helps in healing any kind of ulcers and wound in the mouth. It fades away all the bad or foul smell from the mouth. It promotes the salivary secretion which in turn helps in retarding the bacterial growth. It also helps in promoting digestion in the mouth and maintaining the proper hygienic condition in the mouth. It makes the gums strong and prevents the cavity formation.

Neem (Azadirachta indica) 200 mg: Its powerful microbial action makes it the herb of choice to provide maximum protection to your teeth. It heals up the wounds and also supports the oral cavity by preventing any kind of infection caused. It restores the taste in the mouth. Due to scrapping action it scraps away the plaque formation on the teeth making it appear like pearls. It also promotes the circulation of blood in the mouth keeping gums supplied with the nutrients required to maintain the proper health.

Babool (Acacia arabica) 300 mg: It is used as the mouth freshener because of its unique anti-bacterial and clotting properties. It helps in curbing any kind of infection occurring in mouth.

Khadir (Acacia catechu) 200 mg: It is known as katha, a very famous ingredient of mouth freshener and digestive agent "Paan" is very helpful in restoring the taste in mouth, reduce any foul smell, and also helps in cleaning the tongue coated with debris.

Besides these various oils have also been used that helps in strengthening tooth, fades away the bad breath and also helps you in restoring the lost confidence.

Meswak Toothpaste

Meswak is a scientifically formulated Ayurvedic toothpaste, from pure extract of the Miswak Plant, 'Salvodara persica'.

Ingredients: Calcium carbonate, sorbitol, water, silica, sodium lauryl sulphate, flavour, miswak extract, cellulose gum, carrageenan, sodium silicate, sodium saccharin, formaldehyde.

Action: The astringent and antibacterial properties of meswak help reduce tooth decay, fight plaque and prevent gum disease.

Vicco Vajradanti Ayurvedic Toothpaste

Vicco contains 20 pure herbal extracts, long established by Ayurvedic herbal tradition to be good for teeth, mouth and gums.

Ingredients: Purified chalk, water, sorbitol, sodium lauryl sulphate, gum tragacanth. Extracts of Indian licorice root, Indian almond, common jujube, currant, sarsaparilla, cinnamon, sappan wood, persian walnut, rose apple, medlar, barleria prinoitis, prickly ash, Asian holly oak, bedda nut, bengal madder, bishop's weed, catechu, mayweed and geranium. Menthol, thymol, oils of clove, eucalyptus and peppermint.

Action: Vicco cleans teeth, refreshes your breath, and stimulates your gums.

Babool Toothpowder

Ingredients: The natural active ingredient, babul extract (*Acacia arabica*), neem extract, sorbitol, calcium carbonate, peppermint.

Action: The natural active ingredient, babul extract, in babool toothpowder helps keeps teeth clean and white on the outside and strengthens the roots of the teeth from within.

Evaluation of Toothpastes

1. *Determination of particle size:* Particle size can be determined by using microscopical technique or by sieving technique. Mainly the cleansing nature and the abrasive properties of the dentifrices are depends on the particle size.

2. *Test for abrasive properties:* The cleansing properties of the dentifrices are mainly depends on their abrasive property. The abrasion should not lead to any damage to the enamel hence this test has to carry out on the extracted teeth. The teeth are brushed by mechanical means by paste or powder and the effect of dentifrices on the teeth is studies by comparing the results before and after brushing.

3. *Test for cleansing properties:* The cleansing ability of the dentifrices is done by this test. Tooth pastes or powders are brushed onto a polyester film and the change in reflectance character of the lacquer coating is measured. The *in vivo* method involves brushing of the

Table 4.8: Some of the herbal plants that are use in preparation of dentifrices

Common name	Biological source	Family	Part use	Use
Babul	*Acacia arabica*	Mimosaceae	Bark	Teeth disorders
Neem	*Azadirachta indica*	*Meliaceae*	Leaf	Toothache, antibacterial, dental carries
Vajradanti	*Barleria prionitis*	Acanthaceae	Entire herb	Strengthens teeth, toothache
Cinnamon	*Cinnamomum zeylanicum*	Lauraceae	Bark	It removes bad breath and other oral problems
Clove	*Eugenia caryophyllata*	Myrtaceae	Bud	Toothache, antiseptic
Yastimadhu	*Glycyrrhiza glabra*	*Leguminosae*	Roots	Natural sweetener and flavour
Basil	*Ocimum sanctum*	Labiatae	Leaves	It does not cause bleeding of gums, gingivitis, etc.
Pilu	*Salvadora persica*	Salvadoraceae	Twigs	anti-microbial
Honey plant	*Stevia rebaudiana*	Asteraceae	Leaf	It does not cause cavities, rather it helps in improving dental health, treating bleeding gums and other teeth problems

teeth with dentifrices for 2 weeks and determination of the condition of the teeth before and after brushing and comparing them by means of photographs.

4. **Determination of the consistency of the product:** This test is done to determine the consistency of the product for the maintenance of its flow property all throughout its storage period. Consistency of the product is mainly depends on the rheological properties like particle size, viscosity, etc.

5. **pH determination:** A 10% solution of the paste in water is prepared and the pH of the dispersion is measured using a pH meter. The pH should be in the range of 6.8 to 7.4 for better consistency of the product.

6. **Determination of foaming character:** This test is carried out for only to foaming tooth powders and pastes. Specific amount of the product is mixed with a known amount of water. The solution is then shaken sometimes to produce foam. The foam produced is then collected and studies on its nature, washability and stability are carried out.

7. **Determination of volatile matter and moisture content:** This test is done to determine the volatile matter and moisture content in the product. A specified amount of the product is taken and is kept for drying till a constant weight is obtained.

Table 4.9: Some of the dentrioral care products with their primary evaluation tests

Methods	Products
• Oral malodor	• Toothpaste
• Teeth color determination	• Mouth rinse
• Plaque evaluation	• Tongue cleaner
• Plaque regrowth	• Tooth care chewing gums
• Gingivitis evaluation	• Whitening–products
• Teeth sensitivity	

8. **Determination of heavy metals:** This limit test is done to check the presence of any heavy metals like arsenic, lead in the product, which may lead to toxicity. This test is perform for the raw materials which are used as ingredient in dentifrices.

BIBLIOGRAPHY

1. http://en.wikipedia.org/wiki/Toothpaste
2. http://www.authorstream.com/Presentation/prashantlp-175256-toothpaste-entertainment-ppt-powerpoint/
3. http://www.dabur.com/Products-Health%20Care-Meswak
4. http://www.healthandyoga.com/html/product/toothpaste.html
5. http://www.livestrong.com/article/164327-herbal-toothpaste-ingredients/
6. http://www.the-neem-people.com/miswaktoothpaste.htm
7. http://www.viccotoothpaste.com/
8. Patent 6982077 Issued on Jan 03, 2006, Available from http://www. patentstorm. us/patents/ 6982077.html

MOUTHWASH

Mouthwash or **mouth rinse** is a product used to enhance oral hygiene. It is a solution used as an adjunct to regular oral hygiene methods, like brushing and flossing. Some mouthwash claims that antiseptic and anti-plaque mouth rinse kill the bacterial plaque causing cavities, gingivitis and foul smell. Medicinal herbs are used for dental hygiene and prophylactics of respiratory infections create effective all-natural cleansing mouth-wash with attractive herbal flavor and agreeable taste. They are generally contain flavour, antibacterial compounds, penetrants, astringents, therapeutic or preventive com-pounds and deodorants.

Types of Mouthwashes

a. **Non-medicated:** Mouthwashes are available over-the-counter and can be used regularly in the prevention of

Table 4.10: Some herbs that are used for mouth wash preparations

Common name	Biological source	Family	Part use	Use
Asian holly Oak	Quercus infectoria	Fagaceae	Fruit	Hemostatic, astringent, specific for bleeding gums
Barleria	Barleria prinoitis	Acanthaceae	Bark	Styptic, specific for inflamed, bleeding gums and toothache, natural preventative against tooth cavities
Bedda nut	Terminalia bellerica	Combretaceae	Fruit	Astringent, tonic, analgesic, strengthens gums
Bengal madder	Rubia cordifolia	Rubiaceae	Plant	Astringent, tonic, antiseptic, specific for inflammation of gums, promotes healing of wounds
Catechu	Acacia catechu	Fabaceae	Bark	Astringent, analgesic, hemostatic, treats spongy gums
Common jujube	Zizyphus vulgaris	Rhamnaceae	Bark	Astringent, styptic, antiseptic, specific for inflamed gums
Cinnamon	Cinnamomum zeylanicum	Lauraceae	Bark	It removes bad breath and other oral problems
Chamomile	Matricaria recutita	Asteraceae	Flower	used as a mouthwash against oral mucositis
Clove	Syzygium aromaticum	Myrtaceae	Bud	Aromatic, antiseptic, analgesic, breath purifier, specific for toothache and strengthening of gums
Eucalyptus	Eucalyptus globulus	Myrtaceae	Plant	Purifying, specific for pyorrhea and gum infections
Field mint	Mentha arvensis	Lamiaceae	Plant	Antiseptic, antifungal, deodorant, used for halitosis and tooth decay
Indian almond	Terminalia catappa	Combretaceae	Fruit shell	Demulcent, cleansing, astringent, strengthens gums
Mayweed	Anacyclus pyrethium	Asteraceae	Plant	Stimulant, antiseptic, increases salivation, prevents dry mouth, inflammation of gums and toothache
Medlar	Mimusops elangi	Sapotaceae	Bark	Astringent, antiseptic, hemostatic, strengthens gums, specific for bleeding, spongy gums and loose teeth
Neem	Azadirachta indica	Meliaceae	Bark	Astringent, invigorating, antiseptic, promotes healing of wounds, specific for gum inflammation
Peelu	Salvadora persica	Salvadoraceae	Barks	Anti-inflammatory, gentle abrasive, protects enamel, inhibits tartar build-up
Peppermint	Mentha piperita	Lamiaceae	Plant	Antiseptic, deodorant, antimicrobial, stimulant
Persian walnut	Juglans regia	Juglandaceae	Bark	Astringent, antiseptic, natural detergent, strengthens gums

Contd.

Table 4.10: Some herbs that are used for mouth wash preparations *Contd.*

Common name	Biological source	Family	Part use	Use
Pomegranate	Punica granatum	Lythraceae	Rind	Astringent, specific for bleeding gums
Prickly ash	Zanthoxylum rhetsa	Rutaceae	Fruit	Astringent, aromatic, treats inflamed and bleeding gums
Rose apple	Syzygium jambos	Myrtaceae	Bark	Astringent, stomatic, treats spongy and bleeding gums
Tea tree oil	Melaleuca alternifolia	Myrtaceae	Leaves	healing, cleansing, anti-septic, anti-microbial and disease fighting properties
Tejbal	Zanthoxylum alatum	Rutaceae	Bark	Aromatic, astringent, dentifrice
Sappan wood	Caesalpinia sappans	Fabaceae	Plant	Astringent, hemostatic for bleeding gums
Sarsaparilla	Hemidesmus indicus	Apocynaceae	Root	Demulcent, astringent, anti-inflammatory for gums
Spearmint	Mentha viridis	Labiatae	Plant	Analgesic, antiseptic, deodorant and stimulant
Yastimadhu	Glycyrrhiza glabra	Leguminosae	Roots	Antiseptic, analgesic, healing and freshening

dental diseases and the maintenance of oral health. These mouthwashes usually contain oils such as phenol, thymol, eugenol, etc. (listerine).

b. *Medicated:* Mouthwashes are usually prescribed by the dentist for specific conditions like gum infections, ulcers, or after gum surgery. These may contain Chlorhexidine gluconate having pronounced antiseptic properties. This inhibits the formation of plaque and calculus, thus helping to maintain oral health in those individuals having an excessive tendency to form plaque. These mouthwashes are to be used only for a short time, as prescribed by the dentist, as prolonged use may cause undesirable side effects.

c. *Buffered mouth washes:* Are depends on their action primarily on the pH of the solution. Alkaline preparations may be helpful in reducing stringy saliva or reducing nauseous deposits by dispersion of protein.

Some Formulations

Oregano Mouth Wash

Ingredients: Agrimony (*Agrimonia eupatoria*), Oregano (*Origanum vulgare*), nettle (*Urtica dioica*), marjoram (*Origanum majorana*), rosemary (*Rosemarinus officinalis*), eucalyptus (*Eucalyptus globulus*), thyme (*Thymus vulgaris*), plantain (*Plantago major*), grain alcohol (13–15%), vegetable glycerin, distilled water.

Action: These compounds have been proven to kill the bacteria that can cause bad breath, plaque and gum diseases as well as respiratory infections. Regular use promotes clean teeth, healthy gums and mucous.

Direction: To help combat the throat infections gargle two-three times a day. To promote the effect, dilute 1 teaspoon of salt and 1 teaspoon of baking soda in a glass of warm water and gargle before using oregano. Use after brushing the teeth.

Herbal Mouthwash Sanguino

Ingredients: Bloodroot, calamus, celandine, golden rod, juniper, nettle, plantain, sage, corn silk, soapwort, St. John's wort, grain alcohol (15–17%), vegetable glycerin, distilled water.

Action: It gives natural cleansing and self restorative processes of mucous and promotes healing of gum injuries, especially when used in combination with dental herbal oil.

Direction: After brushing the teeth rinse with water several times until all residuals of the toothpaste are gone. Use one teaspoon to one tablespoon of sanguino per rinse. Rinse until you fill very weak warming sensation in the gums, more pleasant than irritating, usually 1–2 min. Regular 2 times a day during several weeks to achieve beneficial effect. Do not rinse with water after sanguino.

Herbal Dental Oil

Ingredients: St. John's wort oil, sea buckthorn oil, lemon balm (melissa) oil, rose hip oil, plantain oil, cinnamon essence, lemon essence, ginger essence, glove essence, myrrh essence, frankincense essence.

Action: It helps heal gum sores and in general promote healthy gums and teeth by stimulating regeneration of mucous, promote the healing of gum lesions and inhibit bacterial and viral infections.

Evaluation Test for Mouth Wash

1. *Toxic and irritant properties:*

2. *Sensitivity tests:* Tests for cold, hot and sweet taste should perform by clinical study.

3. *Toothache:* They should give comfort of pain from toothache.

4. *Cleansing effect:* They should provide better cleansing effect by removing the food debris's.

5. *Flavoring of mouthwashes:* They should contain pleasant flavoring agents and give mouth freshness.

6. *Mouth freshening effect:* During application it should provide freshness to the mouth and remove the foul smell of the mouth.

7. *Maintaining natural whiteness of the teeth:* They should not cause any damage to the teeth rather improve the whiteness of the teeth.

8. *Antibacterial activity:* They should remove maximum bacterial infection to prevent teeth decay and mouth ulcers.

BIBLIOGRAPHY

1. http://en.wikipedia.org/wiki/Mouthwash
2. http://floraleads.com/OREGANO. HTM
3. http://www.fortune3.com/Auromere/ingredients_mw.html
4. http://www.localharvest.org/herbal-mouthwash-C4707
5. http://your-doctor.com/patient_info/dental_info/dental_disorders/mouth wash. html

TOILETRIES AND PERFUMES

Toiletries and perfumes are products that are mainly concerned with personal hygiene. They are classified differently depending on the gender that uses them and on their uses and as such, there are masculine and feminine perfumes. However, they are not that specific and they can be used by both sexes. Whereas toiletries are used on dirty garments and items, perfumes are used on clean item and garments.

Classification of toiletries: Toiletries are mainly depends on the fragrance. Due to the world turning environmental friendly there has been a production of herbal toiletries that aim at showing how man can use nature in house products. The soaps are mainly classified into powders and non-powdered soaps. The powdered soaps are mainly used in soaking and while washing using machines. This is due to their ability of forming foams faster than the non-powdered soaps. The non-powdered soaps are used in manual washing. It is advisable that the toiletries should be stored in closets and they should be used there. This is hygienic since the children cannot be exposed to them. Some of the toiletries are acidic, they could cause corrosion on the carpets, and they could bleach the skin.

Toiletries include soaps that are used either for washing or for bathing, shampoos that are used to wash hairs, deodorants, bathing creams and lotions. They are mainly used in bathrooms and hence they are given the word toiletries. Perfumes constitute the air freshener sprays, the body sprays, house and car sprays.

Classification of perfumes: Perfumes have the great fragrance that makes people have a lovely smell when they use them. They can also be offered as gifts to loved ones. Perfumes can also used to spray the house to give it a nice smell. They are mainly aerosols and due to this reason, they are sprayed either on the clothes or on the linen and drapery in the house. There are different flavors of perfume in terms of smell and people choose the perfumes whose smell they like. The flavors include lavender, lemon and vanilla. The fragrance varies depending with the brand since different perfume brands have different chemical compositions for their perfumes. There are perfumes that have stronger fragrance than others do and the difference is the composition.

Some Products

Herbal Bath Soap

a. Aloevin bath soap is the original skin care soap with genuine *Aloe vera* gel. *Aloe vera* is a proven moisturizer and revitalize of skin. Regular bathing with this soap is natural way of rejuvenating and keeping skin soft and beautiful.

b. Herbal moisturizer soap, with 100% vegetable oils, Neem Extract, *Aloe vera*, gooseberry and Pomegranate seed oil. No animal fat used. It is helpful on all type of skin problems like; acne, itching, rash, psoriasis, eczema etc.

Action: **Neem** contains soap bestows the skin with a soft glowing texture on regular use. As Neem has blood purifying qualities, the continuous use of Neem enriches oxygen and Nutrients. Soap makes complete use of Neem's antimicrobial powers that helps to contain any illness caused by the microbes. Any protozoa attack conditions can also be treated with the help of Neem based soap. Also dry skin issues and other such dermatological problems are healed effectively.

Table 4.11

Common name	Biological source	Family	Part use	Use
Acacia	Acacia farnesiana	Fabaceae	Flower	It is used in perfume industry
Jasmine	Jasminum sambac	Oleaceae	Flower	Jasmine oil is used for making perfumes and incense. Its oil is also used in creams, shampoos and soaps
Patchouli	Pogostemon cablin	Lamiaceae	Leaves	Patchouli is used widely in modern perfumery and modern scented industrial products such as paper towels, laundry detergents, and air fresheners
Lavender	Lavandula angustifolia	Lamiaceae	Flower	It is used in soaps, shampoos, and sachets for scenting clothes
Raspberry	Rubus idaeus	Rosaceae	Fruit	It is used in cosmetic industries as fragrance in perfume and other preparations
Sandalwood	Santalum album	Santalaceae	Wood	It is used in cosmetic industries as fragrance in soap, perfume and other preparations

Neem act as an antibacterial, antiparasitic, antifungal, antiprotozoa and anti viral.

Amla (gooseberry): Due to its richness of vitamin C, the berry alone has the ability to fight all kinds of free radical damage which can beneficial in a number of ways, such as reducing the signs of aging and prevention hair loss. Its effectiveness as an antioxidant allows it to increase the health of the skin and give you the glowing look that you strive for.

Aloe vera: Aloe vera has antibacterial and antifungal activities, which help in the treatment of skin infections like, eczema, ringworm, psoriasis, cuts, burns, scars, etc.

Pomegranate: It balance skin pH, improve skin elasticity, high moisturizing, anti-inflammatory, protect skin from free radicals can help destroy and prevent cancer cell formation, natural antioxidant, anti-wrinkle and improve the efficacy of the skin. Rich anti-oxidants and benefits eczema, psoriasis, wrinkles and dry skin. It is used to heal and moisturize cracked, dry irritated and mature skin.

Some of the Herbal Plants that are use in Preparation of Perfumes

Amber is fossilized tree resin, which has been appreciated for its color and natural organic beauty. Amber is used as an ingredient in perfumes, as a healing agent in folk medicine, and as jewelry. There are five classes of amber, defined on the basis of their chemical constituents. Because it originates as a soft, sticky tree resin, amber sometimes contains animal and plant material as inclusions.

BIBLIOGRAPHY

1. http://articles.directorym.net/Toiletries _ and_Perfume_Retail_and_Usage_ Laurel_ MS-r956078-Laurel_MS.html#8221098
2. http://www.indiamart.com/jraenterprise/ herbal-products.html
3. http://www.irisherbal.com/perfume. html
4. http://en.wikipedia.org/wiki/Sandalwood
5. http://en.wikipedia.org/wiki/Raspberry

Individual Herbs with their Details in Cosmetics Formulations

Acacia concinna

Plant profile

Kingdom: Plantae	**(unranked):** Angiosperms	**(unranked):** Eudicots
(unranked): Rosids	**Order:** Fabales	**Family:** Fabaceae
Genus: *Acacia*	**Species:** *A. concinna*	

Vernacular names

Hindi: Kochi, Ritha, Banritha, **Bengali:** Rithe, **English:** Shikakai, **Marathi:** Reetha, **Sanskrit:** Bahuphenarasa, Bhuriphena, Carmakansa, **Oriya:** Vimala, **Telugu:** Cheekaya, Chikaya, Gogu, Seekaya, **Tamil:** Shikakai, Sheekay, Cikaikkai, **Kannada:** Sheegae, Shige kayi, Sigeballi, Cige **Malyalam:** Cheeyakayi, Chinik-kaya, Shikai, Cheenikka

Ayurvedic properties

Rasa: Astringent	**Veerya:** Cooling
Vipaka: Pungent	**Guna:** Heavy Dry

INTRODUCTION

Acacia is a climbing, most well-known cosmetic plant. It has thorny branches with brown smooth stripes. Thorns are short, broad-based and flattened. Leaves with caduceus, stipules are not thorn-like. Leaf stalks are 1–1.5 cm long with a prominent gland about the middle. Leaves are double-pinnate, with 5–7 pairs of pinnae, the primary rachis being thorny, velvety. Each pinna has 12–18 pairs of leaflets, which are oblong-lance shaped, 3–10 mm long, pointed, obliquely rounded at base. Inflorescences are a cluster of 2 or 3 stalked rounded flower-heads in axils of upper reduced leaves, appearing paniculate. Stalk carrying the cluster is 1–2.5 cm long, velvety. Flower-heads are about 1 cm in diameter when mature. Flowers are pink, without or with reduced subtending bracts. Pods are thick, somewhat flattened, stalked, 8 cm long, 1.5–1.8 cm wide. Seeds are inside the pods with 6–10 numbers.

Distribution

This woody plant is native of India and widely distributed in tropical Asia and Southern Asia. This plant located in the rain forest, disturbed forest, open grassland, fields, creek sides, in open areas often a sprawling shrub. It grows at an altitude of 50–1050 m.

Chemical Constituents

Pods of *Acacia concinna* contain several saponins including kinmoonosides A-C, triterpenoidal prosapogenols named con-cinnosides A, B, C, D, and E, together with four glycosides, acaciaside (triterpenoid trisaccharide), julibroside A1 ($C_{53}H_{84}O_{22}$), julibroside A3, albiziasaponin C, and their aglycone, acacic acid lactone. The leaves contain oxalic, citric and tartaric acid, tannins, amino acids, proteins and some alkaloids like calyctomine ($C_{12}H_{17}O_3N$) and nicotine. Bark contains saponin, glucose, etc.

Medicinal uses

It is popularly referred as "fruit for the hair" as it has a naturally mild pH that gently cleans the hair without stripping it of natural oils.

1. Shikakai is used to control dandruff, promoting hair growth and streng-thening hair roots.
2. It is a natural conditioner, which removes oil and dirt, and keeps hair free of fungal infections, by detoxifying the blood in the scalp and preventing premature graying of hair.

Julibroside A1

3. It also acts as a de-tangler so no conditioner or rinse is needed after it is used.

4. It is generally considered to be safe. Adverse reactions appear to be mild, with occasional gastrointestinal symptoms.

5. An infusion of the leaves has been used in anti-dandruff preparations. Extracts of the ground pods have been used for various skin diseases.

6. This helps in keeping the body cool, and also prevents the scalp from getting dry.

7. It has a cooling and stimulating effect on the brain and it induces sound sleep.

8. It removes minor eruption over the scalp and keeps hair clean, smooth and silky.

9. It has a natural low pH, is extremely mild, and does not strip hair of natural oils.

Cosmetic uses

The fruit pod, bark and leaves are dried and ground together to make a 'Shikakai powder'. The powder is used to make shampoo. It is slightly acidic and helps to cleanse scalp without stripping it of the natural oils. Removes dandruff, dead cell build-up and dirt stuck at the root of hair. The lather produced by shikakai is not very sudsy. Reetha powder may be added to increase the amount of suds. It is astringent and used to tone-up. Amla powder may be added to restore natural dark hair color and add more nutrients. Shikakai powder mixed with reetha powder and decocted to make soap to wash delicate fabrics such as silk. Shikakai leaves may be made as a tea and used to cleanse skin, wash face with added neem leaves.

Traditional home shampoo preparation: Mix about 2 to 3 tablespoons of shikakai powder (ground pods leaves and bark) with same amount of water to make a runny paste. Let the paste rest overnight. In the morning, wet your hair and apply the paste working it into scalp with fingers. This will help break the dandruff and other stuff stuck to the roots of hair. Let the paste stay in the hair for about 20 minutes. After 20 minutes, work the scalp again with fingers. Wash and rinse hair till water runs clear. Traditionally, after the shampoo one would apply virgin coconut oil to the strands of hair.

Classic ayurveda shampoo preparation: Equal amounts (by weight) of shikakai powder, amla powder (dried fruit without seed), and reetha powder (shell of the nut) are mixed to form a dry powder. Add 3 parts of water to one part of mixed dry powder and boil on low heat for about 4 hours. Cool the solution and dilute with water. In that solution small amount of neem leaves, hibiscus leaves and flower petals, gotu kola extract, and ground fenugreek seeds may add for more effectiveness.

Pharmacological Action

Four saponin fractions were developed from pods of *Acacia concinna*, i.e. acetone fraction (AAC), aqueous fraction (WAC), hydro-methanolic fraction (HAC) and methanolic fraction (MAC) and their haemolytic activities and surface activities were determined in comparison with quillaja saponin (QS). There were no significant differences between the haemolytic activities of MAC and QS. Furthermore, the immunomodulatory effect and the adjuvant potential of MAC on the cellular and humoral immune response of BALB/c mice against ovalbumin were investigated. The finding suggested that MAC might be effect on Th1 and Th2 helper T cells and was indicated that MAC at a dose of 40 µg could be used as vaccine adjuvant to increase immune responses (Kukhetpitakwong et al., 2006).

Antidermatophytic activity of pods of *Acacia concinna* was evaluated against *Trichophyton rubrum, Trichophyton mentagrophytes, Trichophyton violaecum, Microsporum nanum* and *Epidermophyton floccosum* and significant antidermatic activity was established for ethanolic, ethyl acetate and hexane extracts against the dermatophytes studied with the MIC value of 62.5 µg/ ml (Natarajan and Natarajan, 2009).

Antimicrobial activities of *Acacia concinna* pods were evaluated by agar cup diffusion method. Methanol, benzene, chloroform, petroleum ether, butanol and aqueous extracts of pods were prepated and evaluated against bacteria, viz. *Staphylococcus aureus, Klebsiella pneumoniae, Escherichia coli, Pseudomonas aeruginosa, Bacillus subtilis* and fungi like *Aspergillus niger, Penicillium* sp., *Candida albicans,* etc. Finally, result revealed that *Acacia concinna* has antimicrobial activity (Todkar et al., 2010).

The hepatoprotective activity of the ethanolic extract of *Acacia concinna* pods was evaluated against carbon tetrachloride induced hepatotoxicity in rats. Histopathological studies revealed the concurrent administration of the extract with carbon tetrachloride exhibited protective of the liver and finally concluded that the use of ethanolic extract of *Acacia concinna* pods exhibited significant protective from liver damage in CCl_4 induced liver damage model (Kumar et al., 2011).

Toxicological Studies

Bioefficacy of some botanicals was tested against Spodoptera litura, Fabrius on glycine max linn. In present investigation *Acorus calamus, Acacia concinna, Terminalia chebula, Terminalia belerica* are selected for ethanolic extraction of their useful parts. Various dose levels were applied in the field of explore larvicidal potential of phytoextracts. Percentage mortality for various concentrations proved that *Acacia concinna* has effective potential with 85.2% mortality at 50 ppm dose level. Probit analysis gave median dose, i.e. LC_{50}, 25.37 ppm, which can be sustainable solution for eco-friendly management of Spodotera littura on soybean (glycine max), using *Acacia concinna* pod extract (Patil and Chavan, 2010).

Few of Market Formulations

Protein Shampoo

Ingredients:

1. Shikakai (*Acacia concinna*)—47.60 mg
2. Ushira (*Vetiveria zizanioides*)—6.95 mg

3. Musk root (*Nardostachys jatamansi*)—6.95 mg
4. Soapnut (*Sapindus mukorossi*)—2.08 mg
5. Fenugreek (*Trigonella foenum-graeceum*)—2.08 mg.

Action

Fenugreek provides natural proteins for the nourishment and health of your hair. Musk Root promotes hair growth. Shikakai softens the hair and keeps the scalp healthy. Soapnut cleanses hair of excess oily secretions, grime and dust, and leaves the hair soft, silky and healthy.

Direction for application: Wet hair, massage a small amount on hair. Leave for 1–2 minutes and rinse thoroughly. Repeat if necessary. Daily use does not affect the pH balance of hair. It is safe to use on artificially colored or permed hair.

Industry concern: Himalaya drugs.

Hair Shampoo

Ingredients: 1–2 tbsp dried herbs of marigold (*Calendula officinalis*), shikakai (*Acacia concinna*), henna; 8 ounces purified water; 2 ounces liquid glycerin soap; 1 tsp base oil (eliminate or reduce for oily conditions, increase for dry); 15–60 drops essential oils (as conditions require).

Preparation

Prepare a decoction or infusion of all combinations of herbs. For a decoction, place the herbs in water in a stainless steel or glass pan. Bring to boil, reduce the heat, and simmer on low flame for up to 10 minutes. Remove from the heat, let the mixture cool, strain off the liquid, and discard the spent herbs. For an Infusion, bring the water to boil. Place the herbs in a stainless steel or glass pan and pour the boiling water over them. Let steep for at least 20 minutes, then let the mixture cool (if it is not cool already), strain off the liquid, and discard the spent herbs. Mix the herbal infusion or decoction with the soap, base oil, and desired essential oils. Shake well.

Refrigerate the shampoo between uses, for up to a week.

Jasvant Red and Black Shikakai Soap

It increases bounciness and silkiness for hair. Shikakai is mixed with water to make a paste which is worked through the hair. It lathers moderately and cleans hair beautifully. It has a natural low pH, is extremely mild, and does not strip hair of natural oils. Usually, no rinse or conditioner is used since shikakai also acts as a detangler.

Black shikakai soap containing natural ingredients such as shikakai, amla, aritha pure coconut oil and other natural oil.

Industry concern: Jasvant Soap Industries, Gujarat.

Amla Shikakai Shampoo

Ingredients: 100 grams of reetha seeds, 100 grams of amla powder, 100 grams of dried shikakai powder, 20 grams of *Aloe vera* juice and 3 glasses of hot water.

Method: Soak the rita seed in warm water and squeeze out the juice. Add to this amla powder, shikakai powder and *Aloe vera* juice and use as regular shampoo.

Industry concern: Jasvant, Lotus herbals, Gayatri herbals Pvt. Ltd.

General Shikakai Shampoo Natural Herbal Recipe

Ingredients: **Colorants:** Water-6 cups, amla powder-1 tablespoon, cinnamon-2 medium size stick, cloves-12 Cloves, loose black tea leafs-2 teaspoons. **Saponin:** Shikakai powder: 2 cups, select a good quality powder that has the bark, leaves, and the buds. Reetha: Broken shells from 4 reetha nuts, for more lather, substitute more shikakai powder with reetha shell powder. **Nutrients:** Honey-1 tablespoon, almond oil-2 teaspoons, neem oil-2 teaspoons antibacterial for scalp, gotu kola extract: $^1/_4$ teaspoon.

Method

Mix all the colorants. Bring to a boil on low for 20 minutes. Then add shikakai powder and powdered reetha shells. Bring it to a near boil. Let it rest 4 hours or overnight. Add nutrients with stirring. Then cool to room temperature. Transfer to bottles with lids. Store in the refrigerator.

Storage of Products

Shikakai powder: Keep container closed and protect from humidity.

Dosage

Shikakai and amla herbal shampoo can be used daily for washing hair.

Patents

An improved method has developed to isolate and purify 7-hydroxychromones from plant sources like *Acacia concinna, Aloe arborescens, Aloe barbadensis, Aloe cremnophila, Aloe ferox, Aloe saponaria, Aloe vera, Aloe vera* var chinensis. This chemical exhibit potent antioxidant activity.

BIBLIOGRAPHY

1. http://www.101beautysalon.com/body/haircare2.html

2. http://www.5htp5-htp.com/acaciaconcinna.html

3. http://www.agrisources.com/herbs/acaciaconcinna.html

4. http://www.favorfinesse.com/hhproteinshampoo.shtml

5. http://www.flowersofindia.in/catalog/slides/Shikakai.html

6. http://www.indiacurry.com/ayurveda/shikakai.htm

7. http://www.indiacurry.com/women/shikakaishampoorecipe.htm

8. http://www.jasvantsoap.com/jasvant_shikakai_red_soap.html

9. http://www.mysticboard.com/health_and_beauty/43551-more_homemade_remedies_for_haircare_these_r_really_helpful_.html

10. Jai Q and Farrow T.M. 7-Hydroxy Chromones As Potent Antioxidants. Application number: 20100168223. Publication date: 07/01/2010.

http://www. faqs.org/patents/app/20100168223 # ixzz1PoyWUnMQ

11. Kukhetpitakwong R, Hahnvajanawong C, Homchampa P, Leelavatcharamas V, Satra J and Khunkitti W. Immunological adjuvant activities of saponin extracts from the pods of *Acacia concinna*. Int Immunopharmacol. 2006, 6(11): 1729–35.

12. Kumar T.S, Ramakrishnan G, Jaykar K. R. B and Vijaya Kumar D. Hepatoprotective activity of ethanolic extract of pods *of Acacia concinna*. 2011, 03(04): 157–163.

13. Natarajan V and Natarajan S. Anti-dermatophytic Activity of Acacia concinna. Global Journal of Pharmacology. 2009, 3 (1): 06–07.

14. Patil D. S and Chavan N.S. Repellency and toxicity of some botanicals against *Spodoptera litura* fabrius on glycine max. Linn (soybean). The Bioscan. 2010, 5(4): 653–654.

15. Todkar S.S, Chavan V.V and Kulkarni A.S. Screening of secondary metabolites and antibacterial activity of *Acacia concinna*. Res. J. Microbiol. 2010, 5: 974–979.

Acorus calamus

Plant profile

Kingdom: Plantae – Plants   **Subkingdom:** Trachebionta – Vascular plants

Superdivision: Spermatophyta – Seed plants   **Division:** Magnoliophyta – Flowering plants

Class: Liliopside – Monocotyledons   **Subclass:** Arecidae

Order: Arales   **Family:** Acoraceae – Calamus family

Genus: Acorus L. – Sweet flag   **Species:** Acorus calamus L.

Vernacular names

Sanskrit: Vacha, Haimavati, **Hindi:** Bach, **English:** Sweet Flag, **Tamil:** Vasamber, Vashambu, **Marathi:** Vekhand, **Telugu:** Vadaja, **Kannada:** Baje, **Malayalam:** Vayambu

Ayurvedic properties

Rasa: Bitter, pungent, **Veerya:** Heating, **Vipaka:** Pungent, **Gunas:** Light, sharp, subtle

Ayurvedic preparations

Sarasvata choorna, Sarasvatarista, Brahmi prasa, Medhya rasayana, Vaca taila (for nasya).

INTRODUCTION

It is a tall perennial wetland monocot and its grass-like stems growing up to 5-feet tall. This plant has scented leaves and more strongly scented rhizomes. *Acorus calamus* has a single prominent midvein and then on both sides slightly raised secondary veins (with a diameter less than half the midvein) and many fine tertiary veins. This makes it clearly distinct from *Acorus americanus*. The leaves are light to lush green in color, are gladiate, jutting skyward like swords. The leaves are between 0.7 and 1.7 cm wide, with average of 1 cm. The sympodial leaf of *Acorus calamus* is somewhat shorter than the vegetative leaves. The margin is curly-edged or undulate. The spadix, at the time of expansion, can reach a length between 4.9 and 8.9 cm (longer than *A. americanus*). Plants very rarely flower or set fruit, but when they do, the flowers are 3–8 cm long, cylindrical in shape, greenish brown and covered in a multitude of rounded spikes. The fruits are small and berry-like, containing few seeds. The roots are spread horizontally by creeping and can grow to almost 2 m in length.

Distribution

This plant is indigenous to India, but is now found across Europe, in Southern Russia, Northern Asia Minor, Southern Siberia, China, Indonesia, Japan, Burma, Sri Lanka, Australia, as well as Southern Canada and the Northern United States. In India, it is distributed in North Temperate and subtropical regions up to 2200 m altitude in Himalayas but generally it grows between 1000 and 3700 m. It is quite common in Manipur and the Naga hills of India and is found on the edges of lakes and streams.

Chemical Constituents

Main chemical constituents are asarone, alpha pinene, beta asarone, calamenol, calamene, calamenone, eugenol, methyl eugenol, calamone, azulene, sugars, glucosides (10%), saponin (5%) and flavones. Three new sesquiterpenes, 1-beta, 7-alpha (H)-cadinane-4-alpha, 6-alpha, 10-alpha-triol (1), 1-alpha, 5-beta-guaiane-10 alpha-O-ethyl-4-beta, 6-beta-diol (2), and 6-beta, 7-beta (H)-cadinane-1 alpha, 4-alpha, 10-alpha-triol (3), together with 25 known ones, were isolated from the rhizome of *Acorus calamus* Linn.

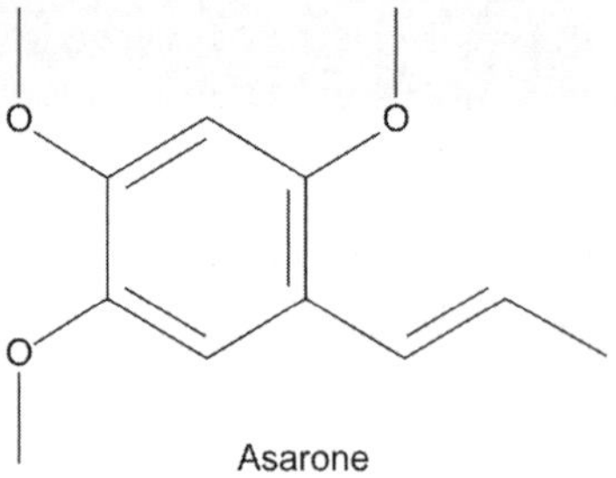

Asarone

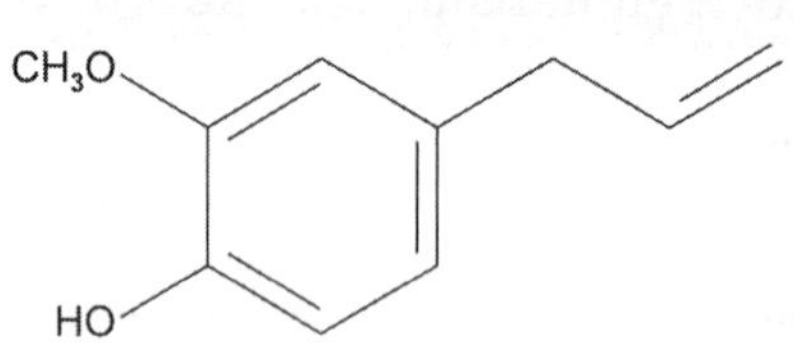

Eugenol

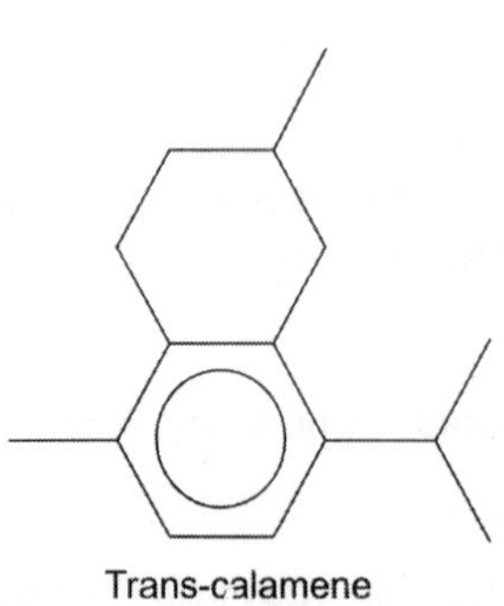

Trans-calamene

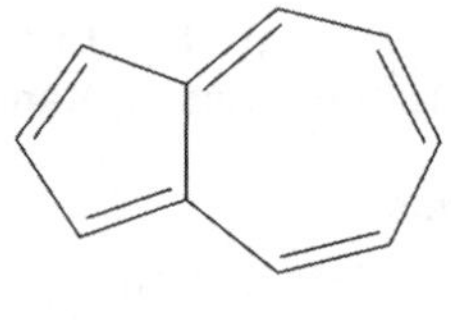

Azulene

It also having phenylpropanoid (produced by plants for defense against herbivores and protection against ultraviolet rays). The oils in the rhizomes of calamus mainly contain eugenol and asarone. Both compounds undergo biotransformation *in vivo*, but asarone probably is the natural precursor to TMA-2 (2,4,5-trimethoxyphenylisopropylamine) and can be converted through an amination-process to TMA-2.

Medicinal uses

Sweet flag is presently classified as an unsafe herb for internal usage by the FDA. It has been used for centuries, however, in Ayurvedic medicine as a renowned rejuvenator of the nervous system for conditions of anxiety, hysteria, insomnia, neurasthenia, and other nervous complaints. It is useful in all conditions of excess vata and is known to enhance awareness and improve memory. A decoction of the root acts as a carminative removing discomfort caused by excess intestinal gas. The root decoction is use in bronchitis and as an aphrodisiac. A small piece of the root is chewed to overcome mental fatigue. The skin of the root is hemostatic. It has been used in dyspepsia, dysentery, headache, gout, and rheumatism. The juice of the root is applied to boils, carbuncles, and painful joints. In large doses it is emetic. The powdered root is used as a snuff to relieve nasal congestion and mental weariness.

Cosmetic uses

Acorus calamus root extract is used as conditioner for eyelash and eyebrow. The root extract is also used as skin conditioner. The oil of *Acorus calamus* is used as an ingredient in flavors, particularly in liquors. It is used a great deal in the formuation of alcoholic drinks and in perfume to give a bitter tang to the former and those special nuances to the perfumes. It is also used in toothpaste preparation. The plant is externally used to treat skin eruptions.

Pharmacological Action

The root is anodyne, aphrodisiac, aromatic, carminative, diaphoretic, emmenagogue, expectorant, febrifuge, hypotensive, sedative, stimulant, stomachic, mildly tonic and vermifuge. It is used internally in the treatment of digestive complaints, bronchitis, sinusitis, etc.

Crude methanolic extract of *Acorus calamus* rhizomes possesses antimicrobial activity against bacteria, fungi and yeast. It shows strong effect against filamentous fungi like *Trichophytum rubrum* and *Microsporum gypseum*, moderate inhibitory effect against yeast and low against bacteria (Phongpaichit et al., 2005). In another experiment, same antimicrobial activity of *Acorus calamus* rhizome and leaf extracts obtained with different solvents, viz. petroleum ether, chloroform, hexane and ethyl acetate was evaluated. Extracts obtained with ethyl acetate among others were found to be highly effective. The results concluded that the rhizomes and leaf ethyl acetate extracts exhibited pronounced antifungal activity (Asha Devi and Ganjewala, 2009).

The study focuses on the effect of *Acorus calamus* on the behavioral, electroencephalographic, and antioxidant changes in $FeCl_3$-induced rat epileptogenesis. Data presented in this study clearly show that *Acorus calamus* possesses the ability for preventing the development of $FeCl_3$-induced epileptogenesis by modulating antioxidant enzymes, which in turn exhibit the potentiality of *Acorus calamus* to be developed as an effective anti-epileptic drug (Hazra et al., 2007).

The study was evaluated the free radical scavenging and anticholinesterase activity of the methanolic extracts of *Acorus calamus* rhizomes *in vitro*. In addition, total phenolics (TP) were also estimated. ACME contained phenolics 23 µg GAE/mg phenolic substances which indicate that total phenolics might be responsible for the observed antioxidant and anticholinesterase activities (Ahmed et al., 2009).

The study was evaluated an anti-inflammatory activity of *Acorus calamus* leaf extract and to explore its mechanism of action on human keratinocyte HaCaT cells. HaCaT cells treated with polyinosinic: Polycytidylic acid (polyI:C) and peptidoglycan (PGN) induced the inflammatory reactions. The anti-inflammatory activities of *Acorus calamus* leaf extract were investigated using RT-PCR, ELISA assay, immunoblotting, and immunofluorescence staining. These results suggest that *Acorus calamus* leaf extract inhibits the production of pro-inflammatory cytokines through multiple mechanisms and may be a novel and effective anti-inflammatory agent for the treatment of skin diseases (Kim et al., 2009).

The present study was investigated insulin releasing and alpha-glucosidase inhibitory effects of different fractions from *Acorus calamus* were detected *in vitro* using HIT-T15 cell line and alpha-glucosidase enzyme. Results revealed that *Acorus calamus* extract may have hypoglycemic effects via mechanisms of insulin releasing and alpha-glucosidase inhibition, and thus improves postprandial hyperglycemia and cardiovascular complications (Si et al., 2010).

Toxicological Studies

Maximum levels depending on the Asarone content, which shall not be exceed 0.01% in the finished cosmetic products.

The primary toxicologic concern focuses on the carcinogenic effect of isoasarone, a major component of the volatile oil of calamus. Feeding studies conducted more than 20 years ago provided evidence for the mutagenic potential of this compound. Subsequently, all calamus-containing products were removed from the US marketplace. However, a recent study found extracts of *A. calamus* to exhibit no mutagenic activity in the salmonella mutagenicity screen. The plant and its extracts continue to find use throughout the world.

The LD_{50} of asarone in mice is 417 mg/kg (oral) and 310 mg/kg (IP). Although *A. calamus* exhibited no mutagenic activity in the *Salmonella* mutagenicity screen, recent experiences showed that calamus oil exhibited genotoxic effects on Swiss mice. Another experiment showed that calamus oil was strongly mutagenic.

In a 2-year study, rats were fed 0, 500 ppm, 1000 ppm, 2500 ppm and 5000 ppm of the oil of calamus (Jammu variety). Dose related toxic effects included moderate to marked growth retardation and degenerative changes in the liver and heart. Malignant tumors developed in the duodenum of rats of all oil of calamus dosages beginning in the 59th week of dosing (Taylor et al., 1967).

Genotoxic activity (strong induction of chromosomal aberrations, slight increase in the rate of sister chromatid exchanges) has been exhibited by 3-asarone in human lymphocyte cultures in the presence of microsomal activation (Abel, 1987).

Toxicokinetics in the roden studies have reported where intragastric administration of beta asarone results in the rapid absorption and elimination of the compound with a peak concentration occurring within about 15 minutes. The serum elimination half life of beta asarone was approximately 1 hour (Wu and Fang, 2004).

Clinical Investigation

The essential oil free alcoholic extract of the rhizome was found to possess sedative and analgesic properties; it has moderate hypotensive and respiratory depressant effects. When administered to experimental animals the oil reduces muscle tone and response to tactile and auditory stimuli. Asarone and beta-asarone are the constituents credited with the sedative and nervine effects. The alcoholic extract has also shown antifungal effects.

Market Formulations
Scavon Vet Cream

This cream is an effective topical antimicrobial, fly repellent and maggoticidal. It is also possessing wound-healing property. The cream has been found to be easily dispersible on application to the moist surfaces, non-irritant, non-staining and possessed a pleasant fragrance. The cream acted as insect repellent.

Ingredients: Ocimum sanctum, Acorus calamus, Cinnamomum camphora, Linum usitatissimum and Eucalyptus globulus.

Action: Ocimum sanctum, Acorus calamus and *Eucalyptus globulus* are known to be anti-bacterial, antimycotic, insecticidal and wound-healing properties. *Linum usitatissimum* is well known for antiseptic, wound healing and antimaggot properties, *Cinnamomum camphora* (Karpura) possesses antibacterial and fungistatic properties. Scavon Vet cream contains Tankana bhasma which has calcinated borax processed by traditional means has antiseptic and wound healing properties.

Industry concern: The Himalaya Drug Company, Bangalore.

Clearwin Cream

It clears stretch marks/minor burn marks, cuts, etc.

Ingredients

Each 100 g contains: **A**-Neem patti (*Azadirachta indica*), Herda (*Terminalia chebula*), Vekh and (*Acorus calamus*), Lodhra (*Symplocus racemosa*), Haldi (*Curcuma longa*), Manjishtha (*Rubia cordifolia*), Kayphal (*Myrica nagi*), Til Tel (*Sesamum indicum*), **B**-Kumari (*Aloe vera-indica*), Til Tel (*Sesamum indicum*), **C**-Siddha Tel **A**, Siddha Tel **B**, Tulsi Ark (*Ocimumo-kilimendi-charicum*), Beeswax, Sangajire (talc powder), Jasad Bhasma (zinc oxide), Tankan Bhasma (borax).

Indications

- Removes the stretch marks and blemishes.
- Useful in a minor burns, cuts, etc.
- Improves complexion.

Dosage: Local application (for external use only).

Skin Support Formula

It is a herbal formula for treating skin problems and diseases.

Ingredients: Nimba Chhal (*Azadirachta indica*), Guduchi (*Tinospora cordifolia*), Vasa (*Adhatoda varica*), Patola (*Trichosaanthes dioica*), Kantakari (*Solanum xanthocarpum*), Shu. Guggule (*Commifora mukul*), Ghrit, Patha, Vidanga (*Emblica ribes*) Devandaru (*Cendrus deodar*), Gajpippali (*Scindapsus officinalis*), Sarjjkshar, Yavakshar, Sunthi (*Zingiber officinalis*), Haridra (*Curcuma longa*), Satapushpa, Cavya (*Piper chaba*), Kushta (*Sausaria leppa*), Jyotishmati (*Celastrus panniculatus*), Marich (*Piper nigrum*), Indrayava (*Holarrhena antidysenterica*), Jirak (*Cuminum Cyminum*), Citrak (*Plunbago zeylanica*), Katuka (*Picrorhiza kurroa*) Shu. Bhallatak (*Senecarpus anacardium*), Vacha (*Acorus calamus*), Pipalimula (*Piper longum*), Manjishta (*Rubia cordifolia*), Haritaki (*Termivalia chebula*), Bibhitaki (*Terminalia bellarica*), Amalaki (*Emblica officinalis*), Yavani.

Indications

- Excellent remedy for skin diseases and problems associated with it like itching, swelling, inflammation, redness, skin rashes, oozing boils, acne and pimples, etc.
- Lowers the incidence of recurrence.

Dosage: As directed by the physician.

Skin Toner Cream

Ingredients: Each gram of cream prepared with 5.3% w/v of the following herbal extracts: *Acorus calamus* (Vacha), *Symplocos racemose* (Lodhra), *Berberis aristata* (Tree-turmeric), *Cinnamomum zeylanicum* (Dalchi, Tawk), *Sida cordifolia* (Bala), in a perfumed cream base q.s. base consist of steric acid, cetosteryl alcohol, white petrolium jelly and glycerin. Sodium benzoate and sodium methyl parabeen (0.2%) is added as preservative.

Industry concern: Ronak Beauty Care Private Limited, Gujarat.

Storage of Products

Acorus calamus oil is kept in the aluminum containers to protect them from the unfavorable conditions.

Cosmetic products should store well closed container and away from sunlight.

Dosage

Rhizome: 1–3 g or by infusion three times daily.

Liquid extract: 1–3 ml (1: 1 in 60% alcohol) three times daily.

Tincture: 2–4 ml (1:5 in 60% alcohol) three times daily.

Patents

An anti-cough, antitussive, and throat soothing synergistic herbal formulation comprising an extract of *Piper cubeba, Glycyrrhiza glabra, Acorus calamus, Alpinia galanga, Zingiber officinale* and pharmaceutically acceptable additives as a syrup, lozenges or chewable tablets for preventing cracking of voice, dryness of mouth and toning of voice, vocal cord; the present invention also provides a method of preparation of this formulation (Pushpangadan et al., 2009).

A conditioner for eyelash and eyebrow includes: Water; hydrolyzed soy protein; erythritol; trehalose; carbomer; arginine; polysorbate 80; niacinamide; panthenol; citric acid; phenoxyethanol; *Acorus calamus* root extract; *Artemisia princes* leaf extract; *Portulaca oleracea* extract; *Polygonum multiflorum* root extract; *Glycyrrhiza glabra* root extract; *Ephedra sinica* extract; *Sophora angustifolia* root extract; *Arctium lappa* seed extract; thuja extract; *Lyceum barbarum* fruit extract; *Ginkgo biloba* leaf extract; *Paeonia suffruticosa* root extract; *Carthamus tinctorius* flower extract; *Bletilla strita* root extract; rheum undulatum root-stalk-stern extract; biotin; methylparaben; and copper tripeptide-1 (Kim, 2010).

BIBLIOGRAPHY

1. Abel G. Chromosome damaging effect on human lymphocytes by Q-asarone. Planta Med. 1987, 251–253.

2. Ahmed F, Narendra Sharath Chandra J.N, Urooj A and Rangappa K.S. *In vitro* anti-oxidant and anticholinesterase activity of *Acorus calamus* and *Nardostachys jatamansi* rhizomes. Journal of Pharmacy Research. 2009, 2(5): 830–833.

3. Asha Devi S and Deepak Ganjewala. Anti-microbial activity of *Acorus calamus* (L.) rhizome and leaf extract. Acta Biol Szeged. 2009, 53(1): 45–49.

4. Available from: http://www. freepaten tsonline.com/y2010/0291018. html

5. Dong W, Yang D and Lu R. Chemical con-stituents from the rhizome of Acorus calamus L. PLanta Med. 2010, 76(5): 454–457.

6. Hazra R, Ray K and Guha D. S. N. Pradhan. Inhibitory role of Acorus calamus in ferric chloride-induced epileptogenesis in rat. Hum Exp Toxicol. 2007, 26(12): 947–953.

7. http://en.wikipedia.org/wiki/Acorus_calamus

8. http://health.indiamart.com/ayurveda/indian-herbs/calamus.html

9. http://plants.usda.gov/java/profile?symbol=ACCA4

10. http://www.entheology.org/edoto/anmv iewer.asp?a=261

11. http://www.indiamart.com/herbosincorps oils.html

12. http://www.motherherbs.com/acorus-calamus-extract.html

13. http://www.niam.com/corp-web/acorus.htm

14. http://www.niam.com/index.php?option =com_content&view=article&id=91& Itemid= 21

15. http://www.nutritional-herbalsupplements.com/herbal-supplements/calamus.html

16. http://www.siddhayu.com/dermatology.HTM

17. Kim A. Conditioner for eyelash and eyebrow. United States Patent Application 20100291018. Application Number: 12/464826. Publication date: 11/18/2010.

18. Kim H, Han T.H and Lee S.G. Anti-inflam-matory activity of a water extract of Acorus

calamus L. leaves on keratinocyte HaCaT cells. J Ethanopharmacol. 2009, 122(1): 149–156.

19. Pujenjob N., Rukachaisirikul V and Ongsakul M. Antimicrobial activities of crude methanol extract of *Acorus calamus*, Songklanakarin J. Sci. Technol., 2005, 27 (2): 517–523.

20. Pushpangadan P, Raghavan G, Madhavan V.K, Mehrotra S, Singh R, Kumar A and Rao C.V. United States Patent 7482031. Application Number: 10/457182. Publication date: 01/27/2009. Available from: http://www.freepatentsonline.com/7482031.html

21. Si M.M, Lou J.S, Zhou C.X, Shen J.N, Wu H.H, Yang B, He Q.J and Wu H.S. Insulin releasing and alpha-glucosidase inhibitory activity of ethyl acetate fraction of *Acorus calamus in vitro* and *in vivo*. J of Ethano-pharmacology. 2010, 128(1): 154–159.

22. Taylor J.M, Jones W.I, Hagan E.C, Gross M.A, Davis D.A and Cook E.L. Toxicity of oil of calamus (Jammu variety). Toxicol. Appl. Pharmacol. 1967, 10: 405.

23. Wu H.B and Yang Y.Q. Pharmacokinetics of beta asarone in rats. Yao. Xue. Xue. Bao. 2004, 39: 836–838. [Chinese].

Alkanna tinctoria

Plant profile

Kingdom: Plantae – Plants

Superdivision: Spermatophyta – Seed plants

Class: Magnoliopsida – Dicotyledons

Order: Lamiales

Genus: Alkanna Tausch – Alkanna

Subkingdom: Trachebionta – Vascular plants

Division: Magnoliophyta – Flowering plants

Subclass: Asteridae

Family: Boraginaceae – Borage family

Species: Alkanna tinctoria L. Tausch– Alkanna

Vernacular names

Hindi: Ratanjot, **English:** Alkanet, dyer's bugloss, **Tamil:** Ratthapaalai.

Other common names

It is also known as orchanet, bugloss of Languedoc, Spanish bugloss, enchusa, lingua bovina, ox tongue, yellow anchusa, and blue bugloss.

INTRODUCTION

Alkanet is a biennial herb cultivated in Central and Southern Europe. It has oblong leaves that grow on a thick hairy stem which rises to approximately 1–3 feet. The species are hispid or pubescent herbs, with oblong, entire leaves, and bracteated racemes, rolled up before the flowers expand. The corolla is rather small, between funnel and salver-shaped; usually purplish-blue, but in some species yellow or whitish; the calyx enlarges in fruit. It has a dark red root of blackish appearance externally but blue-red inside, with a whitish core. The root, which is often very large in proportion to the size of the plant, yields in many of the species a red dye from the rind.

Distribution

It is a grassy plant grown scatteredly in the South, in sandy areas, originating from Northern Africa. It is also grown in the South of France and on the shores of the Levant. Common alkanet originated in the Mediterranean.

Chemical Constituents

Alkanna root contains a mixture of red pigments found in the bark at levels of up to 5 to 6%. These consist mainly of fat-soluble naphthazarin (5, 8-dihydroxy-1, 4-naphthaquinone) components, such as alkannin and related esters. The alkannin esters of beta, beta-dimethylacrylic acid, beta-acetoxy-isovaleric acid, isovaleric acid, and angelic acid have also been isolated from the root. The major compounds detected in the oil, were pulegone (22.27%), 1, 8-cineole (13.03%), α-terpinyl acetate (6.87%), and isophytol (6.83%).

Medicinal uses

It helps the morphy and leprosy. It is used to treat digestive difficulties such as ulcers and also helps liver functions, clearing up jaundice and treating kidney stones. When used to make an ointment, it can treat wounds such as snake bites by either applying topically directly to the site or ingesting orally. It can also relieve skin inflammation, such as smallpox or measles. An ointment made of it is excellent for green wounds, pricks or thrusts. Traditionally used to soften and smooth the skin. Root has astringent and antimicrobial properties. It was used centuries ago to help heal deep wounds and skin ulcers.

Cosmetic uses

It is used by French women as a temporary make-up solution. Some cultures use the root and turn it into a dye which is then utilized in decorations and staining procedures. It is commonly used today as a food coloring. Alkanet root is usually used to produce a natural blue or purple color in the soap. The infused alkanet root oil can then be used in soap or other formulations. Oil made with alkanet is an emollient that is soothing and softening to the skin.

Pharmacological Action

Alkanna root has demonstrated potential antiaging effects. One study found that both monomeric and oligomeric alkannin exhibited high radical scavenging activity. Additionally, an olive oil extract containing *A. tinctoria* possessed free radical scavenging activity at room temperature but with heat, this activity was decreased. Thus, alkanna root's role in cosmetics may extend beyond

Alkannin

Pulegone

Naphthazarin

serving solely as a dye (Assimopoulou and Papageorgiou, 2005).

Several pharmaceutical ointments, such as Helixderm and Histoplastin Red (approved and marketed in Greece) contain alkannin as the active ingredient, and they have been shown to exert activity against gram-positive bacteria, gram-negative bacteria, and fungi. Additionally, alkannin may exert bactericidal action on *Pseudomonas aeruginosa*, a bacteria that forms biofilms against wound healing (Papageorgiou et al., 2008).

This study is designed to examine the chemical composition and *in vitro* antioxidant activity of the hydrodistillated essential oil and the various extracts of alkanet (*Alkanna tinctoria* subsp. *tinctoria*). Gas chromatography (GC) and GC-mass spectrometry (GC-MS) analysis of the essential oil were resulted in the determination of 27 different compounds, representing 93.32% of the total oil (Sabin et al., 2010).

Toxicological Studies

Alkanna root may cause hepatic and/or lung toxicity because of the pyrrolizidine alkaloid components. The pyrrolizidine alkaloid components in alkanna root may cause liver and/or lung toxicity. The most hepatotoxic pyrrolizidine alkaloids include the cyclic diesters, such as retrorsine and senecionine. Fulvine and monocrotaline have been implicated in causing liver and pulmonary toxicities.

Animal Data

A. tinctoria has been studied in male rabbits with partial thickness, severe, and olive oil burns. A solution of *A. tinctoria* 16% was applied twice daily to the left side of the animal. The right side served as a control. Partial thickness burn wounds were completely healed in 7 to 10 days, and olive oil burn wounds were healed in 26 days. However, severe burn wounds were unresponsive to *A. tinctoria*.

Clinical Investigation

The esteric pigments displayed excellent antibiotic and wound-healing properties in a clinical study enrolling 72 patients with ulcus cruris (indolent leg ulcers).

Market Formulations

Eden

Ingredients: Certified organic sucrose (sugar), certified organic *Helianthus annuus* (sunflower) oil and *Alkanna tinctoria* (alkanet) root, honey, *Rosa damascena* (rose) petals, *Rubus idaeus* (raspberry) fruit seeds, *Citrus paradisi* (pink grapefruit) oil, *Pelargonium roseum* (rose geranium) oil, and Vitamin E.

Direction: Apply small amount of scrub to palm of hand. Gently scrub body, feet, and hands to exfoliate. Rinse well. Do not apply to abraded skin.

Mamma Michal's Sensuous Rosy Lips Lip Balm

Ingredients: Red rose petals and alkanet root infused food grade castor oil, olive oil and sweet almond oil, beeswax, vegetable glycerin (100% pure), vitamin E oil and certified organic rose absolute.

Direction: Red rose petals and alkanet root infused food grade castor oil, olive oil and sweet almond oil, beeswax, vegetable glycerin (100% pure), vitamin E oil (natural preservative and fantastic for the skin!) and certified organic rose absolute.

Soap for Goodness Sake Groovy Lavender Patchouli Soap

Ingredients: Saponified organic sunflower oil (*Helianthus annuus* seed oil), saponified organic palm oil (sodium palmate), saponified prganic coconut oil (sodium cocoate), water, *Lavandula officinalis* (lavender) essential oil, *Citrus sinensis* (sweet orange) essential oil, *Pogostemon cablin* (patchouli) essential oil, *Cinnamomum zeylanicum* (cinnamon) leaf oil, ground *Alkanna tinctoria* (alkanet) root.

Direction: Lather, wash and rinse properly.

Carol's Daughter Mango Melange Body Butter (Moisturiser)

Ingredients: Sweet almond oil, coconut oil, jojoba oil, beeswax, shea butter, cocoa butter, fragrance, essential oils of tangerine, lime, and patchouli, alkanet root, annatto seeds, orange, and mango slice. Apply body butter to palm of hand and massage onto body. Remove decoration prior to use.

Direction: Apply body butter to palm of hand and massage onto body.

Company concern: EWG's skin deep, Washington.

Berry glow salt scrub: This exfoliating body scrub softens and moisturizes the skin.

Ingredients

Phase A: *Cocos nucifera* (coconut) oil: 9.0%

Phase B: *Simmondsia chinensis*

(jojoba) oil: 9.0%

Euterpe oleracea fruit oil: 9.0%

Fragaria ananassa

(strawberry) seed oil: 9.0%

Phase C: Sodium chloride: 63.0%

Alkanna tinctoria root

extract: 0.5%

Vaccinium myrtillus seed

oil: 0.5%

Method of Preparation

Heat A gently to 100–110°F until fully melted. Add A to B and blend until homogeneous and cooled. Incorporate C by mixing.

Application: Bath/Shower.

Industry concern: Supplied by Naturex SA.

Storage of Products

All cosmetic preparation should be stored in well closed container and in cool place.

Dosage

As directed by the physician.

Patents

The invention relates to a transparent or translucent colored cosmetic composition for making up the skin, lips and superficial body growths comprising a transparent or translucent cosmetic base and at least one coloring agent in an amount such that the transmission of a 10 µm layer of the final composition, measured at the wavelength of the maximum of one of the absorption peaks of the coloring agent, is between 20 and 80% (Nathalie and Christhope, 2003).

The invention relates to a process for the manufacture of a colored make-up cosmetic composition which makes it possible to produce a transparent or translucent colored coat on the skin, lips or superficial body growths, comprising by selecting a cosmetically acceptable base having bulk opaqueness, translucency or transparency, preparing at least one series of samples of this cosmetic base, each series comprising increasing amounts of a coloring agent dissolved or dispersed in the cosmetically acceptable base and their measurements (Christhope and Nathalie, 2006).

BIBLIOGRAPHY

1. Assimopoulou A.N and Papageorgiou V.P. Radical scavenging activity of *Alkanna tinctoria* root extracts and their main constituents, hydroxynaphthoquinones. Phytother Res. 2005, 19(2):141–147.

2. Chojkier M. Hepatic sinusoidal-obstruction syndrome: toxicity of pyrrolizidine alkaloids. J Hepatol. 2003, 39(3): 437–446.

3. Christhope S.J. and Nathalie J.L. Process for making a colored make-up cosmetic composition with controlled transmittance. United States Patent 7023552. Application Number: 10/203374. Publication date: 04/04/2006.

4. http://plants.usda.gov/java/profile? symbol= ALTI

5. http://www.indianmirror.com/ayurveda/alkanet.html

6. https://www.chagrinvalleysoapandcraft.com/ingrednaturaladd.htm

7. Nathalie J.L and Christhope S.J. Coloured transparent or translucent cosmetic composition. United States Patent Application 20030026772. Application Number: 10/203375. Publication date: 02/06/2003.

8. Ogurtan Z, Hatipoglu F, Ceylan C. The effect of *Alkanna tinctoria* Tausch on burn wound healing in rabbits. Dtsch Tierarztl Wochenschr. 2002,109(11): 481–485.

9. Papageorgiou V.P, Assimopoulou A.N and Ballis A.C. Alkannins and shikonins: a new class of wound healing agents. Curr Med Chem . 2008, 15(30): 3248–3267.

10. Sabin Ozer M, Sarikurkcu C, Tepe B and Can S. Essential oil composition and antioxidant activities of alkanet (*Alkanna tinctoria* subsp. *tinctoria*). 2010, 19(5): 1177–1183.

Aloe vera

Plant profile

Kingdom: Plantae – Plants

Superdivision: Spermatophyta – Seed plants

Class: Liliopsida – Monocotyledons

Order: Liliales

Genus: Aloe L. – Aloe

Subkingdom: Tracheobionta – Vascular plants

Division: Magnoliophyta – Flowering plants

Subclass: Liliidae

Family: Liliacaea – Aloe family

Species: *Aloe vera* (L.) Burm. f. – Barbados aloe

Vernicular name

Bengali and Sanskrit: Ghrita kumari, kumari, **Hindi:** Guarpatha, ghikanvar, **Telugu:** Chinna kalabanda, **Kan:** Lolisara, **Tamil:** Chirukuttali, **Malyalam**: Kattuvala, Kumari, **Oriya:** Kumari, **Marathi:** Korphad

Ayurvedic properties

Rasa, Guna, Veerya, Tikta, Madhura, Gura, Snigda, Seeta.

INTRODUCTION

Aloe Vera is one of the succulent plants or succulents. The commercially significant *aloes* are perennials with 15 to 30 fleshy leaves up to 1.5 feet long, 3 to 4 inches across the base, and with saw teeth marking the margins of the leaves. Plants are growing to 60–100 cm (24–39 in) tall, spreading by offsets. Succulent plants are those that retain water in their stem and leaves. These plants are fleshy and store lots of water in their cells. They are also called as flesh plants. They are capable of surviving in harsh climatic conditions without water. They can also survive in deserted areas. The flowers of these plants have a long stem and their color is crimson red. Their leaves are thick and fleshy and the flowers grow from the centre of the plant. The flowers are produced in summer on a spike up to 90 cm (35 in) tall, each flower pendulous, with a yellow tubular corolla 2–3 cm (0.8–1.2 in) long. The color of the leaves may vary from green to grey-green. Some verities of these plants have white spots on the upper and lower surfaces of the leaves. The margin of these leaves is serrated and they have small white teeth on the margin. The sticky substance inside the *Aloe vera* leaves is called as aloe latex.

Distribution

Aloe vera is a native plant of Africa. It is most commonly found in Africa, Northern America, Egypt, India and Sudan. The species is widely naturalised elsewhere, occurring in temperate and tropical regions of Australia, Barbados, Belize, Nigeria, Paraguay and the US.

Chemical Constituents

The main chemical constituents of *Aloe vera* includes: Amino acids, anthraquinones, enzymes, minerals, vitamins, lignins, monosaccharide, polysaccharides, salicylic acid, saponins, and sterols. Essential amino acids are available in *Aloe vera* and they include isoleucine, leucine, lysine, methionine, phenylalanine, threonine, valine, and tryptophan. Some of the other non-essential amino acids found in *Aloe vera* include alanine, arginine, asparagine, cysteine, glutamic acid, glycine, histidine, proline, serine, tyrosine, glutamine, and aspartic acid. The most important anthraquinones are aloin, aloe-emodin and emodin. The main enzymes found in *Aloe vera* include amylase (breaks down sugars and starches), bradykinase (stimulates immune system, analgesic, anti-inflammatory), catalase (prevents accumulation of water in the body),

Aloin

Emodin

Aloe-emodin

Aloesin

cellulase (aids digestion-cellulose), lipase (aids digestion-fats), oxidase, alkaline phosphatase, proteolytiase (hydrolyses proteins into their constituent elements), creatine phosphokinase (aids metabolism), and carboxypeptidase. It contains vitamins such as A, C, and E plus the minerals, zinc, and selenium. Another component of *Aloe vera* consists of the lignins, a major structural material of cellulose content that allows for penetrative properties. Other constituents of *Aloe vera* would include prostaglandins, tannins, magnesium lactate, resins, mannins, proteins such as lectins, monosulfonic acid and gibberlin. Another constituent of *Aloe vera* includes saponins. The plant sterols or phytosteroids in *Aloe vera* include cholesterol, campesterol, lupeol, and B (beta sign) sitosterol. Flavoniods like aloesin, aloeresin D, etc. also present.

Medicinal uses

It helps to cure regular problems like cough, cold, indigestion, cuts, burns, etc. internal use of *Aloe vera* can cure arthritis, insomnia, ulcers, infection, constipation, heartburn, etc. External use helps in curing sprains, burns, bruises, abrasion, eczema, scrapes, acne, stings, etc. *Aloe vera* juice is helpful to lose weight. It is also generally used as a laxative. *Aloe vera* juice is very helpful to stimulate the body to burn fats and carbohydrates and lose weight. Other than stimulating the metabolic function, it also contains high amounts of proteins and collagen; therefore it also promotes fat loss and muscle development.

Treating a wound or burns: Clean the wound thoroughly and then apply *Aloe vera* pulp to the wound or burned area. Wrap with a bandage and keep the bandage soaked in *Aloe vera* juice. The wound will heal quickly with almost no scars. There would be no infection either. It is very handy and effective in case of kitchen burns.

Remedy for digestive problems: Take 2 spoons of aloe juice thrice a day. This will work like a mild regulator of the bowels, similar to what general tonics do. For bowel regulation, you must use the sap extraction. Sap extraction is the yellow bitter liquid between the leaf skin and pulp. You need to store the peelings of *Aloe vera* in a jar and refrigerate to make the tonic. A few sips once a week is enough to regulate your bowels.

Treating poison ivy, allergies and stings: In case of a cut by poison ivy or oak, or in case of allergies, or in case of bites by ants, wasps, bees, mosquitoes, scorpions, yellow jackets, centipedes and other insects, use *Aloe vera* as a pain inhibitor. Apply *Aloe vera* juice directly on irritated area. It reduces itchiness and heals rashes and sores.

Arthritis treatment: Dice the aloe leaves and put in a jar of water and refrigerate. Drink 4 tablespoons everyday. Within weeks you can see the result. For joint and muscle pains, rub *Aloe vera* on the skin. You can feel the relief in minutes.

Cosmetic uses

Aloe vera is also proved to be a best remedy over the skin disease like psoriasis. It is also helpful to avoid itching and reduce swelling. *Aloe vera* may also be used as a moisturizer for oily skin. Application of this gel to your skin will make it to look radiant. It has cleansing, soothing and nourishing properties. It rejuvenates skin. It removes the dead skin and promotes collagen to prevent wrinkles. If this gel is applied to head, it is good for hair and scalp. It has anti-inflammatory properties. It is also renowned for its anti-fungal and anti-bacterial properties. These properties make it a widely used product. Aloesin modulates melanogenesis via competitive inhibition of tyrosinase, thus holding promise as a pigmentation-altering agent for cosmetic and therapeutic applications (Yagi et al., 2003). The elements in *Aloe vera* such as amino acids, enzymes, vitamins and minerals are essential for human body. The aloe gel if applied to scalp restores pH balance of the scalp and encourages hair growth. Therefore, if used in hair oils, it promotes hair growth and helps to reduce dandruff.

Pharmacological Action

The objectives of the study were to evaluate the inhibitory effect of *aloe* pulp and liquid fraction on the mycelial growth of three phytopathogenic fungi and to determine the extract concentrations that can inhibit mycelial development. Antifungal activity of pulp and liquid fraction was evaluated on the mycellium development of *Rhizoctonia solani*, *Fusarium oxysporum*, and *colletotrichum coccodes* that were isolated from a potato crop by the hyphae point and monosporic techniques. The results showed an inhibitory effect of the pulp of *A. vera* on *F. oxysporum* at 10^4 ml μl^{-1} and over a long period (Rodriuez et al., 2005).

The comparative antimicrobial activities of the gel and leaf of *Aloe vera* were tested against *Staphylococcus aureus*, *Pseudomonas aeruginosa*, *Trichophyton mentagrophytes*, *T. schoenleinii*, *Microsporium canis* and *Candida albicans*. Ethanol was used for the extraction of the leaf after obtaining the gel from it. Antimicrobial effect was measured by the appearance of zones of inhibition. The results of this study tend to give credence to the popular use of both *Aloe vera* gel and leaf (Agarry et al., 2005).

The ethanolic extract of *Aloe vera* leaf skin was fractionated by liquid-liquid partition using hexane, ethyl acetate, chloroform-ethanol and butanol. An antioxidant activity was essayed through some *in vitro* models such as the antioxidant capacity by phosphomolybdenum method, β-carotene bleaching method, radical scavenging activity using 2, 2-diphenyl-1-picryl hydrazyl (DPPH) assay and reducing power assay. The chloroform-ethanol fraction showed the highest total phenolics, the highest scavenging activity and the greatest reducing power, followed by ethyl acetate, butanol and hexane extracts (Miladi and Damak, 2008).

An attempt was made to study the beneficial effects of *Aloe vera* (L.) Burm. fil. in streptozotocin-induced diabetic rats. Excess proliferation of epithelium in the small intestine was observed in diabetic rats, which was reduced after *A. vera* feeding. In diabetic rats and diabetic rats fed with *A. vera*, no change was noticed in the kidney and stomach (Noor et al., 2008).

The present investigation was undertaken to explore the antitumor activity of topical treatment with *Aloe vera* (*Aloe vera*) gel, oral treatment with *Aloe vera* extract, and topical and oral treatment with both gel and extract in stage-2 skin carcinogenesis in Swiss albino mice induced by 7,12-dimethylbenz (a)anthracene (DMBA) and promoted croton (croton tiglium) oil. Results concluded that *Aloe vera* protects mice against DMBA/croton oil-induced skin papillomagenesis, likely due to the chemopreventive activity of high concentrations of antioxidants (Saini et al., 2010).

The immunomodulatory activity of saline extracts of leaves of *Aloe vera* Linn was studied on the albino mice. The saline extract of leaves of *Aloe vera* was administered orally according to their body weight in mice. The results obtained and phytochemical studies the immunostimulant effect of *Aloe vera* could be attributed to the alkaloids content (Chandu et al., 2011).

Toxicological Studies

The toxicity profile of the methanol extract of the *Aloe vera* (*Aloe barbadensis*) gel was studied in Wistar rats. A multiple oral administration of the extract at single dose of 4, 8, 16 g/kg body weights for 14 days did not produce signs of toxicity, behavioral appearances, changes on gross appearance. The sub-acute toxicity was determined by administration of graded doses (1, 2, 4, 8 and 16 g/kg b.wt orally) of the extract daily for 6 weeks and the effects on body weight, organ weight, histology as well as serum biochemical parameters were estimated. Body weight of dosed and control rats increased throughout the duration of treatment. Results concluded that the methanol extract of *Aloe vera* do not produce significant toxic effect in rats during acute and sub-acute treatment in rats. Hence, the extract can be utilized for nutraceuticals formulations (Saritha and Anil kumar, 2010).

Clinical Investigation

The study was conducted to evaluate the effect of cosmetic formulations containing different concentrations of freeze-dried *Aloe vera* extract on skin hydration, after a single and a 1- and 2-week period of application, by using skin bioengineering techniques. Stable formulations containing 5% (w/w) of a trilaureth-4 phosphate-based blend were supplemented with 0.10%, 0.25% or 0.50% (w/w) of freeze-dried *Aloe vera* extract and applied to the volar forearm of 20 female subjects. Skin conditions in terms of the water content of the stratum corneum and of transepidermal water loss (TEWL) (Corneometer CM 825 and Tewameter TM 210) were analysed before and after a single and 1- and 2-week period of daily application. Results show that freeze-dried *Aloe vera* extract is a natural effective ingredient for improving skin hydration, possibly through a humectant mechanism (Dal Belo et al., 2006).

The effect of *Aloe vera* on the reduction of plaque and gingivitis was evaluated in a randomized, parallel and double-blind clinical trial. Subjects were randomly allocated to the test group (n=15) – dentifrice containing *Aloe vera* – or the control group (n = 15) – fluoridated dentifrice. Plaque index (PI) and gingival bleeding index (GBI) were assessed at days 0 and 30. Results concluded that the dentifrice containing *Aloe vera* did not show any additional effect on plaque and gingivitis control compared to the fluoridated dentifrice (de Oliveira et al., 2008).

The aim of this study was to formulate and evaluate herbal cosmetic creams for their improvement of skin viscoelastic and hydration properties. The cosmetic cream formulations were designed by using ethanolic extracts of *Glycyrriza glabra*, *Curcuma longa* (roots), seeds of *Psorolea corlifolia*, *Cassia tora*, *Areca catechu*, *Punica granatum*, fruits of *Embelica officinale*, leaves of *Centella asiatica*, dried bark of *Cinnamon zeylanicum* and fresh gel of *Aloe vera* in varied concentrations. The studies were carried out for 6 weeks on normal subjects (6 males and 12 females, between 22 and 50 years) on the back of their volar forearm for evaluation of viscoelastic properties in terms of extensibility via a suction measurement, firmness using laboratory fabricated instruments such as ball bouncing and skin hydration using electric (resistance) measurement methods. The formulations showed an increase in percentage extensibility, firmness and improved skin hydration and were found more effective compared with the control product after the 6-week study (Ahshawat et al., 2008).

The objective of this study was to determine the effects of skin moisturizers on total antioxidant capacity (TAC) of human skin using epiderm model. Three different skin moisturizers containing antioxidant ingredients (samples 1–3) or *Aloe vera* extract were topically applied to epiderm units and incubated for 2 and 24 hrs to determine acute and longer-term effects of applied samples on TAC and glutathione peroxidase activity in medium and/or homogenized skin tissues. Results of this experiment will help to better understand mechanisms of effects of skin moisturizers containing antioxidant ingredients on skin function at the tissue level and to establish effective strategies for skin protection and clinical treatments of skin disorders and possibly healing wounds (Grazul Bilska et al., 2009).

A research described the design of hydrogels and pluronic-lecithin organogels elaborated as vehicles of *Aloe vera* (*Aloe vera* linné) and *Hydrocotyle asiatica* (*Centella asiatica*) for the treatment of cellulite. The objective of this work was to carry out a complete evaluation of the proposed formulae through the study of the organoleptic and rheological properties of the formulae. Work revealed that, in appearance, hydrogels show better organoleptic characteristics than organogels (Morales et al., 2009).

At least 3 human clinical studies have been conducted on *Aloe vera*'s effect on psoriasis. A randomized, comparative, double-blind, 8-week study on 80 patients found topical *Aloe vera* cream (prepared with 70% mucilage to be more effective than 0.1% triamcinolone acetonide in mild to moderate cases (Choonhakarn et al., 2010).

Market Formulations

Aloe vera products various gel, creams and lotions are available in the market to heal frostbite, burns, insect bites, blisters and allergic reactions, beneficial for cracked and dry skin, eczema, burns, psoriasis, inflammations, wounds, anti-wrinkle creams, facial masks, skin conditioners and lipsticks, lesions and many more the list is never ending.

Aloe vera beauty products are widely spread in the market to make your skin soothing and soft. If you are interested in the Aloe vera cosmetic products you can go through the list mentioned below and benefit from the products available in the market:

- *Aloe vera gel:* Available to heal your burns, scars, insect bites and external problem related to your skin.
- *Skin care:* When we are talking about skin care we will be discussing various *Aloe vera* plant products and various *Aloe vera* beauty products for your use.
- *Body conditioning cream:* To make your body skin smooth.
- *Aloe lotion:* For your skin to give a glowing effect and freshness.
- *Aloe scrub:* Aloe scrub for your face.
- *Aloe sunless tanning lotion:* To protect you from sun rays.
- *Aloe sunscreen:* To protect you from sun rays.
- *Heat lotion:* To protect you from heat.
- *Moisturizing lotion:* Acts as moisturizers for your beautiful skin.
- *Seasonal products:* Seasonal *Aloe vera* products are available in the market to heal your medicinal and season, cosmetic needs.

- *Aloe lip gel:* Aloe lip gel to give you soothing effect in winters.
- *Aloe sun cream:* To keep you away from sun rays and heat.

Dietary supplements: Aloe dietary supplements are available to keep you healthy and fit.

Dietary juice: Aloe dietary juice available for your stomach problem and also to keep you healthy.

Diet packs: Various aloe diet packs are available to keep yourself fit and disease free.

Herbal tea: Herbal tea is available for you free of caffeine.

Luxury skin care: *Aloe vera* plant products and *Aloe vera* cosmetic products available for your extra skin care.

Alluring eyes: Removes dark circle under your eyes.

Aloe body toner: Body toner available to tone your skin.

Aloe Fleur de Jouvence facial treatment: full aloe facial treatment available to give glowing look.

Epiblanc: Available to give you brighter complexion.

Baby care: Various *Aloe vera* products are available for baby care.

Baby bath: Baby bath soap available to make babies beautiful.

Baby calming oil: Aloe oil for babies available to give them soothing feel.

Baby cream: Baby creams available in market.

Baby lotion: Babies lotions for their soft skins.

Baby shampoo: Baby shampoo to make their hair shine.

Baby powder: Aloe powder to give special fragrance to their presence.

Veterinary products: various veterinary products available for animals.

Veterinary cream: Veterinary creams available in market.

Veterinary lotion: Veterinary lotions for their skins.

Veterinary shampoo: Veterinary shampoo to make their hair shine.

Available Forms

Aloe is available commercially in ointments, creams, and lotions. Aloe gel is often included in cosmetic and over-the-counter skin care products as well. Aloe in the form of capsules, tablets, juice, gel, ointment, cream, and lotion are available in market.

Aloe Body Toner

Ingredients: Aloe barbadensis gel (stabilized *Aloe vera* gel), water, C10-18 triglycerides, glyceryl stearate, propylene glycol dicaprylate/dicaprate, butylene glycol, cetyl alcohol, triethanolamine, propylene glycol, *Hedera helix* (ivy) extract, *Spiraea ulmaria* flower extract, *Clematis vitalba* leaf extract, *Fucus vesiculosis* extract, *Equisetum arvense* extract, soluble collagen, hydrolyzed elastin, *Cinnamomum cassia* leaf oil, PEG-100 stearate, polysorbate 20, sorbitan laurate, carbomer, capsicum frutescens resin, ascorbic acid, methylparaben, propylparaben.

Action: Aloe body toner, a wonderfully warming and invigorating cream. It is formulated to be used with cellophane wrap; it contains emulsifiers, moisturizers and humectants, plus warming agents. This specially prepared cream, designed to beautify and firm the body, includes ingredients chosen to provide a rich warming and invigorating feeling. Combination of aloe with cinnamon oil and capsicum provides remarkable warming property.

Supplied by: Forever Living Products, US.

Aloe Vera Sun Block Cream

Ingredients: Aloe vera gel 85%, neem extract 2.5%, cucumber extract 2%, turmeric extract 3.5%, and other herbal extracts.

Action: This one of the product having 75% *Aloe vera* with cucumber, neem and turmeric extracts. It protects face and body from cosmic rays and overexposure of the ultraviolet (UV) rays from sun which seriously threaten the human health. The exposure of UV radiation can cause premature aging, suppression of the immune system and also causes skin cancer. *Aloe vera* sun block cream protect human being from harmful UV sunrays and nourishes the skin and also maintains the face freshness, fairness and shiny. It also blocks both UV-A and UV-B rays. It made with pure, stabilized *Aloe vera* gel, rich moisturizers and humectants; it maintains the skins natural moisture balance. It also prevents skin from darkening. Finally, a sunscreen that combines modern science with natural ingredients to soothe, lubricate and moisturize.

Direction: This cream applies on the face and exposed area of body, neck and arms regularly everyday before getting out. It could not apply around eyes. It is suitable for all types of skin. Dermatologically it was tested. This physically block from harmful sunrays throughout the day including the cloudy days also.

Supplied by: Probia Exim Private Limited, South India.

Clarins after Sun Moisturizer

Ingredients: Vitamins A and E, *Aloe vera*, chamomile, rosemary, linden, shea butter, walnut.

Action: It prevents lasting damage from the sun's rays.

Direction: Put this after sun moisturizer on skin as soon as gets indoors and leave it on for a while before shower. Then apply another layer when after bathing.

Supplied by: Hawaiin Tropic, USA.

Vaseline Aloe Fresh Moisturizing Daily Body Gel

Vaseline aloe fresh moisturizing daily body gel replenishes skins natural moisture balance. This product formulated with 100% natural *Aloe vera* which make instantly refreshes and soothes skin. This gel glides onto skin for a sheer burst of hydration.

Ingredients: Water, pentylene glycol, glycerin, *Aloe barbadensis* leaf juice (*Aloe vera*), *Cucumis sativus* fruit extract (cucumber), triethanolamine, carbomer, PPG 2 Isoceteth 20 acetate, fragrance, disodium EDTA, methylparaben, DMDM hydantoin.

Action: It instantly refreshes skin with a sheer burst of hydration. Fast-absorbing and non-sticky moisture leaves skin feeling cool and healthy by soothes skin.

Supplied by: Hindustan Unilever Company, Mumbai.

Aloe Vera Hair Oil

Ingredients: Sage, *Aloe vera*, rosemary, cornflower, horsetail, camomile, *Ginkgo biloba*.

Action: This oil conditions and detangles your hair with the use of the natural camomile, *Aloe vera* and rosemary extracts. Moisturizes and soften hair and scalp with *Aloe vera*, cornflower extracts. Nourishes, fortifies and helps restore damaged hair, preventing split ends and helps in controlling dandruff with the natural *Aloe vera*, horsetail, *Ginkgo biloba*, rosemary, and camomile extracts, leaving the hair with renewed strengh and radiance. It gives brillance and natural shine enhancing qualities for the hair. It acts as stimulant and tonic for the scalp, providing a healthy and luxury looking hair with rosemary, sage, camomile, and *Aloe vera* extracts. It protects aginst the harsh conditions of nature with the powerful antioxidants properties of rosemary extracts, *Ginkgo biloba*, horsetail, and camomile.

Direction: Massage gently into wet hair, lather and then rinse well with water.

Supplied by: Inmar Est, Hong Kong as brand name Aqua Doux®.

ALOE DEW (Hair Oil)

It is used for moisturizing the scalp, reversing dry scalp and dry hair conditions, providing essential nutrients to the scalp and promoting natural hair growth.

Ingredients: Mainly *Aloe vera*, Henna, hibiscus, Amla.

Action: Effectively cleanses the sebum, bacteria and dirt wrap in hair's root. It helps the hair to breathe and counters the thinning and loss of hair. Moisturizes and regenerates the scalp. Smoothes and softens hair and promotes sheen and luster. Balances scalp's oil level. Prevents dandruff and clears plugged pores through tea tree oil. It also provides essential vitamins to hair to combat dull hair, breakage and damage, Treats irritated stressed scalp also.

Direction: Regular continuous application is the best way to achieve significant improvement.

Supplied by: SRP Global Exports, India.

Nyle Herbal Moisturizing Shine Shampoo

Ingredients: Water, SLES, CAPB, sodium cocoamphoacetate, alkylpolyglucoside, PEG7 glyceryl cocoate, CMEA, PPG2 hydroxyethyl derivatives, silicone quaternium-20, extracts of Amla, Tulsi, aloe green tea, GHTC, PQ-10, PQ-7, DMDM hydantoin, benzylalcohol, sulphonated benzotriazole derivative, fragrance, BHT C147000, C161565, sodium chloride.

Application: Apply directly on wet hair. Rinse thoroughly with water.

Industry concern: Cavin Kare Pvt Ltd, Haridwar.

Method of Preparation Hair Oil

To prepare *Aloe vera* hair oil you can follow the following steps:

1. First take 1/4 cup of aloe gel and then mix it with vegetable oil.

2. Then heat the oil for 10 minutes and cool it

3. Store it in a cool and dry place.

4. Keep it away from sunlight.

Aloe Hair Oil for Split Ends in the Hair

Ingredients: Four big leaves of aloe and 250 g of coconut oil.

Method: Slit open the leaves and scrape out the juicy pulp. Use a liquidizer and blend to make it into a smooth paste. Heat coconut oil and add the aloe pulp into that. Keep stirring until the oil floats on the top. Strain the oil and keep in the air tight bottle.

Direction: Take two tablespoons of oil and massage it well into the scalp to the end of hair. Leave on for one hour.

Shyamal Hair Oil

Ingredients: Sesamum indicum oil 16%, non-greasy base q.s, extract of *Eclipta alba* 1.6%, *Centella asiatica* 1.6%, *Aloe vera* 1.6%, *Azadirachta indica* 1.6%, *Emblica officinalis* 1.6%, *Indigofera tinctoria* 0.8%, *Jasminum officinate* 0.8%, *Abrus precatorius* 0.8%, *Glycyrrhiza glabra* 0.8%, *Cyperus rotundus* 0.8%, colour-sunset yello-FCF.

Application: Gentle massage on hair and scalp or as directed by physician.

Indication: Prevent hair fall, dandruff, premature graying of hair, splitends.

Industry concern: Vasu Healthcare Pvt. Ltd., Vadodara.

Storage of Products

During processing for the collection of the aloe gel and bitter liquid, harvested leaves are cut and drained slowly and placed in jars for storage or for use when necessary.

Herbal ointment prepared by aloe, can then be stored using clean jars and placed in a cool place and this ointment can be used like the fresh leaves as and when necessary to treat the same types of disorders and conditions normally treated using the gel from fresh leaves.

All the products can store at room temperature and protect from oxidation.

Dosage

Dosage of the herbal remedy made from aloe can differ for one disorder to another, and depends on the person as well, taking one 50–200 mg capsule of the aloe latex a day for a treatment period of ten days can aid in alleviating constipation in the person. The stabilized aloe gel can be applied topically for the treatment of minor burns, the gel can be used as a topical cream to be applied along the affected area of the skin three to five times daily for the treatment of all external injuries in affected individuals. A professional and qualified health care worker must be consulted before any attempt is made to treat more serious burns on the skin using the aloe gel alone. The preferred dose of the *Aloe vera* gel by most people is, for internal purposes, about 30 ml of the aloe gel taken thrice daily during the entire length of the treatment.

The fresh form of the aloe gel can be used internally as a herbal remedy as well, and dosage can be a maximum of 2 tsp of the fresh gel in a glass of water or any fruit juice, this can be taken thrice daily, as a general herbal tonic for the treatment of various internal disorders in the body of the affected patients.

Patents

The present invention discloses aqueous compositions comprising liposomes of phospholipids, and at least one competitive inhibitor of an enzyme for the synthesis of melanin, in combination with at least one non-competitive inhibitor of an enzyme for the synthesis of melanin. The invention also includes the use of the compositions of this invention for the de-pigmentation of skin (Kenneth et al., 2000).

A product and process for stabilizing *Aloe vera* gel is disclosed. The process includes the steps of rapidly heating the *Aloe vera* gel to a temperature in the range of from about 35°C. to about 80°C., adding to the heated *Aloe vera* gel one or more stabilizing antioxidants, and rapidly cooling the heated *Aloe vera* gel to a temperature in the range of from about 20°C. to about 30°C. The stabilizing antioxidants may be a tocotrienol/tocopherol blend, rosmarinic acid, polyphenols, or any combination thereof (Maughan et al., 2005).

Methods for providing skin enhancement and pain relief to an individual in need of treatment by administering to such an individual an effective amount of an immunostimulatory *Aloe vera* derived composition. Oral and topical methods and compositions are provided. *Aloe vera* derived immunostimulatory compositions and methods of producing such compositions are provided (Denise et al., 2010).

BIBLIOGRAPHY

1. Agarry O.O, Olaleye M.T, Bello-Michael C.O. Comparative antimicrobial activities of *aloe vera* gel and leaf. African Journal of Biotechnology. 2005; 4 (12): 1413–1414.

2. Ahshawat MS, Saraf S, Saraf S. Preparation and characterization of herbal creams for improvement of skin viscoelastic properties. Int J Cosmet Sci. 2008; 30 (3): 183–93.

3. Chandu AN, Kumar S.C, Bhattacharjee S, Debnath S, Kannan K.K. Studies on immuno-modulatory activity of *Aloe vera* (Linn). International Journal of Applied Biology and Pharmaceutical Technology. 2011; 2(1): 19–22.

4. Choonhakarn C, Busaracome P, Sripanidkulchai B, Sarakarn P. A pro-spective, randomized clinical trial comparing topical *aloe vera* with 0.1% triamcinolone acetonide in mild to moderate plaque psoriasis. J Eur Acad Dermatol Venereol. 2010; 24(2): 168–172.

5. Dal'Belo SE, Gaspar LR, Maia Campos PM. Moisturizing effect of cosmetic formulations containing *Aloe vera* extract in different concentrations assessed by skin bioengineering techniques. Skin Res Technol. 2006; 12(4): 241–246.

6. de Oliveira SM, Torres TC, Pereira SL, Mota OM, Carlos MX.. Effect of a dentifrice containing *Aloe vera* on plaque and gingivitis control. A double-blind clinical study in humans. J Appl Oral Sci. 2008; 16(4):293–296.

7. Denise D, Stanley PD, Ceri PND, Alex P, Alvin N. Aloe preparation for skin enhancement. United States Patent 20100255130. Application Number: 12/599532, Publication Date:10/07/2010. http://www. freepaten tsonline.com/y2010/0255130.html

8. Grazul Bilska AT, Bilski JJ, Redmer DA, Reynolds LP, Abdullah KM, Abdullah A. Antioxidant capacity of 3D human skin EpiDerm model: effects of skin moisturizers. Int J Cosmet Sci. 2009; 31(3):201–208.

9. http://www.beautyandgroomingtips.com/2007/06/aloe-oil-for-split-ends-in-hair.html

10. http://www.herbs2000.com/herbs/herbs_aloe_vera.htm

11. http://www.thealoevera.com/

12. Kenneth J, Abeysinghe P, Ernst ET, Gabriele B. Aqueous composition comprising active ingredients for the de-pigmentation of the skin. United States Patent 6123959. Application Number: 09/066161, Publication Date: 09/26/2000. http://www.freepatentsonline.com/6123959.html

13. Maughan RG, Roger PA, Phan BV. Product and process for stabilizing *Aloe vera* gel. United States Patent 6869624. Application Number: 10/447970, Publication Date:03/22/2005. http://www. freepatentsonline.com/6869624. html

14. Miladi S, Damak M. *In vitro* antioxidant activities of *Aloe vera* leaf skin extracts. Journal de la Société Chimique de Tunisie. 2008; 10: 101–109.

15. Morales ME, Gallardo V, Clars B, Garcia MB, Ruzi MA. Study and description of hydrogels and organogels as vehicles for cosmetic active ingredients. J cosmet Sci. 2009; 60(6): 627–636.

16. Noor A, Gunasekaran S, Manickam A.S, Vijayalakshmi M.A. Antidiabetic activity of *Aloe vera* and histology of organs in streptozotocininduced diabetic rats. Current Science. 2008; 94(8): 1070–1076.

17. Rodriuez DJ, Castillo DH, Garci R.R, Sachez J.L.A. Antifungal activity *in vitro* of *Aloe vera* pulp and liquid fraction against plant pathogenic fungi. Industrial Crops and Products. 2005; 21: 81–87.

18. Saini M, Goyal P.K, Chaudhary G. Anti-tumor activity of *Aloe vera* against DMBA/croton oil-induced skin papillomagenesis in Swiss albino mice. J Environ Pathol Toxicol Oncol. 2010; 29(2):127–35.

19. Saritha V, Anilakumar K.K. Toxi-cological evaluation of methanol extract of *Aloe vera* in rats. International Journal on Pharmaceutical and Biomedical Research. 2010; 1(5): 142–149.

20. Yagi A, Kabash A, Mizuno K, Moustafa SM, Khalifa TI, Tsuji H. Radical scavenging glycoprotein inhibiting cyclooxygenase-2 and thromboxane A2v synthase from *Aloe vera* gel. Planta Med. 2003; 69: 269–271.

Azadirachta indica

Plant profile

Kingdom: Plantae – *Plants*

Subkingdom: Tracheobionta – Vascular plants

Superdivision: Spermatophyta – Seed plants

Division: Magnoliophyta – Flowering plants

Class: Magnoliopsida – Dicotyledons

Subclass: Rosidae

Order: Sapindales

Family: Meliaceae – Mahogany family

Genus: Azadirachta *A. Juss.* – Azadirachta

Species: *Azadirachta indica A. Juss.* – Neem

Vernacular names

San: Arishta, Nimba, Minbaka, **Assam:** Nim, **Beng:** Nim, Neemgachh, Neem, **Eng:** Neem, Indian Lilac, Margosa Tree, Nim, Crackjack, Paradise Tree, White Cedar, Chinaberry, **Guj:** Limba, Limbado, Leemgo, Danu-jhada, Kohalu Limdo, **Hin:** Nim, Nimb, Bal-nimb, Neem, Nind, **Kan:** Bevinamara, Bevu, Hebbevu, Kiri-bevu, **Konkani :** Nim, **Mal:** Aryaveppu, Veppu, Aryaveshnu, Rajavedhu, Vepe, **Marathi:** Kadukhajur, Limba, Limb, Nimbay, Bakayan, Balanti-limb, **Oriya:** Kakopholo, Limbo, Nimbu, Nimo, **Telugu:** Nimbanuv, Vepa, Yeppa, Yapa, Vepachettu, Yapachattu.

Ayurvedic properties

Guna: Laghu, **Rasa:** Tita, kashaya, **Veerya:** Sheeta, **Vipaka:** Katu

Ayurvedic action

According to Ayurveda principles vitiated kapha and pitta cause skin diseases. Neem pacifies vitiated kapha and pitta, thus helps to cure skin ailments.

INTRODUCTION

A moderate to large sized tree grows up to 30 meters in height. Leaves are imparipinnately compound, leaflets opposite, serrate and acuminate. The pinnate leaves are 20–40 cm (8 to 16′) long, with 20 to 31 medium to dark green leaflets about 3–8 cm (1 to 3′) long. The terminal leaflet is often missing. The petioles are short. Flowers are arranged axillary, normally in more-or-less drooping panicles which are up to 25 cm (10′) long. The inflorescences, which branch up to the third degree, bear from 150 to 250 flowers. An individual flower is 5–6 mm long and 8–11 mm wide. Protandrous, bisexual flowers and male flowers exist on the same individual. Fruits are elliptical one-seeded drupes yellowish when ripe are 1.4–2.8 × 1.0–1.5 cm.

Distribution

It is native to Asian countries like Burma, India and Pakistan. It grows in tropical and semitropical regions and is widely found everywhere. Throughout India, it growing deciduous forests also cultivated along road sides and farm boundaries. Normally it thrives in areas with sub-arid to sub-humid conditions, with an annual rainfall between 400 and 1200 mm. It can grow in regions with an annual rainfall below 400 mm, but in such cases it depends largely on ground water levels. Neem can grow in many different types of soil, but it thrives best on well drained deep and sandy soils. It is a typical tropical to subtropical tree and exists at annual mean temperatures between 21 and 32°C.

Chemical Constituents

Neem tree has numerous medicinal properties by virtue of its chemical compounds. Seeds of the neem tree contain the highest concentration of azadirachtin. Diterpenoids and triterpenoids compounds contain protomeliacins, limonoids, azadirone, gedunin. C-secomeliacins such as nimbin, salannin, azadirachtin are also present.

Azadirachtin

Nimbin

Salannin

Medicinal uses

The Neem tree has many medicinal uses. The chemical compounds present in Neem have anti-inflammatory, antiarthritic, antipyretic, hypoglycaemic, antifungal, spermicidal, antimalarial, antibacterial and diuretic properties. Flower, leaves, bark and seeds of Neem are used in home remedies and in preparation of medicines. Bark of Neem acts as antipyretic and helps to reduce fever. Flowers are used in intestinal disorders. Juice from fresh leaves is very helpful in treating skin diseases, wounds and obesity. Oil from Neem seeds is used in arthritis, skin diseases and muscular sprains. Neem is very effective in treating gum diseases. The Neem is proved to be beneficial in treating skin diseases because of its antibiotic, antifungal and blood purifying properties. Due to its detoxifying properties it helps to keep organ systems healthy, especially circulatory, digestive, respiratory and urinary systems. Neem acts as antidiabetic, reduces blood cholesterol level and keeps the heart healthy.

Nimbolide also isolated from a major crude bitter principle extracted from oil of seed kernels and known to have antimalarial activity by inhibiting growth of *Plasmodium falciparum*. Nimbolide also shows antibacterial activity against *S. aureus* and *S. coagulase*.

Gedunin was isolated from Neem seed oil. It is known to possess antifungal and antimalarial activity. Gedunin found effective against malaria which can be treated through formation of tea and infusion of the leaves.

Mahmoodin is antibacterial, isolated from Neem seed oil. It is a deoxygedunin and shows antibacterial activity against human pathogenic bacteria.

Cosmetic uses

It promotes wound healing as it is antibacterial and astringent. In psoriasis it reduces itching, irritation, roughness of skin and heals the psoriatic patches. In same way it heals eczema too. It reduces infection and inflammation of acne. Neem helps to maintain the health of scalp skin and prevents dandruff.

Nimbidin also found effective against many skin diseases such as pyorrhea, bleeding gums and sore throat. It is also used in different lotions, paste and toothpastes which are prepared according to usual prescription and have effective against pyorrhea (diseases of gum, tooth, and sockets).

Neem oil offers moisturizing, regenerative and restructuring properties and is now widely used in the cosmetic industry along with other skin care herbs.

Home Remedies with Neem

- Apply crushed fresh leaves of Neem on acne. In case of body acne mix fine paste of fresh neem leaves in little water and smear this mixture on back, chest and shoulders.
- In itching, application of neeem oil on affected areas helps. Boil neem leaves in a big bowl of water and mix this in bathing water. This reduces body itch.
- Massaging Neem oil to scalp removes head lice and prevents formation of dandruff.
- Mix dry Neem powder, shikakai and Amla in water and apply this as pack on head. This pack has to be kept for 45 minutes and washed off later. This prevents hair loss and dandruff. Fresh Neem leaves can also be used instead of dry Neem powder.
- A freshly prepared paste of turmeric, Neem and sesame seeds is recommended in Ayurveda for fungal infection between toes.

Pharmacological Action

The present study evaluates its hepatoprotective role. Fresh juice of tender leaves of *Azadirachta indica* (200 mg/kg body wt. p.o.) inhibited paracetamol (2 g/kg body wt. p.o.)-induced lipid peroxidation and prevented depletion of sulfhydryl groups in liver cells. There was an increase in serum marker enzymes of hepatic damage (aspartate transaminase, alanine transaminase and alkaline phosphatase) after paracetamol administration. *Azadirachta indica* pretreatment stabilized the serum levels of these enzymes (Yanpallewar et al., 2003).

The present study revealed that nimbidin significantly inhibited some of the functions of macrophages and neutrophils relevant to the inflammatory response following both *in vivo* and *in vitro* exposure. Oral administration of 5–25 mg/kg nimbidin to rats for 3 consecutive days significantly inhibited the migration of macrophages to their peritoneal cavities in response to inflammatory stimuli and also inhibited phagocytosis and phorbol-12-myristate-13-acetate (PMA) stimulated respiratory burst in these cells. The results suggest that nimbidin suppresses the functions of macrophages and neutrophils relevant to inflammation. Thus nimbidin can be valuable in treating inflammation/inflammatory diseases (Kaur et al., 2004).

Present study was revealed the effect of *Azadirachta indica* to possess antioxidant, anti-inflammatory and anxiolytic properties, on cerebral reperfusion injury and long-term cerebral hypoperfusion. *A. indica* pretreatment (500 mg/kg/day × 7 days) attenuated the reperfusion induced enhanced lipid peroxidation, SOD activity and prevented fall in T-SH groups. Moreover, *A. indica* per sec increased brain ascorbic acid level, which was unchanged during reperfusion insult. Long-term cerebral hypoperfusion induced by permanent BCCA occlusion has been reported to cause behavioral and histopathological abnormalities. Reactive changes in brain histology like gliosis, perivascular lymphocytic infiltration, recruitment of macrophages and cellular edema following long-term hypoperfusion were also attenuated effectively by *A. indica* (Yanpallewar et al., 2005).

The research was conducted to investigate *in vitro* antiviral, antitumor and cytotoxic activities of crude aqueous extracts of seeds and leaves of Neem plant. Results show the crude acidic extract of leaves and seeds (90.38, 65.38%, respectively) and alkaline extract of seeds (77.61%) possesses very remarkable

antiviral activity against herpes simplex virus type 1 compared with acyclovir (54.33%) at concentration 20 μg/ml. Furter cytotoxicity and cell cancer viability test was measured microscopically and by viable cell counting. The cytotoxic activity of crude aqueous extracts from seeds and leaves revealed significant inhibition of Ehrlich ascites carcinoma cell line growth and had anticancer activity at different concentrations (250, 500, 750 and 1000 μg/ml) of crude aqueous extracts (Hassan et al., 2010).

Pharmacological hypoglycemic action of *Azadirachta indica* was evaluated in diabetic rats. After treatment for 24 hrs, *Azadirachta indica* 250 mg/kg (single dose study) reduced glucose (18%), cholesterol (15%), triglycerides (32%), urea (13%), creatinine (23%), and lipids (15%). Multiple dose study for 15 days also reduced creatinine, urea, lipids, triglycerides and glucose. Hence it serves as an important alternative source in the management of diabetes mellitus involved in reducing increased blood glucose during diabetes which should be examined further by oral hypoglycemic therapy (Dholi et al., 2011).

Toxicological Studies

The acute oral toxicity in rats fed technical grade azadirachtin ranged from greater than 3,540 mg/kg to greater than 5,000 mg/kg, the highest dose tested when administered undiluted to albino rats (Thomson, 1992, US Environmental Protection Agency, 1993). The acute inhalation toxicity study in rats exposed to technical azadirachtin showed that the acute inhalation LD_{50} is greater than 2.41 mg/l per animal, the highest dose tested. Although this figure is below the 5.0 mg/l limit test dose for an acute inhalation study, the reported concentration was the maximum dose possible under the test conditions. No deaths occurred during the course of the study. Azadirachtin was given a toxicity classification of Category III (U.S. Environmental Protection Agency, 1993).

Acute toxicity of the seed oil of *Azadirachta indica* (Neem oil) was documented in rats and rabbits by the oral route. Dose-related pharmacotoxic symptoms were noted along with a number of biochemical and histopathological indices of toxicity. The 24-h LD_{50} was established as 14 ml/kg in rats and 24 ml/kg in rabbits. Prior to death, animals of both species exhibited comparable pharmacotoxic symptoms in order and severity, with lungs and central nervous system as the target organs of toxicity. Edible mustard seed oil (80 ml/kg) was tested in the same manner to document the degree to which the physical characteristics of an oil could contribute to the oral toxicity of Neem oil (Gandhi et al., 1988).

Acute toxicity of *Azadirachta indica* (Neem) leaf aqueous extract was tested on chickens; and the median lethal dose (LD_{50}) was calculated. Both clinical signs, gross lesion and microscopic lesions at post-mortem were recorded. Five experimental groups of chickens were used in this study. Four of the groups were respectively given intraperitoneal doses of 800 mg/kg, 1600 mg/kg, 3200 mg/kg and 6400 mg/kg of the extract, while group five was control. The LD_{50} was calculated. Clinical signs and tissue post-mortem and microscopic lesions recorded. Results revealed the calculated median lethal dose (LD_{50}) was 4800 mg/kg. The severity of the clinical signs, gross and microscopic lesions of tissues observed at post-mortem was dose-dependent (Biu et al., 2010).

Chronic toxicity of Neem (*Azadirachta indica*) leaf aqueous extract in chicken (*Gallus gallus domesticus*) was studied by daily intraperitoneal administration of graded doses of 500, 1000, and 2000 mg/kg of the extract for a period of 18 days. Clinical symptoms of ruffled feathers, weakness, anorexia, depression, and dyspnoea were observed at 20 minutes postexposure extract-dose-dependent, and death intervened on day 6 postexposure. Microscopic lesions were also extract-dose-dependent and included pulmonary congestion and edema with interstitial mononuclear cell infiltration, hepatic vacuolar degeneration with Kupffer cell proliferation, sinusoidal congestion, haemorrhage and necrosis with foci of bacterial

colonies, nephritis, congestion of vasa vasori, spongiosis, matted and stunted intestinal villi with goblet cell hyperplasia (Biu et al., 2010).

Clinical Investigation

A recent study proved that, when a patient took either Neem leaf extract or Neem capsules for a month, her high cholesterol levels fell subsequently. In another study, alcoholic extract of Neem leaves reduced serum cholesterol by approximately 30 percent two hours after its administration. The cholesterol level stayed low for an additional four hours until testing ceased. In 1993, in a preliminary study, the National Institutes of Health reported positive results from *in vitro* tests where Neem bark extracts killed the AIDS virus.

Oral doses of neem were tested at least one year on fifteen patients who had the vitiligo. They also applied a cream made up of several herbs to patched, which were then exposed to the sun. After ninety days, 25 percent of the patients showed complete relief. No adverse reactions were shown by any participants. Those who stayed on the treatment the longest showed the most improvement. The dosage was four grams of Neem leaves three times a day, ideally taken before each meal.

In clinical studies Neem extracts and oil were found to be as effective as coal tar and cortisone in treating psoriasis. However, there were none of the usual side effects accompanying the use of Neem as there was with coal tar and cortisone. When applied to the skin, Neem extracts and oil removed the redness and itching while improving the condition of the skin for the duration of the treatment.

The present study was revealed the response of the drug "Neem Guard" in skin diseases in general with a motto to study the specific effect of the drug in particular skin diseases, the immunoresponse of the drug particularly on IgE and the safety and toxicity of the drug. Total 57 patients of skin diseases were registered during the course of the trial. The study consists clinical patterns in skin disease and management of the disease with Neem

Guard capsule. Out of 57 cases 35 cases had completed the total tenure of the treatment and rest of cases had discontinued at different point of time. The product "Neem Guard" capsule has been found to be effective on lowering IgE level particularly in case of skin diseases as revealed from the study. It has also been found to be an effective product in some selective skin diseases like acne, contact dermatitis, urticaria, allergic dermatitis, wet eczema where the product "Neem Guard" capsule can safely be prescribed.

The use of Neem oil and chewing on Neem twigs have demonstrated varying efficacy versus oral flora and microorganisms responsible for dental caries (Pai et al., 2004; Prashant et al., 2007; Sharma et al., 2008).

Market Formulations

Neem bark: Neem bark is extensively use to manufacture and export quality herbal and natural cosmetics and personal hygiene products. These ecofriendly and organic products suit almost all skin types and do not have any harmful effects on the skin. It can also be used in blemish removing creams and lotions.

Neem Leaves: Leaves in powdered form or as extracts are used to manufacture face and body lotions and creams, they are very effective, it is also used in certain fairness creams. It is essential for maintaining a healthy and radiant skin. Neem leaf extracts are sometimes topically added to shampoo as a measure to control dandruff and increase their strength. Leaf granules are used especially in herbal face packs. Due to its anti-fungal and anti bacterial properties, the extract can also be used as skin and hair soother lotion.

Neem Oil: After the process of refining, neem oil loses its unpleasant smell and can be used to manufacture cosmetics. It is used for damaged and delicate hair, hand creams, facial care and sun screen products.

Himalaya Herbal Health Care, Organique, Toothpaste

Ingredients: Vegetable glycerin (from vegetable oils), xylitol (Himalayan birch wood), water (aqua), hydrated silica, calcium carbonate, lauryl glucoside (from coconut oil), sodium cocoyl glutamate and disodium cocoyl glutamate (from amino acids and vegetable oil), *Stevia rebaudiana* (stevia) extract, natural flavor (using essential oils and botanicals), potassium sorbate (sorbus cashmiriana berries), menthol (from peppermint leaves), *Chondrus crispus* (carrageenan), xanthan gum (natural thickener), sodium chloride (sea salt), thymol (from thyme leaf oil), *Azadirachta indica* (Neem) leaf extract, *Punica granatum* (pomegranate) fruit extract, *Terminalia chebula* (chebulic myrobalan) fruit extract, *Terminalia bellerica* (belleric myrobalan) fruit extract, *Phyllanthus emblica* (Amla) fruit extract, *Embelia ribes* (vidanga) fruit extract, *Acacia nilotica* (Indian gum arabic tree) bark extract.

Action: This minty fresh toothpaste cleans your teeth and promotes healthy gums. Xylitol is used as a natural sweetener and helps fight plaque and enhances sparkling smile. This paste is gluten, fluoride, saccharin and sodium lauryl sulphate free.

Storage of products: Store at room temperature.

Industry concern: Himalaya Drug Company, Bangalore.

Neemaura Naturals Neem Skin Salve

Ingredients: Each 1.0 ounce contains a proprietary blend of organic Neem leaf (*Azadirachta indica*), almond oil, safflower oil, beeswax and vitamin E oil.

Direction: Keep out of the reach of children. (For external use only).

Neemaura Naturals Inc Concentrated Neem Cream

Ingredients: Neem aura herbal cream complex (*Aloe vera* liquid, neem leaf, St. Johns wort, chamomile leaf, white willow bark, grape seed, arnica flower, horse chestnut leaf, thyme leaf, goldenseal root, barberry and ivy leaf extracts), glyceryl stearate (from coconut), rice bran oil, glycerin, cetyl alcohol (from coconut), behentrimonium methosulfate, cetearyl alcohol, arachidyl alcohol, behenyl alcohol, arachidyl glucoside, Neem oil, locust bean gum, jojoba oil, *Hamamelis virginiana* (witch hazel), wheat protein, ginger ext., citrus aurantium dulcis seed ext., allantoin (from Comfrey), dL panthenol (pro-vitamin B5), tocopherol acetate (vitamin E), melissa oil, chamomile oil, vitamin A palmitate and lavender oil.

Action: Neem's active ingredients absorb into the skin, restoring, rejuvenating and repairing damage. Neem leaf extract and Neem oil are provided in extra-potency strength. This highly concentrated cream goes a long way so apply sparingly.

Neem Herbal Skin Conditioning Spray

Ingredients: Neem aura moisturizing complex (*Aloe vera*, and Neem leaf, goldenseal root, barberry plant, chamomile leaf and thyme leaf extracts), ethyl alcohol (from grain), vegetable glycerin, coconut soap, Neem oil, decyl glucoside (from coconut), behen-trimonium methosulfate (from nuts), orange oil, lemongrass oil, cedarwood oil, rhodium wood oil, citronella oil, myrrh gum extract, lavender oil and citrus aurantium dulcis seed extract.

Action: Neem's active ingredients absorb into the skin, restoring and rejuvenating damage. It protects the skin and help repair damage from sun or wind. It also recommended for sunburned skin and minor cuts or abrasions.

Direction: Use this spray before and after exposure to irritating elements such as wind and sun and to reduce irritation resulting from insect bites. Neem aura can soothe the effects of irritated skin no matter what the cause, whether too much heat, allergies or insect bites. Do not spray in your eyes.

Neem Face Wash

Ingredients: It is mainly content Neem and turmeric.

Action: Neem having antibacterial properties which helps to control acne and pimples. Turmeric helps in acne causing bacteria.

Direction: Moisten the face and then apply small amount of the product and gently rub with circular motion. Wash off and pat dry.

Industry concern: Himalaya Drugs Company, Bangalore.

Acnovin Cream

This cream is used to eliminate causative bacteria, promotes the collagen formation. It is effectively get rid of blemishes and scars and bring a glow to the skin.

Ingredients: Kumkumadya Tailam with herbs like Neem and Haridra.

Indications: Acne vulgaris, pimples, blemishes on skin, blackheads, chaffed skin.

Method of application: Wash face with the lukewarm water; also dry with clean towel, and gently apply Acnovin cream, rubbing evenly till it penetrates the skin or as directed by physician. Apply 2–3 times in a day.

Duration: Up to one month after face is totally cleared of acne problems. Simultaneous use of the Acnovin capsules gives better results.

Patents

There are provided topical cosmetic compositions for improving the aesthetic appearance of skin and remediating the effects of aging, and methods of use thereof. One composition is a blend of neem seed cell broth and one or more additional botanical ingredients. Another composition has pomegranate fruit extract and, optionally, one or more additional botanical ingredients (Michelle et al., 2003).

The present invention relates to the preparation and use of compositions for the treatment of skin disorders itchy and/or infected skin such as impetigo, acne (on face, forehead scalp and on the back of the body) and fungal infection of skin and nails. Whether the infection may be acute or chronic or sub-acute or acute on chronic. The compositions are based on extracts from the plants *Cassia tora*, *Centratherum anthelminticum* and/or *Melia azadirachta*. A variety of other herbal extracts may be included and the composition may take the form of a freeze dried or a spray dried powder or presence of this in a cream or ointment based on ghee, or in a cream or ointment form developed with any other vehicle, or they may be in a powdered form without spray drying (Shah, 2007).

This invention relates to novel dental compositions and methods for preventing dental plaque and caries formation and generally for inhibiting tooth decay and brightening/whitening teeth. The compositions of this invention comprise herbs such as *Citrus karna* raf., *Zanthoxylum armatum* DC and *Azadirachta indica* A. Juss. thereof which can be combined with pharmaceutically acceptable carriers or diluents to be administered in the form of conventional dental compositions. The compositions of the present invention, also preferably, contain *mint* (Pushpangadan et al., 2007).

BIBLIOGRAPHY

1. http://www.azadirachta-indica.com/

2. http://www.ayurhelp.com/plants/neem.htm

3. http://www.isrj.net/may/2011/Chemistry_NATURES_DRUG_ STORE. aspx

4. Yanpallewar SU, Sen S, Tapas S, Kumar M, Raju SS, Acharya SB. Effect of Azadirachta indica on paracetamolinduced hepatic damage in albino rats. Phytomedicine 2003; 10(5):391–396.

5. Kaur G, Sarwar Alam M, Athar M. Nimbidin suppresses functions of macrophages and neutrophils: relevance to its antiinflammatory mechanisms. Phytother Res. 2004;18(5): 419–424.

6. Yanpallewar S, Rai S, Kumar M, Chauhan S, Acharya SB. Neuroprotective effect of Azadirachta indica on cerebral postischemic reperfusion and hypoperfusion in rats. Life Sci. 2005; 76(12):1325-1338.

7. Hassan A, Wafaa AH, Hanan AAT. In Vitro antitumor and antiviral activities of seeds and leaves neem (Azadirachta Indica) extracts.

International journal of Academic Research 2010; 2(2): 47–51.

8. Dholi SK, Raparla R, Mankala SK, Nagappan K. Invivo Antidiabetic evaluation of Neem leaf extract in alloxan induced rats. Journal of Applied Pharmaceutical Science 2011; 01 (04): 100–105.

9. Thomson, W.T. Agricultural Chemicals. Book I: Insecticides. 1992. Thomson Publications, Fresno, CA.

10. U.S. Environmental Protection Agency. 1993. Azadirachtin: Tolerance Exemption. Federal Register. Vol. 58, No. 30. Rules and Regulations. Wednesday, February 17, 1993.

11. Gandhi M, Lal R, Sankaranarayanan A, Banerjee CK, Sharma PL. Acute toxicity study of the oil from Azadirachta indica seed (neem oil). Journal of Ethnopharmacology 1988; 23 (1): 39–51.

12. Biu AA, Yusufu SD, Rabo JS. Acute toxicity study on neem, (Azadirachta indica) leaf aqueous extract in chicken (Gallus gallus domesticus). African Scientist 2010; 11(4): 241–244.

13. Biu AA, Yusufu SD, Rabo JS. Chronic toxicity study on neem, (Azadirachta indica) leaf aqueous extract in chicken (Gallus gallus domesticus). African Scientist 2010; 11(4): 245–247.

14. http://www.herbalremedy.in/product_details.php?item_id=142&currency_ code= ZAR

15. http://rawandnatural.org/cosmetics-4/neem-oil-health-benefits/

16. http://www.allayurveda.com/research_neemguard.asp

17. Pai MR, Acharya LD, Udupa N. Evaluation of antiplaque activity of Azadirachta indica leaf extract gel—a 6-week clinical study. J Ethnopharmacol 2004; 90(1): 99–103.

18. Prashant GM, Chandu GN, Murulikrishna KS, Shafiulla MD. The effect of mango and neem extract on four organisms causing dental caries: Streptococcus mutans, Streptococcus salivavius , Streptococcus mitis, and Streptococcus sanguis: an in vitro study. Indian J Dent Res 2007; 18(4):148–151.

19. Sharma S, Saimbi CS, Koirala B, Shukla R. Effect of various mouthwashes on the levels of interleukin-2 and interferon-gamma in chronic gingivitis. J Clin Pediatr Dent 2008; 32(2):111–114.

20. Michelle L, Duggan M, Menon GK, Theophilus EH, Dokka S, Wang H. Topical cosmetic composition with skin rejuvenation benefits. United States Patent Application 20030091665. Application number: 10/039746, Publication date: 05/15/2003. http://www. freepaten tsonline. com/y2003/0091665.html

21. Shah EM. Novel base material for pharmaceutical and/or cosmetic cream (herbal composition for itchy or infected skin). United States Patent Application 20070014749. Application number: 11/517986, Publication date:01/18/2007. http://www. freepat entsonline. com/y2007/0014749. html

22. Pushpangadan P, Rao CV, Ojha SK, Nair KPN, Pandey MM, Rawat AKS, Mehrotra S. Herbal oro-dental care composition and process for preparing the same. United States Patent 7279151. Application number: 10/810011, Publication date: 10/09/2007. http://www.freepatentsonline. com/279151. html.

Berberis aristata

Plant profile

Kingdom: Plantae – Plants
Class: Magnoliopsida
Genus: Berberis L.

Division: Magnoliophyta – Flowering plants
Order: Ranunculales
Species: *Berberis aristata*

Family: Berberidaceae

Vernacular names

Sanskrit: Daruharidra, Darvi, Katankateri, Peetita, Sthiraraga, Kamini, **Hindi:** Daruhaldi, Jharihaldi, Kashmal, Chitra, **English:** Indian Berberi, Tree Turmeric, **Malayalam:** Maramanjal. **Bengali:** Darhaldi, **Marathi:** Daruhald, **Kerala:** Maradarisina, Maramanjal, **Tamil:** Masamangal, Mullukale, **Punjabi:** Sumlu, Simlu

Ayurvedic properties

General properties like, **Rasa:** Tikta, Kashaya, **Guna:** Lakhu, Rooksha, **Veerya:** Ushna.
Caraka has categorized as **stanyasodhana** – lactode purant, **lekhana** – a reducing herb, **arsoghna** – antihaemorrhoidal, **kandughna** – antihaemorrhoidal, **kandughna** – antipruritic and as **svedala** – promotes sweating, **rasayana**–rejuvenative. Susruta have mentioned as **ropana** – a wound healer

Ayurvedic preparations

Darvyadi kavatha; Darvyadi leha; Darvyadi taila; Rasanjana; Dasanga lepa.

INTRODUCTION

Berberis aristata is an evergreen small shrub growing to 3.5 m (11' 6") at a medium rate. It is hardy to zone 6 and is not frost tender. The leaves are strong with fine whorled venation, straight, but with dentate margins. Leaves are 4.9 cm long, 1.8 cm broad, deep green on the dorsal surface and light green on the ventral surface. Inflorescence 5 to 8 cm long, with large yellow colored flowers. Flowers are actinomorphic caducous, 4 to 5 mm long. Corolla is polypetalous, with 6 petals, yellow, actinomorphic, 4 to 5 mm long. Androecium is polyandrous, with 6 stamens. Adnate is 5 to 6 mm long. Gynoecium is one, 4 to 5 mm long, with a short style and a broad stigma. The fruits are bluish purple and small. The fruit is about 7 mm × 4 mm and it can be up to 10 mm long. It is in flower in May. The flowers are hermaphrodite (have both male and female organs) and are pollinated by insects, self. The plant is self-fertile. Seeds are 2 to 5 in number, varying in color from yellow to pink. Each seed is weighing 25 mg and being 29 microlitres in volume.

The plant prefers light (sandy), medium (loamy) and heavy (clay) soils and can grow in heavy clay and nutritionally poor soils. The plant prefers acid, neutral and basic (alkaline) soils. It can grow in semi-shade (light woodland) or no shade. It requires dry or moist soil.

Distribution

Berberis aristata is native to the Himalayas in India and in Nepal. The plant is distributed through whole range of Himalaya mountains at an elevation 2000 to 3500 meters. It also occurs in Nilagiri range in Southern India.

Berberine

Berbamine

Aromoline

Oxyacanthine

Chemical Constituents

The root bark of *Berberis aristata* contains alkaloids, such as berberine, berbamine, aromoline, karachine, palmatine, hydrastine, umballiatine, oxyacanthine and oxiberbarine. Fruit contains citric and malic acid. Apart from that fruit also contains protein: 2.3 g, carbohydrate: 12 g.

Medicinal uses

The fruit, root bark, stem, wood and extract are used for medicinal purpose like antiperiodic, deobstruent, diaphoretic, laxative, ophthalmic and tonic (bitter). Externally, the wounds are dressed with the medicated oil of the decoction of daruharidra. It reduces the pruritus also. The paste applied on painful and swollen parts effectively mitigates the symptoms. The popular preparation of daruharidra, called Rasanjana is a crude extract, prepared from the root bark. The decoction of root bark, mixed with equal quantity of milk is then beated until it forms a solid substance which is useful in unfections of the ear like otitis media. It is also useful as a wash for piles. An infusion is used in the treatment of malaria, eye complaints, skin diseases, menorrhagia, diarrhoea and jaundice. Berberine has antioxidant properties that allow it to act in anti-inflammatory, antimicrobial, antitumor and antidiabetic activity. Since it is not appreciably absorbed by the body, it is used orally in the treatment of various enteric infections, especially bacterial dysentery. It is also reported to be a mild laxative, a tonic and is useful in curing ulcers and fevers.

Cosmetic uses

Its ointment made with camphor and butter is applied to pimples and boils. Externally, the decoction of the root bark is used as a wash for unhealthy ulcers to improve their appearance and cicatrisation. Rasanjana mixed with honey is a useful application to aphthous sores, abrasions and ulcerations of the skin. A decoction is used as mouthwash for treating swollen gums and toothache. It is also used to treat infections, eczema,

parasites, psoriasis, and vaginitis. For eye problems, the root powder of the plant is boiled in water until a little amount of the water is left, filtered, mixed with butter and alum, or with opium and lime-juice, and or alone, which is then applied externally over the eyelids to cure conjunctivitis. Externally, the root powder is used for healing wounds, swellings and inflammations.

Pharmacological Action

Present study was investigated the anti-inflammatory effect of topical application of *Curcuma longa* (*C. longa*) and *Berberis aristata* (*B. aristata*) aqueous extracts on experimental uveitis in the rabbit and was evaluated by grading the clinical signs and histopathologic changes and estimating the inflammatory cell count, protein, and TNF-α levels in the aqueous humor. Finally, results concluded their anti-inflammatory activity (Gupta et al., 2008).

In a scientific study of the antidiabetic activity of the plant, diabetic rats treated with the ethanol extract of root of *Berberis aristata* showed a significant reduction of serum glucose level, however, it also showed a significant increase in the level of HDL cholesterol (Semall et al., 2009).

The present study was evaluated the antihyperglycemic and antioxidant potential of 50% aqueous ethanolic root extract of *Berberis aristata* in alloxan induced diabetic rats. *Berberis aristata* root extract (250 mg/kg) was administered to diabetic rats and standard drug glybenclamide (0.6 mg/kg) to group serving as positive control. Effect of extract on antioxidant and carbohydrate metabolism regulating enzymes of liver was studied in diabetic rats along with its safety parameters. Results concluded that the extract of *Berberis aristata* (root) has strong potential to regulate glucose homeostasis through decreased gluconeogenesis and oxidative stress (Singh and Kakkar, 2009).

The methanolic extract of the stems of *Berberis aristata*, was investigated against human colon cancer cell line (HT29) to explore its anticancer potential. The effect of *Berberis aristata* methanolic extract on proliferation of HT29 cancer cell line was determined by 3-[4,5-dimethylthiazol-2-yl]-2,5-diphenyl tetrazolium bromide (MTT) microculture tetrazolium viability assay. The cells were exposed to different concentrations (100, 50, 25, 12.5, 6.25, 3.125 and1.5 µg/ml) of *Berberis aristata* methanolic extract. It was observed that *Berberis aristata* methanolic extract induces a concentration dependent inhibition of HT29 cells, with an IC value of 1.8964 µg/ml after 72 hrs of incubation (Das et al., 2009).

The aqueous and alcoholic extract of fresh *Berberis aristata* DC roots, as well as aqueous extract of dried roots were compared for their antibacterial and antifungal activities by the disc diffusion method. All three extracts showed wide antibacterial activity against Gram-positive bacteria (Shahid et al., 2009).

A selected polyherbal formulation composed of 7 herbal extract mixtures such as:

Phyllanthus niruri, Eclipta alba, Cichorium intybus, Boerhaavia diffusa, Embelia ribes, Berberis aristata and *Picrorhiza kurroa* and the antioxidant activity was compared with ascorbic acid (ASC) and rutin as standard. The hepatoprotective activity in carbon tetrachloride induced hepatotoxicity was studied. The study was revealed the formulation exhibit significant activity (Arsul et al., 2010).

The antimicrobial and MIC of *B. aristata* leaves and root extracts were evaluated by agar well diffusion method against six ear pathogens. Acetone root extracts and aqueous leaves extracts showed highest MIC and finally concluded that organic extract of *B. aristata* showed broad-spectrum antimicrobial activity which help in the treatment of ear infections (Sharma et al., 2011).

Toxicological Studies

The oral LD in mice of berberine is 329 mg/kg. Oral doses of up to 100 mg/kg of berberin

sulphate have been well tolerated in animal studies without lasting effects. However, prolonged administration caused organ damage and death after 8–10 days.

The oral LD of hydrastis canadensis extract in mice is 1620 mg/kg.

The LD of berberine sulphate in mice by intraperitoneal route is 24.3 mg/kg. In high doses, berberine caused hemorrhagic nephritis and eventually death by respiratory failure.

Acute toxicity study of methanoilc extract of *Berberis aristata* DC stem was determined as per the OECD guideline No. 423 (acute toxic class method). It was observed that test extract was not lethal to the rats even at 2500 mg/kg dose. Hence, 1/10th (250 mg/kg) and 1/5th (500 mg/kg) of this dose were selected for further study (Upwar et al., 2011).

Clinical Investigation

Fifty-one patients with clinically active trachoma lesions (stages 1 and 11) were treated for 8 weeks with eyedrops of either 0.2% berberine chloride or the antitrachoma drug, sulfacetamide eyedrops gave the best clinical results but the infective agent (*Chlamydia trachomatis*) remained present in the conjunctiva and relapses of symptoms occurred. Berberine-treated patients showed only very mild ocular symptoms and were negative for infective agent. Also, no relapses occurred among these patients.

A double-blind placebo-controlled clinical trial was conducted on 96 children with trachoma stage 11a or 11b over a period of 3 months. Berberine eyedrops (0.2%) were compared with berberine plus neomycin, sulfacetamide and a placebo. In patients treated with berberine alone, 84% were clinically cured (p < 0.001) but only 50% were microbiologically cured. The response rate was higher in those treated with berberine and neomycin and lower in the sulfacetamide group. Berberine treatment was better tolerated than dulfacetamide.

The study was conducted to evaluate aqueous and alcoholic extract ointments of *Adhatoda vasica* and *Berberis aristata* in the treatment of open wounds in 30 Barbari adult male goats. The healing of wounds was evaluated on the basis of clinical observations, rate of healing and changes in histomorphological features. The study revealed that wound contraction, epithelialization and percent healing were excellent in wounds treated with alcoholic extract ointment of *A. vasica* when compared with either alcoholic extract ointment of *Berberis aristata* or aqueous extract ointments of *A. vasica* and *B. aristata*. It was evident from the results that the alcoholic extract ointment of *A. vasica* showed better healing effect (Rajput et al., 2004).

Market Formulations

Rakta Nikhar Capsule

Ingredients: *Azadirachta indica:* 60 mg, *Swertia chirata:* 120 mg, *Rubia munjista:* 30 mg, *Glycyrrhizae radix:* 60 mg, *Berberis aristata:* 20 mg, *Tinospora cordifolia:* 90 mg, *Embelia ribes:* 60 mg, *Picrorrhiza kurroa:* 60 mg.

Action: It serves to cure skin diseases, neutralizes toxins responsible for skin allergies, helps to cure acne, pimples, itching, eczema and improves circulation to all tissues. It normalizes melanin distribution pattern, thus improving the fairness. It also cures and prevents anemia and thus adds glow to the skin.

Dosage: One capsule twice a day with water.

Supplied by: Ayurved Vikas Sansthan, Uttarakhand.

Body Lotion

Ingredients: Aqueous extract of *Aloe barbadensis, Santalum album, Azadirachta indica, Berberis aristata, Glycyrrhiza glabra*, oil of olive, almond, wheatgerm and coconut.

Action: Used for moisturizing lotion.

Supplied by: D'pure pharmaceuticals, West Bengal.

Stretch Nil Lotion

Stretch nil is a herbal preparation which is to be used for minimization of stretch marks. It is a herbal lotion with natural ingredients that have stood the test of time. Each of these ingredients have been advocated for skin related disorders in ancient treatises on Ayurveda.

Ingredients: *Rubia cordifolia, Azadirachta indica, Ocimum sanctum, Berberis aristata, Glycyrrhiza glabra, Trichosanthes diocia, Sesamum indicum.*

Action: *Rubia cordifolia* improving the complexion and it helps stop the spread of skin diseases. *Azadirachta indica* help the localization of abscesses and boils, hence useful in averting secondary infection. *Ocimum sanctum* is used in treating chronic skin diseases and bleeding disorders. *Berberis aristata* helps in treating ulcers and itchy skin conditions. *Glycyrrhiza glabra* improves the body's resistance to infection and improves the complexion.

A combination of the above ingerdients provides anti-inflammatory, healing, skin tonning and complexion improving properties.

Method of Application Stretch Nil

Apply liberally over stomach, abdomen and especially on flanks, twice a day from the 4th month of pregnancy till delivery. Rub gently till it disappears on skin.

Duration of use

To be applied twice a day from the 4th month of pregnancy till delivery.

Supplied by: Gufic Biosciences Limited, Coimbatore.

Breast Nat

Ingredients: *Asparagus racemosus* 100 mg, *Withania somnifera* 100 mg, *Glycyrrhiza glabra* 60 mg, *Pueraria tuberosa* 60 mg, *Berberis aristata* 60 mg, *Cissampelos pareira* 60 mg, Lauh bhasm

60 mg, *Tinospora cordifolia* 60 mg, *Cuminum cyminum* 50 mg, *Trigonella foenum graecum* 50 mg.

Action: Breast nat contains phytoestrogens (naturally occurring non-hormonal plant estrogens). These phytoestrogens stimulate your body to support the growth of new breast tissue. It includes a unique formula of finest ingredients like *Withania somnifera* providing a rich amount of phytoestrogens that play the same role as natural hormones allowing further enlarging process and our body's breast tissues respond to them perfectly. *Asparagus racemosus* is renowned for its rejuvenating effect that maintains the temperature, strengthens and nourishes the tissues. *Tinospora cordifolia* is used to improve the immune system and the body's resistance power. It is also known for its use in dyspepsia, fever and urinary disease.

Direction: Take one capsule twice a day with milk, and especially for the first month, for most effective results, be very disciplined and do not miss any dose. You will notice your breasts becoming bigger and firmer, within the first week. Each bottle contains 60 capsules.

Supplied by: Fita Pharma, Reston.

Clarina Cream

Ingredients: Berberis aristata, Prunus amygdalus, Aloe vera and Yashad bhasma.

Action: Berberis aristata and is an important ingredient in various Ayurvedic preparations used to treat burns and wounds. *Prunus amygdalus* has emollient and demulcent properties which are useful in skin diseases. *Aloe vera* is useful in skin diseases, and is used for local application in painful inflammation and chronic ulcers. Yashad bhasma is a preparation of calcinated zinc containing chiefly zinc oxide. Zinc oxide is well known for its mild astringent and antiseptic properties and accelerates wound-healing.

Use: It is use to treat acne lesions effectively without any side effects and contraindications.

Storage of Products

Stored in a cool and dry well-closed container, keep away from moisture and strong light/ heat.

Dosage

Root bark juice: 12–24 gm. Decoction: 50–100 gm.

Rasanjana/Rasauta (wood extract): 0.5–1.0 gm. (in Vishama jwara 1.0–2.0 gm)

Fruits: 6–12 gm.

Patents

The present invention pertains to a herbal formulation with highly potent wound healing properties, in humans and animals. The composition consists of aqueous extracts of *Azadirachta indica*, in a mixture of natural oils along with herbs, viz. *Berberis aristata, Curcuma longa, Glycyrrhiza glabra, Jasminum officinale, Picrorhiza kurroa, Pongamia pinnata, Rubia folia, Saussurea lappa, Terminalia chebula, Trichosanthes dioica,* Capsicum and Stellata wild in well-defined ratios. The invention also includes a process for preparing the formulation by extracting the water-soluble components from bark of *Azadirachta indica* (Saxena, 2010).

The present invention relates to a synergistic antipyretic formulation comprising extracts of plants *Berberis aristata, Tinospora cordifolia, Alstonia scholaris, Andrographis paniculata* and *Hedychium spicatum*, optionally along with pharmaceutically acceptable additives; its use in treating fever, and lastly, a process for the preparation said synergistic antipyretic formulation (Pushpangadan et al., 2010).

The present invention pertains to a herbal formulation with highly potent wound healing properties, in humans and animals. The composition consists of aqueous extracts of *Azadirachta indica*, in a mixture of natural oils along with herbs, viz. *Berberis aristata, Curcuma longa, Glycyrrhiza glabra, Jasminum officinale, Picrorhiza kurroa, Pongamia pinnata,*

Rubia folia, Saussurea lappa, Terminalia chebula, Trichosanthes dioica, Capsicum and Stellata wild in well-defined ratios. The invention also includes a process for preparing the formulation by extracting the water-soluble components from bark of *Azadirachta indica* (Saxena, 2010).

BIBLIOGRAPHY

1. Parmar C, Kaushal MK. *Berberis aristata*. In: Wild Fruits. Kalyani Publishers, New Delhi, India. 1982; p.10–14. http://www. hort. purdue.edu/newcrop/parmar/03. html

2. http://www.herbalcureindia.com/herbs/ daruharidra.htm

3. http://www.pfaf.org/user/Plant.aspx? LatinName=Berberis%20aristata

4. Gupta SK, Agarwal R, Srivastava S, Agarwal P, Agarwal SS, Saxena R, Galpalli N. The Anti-inflammatory Effects of *Curcuma longa* and *Berberis aristata* in Endotoxin-Induced Uveitis in Rabbits. Invest. Ophthalmol. Vis. Sci. 2008; 49(9): 4036–4040.

5. Semall C, Bhupesh, Gupta J, Singh S, Kumar Y, Giri M. Antihyperglycemic activity of root of *Berberis aristata* D.C. in alloxan-induced diabetic rats. International Journal of Green Pharmacy 2009; 259–262.

6. Singh J, Kakkar P. Antihyperglycemic and antioxidant effect of *Berberis aristata* root extract and its role in regulating carbo-hydrate metabolism in diabetic rats. J of Ethanopharmacology 2009; 123 (1): 22–26.

7. Das S, Das MK, Mazumder PM, Das S, Basu SP. Cytotoxic Activity of Methanolic Extract of *Berberis aristata* DC on Colon Cancer. Global Journal of Pharmacology 2009; 3 (3): 137–140.

8. Shahid M, Rahim T, Shahzad A, Tajuddin, Latif A, Fatma T, Rashid M, Raza A, Mustafa S. Ethnobotanical studies on *Berberis aristata* DC. Root extracts. African Journal of Biotechnology 2009; 8 (4): 556–563.

9. Arsul V, Ganjiwale RO, Yeloe PG. Phyto-chemical and pharmacological standardisation of polyherbal tablets for hepatoprotective activity against carbon tetrachloride induced hepatotoxicity. International Journal of Pharmaceutical Sciences and Drug Research 2010; 2(4): 265–268.

10. Sharma C, Aneja KR, Kasera R. Screening of *Berberis aristata* DC. for antimicrobial potential against the pathogens causing ear infection. Int. J. Pharmacol 2011; 7(4): 536–541.

11. http://www.hillgreen.com/pdf/BERBE RISARISTATA.pdf

12. Upwar NK, Patel R, Waseem N, Mahobia NK. Hypoglycemic effect of methanolic extract of *Berberis aristata* DC stem on normal and streptozotocin induced diabetic rats. International Journal of Pharmacy and Pharmaceutical Sciences 2011; 3(1): 222–224.

13. Rajput N, NigampJM, Srivastava DN, Katiyar AK. Evaluation of *Adhatoda vasica* and *Berberis aristata* as wound healing agents in goats. Indian Journal of Veterinary Surgery 2004; 25(2): 89–93.

14. Saxena M. Herbal formulation for wound healing. United States Patent Application 20100178367. Application Number: 12/519137. Publication date: 07/15/2010. http://www. freepatentsonline.com/y2010/0178367. html

15. Pushpangadan P, Rawat AKS, Rao CV, Srivastava SK, Khatton S. Synergistic anti-pyretic formulation. US Patent 7658954, Application Number: 11/025757. Publication date: 02/09/2010. http://www.docstoc. com/docs/78013459/Synergistic-Antipyretic-Formulation—Patent-7658954

16. Saxena N. Herbal Formulation For Wound Healing. Patent application number: 20100178367. Publication date: 07/15/2010. http://www.faqs.org/patents/app/ 20100178367.

Bixa orellana

Plant profile

Kingdom: Plantae – Plants
Superdivision: Spermatophyta – Seed plants
Class: Magnoliopsida – Dicotyledons
Order: Violales
Genus: Bixa L. – Bixa

Subkingdom: Trachebionta – Vascular plants
Division: Magnoliophyta – Flowering plants
Subclass: Dilleniidae
Family: Bixaceae – Lipstick-tree family
Species: *Bixa orellana* L.– Lipstick-tree

Vernacular names

Sanskrit: Karachhada, Raktabija, Raktapushpa, Shonapushpi, Sindurapuspi, **Hindi:** Gowpurgee, Latkan, Latkhan, Senduria, **English:** Arnotta plant, Annatto, **Tamil:** Amudadaram, Amuttaram, Aruna, Avam, Camankalikam, Cappira, **Bengali:** Latkan, Belatihaldi, Utkana, **Kannada:** Aarnatu, Arnattu, Bangaara Kaayi, Bangarakayi, **Malayalam:** Korangumunga, Korungoomungal, Kuppamanjal, **Marathi:** Kaesari, Kesari, Keshri, Kesui, Kesuri, **Telugu:** Jaabara, Jaabura, Jaapharaa Chettu, Jabaru Kaya

Ayurvedic properties

Rasa: Kashaya, Tikta, **Guna:** Lakhu, Rooksha, **Veerya:** Seeta.

INTRODUCTION

It is a small evergreen tree, grows up to 2–8 m tall, generally with branches from near the ground. Leaves are 10–20 cm long, ovate, cordate and acuminate. Leaves are arranged spirally, simple stipulate. The petiole is 4.5–1 cm long, cylindrical and thickened at both ends. The flowers are borne in terminal 8–50 flowered panicles, fragrant and measure 4–6 cm across. Flowers are 5 cm in diameter, white or pink in terminal panicles. Capsule 3.7 cm long, ovoid or subglobose, clothed with long soft prickles. Fruit contains 15–20 trigonous seeds and are covered with red pulps. The bark is light to dark brown, tough and smooth. The inner bark is filled with orange sap.

Distribution

It is native to Central America and tropical South America. It is widely planted in the tropical region of the world including South-East Asia. The plant grows equally well in lowlands and mountainous regions or areas of higher elevation. It is found in largest quantities from Mexico to Ecuador, Brazil, and Bolivia. This plant is cultivated in warm regions of the world, such as India, Sri Lanka, and Java mainly for the dye which the seeds yield. Bixa has been found in and about the towns of the Philippines, Southeastern Africa and Dominica and is commonly planted in Florida as an ornamental. There are a variety of common names for *B. orellana* because it flourishes in a variety of places.

Chemical Constituents

Roots contain triterpenoid and tomentosic acid. Leaves yield essential oil (0.4%), which contains sesquiterpene, bixaghanene (ishwarane). Flavonoids, 7-bisulphates of apigenin, luteolin and hypolaetin have also been isolated from leaves. Analysis of annatto seeds indicates that they contain 40% to 45% cellulose, 3.5% to 5.5% sucrose, 0.3% to 0.9% essential oil, 3% fixed oil, 4.5% to 5.5% pigments and 13% to 16% protein, as well as alpha- and beta-carotenoids and other constituents. Annatto oil is extracted from the seeds and is the main source of pigments named bixin and norbixin, which are classified as carotenoids. In addition to bixin and norbixin, annatto contains bixein, bixol, crocetin, ellagic acid, isobixin, phenylalanine, salicylic acid, threonine, tomentosic acid, and tryptophan.

infusion of root is used for diabetes, flu, venereal diseases. Gum made from crushed leaves mixed with water (ingest) is used as diuretic, purgative, treats gonorrhea. Root decoction is used for diuretic, treatment of jaundice, diabetes, influenza and venereal diseases. Oil derived from seeds is uded for leprosy treatment.

Cosmetic uses

The cosmetic industry increasingly uses it as a natural red color additive due to the swelling number of restrictions being made against unnatural, synthetic pigment ingredients. It is particularly popular with body lotion/cream and shampoo/conditioner because it can give the skin and hair a natural "sunny" glow. Natural colour isolated from the seeds of the annatto tree (*Bixa orellana*). Annatto is

Bixin

cis-norbixin

Medicinal uses

Decoction of leaves and red coloring matter is used for ailments of the womb or uterus and acts as a female aphrodisiac. Dye is used for antidote to prussic acid poisoning. Decoction of leaves is used for asthma, pleurisy, labored breathing and anti-inflammatory. The steroid presence in the leaves, suggests they are responsible for its anti-rheumatic and inflammatory reliever properties that popular medicine attributes in the treatment of rheumatism and prostate inflammation. Leaves are also used for treatment of snake bites, sore throat and liver trouble. Pulp surrounding seeds are used as astringent. Seed pulp dissolved in warm water are used to bring out measles quickly, relieves stomachaches and asthma. Seeds ground and boiled in oil are used for burn treatments. Root bark is used as antipyretic and antiperiodic. An aqueous and rum

the name of the crude extract, whereas bixin is the fat-soluble color and norbixin the water-soluble color. Mixture of red pulp and oil is used for healing of minor wounds and burns, prevents scaring and blisters by applied on the skin. Red, resinous substances of seeds are used for certain skin diseases. Seeds ground and boiled in oil are used for burn treatment. In addition, because annatto oil is an emollient rich in carotenoids, it may also function as a powerful antioxidant. For the most part, this ingredient is used in products such as facial moisturizer, permanent waves, bar soap, hair dye, lipstick, shampoo/conditioner, powder and tanning products.

Pharmacological Action

The ethanolic extracts of the leaves and seeds of *Bixa orellana* showed a broad spectrum of antimicrobial activity. The activity of the leaf extract was more pronounced (Fleischer et al., 2003).

Preliminary pharmacological studies were performed on the methanol extract of *Bixa orellana* L. (Bixaceae) leaves to investigate neuropharmacological, anticonvulsant, analgesic, antidiarrheal activity and effect on gastrointestinal motility. All studies were conducted in mice using doses of 125, 250 and 500 mg/kg of body weight. The extract significantly and dose-dependently reduced the writhing reflex in the acetic acid-induced writhing test. Antidiarrheal activity was supported by a statistically significant decrease in the total number of stools (including wet stools) in castor oil-induced diarrhea model. A statistically significant delay in the passage of charcoal meal was observed at 500 mg/kg in the gastrointestinal motility test. The extract was further evaluated *in vitro* for antioxidant and antibacterial activity. It revealed radical scavenging properties in the DPPH assay (IC_{50} = 22.36 µg/ml) and antibacterial activity against selected causative agents of diarrhea and dysentery, including *Shigella dysenteriae* (Shilpi et al., 2006).

This present investigation sought to determine the effects of the glucose-lowering extract on C-peptide and streptozotocin-induced diabetic dogs. This annatto extract was found to decrease blood glucose levels in fasting normoglycaemic and streptozotocin-induced diabetic dogs. It was concluded that *Bixa orellana* (annatto) lowered blood glucose by stimulating peripheral utilization of glucose, and it is possible that this glucose-lowering extract might be of pharmacological importance (Russell et al., 2008).

The relationship between the aqueous annatto extract and its influence on lipid profile in animals was reported. Male fisher rats were divided into three groups (n = 12): C group, fed standard diet and water; H group, fed high-lipid diet and water and; HU group, with high-lipid diet and aqueous annatto extract for 60 days. The treatment with annatto extract in animals fed with the high-lipid diet lowered the LDL and total cholesterol and raised the HDL-cholesterol, suggesting a hypocholesterolemic effect. Neither high fat diet nor aqueous annatto extract had any significant effect on serum levels of albumin or serum activities of transaminases which suggested that no liver injury was induced (Paula et al., 2009).

Diuretic activity of the *Bixa orellana* leaves was evaluated. Among all the different extracts methanolic extract showed significant diuretic activity at dose of 500 mg/kg body weight by increased the volume of urine and level of sodium, potassium and chloride in urine as compared to standard frusemide (Radhika et al., 2010).

Toxicological Studies

Germano et al. (1997) showed good cutaneous tolerability (as shown by histological examination of skin and hair) of extracts of *Bixa orellana*, after single or repeated application to the skin of rabbits, although these studies are not relevant for oral food use. The acute toxicity of annatto when administered orally is low.

In a study, pure norbixin or annatto extract containing 50% norbixin was given for 21 days at a dose of 0.8, 7.6, 66 or 274 mg/kg bw. Norbixin induced an increase in plasma alanine aminotransferase (ALT) while both norbixin and annatto induced a decrease in plasma total protein and globulins ($p < 0.05$). No signs of toxicity were detected by histopathological analysis of the liver, and no enhancement in DNA strand breakage was detected in liver or kidney from mice treated with annatto pigments, as evaluated by the comet assay (Fernandes et al., 2002). In another study they have showed absence of toxicity when rats were given drinking water containing either annatto extract consisting of 50% norbixin (at a dose of 0.8, 7.5 or 68 mg/kg bw per day) or pure norbixin (at a dose of 0.8, 8.5 or 74 mg/kg bw per day) was for 21 days. No toxic effects on plasma clinical chemistry parameters were detected. A significant hyperglycaemic effect was reported.

In the study we evaluated the developmental toxicity of annatto (28% of bixin). Annatto (0, 31.2, 62.5, 125, 250 and 500 mg/kg body weight/day) was given by gavage to Wistar rats on days 6–15 of pregnancy. No increase in embryolethality and no reduction of fetal body weight were observed among annatto-exposed rats. Annatto did not induce any increase in the incidence of externally-visible, visceral or skeletal anomalies in the exposed offspring. These findings suggest that annatto was neither maternally toxic nor embryotoxic in the rat. Therefore, the no-observed-adverse-effect-level (NOAEL) for annatto-induced maternal and developmental toxicity was 500 mg/kg body weight/day or greater (or > or = 140 mg bixin/kg body weight/day) by the oral route (Paumgartten et al., 2002).

A subchronic oral toxicity study of annatto extract (norbixin), was conducted. Groups of 10 male and 10 female Sprague-Dawley rats were fed annatto extract at dietary levels of 0, 0.1, 0.3 and 0.9% for 13 weeks. There were no treatment-related adverse effects on body weight, food and water consumption, ophthalmology and hematology data. Marked elevation in absolute and relative liver weights was also found in both sexes of the 0.9% and 0.3% groups, but not the 0.1% group. Hepatocyte hypertrophy was evident and an additional electron microscopic examination demonstrated this to be linked to abundant mitochondria after exposure to a dietary level of 0.9% annatto extract for 2 weeks. Thus, the no-observed-adverse-effect-level (NOAEL) was judged to be a dietary level of 0.1% (69 mg/kg body weight/day for males, 76 mg/kg body weight/day for females) of annatto extract (norbixin) under the present experimental conditions (Hagiwara et al., 2003).

The toxic effects of annatto powder (bixin 27%) have been assessed following administration of a subacute regimen (4 weeks, 20 doses) in Wistar male and female rats. A full study with three dose levels was considered unnecessary since no sign of toxicity had been noted in a preliminary experiment with 1000 mg/kg body weight/day. In the study, annatto administered by gavage at a dose level of 2000 mg/kg/day decreased male body weight gain, but had no effect on either food intake or food conversion efficiency. Hematological and plasma biochemical examination as well necropsy performed at the end of administration (29th day) and observation (43rd day) periods revealed no alterations related with annatto administration. These findings suggest that annatto was no toxic to the rat (Bautista et al., 2004).

Clinical Investigation

The FDA lists *annatto* as a color additive exempt from certification, and they, along with the EU cosmetics directive, deem it as safe to use in food and OTC cosmetic products but no epidemiological studies or case reports investigating the association of exposure to annatto, bixin, or norbixin and cancer risk in humans were identified in the available literature.

Annatto can elicit an allergic response in some individuals. Fifty-six patients with urticaria or angioedema were orally challenged with annatto extract, and 26% had a positive test (Mikkelson et al., 1978). In a study of 112 patients with recurrent urticaria, 10% orally challenged with annatto dye, had a positive response and 14% had an uncertain challenge (Juhlin, 1981). Neither of the above studies was double blinded or placebo controlled (Hallagan et al., 1995). A patient developed urticaria, angioedema, and severe hypotension within 20 minutes following ingestion of milk and Fiber One™ cereal. Skin tests to milk, wheat, and corn were negative, but the patient had a strong positive skin test to annatto dye. The nondialyzable fraction of annatto dye displayed two protein staining bands in the 50 kD range. Immunoblotting demonstrated patient IgE specific for one of these bands (Nish et al., 1991).

Market formulations: There are many formulations composed of annatto dyes as natural color agent. It is mainly used as in preparation of lipsticks, bath soap, and some skin preparations.

Lip Liquid

Ingredients: Blend of annatto seed, sunflower oil and castor oil. Annatto is approved by the FDA for general cosmetic use, including the eye area.

> *Use in soap:* Stable and bleeding.

> *Direction:* Wash hands after handling.

> *Supplied by:* TKB Trading, LLC.

Annato Soap

Ingredients: Annatto, coconut oil, palm oil, castor oil, sesame oil, Neem oil, natural essentail oil and natural herbal powders.

> *Action:* It has healing and soothing properties and excellent for sun damaged or matures skin. It has antibacterial, anti-inflammatory, antioxidant, antiseptic properties.

> *Supplied by:* Auronature, Puducherry.

Method of Preparation

Annatto seeds are washed and dried separately from the pulp of the seed pod for culinary use. They may be added directly to a cooking liquid or infused in hot water until the desired color is obtained and then used for stocks or coloring rice. It is also common to fry the seeds in oil for a few minutes (best done in a covered pan as the hot seeds jump), then discard the seeds and use the oil. Try using one teaspoon of seeds to 4 tablespoons of oil.

Storage of Products

Annatto seeds should be kept out of light in an airtight container.

Dosage

Standard decoctions of annatto leaves are taken by the half-cupful two or three times daily for prostate and urinary difficulties as well as for high cholesterol and hypertension. Ground annatto seed powder is also used in small dosages of 10–20 mg daily for high cholesterol and hypertension.

- A typical dose of annatto is one to two grams of powdered leaf in tablets or capsules taken by mouth twice daily.
- Tea made from Bixa leaves or seeds have been taken by mouth by the half-cupful two to three times daily for prostate and urinary conditions, high cholesterol, and high blood pressure.
- Two to four milliliters of a 4:1 tincture (alcohol extract), of Bixa twice daily has been used, based on anecdotal evidence.
- For urinary disorders (benign prostatic hyperplasia), capsules containing 250 milligrams of dried Bixa leaf have been taken by mouth three times daily for 12 months.

Patents

A method of eliciting a therapeutic health effect in a human or non-human animal by adding an effective amount of at least one bixin compound to a pharmaceutical compound is also provided. The beneficial or therapeutic health effect may be further enhanced by the addition of at least one carotenoid compound or carotenoid-containing extract to the feed, food product, nutritional supplement, or pharmaceutical composition. These methods are particularly suited to companion animals, such as cats and dogs, and to farmed animals, such as layer hens, broiler chickens, and pigs (Levy and Levy, 2003).

The present invention deals with a composition for evidencing bacterial plaque based on natural colorants comprising at least one concentrated solution of natural colorant selected from the group consisting of colorants extracted from *Euterpe oleracea* and colorants extracted from urucum (*Bixa orellana*), in a pharmaceutically acceptable vehicle and, optionally, acceptable pharmaceutical additives. Preferentially, in the case of *Euterpe oleracea* extracts, the concentration varies from 60 to 95% of concentrated extract, whilst for the urucum solution, the concentration varies between 56 and 90% of concentrated extract (Ribeiro et al., 2007).

BIBLIOGRAPHY

1. http://www.rain-tree.com/annato.htm

2. http://www.mpbd.info/plants/bixaorellana.php

3. http://www.globinmed.com/index.php?option=com_content&view=article&id=79146:bixa-orellana-l&catid=366:b

4. Honychurch, P. N. Caribbean Wild Plants and Their Uses in Canada. 1980; 16.

5. Wolf M A. 1997. http://home. braunschweig.netsurf.de/~andree.wolf /urucum.html

6. Williamson J. Useful Plants of Nyasaland. Zomba, Nyasaland: 1955; 23.

7. http://www.virtualherbarium.org/gl/bixa/bixaorellana.htm

8. Fleischer TC, Ameade EPK, Mensah MLK, Sawer IK. Antimicrobial activity of the leaves and seeds of *Bixa orellana*. Fitoterapia 2003; 74(1–2): 136–138.

9. Shilpi JA, Rahman TU, Uddin SJ, Alam, Sadhu SK, Seidel V. Preliminary pharmacological screening of *Bixa orellana* L. leaves. J of Ethanopharmacology. 2006; 108 (2): 264–271.

10. Russell KR, Omoruyi FO, Pascoe KO, Morrison EY. Hypoglycaemic activity of *Bixa orellana* extract in the dog. Methods Find Exp Clin Pharmacol. 2008; 30(4): 301–305.

11. Paula HD, Pedrosa ML, Júnior JVR, Haraguchi FK, dos Santos RC, Silva ME. Effect of an Aqueous Extract of Annatto (*Bixa orellana*) Seeds on Lipid Profile and Biochemical Markers of Renal and Hepatic Function in Hipercholesterolemic Rats. Brazilian archives of biology and technology. 2009; 52(6): 1373–1378.

12. Radhika B, Begum N, Srisailam K, Reddy VM. Diuretic activity of *Bixa orellana* Linn. Leaf extracts. Indian J of Natural products and resources. 2010; 1(3): 353–355.

13. Germano MP, DePasquale R, Rapisarda A, Monteleone D, Keita A, Sanogo R. Drugs used in Africa as dyes: 1. Skin absorption and tolerability of *Bixa orellana* L. *Phytomedicine*, 1997; 4: 129–131.

14. Fernandes ACS, Almeida CA, Albana F, Laranja GAT, Felzenszwalb I, Lage CLS, de Sa CCNF, Moura AS, Kovary K. Norbixin ingestion did not induce any detectable DNA breakage in liver and kidney but caused a considerable impairment in plasma glucose levels of rats and mice. *J. Nutr. Biochem.* 2002; 13: 411–420.

15. Paumgartten FJ, De-Carvalho RR, Araujo IB, Pinto FM, Borges OO, Souza CA, Kuriyama SN. Evaluation of the developmental toxicity of annatto in the rat. Food Chem Toxicol. 2002; 40(11): 1595–601.

16. Hagiwara A, Imai N, Ichihara T, Sano M, Tamano S, Aoki H, Yasuhara K, Koda T, Nakamura M, Shirai T. A thirteen-week oral toxicity study of annatto extract (norbixin), a natural food color extracted from the seed coat of annatto (*Bixa orellana* L.), in SpragueDawley rats. Food and Chemical Toxicology. 2003 41(8): 1157–1164.

17. Bautista AR, Moreira EL, Batista MS, Miranda MS, Gomes IC. Subacute toxicity assessment of annatto in rat. Food Chem Toxicol. 2004; 42(4): 625–629.

18. Mikkelsen H, Larsen JC, Tarding F. Hypersensitivity reactions to food colors with special reference to the natural color annatto extract. Arch. Toxicol. Suppl. 1978; 1: 141–143.

19. Juhlin L. 1981. Recurrent urticaria: clinical investigation of 330 patients. Br. J. Dermatol. 1981; 104: 369–381.

20. Hallagan J B, Allen D C, Borzelleca JF. The safety and regulatory status of food, drug and cosmetics colour additives exempt from certification. Food. Chem. Toxicol. 1995; 33: 515–528.

21. Nish W, Whisman B, Goetz D, Ramirez D. Anaphylaxis to annatto dye: a case report. Ann. Allergy 1991; 66: 129–131.

22. Levy PE, Levy LW. Supplements containing annatto extracts and carotenoids and methods for using the same. United States Patent Application 20030104090. Application number: 10/309732, Publiucation date: 06/05/2003. http://www.freepatentsonline.com/y2003/0104090.html

23. Ribeiro de NRF, Emmi DT, Barroso RFF, Oliveira Da RP. Bacterial plaque evidencing composition based on natural colorants. United States Patent Application 7182935. Application number: 10/173844, Publiucation date: 02/27/2007.

Calendula officinalis

Plant profile

Kingdom: Plantae – Plants

Subkingdom: Trachebionta – Vascular plants

Superdivision: Spermatophyta – Seed plants

Division: Magnoliophyta – Flowering plants

Class: Magnoliopsida– Dicotyledons

Subclass: Arecidae

Order: Asterales

Family: Asteraceae

Genus: Calendula L. – Marigold

Species: *Calendula officinalis* L. – Pot marigold.

Vernacular names

Sanskrit: Zendu, **Hindi:** Genda, **English:** Marigold, **Punjabi:** Gulsarfi, **Bengali:** Genda ful, **Gujarati:** Galgotta, Dipmal, **Marathi:** Zendu.

INTRODUCTION

Calendula officinalis, or pot marigold, is an herbaceous annual or perennial plant related to daisies and asters. This marigold has dark orange daisy-like flowers with reddish-brown tinted petals. It is a short-lived aromatic plant, growing to 80 cm tall, with sparsely branched lax or erect stems. The leaves are oblong-lanceolate, 5–17 cm long, hairy on both sides, and with margins entire or occasionally waved or weakly toothed. The inflorescences are yellow, comprising a thick capitulum or flowerhead 4–7 cm diameter surrounded by two rows of hairy bracts; in the wild plant they have a single ring of ray florets surrounding the central disc florets. The disc florets are tubular and hermaphrodite, and generally of a more intense orange-yellow color than the female, tridentate, peripheral ray florets. The flowers may appear all year long where conditions are suitable. The fruit is a thorny curved achene.

Distribution

Calendula is a native of Southern Europe, but has become naturalized throughout temperate regions of the world. These days, it is cultivated in many temperate regions of the world for use in many processes and is naturalized in temperate North America and Asia. Ideal soil profiles for the growth of the Calendula are light to sandy and moderately rich soils. The soil must be fairly moist with a good drainage without water logging. The Calendula tolerates a pH range from an acidic 4.5 to a very alkaline 8.3.

Chemical Constituents

The flowers of *Calendula officinalis* contain flavonol glycosides, triterpene oligoglycosides, oleanane-type triterpene glycosides and a sesquiterpene glucoside. Essential oil contains carotenoids (carotene, calenduline and lycopine), a saponin, resin and bitter principle.

Flavonoids: Pharmacopoeial standard not less than 0.4% flavonoids. Flavonol (isorhamnetin, quercetin) glycosides including isoquercitrin, narcissin, neohesperidoside, and rutin.

Polysaccharides: Three polysaccharides PS-I, -II and –III have a (1→3)-β-D-galactan backbone with short side chains at C-6, comprising α-araban-(1→3)-araban, α-L-rhamnan-(1→3)-araban or simple α-L-rhamnan moieties.

Terpenoids: Many components, including α- and β-amyrin, lupeol, longispinogenin, oleanolic acid, arnidiol, brein, Calenduladiol, erythrodiol, faradiol, faradiol-3-myristic acid ester, faradiol-3-palmitic acid ester, helantriols A1, B0, B1 and B2, lupeol, maniladiol, urs-12-en-3, 16, 21-triol, ursadiol; oleanolic acid saponins including calendulosides C-H; campesterol, cholesterol, sitosterol, stigmasterol and taraxasterol (sterols).

Volatile oils: Terpenoid components include menthone, isomenthone, caryophyllene and an epoxide and ketone derivative, pedunculatine, α- and β-ionone, a β-ionone epoxide derivative, dihydroactinidiolide.

Other constituents: Bitter (loliolide), arvoside A (sesquiterpene glycoside), carotenoid pigments and calendulin (gum).

Medicinal uses

Marigold has always been used from time immemorial in the treatment of headaches, toothache, and swelling and also for strengthening of the heart. Its action is stimulant and diaphoretic. Given internally, it assists local action and prevents suppuration. The infusion of 1 ounce to a pint of boiling water is given internally, in doses of a tablespoonful, and externally as a local application. It is useful in chronic ulcer, varicose veins, etc. It has been asserted that a marigold flower, rubbed on the affected part, is an admirable remedy for the pain and swelling caused by the sting of a wasp or bee. A lotion made from the flowers is most useful

Beta-carotene

Lycopine

for sprains and wounds, and water distilled from them is good for inflamed and sore eyes. An infusion of the freshly-gathered flowers is employed in fevers, as it gently promotes perspiration and throws out any eruption—a decoction of the flowers is much in use in country districts to bring out smallpox and measles. The leaves when chewed at first communicate a viscid sweetness, followed by a strong penetrating taste, of a saline nature. The expressed juice, which contains the greater part of this pungent matter, has been given in cases of costiveness and proved very efficacious. Snuffed up the nose it excites sneezing and a discharge of mucous from the head. The leaves, eaten as a salad, have been considered useful in the scrofula of children, and the acrid qualities of the plant have caused it to be recommended as an extirpator of warts.

Cosmetic uses

Treating skin conditions is the most frequent use of *Calendula officinalis*. The herb has been found to be effective in treating inflammation of the skin. This includes conditions such as varicose veins, hemorrhoids, ulcerations, cysts and mastitis. It is very effective in treating eczema, a skin condition where the skin is irritated by constant itching. The infused oil or cream of marigold flowers is used extensively in aromatherapy to treat eczema, scars, cracked skin, rashes, inflammation and viral infections. Acne is also helped by the use of Calendula lotions. The flowers help relieve the inflammation due to menstruation. This is because Calendula is believed to have antispasmodic properties. It is used in the healing of wounds and burns. Rubbing some Calendula on a burn can take some of the pain away. This even works on sunburns. There have been several animal studies done to show that Calendula helps increase the speed with which wounds heal. It is believed that the herb works by increasing the amount of blood to the wound. Also, because Calendula has anti-inflammatory properties, it helps to

sooth wounds and reduces swelling. Women who are breastfeeding find that rubbing Calendula on sore nipples helps relieve the pain and irritation. Because of the anti-inflammatory properties of Calendula, it is often used to treat sore throats and mouth sores. It makes a great antioxidant, which may help to increase the immune function of the body. It is said that Calendula mixed into a lotion will help control acne, diaper rash, and razor burns. The plant has been used to treat conjunctivitis and other eye problems. It helps to reduce any swelling in the eye. The petals of this summer flower are used to make an eye wash.

Pharmacological Action

The aim of this study was to investigate the relationship between the beneficial properties of this plant and its antioxidant action. The butanolic fraction (BF) was studied because it is non-cytotoxic and is rich in a variety of bioactive metabolites including flavonoids and terpenoids. Superoxide radicals ($O_2^{\bullet-}$) and hydroxyl radicals ($HO^{\bullet}$) are observed in decreasing concentrations in the presence of increasing concentrations of BF with IC(50) values of 1.0 +/− 0.09 mg/ml and 0.5 +/−0.02 mg/ml, respectively, suggesting a possible free radical scavenging effect. The results obtained suggest that the butanolic fraction of *C. officinalis* possesses a significant free radical scavenging and antioxidant activity and that the proposed therapeutic efficacy of this plant could be due, in part, to these properties (Cordova et al., 2002).

Calendula officinalis flower extract possessed significant anti-inflammatory activity against carrageenan and dextran-induced acute paw edema. Oral administration of 250 and 500 mg/kg body weight Calendula extract produced significant inhibition (50.6 and 65.9% respectively) in paw edema of animals induced by carrageenan and 41.9 and 42.4% respectively with inflammation produced by dextran. In chronic anti-inflammatory model using formalin, administration of 250 and 500 mg/kg body weight Calendula extract produced an inhibition of 32.9 and 62.3% respectively compared to controls. The results showed that potent anti-inflammatory response of *C. officinalis* extract may be mediated by the inhibition of proinflammatory cytokines and Cox-2 and subsequent prostaglandin synthesis (Preethi et al., 2009).

The present work was planned to evaluate the potential hepatoprotective effects of *Morus alba* and *Calendula officinalis* extracts against cytotoxicity and oxidative stress induced by carbon tetrachloride (CCl_4) in isolated primary rat hepatocytes. Preincubation of hepatocytes with either *Morus alba* and *Calendula officinalis* extracts ameliorated the hepatotoxicity and oxidative stress induced by CCl_4, as indicated by significant improvement in cell viability and enzymes leakages (ALT, AST and LDH). Also, significant improvement of GSH content and significant decrease in TBARS formation as compared to CCl_4 treated cells. The present study was indicated the *Morus alba* and *Calendula officinalis* extracts possess a highly promising hepatoprotective effects against CCl_4- induced hepatotoxicity (Hussein et al., 2010).

Seven microorganisms were used to study the antimicrobial effects of *Calendula officinalis* extracts. Extracts of *Calendula officinalis* leaf, stem, root and flower made in n-butanol, ethanol and distilled water were tested for their antimicrobial activity. The extracts made in ethanol, water and n-butanol were found to be effective against most of the human pathogenic microorganisms. Leaf, stem and flower part of *Calendula officinalis* showed very effective results against pathogenic microorganisms (Goyel and Mathur, 2011).

The aim of the study was the formulation of suitable medicinal form for the topical application of *Calendula officinalis* L. This extract was proved to be an effective scavenger of H_2O_2 radicals in *in vitro* studies with the mitochondria of rat cardiac muscles. Subsequently, Calendula extract was incorporated, and the concentration of extract which provided significant antioxidant effect ($p < 0.05$), has been determined. Antioxidant activity of the cream with Calendula extract was due to the content of carotenoids, polyphenols and flavonoids. Cream with the best properties (0.9% of Calendula extract) contained 0.73 ± 0.04 mg/100 g of total carotenoids expressed as β-carotene. Achieved results suggest that developed cream with Calendula extract poses the good quality emulsion system warranting the stability of carotenoids, and thus the therapeutic, namely antioxidant activity of preparation (Bernatoniene et al., 2011).

Toxicological Studies

The hydroalcohol extract (HAE) of *Calendula officinalis* L. was evaluated for its acute toxicity by the oral route in rats and mice and for the subacute effect on hematological, biochemical and morphologic parameters in rats. In the acute toxicity test, HAE failed to cause death in the animals after administration of oral doses up to 5.0 g/kg. Oral treatment with HAE at 0.025, 0.25, 0.5 and 1.0 g/kg did not induce hematological alterations when compared with the control group (Silva et al., 2007).

The purpose of the paper was to determine the safe use of Calendula flower (*Calendula officinalis*) (CF) and *C. officinalis* extracts (CFE) based upon the threshold of toxicological concern (TTC) class for each of its known constituents. For each constituent, the concentration in the plant, the molecular weight, and the estimated skin penetration potential were used to calculate a maximal daily systemic exposure which was then compared to its corresponding TTC class value. The paper has demonstrates the utility and practical application of the TTC concept when used as a tool in the safety evaluation of botanical extracts (Re et al., 2009).

The acute and subchronic oral toxicities of *Calendula officinalis* extract in male and female Wistar rats were studies. A single acute *C. officinalis* extract dose of 2000 mg/kg dissolved in distilled water was administered by oral gavage for acute toxicity. Subchronic doses of 50, 250 and 1000 mg/kg/day were administered in drinking water. The major toxicological endpoints examined included animal body weight, water and food intake, selected tissue weights, and histopathological examinations. In the acute study, there were no mortality and signs of toxicity. In the subchronic study, several of the blood elements were significantly affected in males and females after 90 days; hemoglobin, erythrocytes, leukocytes and blood clotting time. For blood chemistry parameters, ALT, AST and alkaline phosphatase were affected. Histopathological examination of tissues showed slight abnormalities in hepatic parenchyma that were consistent with biochemical variations observed. These studies indicate that the acute and subchronic toxicities of *C. officinalis* extract are low (Lagarto et al., 2011).

Clinical Investigation

The goal of the study was to compare the effectiveness of *Calendula officinalis* with that of trolamine with 254 patients who had been operated on for breast cancer and who were to receive postoperative radiation therapy were randomly allocated to application of either trolamine (128 patients) or Calendula (126 patients) on the irradiated fields after each session. The primary end point was the occurrence of acute dermatitis of grade 2 or higher. Prognostic factors, including treatment modalities and patient characteristics, were also investigated. Secondary endpoints were the occurrence of pain, the quantity of topical agent used, and patient satisfaction. The results concluded that Calendula was highly effective for the prevention of acute dermatitis of grade 2 or higher and should be proposed for patients undergoing postoperative irradiation for breast cancer (Pommier et al., 2004).

The study was conducted to determine the effect of *Calendula officinalis* cream in healing pressure sores. The comparative clinical trial done on 20 patients with pressure sores who received the recommended treatment. The condition of patients such as, duration of pressure sore, and the extent of the sore were recorded. The sore was washed with normal saline, dried, followed by applying Calendula cream three times a day for a duration of four weeks. Each week, the extent and rate of healing was observed and recorded. Results revealed the cream can be used in the treatment of pressure sores (Esmaili et al., 2008).

A liquid extract of marigold is prepared through percolation and was used for the preparation of gel, based on methylcellulose. The transparent gel was applied on 24 men and women with acne. A standard questionnaire of 10 questions was applied. Results revealed the relative share of women is twice as higher as that of men. After processing and analysis it was found that, the young people with acne had sought medical advice with a dermatologist (54%), GP (29%) and with medical cosmetics expert (17%). Results concluded that more than half of the young people have successfully solved this problem, there is no effect in only two of them, and in 1/3 there is a partial improvement, and longer therapy is required (Ibrahim et al., 2010).

The aim of this study was to determine the effects of newly formulated w/o emulsion (cream) of Calendula versus its vehicle (base) as control on skin pH, skin melanin, skin erythema, skin moisture content and transepidermal water loss (TEWL). Hydroalcoholic extract of Calendula plant was entrapped in the inner aqueous phase of the w/o emulsion. Base without active and formulation having 3% extract of Calendula in the aqueous phase were prepared and were applied to the cheeks of 21 healthy human volunteers for a period of eight weeks. Both creams were esthetic with respect to sensory evaluation. The topical non-invasive application of *Calendula officinalis* cream showed a positive rejuvenating effect on human skin. This study will encourage more attention towards research and more conviction towards utilization of herbal medicines (Akhtar et al., 2011).

The aim of the study was the formulation of suitable medicinal form for the topical application of *Calendula officinalis* L. Dry Calendula extract was used as an active compound. This extract was proved to be an effective scavenger of H_2O_2 radicals in *in vitro* studies with the mitochondria of rat cardiac muscles. Several compositions of the cream base were evaluated and the hydrophilic cream containing complex emulsifier was chosen as the delivery system. Achieved results suggest that developed cream with Calendula extract poses the goodquality emulsion system warranting the stability of carotenoids, and thus the therapeutic, namely antioxidant activity of preparation (Bernatoniene et al., 2011).

Market Formulations

The most common use of Calendula is topical. It can be made into lotions, compresses, or poultices. It works really well to heal ugly skin conditions, such as acne and broken capillaries. Taking sap from the stem of the plant helps with warts and calluses. The dried flowers can be rubbed on wounds to help speed healing.

Chamomile and Calendula Calming Cream

Ingredients: 50 ml Calendula and chamomile infusion (steep a tablespoon of each herb in a cup of freshly boiled water, strain and measure out required amount). 25 ml *Aloe vera* gel (also calming and healing), 1/2 teaspoon vegetable glycerine, 10 g beeswax, 20 g coconut oil (considered cooling and calming in Ayurvedic medicine), 25 ml Calendula infused oil, 25 ml chamomile infused oil, 2 ml vitamin E, 5 drops vitamin A, 10 drops lavender essential oil, 4 drops roman chamomile essential oil.

Action: It is use on areas of dry or irritated skin, sunburn, insect bites, allergies, scars or just as a lovely moisturiser that is great for sensitive skins. It also makes a perfect cream for mother and baby and can be used to help a range of problems such as nappy rash, cradle cap and sore nipples, etc.

Vitamins A and E also essential oils can act as preservatives.

Method of preparation: Melt the wax and coconut oil in a bain marie or double boiler on a low heat, adding the Calendula and chamomile infused oils when liquid and stirring a little if the waxes start to solidify. In a separate container mix the herbal infusion with the *Aloe vera* and glycerine. Take the oils off the heat and allow to cool slightly before adding the vitamins A and E. Whilst this is cooling you can warm the water mixture by placing the container in the pan of hot water over which the oils were melting. It is important to get the oils and waters to roughly the same temperature to enable them to mix properly, otherwise your cream will separate. Begin to blend/whisk the water mixture and slowly add in the oils, a drizzle at a time. Continue to blend until you have a nice smooth, even, creamy consistency. Spoon into a jar or jars and stir in the essential oils. Pop in the fridge for a short while to cool.

Storage of products: It has a shelf life of about 3 months out of the fridge or 6 month in.

Calendula Herbal Oil

Ingredients: Calendula flowers *(Calendula officinalis)*, infused in certified organic olive oil and a pinch of vitamin E oil.

Action: Calendula oil is the most successful oil for assisting us with dry and damaged skin, skin inflammations, rashes, diaper irritations, and other skin disorders.

Restorative Skin Oil

Ingredients: Organic Calendula oil, organic Rosehip seed oil, vitamin E, and a blend of pure essential oils including neroli and helichrysum.

Action: This skin oil, this is currently the much revered formula for dry and mature skin.

Natrabio the Calendula Rub

Ingredients: Calendula officinalis (garden marigold) 1X 5%, *Arnica montana* (leopard's bane) 1X, belladonna (nightshade) 1X,

Symphytum officinale (comfrey) 1X, *Hypercum perforatum* (St. John's wort) 1X, *Ledum palustre* (wild rosemary) 1X, *Grindelia robusta* (wild sunflower) 1X.

Action: Safe and natural temporary relief for minor skin injuries, cuts, scrapes, rashes, insect bites, sunburn and minor burns. Reduces pain, inflammation and redness, helping to speed healing.

Direction: For external use only. Adults and children over 2 years of age: Apply generously with gentle rubbing to the affected area 2–3 times daily or as needed. Apply a bandage or band-aid to the affected area if desired. For children under 2 years of age consult a physician.

Supplied by: Northwest Cosmetic Laboratories, USA.

Storage

All cosmetic preparations of Calendula must be protected from light, moisture and heat.

Dosage

It is safe to use Calendula long-term. For internal purposes it should be taken three times a day. For external purposes it should be applied three times a day.

A tea of Calendula can be made by pouring 200 ml of boiling water over 1–2 teaspoons of the flowers, which is steeped, covered for 10–15 minutes, strained, and then drunk. At least 3 cups of tea are generally drunk per day. Tincture is similarly used three times a day, taking 1–2 ml each time. The tincture can be taken in water or tea.

Prepared ointments are often useful for skin problems, although wet dressings made by dipping cloth into the tea (after it has cooled) are also effective. Home treatment for eye conditions is not recommended, as absolute sterility must be maintained.

Juice: Take 1 tsp at a time, always freshly pressed.

Tincture: To make, soak a handful of flowers in 0.5 quart rectified alcohol or whiskey for 5 to 6 weeks. A dose is 5 to 20 drops.

Salve: Boil 1 oz dried flowers or leaves, or 1 tsp fresh juice, with 1 oz of lard.

Patents

Inventors were claimed that therapeutical composition contains an extract of the medicinal herbs *Calendula officinalis* and *Hypericum perforatum* in a ratio from 1:1 to 3:1 parts by weight is use for the treatment of dermatosis, in particular psoriasis. The extract contains low-molecular peptides and free amino acids to a total amount of 0.5–6.0% of the extract dry mass, more specifically histidine, cystine, serine, alanine, valine, leucine, isoleucine, phenylalanine, glycine, lysine, proline, tyrosine, glutamine, asparagine and methionine; unsaturated fatty acids 1.0–10.0%, more specifically linoleic acid, linolenic acid and oleic acid; vitamins 1.0–6.0%, including retinoides 0.6–3.0%; and minerals and microelements 1.0–6.0% of the extract dry mass (Yosifov, 2006).

A unique, synergistic formulation for the treatment of first-degree skin burns is provided for topical use and alleviates the full spectrum of symptoms and concerns associated with epidermal burns, including pain, blistering, redness, swelling, compromised skin integrity, risk of infection and scarring. The formulation contains natural homeopathic extracts combined with a pharmaceutically acceptable carrier suitable for topical administration. The formulation includes the following homeopathic extracts: *Cantharis vesicatoria, Echinacea angustifolia, Calendula officinalis* and *Hypericum perforatum* (Manickam, 2011).

BIBLIOGRAPHY

1. Akhtar N, Shahiq-uz-zaman, Khan BA, Haji M, Khan S, Ahmad M, Rasool F, Mahmood T, Akhtar R. Evaluation of various functional skin parameters using a topical cream of *Calendula officinalis* extract. African Journal of Pharmacy and Pharmacology, 2011; 5(2): 199 – 206.

2. Bernatoniene J, Masteikova R, Davalgiene J, Peciura R, Gauryliene R, Bernatoniene R et

al., Topical application of *Calendula officinalis* (L.): Formulation and evaluation of hydrophilic cream with antioxidant activity. Journal of Medicinal Plants Research. 2011; 5(6): 868–877.

3. Bernatoniene J, Masteikova R, Davalgiene J, Peciura R, Gauryliene R, Bernatoniene R et al., Topical application of *Calendula officinalis* (L.): Formulation and evaluation of hydrophilic cream with antioxidant activity. Journal of Medicinal Plants Research. 2011; 5(6): 868–877.

4. Cordova CA, Siqueira IR, Netto CA, Yunes RA, Volpato AM, Cechinel Filho V, Curi-Perrosa R, Creczynski Pasa TB. Protective properties of butanolic extract of the *Calendula officinalis* L. (marigold) against lipid pero-xidation of rat liver microsomes and action as free radical scavenger. Redox Rep. 2002; 7(2):95–102.

5. Esmaili R, Khalilian AR, Nasiri E, Dehghani HJO, Alipour S. Study regarding the effect of *Calendula officinalis* cream in healing of pressure sores. J of Mzandaran University of Medicial Sciences. 2008; 18 (66): 19–25.

6. Goyel M, Mathur R. Antimicrobial effects of *Calendula officinalis* against human pathogenic microorganisms. Journal of Herbal Medicine and Toxicology 2011; 5 (1): 97–101.

7. http://en.wikipedia.org/wiki/Calendula_officinalis

8. http://www.anniesremedy.com/herb_detail145.php

9. http://www.herbalist.com/wiki. details/7/category/8/start/0/

10. http://www.wildcrafted.com.au/Botanicals/Calendula.html

11. Hussein MS, Tawil OSE, Yassin NEH, Abdou KA. The protective effect of *Morus alba* and *Calendula officinalis* plant extracts on carbon tetrachloride- Induced hepatotoxicity in isolated rat hepatocytes. Journal of American Science 2010;6(10):762–773.

12. Ibrahim ZH, Dimitrova ZI, Georgiev ST, Madzharov V, Titeva ST. Use of cosmetic products containing extract of marigold (*Calendula officinalis*) in cases of acne problem at the educational and beauty center "Top Beauty" – Sofia. Trakia J of Sciences 2010; 8(2): 353-357.

13. Lagarto A, Bueno V, Guerra I, Valdes O, Vega Y, Torres L. Acute and subchronic oral toxicities of *Calendula officinalis* extract in Wistar rats. Experimental and Toxicological Pathology. 2011; 63(4): 387–391.

14. Manickam SS. Antipyrotic formulation for the treatment of Epidermal burns. Patent No: 7959955. Application Number: 12/137026. Publication Date: 06/14/2011. http://www.docstoc.com/docs/82284862/Antipyrotic-Formulation-For-The-Treatment-Of-Epidermal-Burns—Patent-7959955.

15. Pommier P, Gomez F, Sunyach MP, D' Hombres A, Carrie C, Montbarbon X. Phase III Randomized Trial of *Calendula Officinalis* Compared With Trolamine for the Prevention of Acute Dermatitis During Irradiation for Breast Cancer. Journal of Clinical Oncology. 2004; 22 (8):1447–1453.

16. Preethi KC, Kuttan G, Kuttan R. Anti-inflammatory activity of flower extract of Calendula officinalis Linn. and its possible mechanism of action. Indian J Exp Biol. 2009; 47(2): 113–20.

17. Re TA, Mooney D, Antignac E, Dufour E, Bark I, Srinivasan V, Nohynek G. Application of the threshold of toxicological concern approach for the safety evaluation of Calendula flower (*Calendula officinalis*) petals and extracts used in cosmetic and personal care products Food and Chemical Toxicology. 2009; 47(6): 1246–1254.

18. Silva EJ, Gonçalves ES, Aguiar F, Evêncio LB, Lyra MM, Coelho MC, Fraga Mdo C, Wanderley AG. Toxicological studies on hydroalcohol extract of *Calendula officinalis* L. Phytother Res. 2007; 21(4): 332–336.

19. Yosifov NV. therapeutical composition for the treatment of dermatosis com-prising an extract of *Calendula officinalis* and *Hypericum perforatum*. Application Number: EP/EP04802161. Publication Date: 10/11/2006. http://www.patsnap. com/patents/view/EP1708725a1.html

Centella asiatica

Plant profile

Kingdom: *Plantae* – Plants

Subkingdom: *Tracheobionta* – Vascular plants

Superdivision: *Spermatophyta* – Seed plants

Division: *Magnoliophyta* – Flowering plants

Class: *Magnoliopsida* – Dicotyledons

Subclass: *Rosidae*

Order: *Apiales*

Family: *Apiaceae* – Carrot family

Genus: *Centella* L. – Centella

Species: *Centella asiatica* (L.) Urban – Spadeleaf

Vernacular name

San: Mandukaparni, **Beng:** Thankuni, **Guj:** Moti Brahmi, **Hin:** Khulakhudi, Brahma-manduki, **Kan:** Brahmisoppu, **Mal:** Kodangal, Muyalchevi, **Ori:** Thalkudi, **Tam:** Vallarei, **Tel:** Brahmi, Saraswataku

Ayurvedic properties

Rasa: Tikta, **Vipaka:** Laghu, **Veerya:** Shita, **Karma:** Rasayanam, Medhyam, Mutra Virecana, **Prabhava:** Rasayanam, Medhyam, Mutra Virecana.

INTRODUCTION

It is a small herbaceous annual plant. The stems are slender, creeping stolons, green to reddish green in color, interconnecting one plant to another. It has long-stalked, green, reniform leaves with rounded apices which have smooth texture with palmately netted veins. The leaves are borne on pericladial petioles, around 2 cm. The rootstock consists of rhizomes, growing vertically down. They are creamish in color and covered with root hairs. The flowers are pinkish to red in color, born in small, rounded bunches (umbels) near the surface of the soil. Each flower is partly enclosed in two green bracts. The hermaphrodite flowers are minute in size (less than 3 mm), with 5–6 corolla lobes per flower. Each flower bears five stamens and two styles. The fruit are densely reticulate, distinguishing it from species of hydrocotyle which have smooth, ribbed or warty fruit. Seeds are present solitary in each mericarp, pendulous embryo, laterally compressed.

Distribution

The plant is native to India, Sri Lanka, Northern Australia, Indonesia, Iran, Malaysia, Papua New Guinea and other parts of Asia. It is found throughout tropical and subtropical regions of India up to an altitude of 600 m. The plant is reported to occur also at higher altitudes of 1550 m in Sikkim and 1200 m in Mount Abu. It grows widely in regions of East India, China, Japan and also in South Africa.

Chemical Constituents

Main chemical constituents are vallarine, asiaticoside, hydrocotylin, pectic acids, steroids, hersaponin, bacogenin, monnierin, triterpene, tannin, etc. Indian brahmi herb contains the following glycosides: Indocentelloside, brahmoside, brahminoside, asiaticoside, thankuniside and isothankuniside. The corresponding triterpene acids obtained on hydrolysis of the glycosides are indocentoic, brahmic, asiatic, thankunic and isothankunic. These acids, except the last two, are also present in free form in the plant apart from isobrahmic and betulic acids. The presence of mesoinositol, a new oligosaccharide, 'centellose', kaempferol, quercetin and stigmasterol, have also been reported. Also contains a green, strongly volatile oil composed of unidentified terpene acetate, camphor, cineole, and other essential oils.

Vallarine: A bitter principle which is present in the leaves and roots of this plant is an oily liquid of a pale color, soluble in alcohol, ether and caustic ammonia. It is believed that it is active properties reside in this principle.

Asiaticoside: Asiatoicside has been shown to be active in the treatment of leprosy. It probably acts by dissolving the waxy covering of *Bacillus leprae*. The bacillus thus become fragile and may easily be destroyed. Leprosy nodules are broken down, diffuse infiltrations do disappear. Perforating ulcers and lesions on the fingers heal and most remarkable of all, eye lesions are cured if treatment is given before the posterior chamber of eye

Asiaticoside

R1= OH, R2= H: Asiatic acid
R1= OH, R2= O: Madecassic acid
R1= O-glu-gluc-gluc, R2= OH: Madecasside

Asiatic acid derivates

is involved. These improvements are seen in cases of anaesthetic leprosy and it is not really evident in tubercular leprosy.

Sitosterol, tannin: From the alcoholic extract of the herb, an essential oil green in color possessing the strong odor of herb which is known to have antiprotozoal and spasmolytic effect have been reported.

Medicinal uses

Traditionally, healers have used *Centella asiatica* to treat a variety of different ailments including colds, respiratory infections, fevers, diarrhea, syphilis, hepatitis, stomach ulcers, epilepsy, scleroderma, insomnia, cellulite, scars, cancer, gastrointestinal problems and asthma. Other traditional, though unproven, uses of the herb include improving mental clarity, enhancing memory, increasing longevity and reducing mental fatigue. It can treat mental decline that occurs in old age because of its ability to protect blood vessels carrying oxygen to the brain. It also can help improve memory and concentration. It is helpful for varicose veins because glycosaminoglycans are also responsible for producing and stabilizing the connective tissue that encases the veins of the legs. Additionally, gotu kola boosts the amount of oxygen being carried through the veins. The isolated steroids from the plant have been used to treat leprosy. It is a mild adaptogen, is mildly antibacterial, antiviral, anti-inflammatory, anti-ulcerogenic, anxiolytic, a circulatory stimulant. There have been many reports showing the medicinal properties of *C. asiatica* extract in a wide range of disease conditions like diabetic microangiopathy, edema, venous hypertension, etc.

Cosmetic uses

The use of ointments and creams containing *Centella asiatica* to treat wounds, burns, scrapes and minor skin lesions. The herb contains triterpenoids—substances that expedite wound healing by boosting antioxidants in the wound, strengthening the skin and increasing blood supply to the wounded area. Low doses of chemical compounds found in *Centella asiatica* known as asiaticosides enhanced burn-wound healing. The plant is used in skin eczema. It also can help heal the skin after a burn or skin graft.

The flavonoids may also be used to effect in hair care products, where stimulation of the peripheral circulation of the scalp will promote healthy scalp condition and prevent hair loss. It is also reported to aid capillary growth in psychosomatic alopecia in the case that the piliferous papillae are not atrophied. Certain of the constituents are responsible for accelerated growth of hair and nails.

Action: The asiaticosides and triterpenes extracted from the plant have modulating properties on the development and meta-bolism of connective tissue, improve the synthesis of collagen and other tissue proteins by modulating the action of fibroblasts in the vein wall, and stimulate collagen remodeling in and around the venous wall. They, therefore, improve wound repair, with a better re-epithelialization and a normalization of perivascular connective tissue, thus allowing an improvement of the venous wall tone and elasticity. It is found that in poor connective tissue conditions, the triterpenes in *Centella asiatica* are able to renew the collagen, in quantity and quality, and restore tissue firmness and skin elasticity, improving skin appearance and comfort. Apart from this it also has anti-psoriatic properties. These saponins also help prevent excessive scar formation, by modulating and slowing down the excessive production of collagen (the material that makes up connective tissue), at the wound site. Various studies have supported these actions of centella and furthermore have been shown it to be effective in treating gangrene, traumatic injuries, burns, chronic skin lesions, ulcers and even leprosy wounds. *Centella asiatica* is used in the following products: rejuvenating face wash, rejuvenating toner, rejuvenating serum, rejuvenating day and night creams, eye cream, neck cream, face scrub and mask. It is used for purifying face mask, as skin tonner, skin spot treatment, as cellumend anticellulite cream, etc.

Pharmacological Action

Bacoside A assists in release of nitric oxide that allows the relaxation of the aorta and veins, to allow the blood to flow more freely through the body. Therefore, Brahmi is revered for strengthening the immune system, improving vitality and performance and promoting longevity. Bacoside B is a protein valued for nourishing the brain cells, as a result Brahmi improves mental clarity, confidence, intelligence and memory recall. Asiaticosides stimulate the reticuloendothelial system where new blood cells are formed and old ones destroyed, fatty materials are stored, iron is metabolized, and immune responses and inflammation begin. The primary mode of action of Centella appears to be on the various phases of connective tissue development, which are part of the healing process.

Present investigation was planned to evaluate the nootropic effect of *Centella asiatica*. Three months old male Swiss albino mice were injected orally with graded doses (200, 500, 700, 1000 mg/kg body weight) of *C. asiatica* aqueous extract for 15 days to select an effective dose for nootropic studies. Animals were tested in radial arm maze to assess the learning and memory performance. Based on these results, mice were treated orally with 200 mg/kg of *C. asiatica* for 15 days from day 15 to day 30 postpartum (p.p.) and the nootropic effect was evaluated on the 31st day and 6 months p.p. Results of the present investigation show that treatment during postnatal developmental stage with *C. asiatica* extract can influence the neuronal morphology and promote the higher brain function of juvenile and young adult mice (Rao et al., 2005).

The study was designed to determine whether extract of *Centella asiatica*, an antioxidant, when administered orally (300 mg/kg body weight/day) for 60 days would prevent age-related changes in antioxidant defense system, lipid peroxidation (LPO) and protein carbonyl (PCO) content in rat brain regions such as cortex, hypothalamus, striatum, cerebellum and hippocampus. Aged rats elicited a significant decline in the antioxidant status and increased the LPO and PCO as compared to control rats in all five regions studied. Supplementation of *C. asiatica* was effective in reducing brain regional LPO and PCO levels and in increasing the antioxidant status. Thus, *C. asiatica* by acting as a potent antioxidant exerted significant neuroprotective effect and proved efficacious in protecting rat brain against age related oxidative damage (Subathra et al., 2005).

The objective of the study was to evaluate the wound healing potential of the ethanolic extract of the plant in both normal and dexamethasone-suppressed wound healing. The study was done on Wistar albino rats using incision, excision, and dead space wounds models. The extract of *C. asiatica* significantly increased the wound breaking strength in incision wound model compared to controls ($p < .001$). The results indicated that the leaf extract promotes wound healing significantly and is able to overcome the wound healing suppressing action of dexamethasone in a rat model. These observations were supported by histology findings (Shetty et al., 2006).

The ayurvedic medicinal plant Gotukola (*Centella asiatica*) was evaluated for its anxiolytic properties. Various paradigms were used to assess the anxiolytic activity, including the elevated plus maze (EPM), open field, social interaction, locomotor activity, punished drinking (Vogel) and novel cage tests. The EPM test revealed that Gotukola, its methanol and ethyl acetate extracts as well as the pure asiaticoside, imparted anxiolytic activity. Furthermore, the asiaticoside did not affect locomotor activity, suggesting these compounds do not have sedative effects in rodents (Wijeweera et al., 2006).

An experiment was carried out to study the antimicrobial activity of petroleum ether, ethanol and water extract of *Centella asiatica* plant by agar diffusion method. Zone of inhibition produced by petroleum ether, ethanol and water extract in dose of 62.5, 125, 250, 500 and 1000 µg/ml against some selected strains was measured and compared with standard antibiotics ciprofloxacin (10 µg/ml). The present study demonstrated that the ethanolic extract of *Centella asiatica* has higher antimicrobial activity than petroleum ether and water extract (Jagtap et al., 2009).

The present study was performed to evaluate the anti-ulcerogenic activity of ethanol extract of *Centella asiatica* against ethanol-induced gastric mucosal injury in rats. Five groups of adult *Sprague Dawley* rats were orally pre-treated respectively with carboxymethyl cellulose (CMC) solution (ulcer control group), Omeprazole 20 mg/kg (reference group), and 100, 200 and 400 mg/kg *C. asiatica* leaf extract in CMC solution (experimental groups), one hour before oral administration of absolute ethanol to generate gastric mucosal injury. Histological studies revealed that ulcer control group exhibited severe damage of gastric mucosa, along with edema and leucocytes infiltration of submucosal layer compared to rats pre-treated with *C. asiatica* leaf extract which showed gastric mucosal protection, reduction or absence of edema and leucocytes infiltration of submucosal layer. Acute toxicity study did not manifest any toxicological signs in rats. The present finding suggests that *C. asiatica* leaf extract promotes ulcer protection as ascertained grossly and histologically compared to the ulcer control group (Abdulla et al., 2010).

The present study has been conducted in order to explore the possible role of *Centella asiatica* against D-galactose induced cognitive impairment, oxidative and mitochondrial dysfunction in mice. Chronic administration of D-galactose (100 mg/kg s.c.) for a period of six weeks significantly impaired cognitive task (both in both Morris water maze and elevated plus maze) and oxidative defense (increased lipid peroxidation, nitrite concentration and decreased activity of superoxide dismutase, catalase and non-protein thiols)

and impaired mitochondrial complex (I, II and III) enzymes activities as compared to sham group. Six weeks *Centella asiatica* (150 and 300 mg/kg, p.o.) treatment significantly improved behavioral alterations, oxidative damage and mitochondrial enzyme complex activities as compared to contro l (D-galactose). *Centella asiatica* also attenuated enhanced acetylcholine esterase enzyme level in D-galactose senescence mice. Present study highlights the protective effect of *Centella asiatica* against D-galactose induced behavioral, biochemical and mitochondrial dysfunction in mice (Kumar et al., 2011).

Toxicological Studies

The acute toxicity study of *Centella asiatica* was studied on Swiss mice with a dose of 3, 5 and 7 g/kg body weight orally. The single administration exposure of the whole plant powder in the form of aqueous slurry on Swiss mice was carried out and the exposure route was oral with water as a vehicle. The observations of changes in body weight, food and water intake as well as cage side observations were reported. The whole plant powder was found to be nontoxic.

Alcoholic extracts of Gotukola have shown no toxicity at doses of 350 mg/kg when given i.p. to rats. Madecassol is an extract of *Centella asiatica*. It contains madecassic acid, asiatica acid and asiaticoside. It has been used as a wound healing agent and for the prevention of cicatrization. Report contact dermatitis due to madecassol and a control study with its individual ingredients. Three women (61, 52 and 49 years old) who developed jaundice after taking *Centella asiatica* for 30, 20 and 60 days and recovered on discontinuation of the herb.

An Acute toxicity study was carried out in which the animals were treated with the rhizome extract at a dose of 2 and 5 g/kg of *C. asiatica* leaf extracts and were kept under observation for 14 days. All the animals remain alive and did not manifest any significant visible signs of toxicity at these doses. There were no abnormal signs, behavioral changes, body weight changes, or macroscopic finding at any time during the observation period. From these results it is concluded that the extract is quite safe even at these higher doses and had no acute toxicity and the oral lethal dose (LD_{50}) for the male and female rats was greater than 5 g/kg body weight (Abdulla et al., 2010).

Clinical Investigation

Centella is used in Ayurvedic medicine for the treatment of anxiety. After assessing baseline measurements of acoustic startle response (ASR), mood self-rating scale, heart rate, and blood pressure, 40 healthy subjects (21 males, 19 females; ages 18–45 years) were randomized to receive 12 g non-standardized Centella dissolved in 300 ml grape juice or placebo. Evaluations were recorded at 30, 60, 90, and 120 minutes after beginning therapy. Centella significantly decreased ASR amplitude compared to placebo at 30 ($p < 0.02$) and 60 ($p < 0.001$) minutes; heart rate, blood pressure, and mood did not change (Bradwejn et al., 2000).

Forty patients (21 males, 19 females; mean age 48 years) with severe venous hypertension, ankle swelling, and lipodermatosclerosis, were randomized to receive total triterpenoid fraction of *Centella asiatica* (TTFCA) 60 mg twice daily or placebo for eight weeks; patients in the study did not wear compression stockings. After trial conclusion, patients taking the herbal extract experienced a significant decrease in skin flux and rate of ankle swelling compared to baseline values ($p < 0.05$). In addition, patients in the active group reported rapid clinical improvement, reflected by a reduction in the analogue scale line score (e.g. symptoms of edema, pain, restless limbs, swelling, and change in skin condition/color) from 9.5 at baseline to 4.5 after eight weeks (Cesarone et al., 2001). In another study using Laser Doppler evaluation, subjects taking 60 mg TTFCA twice daily for six weeks demonstrated a 29-percent decrease in resting flux ($p < 0.05$), 52 percent increase in venoarteriolar response ($p < 0.05$), and 66-ml reduction in leg volume. Similarly, those utilizing the herbal extract demonstrated 7.2 percent increase in pO_2 and 9.6 percent reduction in pCO_2 ($p < 0.05$) (Cesarone et al., 2001).

Asymptomatic patients (49 men, 38 women; mean age 56) with high-risk, echolucent carotid artery plaques were randomized to receive 60 mg TTFCA or placebo three times daily for one year; patients also took platelet anti-aggregating medication throughout the trial. After 12 months, sonographic evaluation indicated a significant decrease in plaque echolucency in the TTFCA group. Incidence of positive MRI images indicating cerebral ischemic lesions was seven percent in the TTFCA group and 17 percent in the control group ($p < 0.05$) (Cesarone et al., 2001). In another study testing a similar dose of Gotukola, patients with femoral plaques demonstrated a decrease in plaque echolucency after 12 months of therapy, compared to no change in the control group. Degree of stenosis and walking distance did not change in the two groups (Incandela et al., 2001).

The cosmetic cream formulations were designed by using ethanolic extracts of *Glycyrriza glabra*, *Curcuma longa* (roots), seeds of *Psoralea corylifolia*, *Cassia tora*, *Areca catechu*, *Punica granatum*, fruits of *Emblica officinale*, leaves of *Centella asiatica*, dried bark of *Cinnamon zeylanicum* and fresh gel of *Aloe vera* in varied concentrations (0.12–0.9% w/w) and characterized using physicochemical and physiological measurements. The studies were carried out for 6 weeks on normal subjects (6 males and 12 females, between 22 and 50 years) on the back of their volar forearm for evaluation of viscoelastic properties in terms of extensibility via a suction measurement, firmness using laboratory fabricated instruments such as ball bouncing and skin hydration using electric (resistance) measurement methods. The formulations showed an increase in percentage extensibility, firmness and improved skin hydration and were found more effective compared with the control product after the 6-week study (Ahshawat et al., 2008).

Market Formulations

The proven clinical efficacy of *Centella asiatica* selected triterpenes has led to its incorporation in numerous cosmetic products including soothing creams and milks, repairing and regenerating creams, after-sun creams, after-shave products, creams for chapped hands and cellulite. The potential cosmetic uses of this herb are varied, given its skin-tightening and regenerative capacity. It is included in some products for its antiwrinkling properties and for its ability to reduce acne-induced blemishes.

Anti-cellulite balm: It is an oil-based balm that is blended with soya liposomes to enhance penetration.

Ingredients: Gotukola, garcinia, chebulic myrobalan, essential oils of ylang ylang (*Cananga odorata*), rosemary, patchouli.

Action: Helps guard against cellulite, adiposites, and 'orange dimple skin'. It motivates the lymphatic system to eliminate extracellular sebum and toxins.

Direction: Massage the body in upward, circular motions until the balm is absorbed by the skin. Follow with hot bath.

Supplied by: Sadatan Lab, Nagpur, India.

Pure heal gel: This gel cleanses and supports the skin after everyday common cuts and scrapes. This product contains absolutely no parabens, sodium laurel sulfate (SLS) or harmful petrochemicals. It does not contain animal products, gluten, artificial colors, flavors or preservatives.

Ingredients: *Calendula officinalis*, *Melaleuca alternifolia* (tea tree), *Lavandula officinalis*, *Centella asiatica* and *Aloe ferox*.

Action: Calendula will help to maintain healthy skin and tissue health, making Calendula an ideal topical skin remedy. Calendula also acts as an excellent skin tonic and has been studied for its ability to help soothe the skin. The skin-supporting qualities of tea tree are enhanced by its ability to penetrate the skin. Lavender is valued for its soothing action on the skin. In addition, the aromatherapy properties of lavender will also help to comfort and soothe the individual. This herb supports healthy circulation to the skin thereby maintaining skin health and has even been researched for its soothing properties on the most delicate of areas. *Centella asiatica* contains compounds called triterpenoids which have an excellent reputation in supporting skin health, increasing the concentration of anti-oxidants and maintaining healthy blood supply to the affected area. *Aloe ferox* has skin supportive properties.

Direction: With clean hands, squeeze a small amount of gel onto fingertips and apply to affected area 2–3 times daily.

Supplied by: Native Remedies, USA.

Nourishing Skin Cream

Ingredients: *Aloe vera* (Barbados Aloe, Kumari), *Pterocarpus marsupium* (Indian kino tree, Bijaka), *Withania somnifera* (winter cherry, Ashvagandha), *Centella asiatica* (Indian Pennywort, Mandukaparni).

Action: It provides all day moisturizing, nourishment and protection. It is enriched with *Aloe vera* which nourishes and moisturizes, and winter cherry and Indian kino tree extracts, that protect skin from pollution and dry weather.

Direction: Apply gently over face and neck twice daily. It can also be used as a base for make-up.

Supplied by: The Himalaya Drug Company, Bangalore, India.

SNP Control Cream

Ingredients: Mink oil, tocopheryl acetate, *Centella asiatica* extract, allantoin.

Action: SNP control cream contains mink oil similar to human serum, which improves skin environment by controlling oil, water balance of skin. Thus, it prevents the evaporation of moisture from the skin and forms a protective layer, which makes your skin moist and soft. It improves skin's moisture maintenance function with its containment of such ingredients as tocopheryl acetate, *Centella asiatica* extract and allantoin, improves drying and tough phenomenon, a basic cause of skin aging, and helps to make skin moist, soft and elastic by stimulating skin elasticity. Also, SNP control cream, which provides dried and tough skin with well-moisturizing condition 24 hours, helps undiluted mink oil (99%), a skin care oil similar to human serum ingredient, control oil, water balance and use it silky without being sticky or shiny.

Direction: At the final stage of basic make up, apply a suitable amount of cream to the whole face evenly for absorbing.

Supplied by: SD Biotechnologies Co. Ltd., Korea.

Storage of Products

Store in cool and dry place. Keep away from direct strong light and heat.

Dosage

Typical dosage recommendations are in the range of 60–180 mg/day, usually consumed in divided doses (20~60 mg) three times daily for at least 4 weeks.

Pediatric

Gotukola is not recommended for those under 18 years old.

Adult

The adult dosage of Gotukola varies depending on the condition being treated. The standard dose of Gotukola (*Centella asiatica*) varies depending on the preparation. Most studies have used standardized extracts:

- Dried herb—you can make a tea of the dried leaf, three times daily.

- Powdered herb (available in capsules)—1,000–4,000 mg, three times a day.

- Tincture (1:2 w/v, 30% alcohol) 30–60 drops (equivalent to 1.5–3 ml—there are 5 ml in a teaspoon), three times daily.

- Standardized extract—50–250 mg, two to three times daily. Standardized extracts should contain 40% asiaticoside, 29–30% asiatic acid, 29–30% madecassic acid, and 1–2% madecassoside. Doses used in studies mentioned in the treatment section range from 20 mg (for scleroderma) up to 180 mg (in one study for venous insufficiency, although most of the studies for this condition were conducted using 90–120 mg daily).

Patents

The present invention relates to cosmetic formulations for topical application containing extracts from *Phyllanthus emblica* and *Centella asiatica* and/or *Bacopa monnieri*, and the use of such formulations for the care of the human skin. In particular, the present invention relates to cosmetic formulations for topical application containing extracts from *Phyllanthus emblica* and *Centella asiatica* and/or *Bacopa monnieri* in addition to *per se* known adjuvants and expedients (Singh-Verma, 2001).

The present invention relates to novel antiacne compositions containing extract of Krameria, or a pure compound from the extract, at least one of ximeninic acid or lauric acid, and an anti-inflammatory saponin extracted from *Olax dissitiflora, Aesculus hippocastanum, Centella asiatica, Terminalia sericea, Glycyrrhiza glabra,* or mixtures thereof (Bombardelli et al., 2001).

Compositions and methods are disclosed for use in the treatment or prevention of hair loss and/or the promotion of hair growth. The compositions comprise in combination *Centella asiatica* extract, or one or more active principles thereof, green coffee extract, or one or more active principles thereof, and one or more antioxidants, and a dermatologically acceptable carrier (Long, 2010).

Disclosed is a composition comprising at least two of the following ingredients: *Magnolia* extract, honokiol, *Humulus lupulus* extract, hesperidin methyl chalcone, Gotukola, dipeptide valyl-tryptophan, palmitoyl tetrapeptide-3, *Corylus avellana* bud extract, *Centella asiatica* extract, *Cucumis sativa* extract, *Morus alba* extract, *Hibiscus sabdariffa* flower extract, *Vitis vinifera* extract, ascorbyl glucoside, *Citrus medica limonum* extract, *Avena sativa* kernel extract, hydrolyzed soy protein, aniseed myrtle extract, *Tasmania lanceolata* leaf extract, *Artemisia abrotanum* extract, or *Citrus grandis* fruit extract or any combination thereof. Also disclosed are methods of treating skin conditions by topically applying the composition to skin (Faller et al., 2010).

Combinations of vasoactive substances which act at venous or arterial level with phosphodiesterase inhibitors including phosphodiesterase V, in particular: Visnadin or esculoside; at least one compound selected from icarin or derivatives thereof or extracts containing it, *Ginkgo biloba* dimeric flavones either in the free form or complexed with phospholipids, amentoflavone; at least one compound selected from escin, escin beta-sitosterol complexed with phospholipids, sericoside, sericoside complexed with phospholipids or *Centella asiatica* extract in the free form or complexed with phospholipids. The formulations are useful in reducing pharmiculopathy and problems associated with venous insufficiency of the lower limbs (Bombardelli, 2011).

BIBLIOGRAPHY

1. Abdulla MA, Al-Bayaty FH, Younis LT, Abu Hassan MI. Anti-ulcer activity of *Centella asiatica* leaf extract against ethanol-induced gastric mucosal injury in rats. Journal of Medicinal Plants Research. 2010; 4(13): 1253–1259.

2. Ahshawat MS, Saraf S, Saraf S. Preparation and characterization of herbal creams for improvement of skin viscoelastic properties. Int J Cosmet Sci. 2008; 30 (3): 183–193.

3. Bombardelli E, Morazzoni P, Cristoni A, Seghizzi R. Pharmaceutical and cosmetic formulations with anti-microbial activity. United States Patent 20010046525. Application No: 09/884939. Publication date: 11/29/2001. http://www. freepatentsonline. com/y2001/0046525. html

4. Bombardelli E. Combinations of vasoactive agents, their use in the pharmaceutical and cosmetic field, and formulations containing them. United States Patent 7976881. Application No: 10/563965. Publication date: 07/12/2011. http://www.freepatentsonline.com/7976881.html.

5. Bradwejn J, Zhou Y, Koszycki D, Shlik J. A doubleblind, placebocontrolled study on the effects of Gotu Kola (*Centella asiatica*) on acoustic startle response in healthy subjects. J Clin Psychopharmacol 2000; 20: 680–684.

6. Cesarone MR, Belcaro G, De Sanctis MT, et al. Effects of the total triterpenic fraction of *Centella asiatica* in venous hypertensive microangiopathy: a prospective, placebo-controlled, randomized trial. Angiology. 2001; 52:S15-S18.

7. Cesarone MR, Belcaro G, Nicolaides AN, et al. Increase in echogenicity of echolucent carotid plaques after treatment with total triterpenic fraction of *Centella asiatica*: a prospective, placebo-controlled, randomized trial. Angiology. 2001; 52: S19-S25.

8. Cesarone MR, Belcaro G, Rulo A, et al. Microcirculatory effects of total triterpenic fraction of *Centella asiatica* in chronic venous hypertension: measurement by laser Doppler, $TcPO_2$–CO_2, and leg volumetry. Angiology. 2001; 52: S45-S48.

9. Faller J, Gan D, Hines M, Mangos L. Magnolia extract containing compositions. United States Patent 7744932. Application No: 12/048953. Publication date: 06/29/2010. http://www.freepatentsonline. com7744932. html.

10. http://www.centellaasiatica.com/

11. http://www.dermaxime.com/centella-asiatica.htm

12. http://www.livestrong.com/article/121057-uses-centella-asiatica/#ixzz1U8w RnSl1

13. http://www.mdidea.com/products/her bextract/gotukola/data07.html

14. http://www.umm.edu/altmed/articles/ gotukola-000253.htm

15. Incandela L, Belcaro G, Nicolaides AN, et al. Modification of the echogenicity of femoral plaques after treatment with total triterpenic fraction of *Centella* asiatica: a prospective, randomized, placebocontrolled trial. Angiology. 2001; 52: S69-S73.

16. Jagtap NS, Khadabadi SS, Ghorpade DS, Banarase NB, Naphade SS. Antimicrobial and Antifungal Activity of *Centella asiatica* (L.)Urban, Umbeliferae. Research J. Pharm. and Tech. 2009; 2 (2): 328–330.

17. Kumar A, Prakash A, Dogra S. *Centella asiatica* Attenuates D-Galactose-Induced Cognitive Impairment, Oxidative and Mitochondrial Dysfunction in Mice. Int J Alzheimers Dis. 2011; Article ID 347569: 1–9.

18. Long SP. Hair care compositions and methods. United States Patent 7754249. Application No: 11/629773. Publication date: 07/13/ 2010. http://www.freepatentsonline.com/ 7754249. html

19. Rao SB, Chetana M, Uma Devi P. *Centella asiatica* treatment during postnatal period enhances learning and memory in mice. Physiol Behav. 2005; 86(4): 449–457.

20. Shetty BS, Udupa SL, Udupa AL, Somayaji SN. Effect of *Centella asiatica* L (Umbelliferae) on Normal and Dexamethasone-Suppressed Wound Healing in Wistar Albino Rats. International Journal of Lower Extremity Wounds. 2006; 5(3): 137–143.

21. Singh-verma SB. Cosmetic preparations containing extracts from *phyllanthus emblica* and *Centella asiatica* and/or *bacopa monnieri*. United States Patent 6261605. Application No: 09/331791. Publication date: 07/17/ 2001. http://www. freepatentsonline.com/ 6261605. html

22. Subathra M, Shila S, Devi MA, Pan-neerselvam C. Emerging role of *Centella asiatica* in improving agerelated neurological anti-oxidant status. Exp Gerontol. 2005; 40(8-9): 707–715.

23. Wijeweera P, Arnason JT, Koszycki D, Merali Z. Evaluation of anxiolytic properties of Gotukola–(*Centella asiatica*) extracts and asiaticoside in rat behavioral models. Phytomedicine. 2006; 13 (9–10): 668–676.

Chamomilla recutita

Plant profile

Kingdom: *Plantae* – Plants

Superdivision: *Spermatophyta* – Seed plants

Subkingdom: *Tracheobionta* – Vascular plants

Division: *Magnoliophyta* – Flowering plants

Class: *Magnoliopsida* – Dicotyledons

Order: *Asterales*

Genus: *Chamomile*

Subclass: *Asteridae*

Family: *Asteraceae*

Species: *Chamomilla recutita*

Other common names

Mayweed, scented mayweed, pinheads, pineapple weed, german-hundskamille.

Eng: German Chamomile.

INTRODUCTION

Chamomilla is an erect annual, up to 60 cm in height, with wispy 2–3 leaves and terminal peduncles supporting single flowerheads. The leaves are light green and feathery with a bipinnate pattern. The entire plant has a sweet, pineapple scent. Yellow tubular florets without membranous bracts are implanted on a raised and hollow receptacle. This is surrounded by a single row of white ligulate florets which are often bent backwards. The flowers are small (about 1 inch across) and daisy-like with a white collar around a cone shaped, yellow center. The flower heads are collected when they are mature and expanded, from June to August. They should be dried carefully in the shade and stored in a cool dark place.

Distribution

Chamomile is a member of the daisy family and is native to Europe and West Asia. Its name comes from the Greek word "kamai melon," which means "Ground apple," due to chamomile's pleasant apple aroma and flavor. The plant grows low to the ground, thrives in sunny areas, and produces a bright yellow disk with white florets flower. It is also grown in Germany, Hungary, France, Russia, Yugoslavia, and Brazil. In India, it is grown in Punjab, Uttar Pradesh, Maharashtra, and Jammu and Kashmir. The plants can be found in North Africa, Asia, North and South America, Australia, and New Zealand.

Chemical Constituents

Chamomilla contains 0.3–2% volatile oil (α-bisabolol, β-bisabolol oxides A, B, and C);

Anthemidin

Apegenin

α-(-)-Bisabolol

β-Bisabolol

Chamazulene

bitter glycosides (anthemic acid); flavone glycosides (anthemidin, apegenin, luteolin), coumarins (including umbelliferon and herniarin), phenolic carboxylic acids, polysaccharides, mucilage, choline, amino acids, tannins, malic acid. Blue chamazulene is formed (1–15%) from the sesquiterpene lactone matricin during steam distillation.

Medicinal uses

Chamomile is used in homoeopathic medicine for inner turmoil, anxiety, anger, convulsions, throbbing headache, earache, teething, hacking coughs, dysmenorrhea and diarrhea. The root was traditionally chewed to relieve toothache.

Chamomilla has a wide range of actions. It is used in the treatment of insomnia, anxiety and nervous tension, for the relief of spasmodic pain such as dysmenorrhea or migraine, and is a safe remedy for children's problems with a nervous component. This spasmolytic action is due to the presence of flavones, bisabolol and other constituents of the volatile oil. This herb is particularly suited to digestive problems such as nervous dyspepsia and colic. The dicyclic ether in the volatile oil relaxes the smooth muscle, regulating peristalsis, while the carminative volatile oil reduces flatulence and irritation of the gut wall. The bitter glycosides stimulate the appetite and digestive activity, and the herb also helps relieve inflammatory conditions of the upper digestive tract.

Chamazulene and bisabolol directly reduce inflammation in tissues with which they come into contact, stimulate the formation of granulation tissue, and have an anti-bacterial action. Bisabalol is also protective against ulcers. The polysaccharides are having an immunostimulant action, activating macrophages and B-lymphocytes, thus demonstrating a scientific basis for the use of the herb in the topical treatment of wounds and ulcers. Chamomilla has a traditional use on the Continent in the treatment of asthma and hayfever, probably due to the herb's action on the mucous membranes of the upper respiratory tract. It is thought to reduce the reaction to allergens such as pollen or dust in sensitive individuals. Chamomile decreases spasms of smooth and skeletal muscles, It can be used alongside other herbs for the relief of stomach cramps, bloating and indigestion.

Flowers: Sedative, anti-oxidant, antidepressant, antihistaminic, diaphoretic.

Flower extracts (volatile oil): Used as bactericidal and fungicidal, antispasmodic, anti-inflammatory, analgesic, antiseptic, antipyretic, and antianaphylactic.

Cosmetic uses

It is used topically for skin conditions as a cream or an ointment, or used as a mouth rinse. *Chamomilla recutita* (Matricaria) flower Extract is used as ingredient in variety of cosmetic products. It helps to reduce redness of the skin and to reduce irritation. The volatile oil is significantly anti-inflammatory and chamomile is useful as a wash in cases of sunburn, or for cuts and scrapes. Research has demonstrated an immune stimulating action which may also help in the treatment of wounds with chamomile. It is used against irritations of the skin and mucosa (skin cracks, bruises, frostbite, and insect bites), including irritations and infections of the mouth and gums, and haemorrhoids. The essential oil present in the flower heads contains azulene and is used in perfumery, cosmetic creams, hair preparations, skin lotions, toothpastes, and also in fine liquors. The dry flowers of chamomile are also in great demand for use in baby massage oil. An infusion of chamomile flowers is used as a hair shampoo, especially for fair hair. An essential oil from the whole plant is used as a flavoring and in making perfume. The dried flowers are used as an insect repellent. The oil also gets rid of mouth infections and wards off bad breath and acid reflux.

Chamomile tea is used for burns, scrapes and wounds, skin problems such as eczema, chickenpox, diaper rash and even psoriasis, gum inflammation, lightens skin tone,

Reduces dark circles around the eyes, acne and sunstroke, etc.

Pharmacological Action

The inhibitory effect of chamomile essential oil and its major constituents on four selected human cytochrome P450 enzymes (CYP1A2, CYP2C9, CYP2D6 and CYP3A4) was investigated. Increasing concentrations of the test compounds were incubated with individual, recombinant CYP isoforms and their effect on the conversion of surrogate substances was measured fluorometrically in 96-well plates; enzyme inhibition was expressed as IC_{50} and K_i value in relation to positive controls. Finally, results concluded that chamomile preparations contain constituents inhibiting the activities of major human drug metabolizing enzymes (Ganzera et al., 2006).

Several essential oil samples from dried flower-heads of *Chamomilla recutita* (L.) Rauschert, were analyzed by GC-MS and the chemical constituents were compared. The local grown herbs showed levels of essential oil below the standards recommended by the Brazilian Pharmacopoeia. Also, differences in their composition as well as in the quantity of several components were found such as the unexpected inversion of the relative constitution of the A and B α-bisabolol oxides. Of particular interest was the striking effect of the chamomile extracts upon human leukocyte chemotaxis, a biological anti-inflammatory activity not reported before, in which cell migration was *in vitro* inhibited at the same level as showed by dexamethasone (Presibella et al., 2006).

The antibacterial activity of an oil extract of *Chamomilla recutita* flowers against *Helicobacter pylori* (*H. pylori*) was evaluated by the agar dilution method using Colombia agar with 10% sheep blood, an inoculum of McFarland 0.5 and incubation in an anaerobic atmosphere at 37°C for 3 days. The MIC(90) (minimal inhibitory concentration) and MIC (50) were 125 mg/ml and 62.5 mg/ml, respectively. It was shown that the *Chamomilla recutita* oil extract inhibited the production of urease by *H. pylori*. In addition, it was found that the morphological and fermentative properties of *H. pylori* were affected by application of the *Chamomilla recutita* oil extract (Shikov et al., 2008).

Recent study revealed the effects of anti-allergic activity of *Matricaria recutita* L. on mast cell mediated allergic models. The methanol extract of *Matricaria recutita* L. showed inhibitory effects on anaphylaxis induced by compound 48/80 and significant dose dependent anti-pruritis property was observed by inhibiting the mast cell degranulation. The results suggest that the methanol extract of *Matricaria recutita* showed potent anti-allergic activity by inhibition of histamine release from mast cells (Chandrashekhar et al., 2011).

The purpose of the study was to evaluate the effect of *Chamomilla recutita* on the healing of ulcers in rats. A 5-mm wound was inflicted on the tongue of 36 rats. Treatment group animals were treated topically with 0.04 ml/day of chamomile ointment, whereas control group animals were not treated. Animals treated with chamomile showed the best results regarding epithelialization and percentage of collagen fibers after 10 days. As expected, time had a statistically significant effect ($p < 0.05$) on fibroblast count, epithelialization, inflammation and wound size; animals sacrificed at 3 days showed the worst results. Hence, it was concluded that chamomile stimulated re-epithelialization and the formation of collagen fibers after 10 days of treatment; it did not, however, influence inflammation or fibroblast count (Fonseca et al., 2011).

Toxicological Studies

Chamomile is generally considered safe and nontoxic. Side effects are extremely rare for health users.

The acute oral LD_{50} of German or Roman chamomile in rats is > 5 g/kg; oral LD_{50} of German chamomile oil in mice is 2.5 ml/kg. Long-term oral toxicity studies of German chamomile in dogs and rats found no toxicity and no changes in pups from prenatal dosing; also, no teratogenic effects in rats after long-term administration; no toxicity from 3-week topical application on rabbits; no toxic effects from the oil applied to the backs of hairless mice and limited irritation applied for 24 hrs to the skin of rabbits.

Clinical Investigation

This study evaluates the fluid extract from *Chamomilla recutita's* safety and effectiveness in pain relief from aphthous stomatitis and other painful ulcers of the oral mucous membrane. The analgesic effect was considered excellent by 82% and good by 18% of the patients, as demonstrated with the analogical visual scale for chronic and experimental pains after 5, 10, and 15 minutes. Tolerance was evaluated as excellent by 97% and good by 1% of the subjects. The fluid extract from *Chamomilla recutita*, due to its analgesic effect, may give these patients a better quality of life (Silva et al., 2006).

Recent studies have shown several possible non-sesquiterpene lactone allergens in tea (infusions) from the *Chamomilla recutita* plant. The aim of this study was to report the results of patch testing with herniarin (7-methoxycoumarin), which is one of the possible coumarin allergens in chamomile. Selected patients with known or suspected compositae contact allergy were patch tested with herniarin 1% petrolatum. Among 36 patients tested, there was one positive and three doubtful positive reactions to herniarin. All 4 patients had a relevant contact allergy to German chamomile, whereas the majority of the remaining 32 patients had chamomile allergy of unknown relevance. The clinical results suggest that herniarin indeed is one of the non-sesquiterpene lactone sensitizers in German chamomile and that sensitization may occur through, for example, external use of chamomile tea or use of chamomile-containing topical herbal remedies (Paulsen et al., 2010).

Market Formulations
Miessence Balancing Moisturizer for Normal and T-Zone Skin

Miessence certified organic skin moisturizer is the best non-toxic, healthy solution to moisturizing skin. It is gentle, all natural and certified organic skin moisturizing ingredients create a protective barrier against free radicals and moisture loss (one of the main causes of wrinkles). The special ingredients in the unique organic base of this certified organic moisturizer have lasting effects on skin hydration and smoothness.

Ingredients: Plant phospholipids, organic seed butters, oils, organic herbs and flowers. It also contains organic rosehip seed oil, jojoba, Calendula, chamomile, olive leaf, marshmallow and lavender.

Recommended skin type: Balancing moisturizer is ideal for people with normal or combination skin who have even skin tone, fine to average pore size, some enlarged pores, some T-zone oiliness, occasional breakouts or some blackheads.

Direction: Twice daily in the morning and evening after conditioning (and treatment, if required), apply one to two pumps of the moisturizer to the palm of your hand then pat and press gently onto your face, throat and decolletage.

Supplied by: Organic and Natural Enterprises Group (ONE Group), Australia.

Fragrance-free Moisturizer

Ingredients: Purified water, helianthus annuus (sunflower) seed oil, stearic acid, C12-C15 alkyl benzoates, glyceryl stearate, stearyl alcohol, cetyl alcohol, persea gratissima (avocado) oil, butyrospermum parkii (shea butter) fruit, emulsifying wax, PEG-7 glyceryl cocoate, cetyl palmitate, PEG-100 stearate, sodium hydroxymethylglycinate, coconut triglycerides, tetrahydroxypropyl ethylenediamine, yeast extract (phyto hyaluronic acid), *chamomilla recutita* (German chamomile) flower, arnica montana (arnica) flower extract, steareth-5, oleth-5, sodium PCA, gluconolactone, PEG-40 prunus dulcis (almond) triglycerides, sodium benzoate, gamma-aminobutyric acid (GABA), phytonadione (vitamin K).

Directions: After cleansing and toning, warm a small amount between palms and smooth over face and neck.

Green Tea Gel Masque

This ultrasoothing oil-free gel mask is a must for those with sensitive, blotchy skin, rosacea-prone skin or any complexion that reddens or irritates easily. This masque is a blue gel natural color from azulene.

Ingredients: Purified water, glycerine, steareth-20, *Carbomer, allantoin, Boswellia serrata* extract, *Chamomilla recutita* (German chamomile) extract, tetrahydroxypropyl ethylenediamine, *Calendula officinalis* (Calendula) flower extract, *Camellia sinensis* (green tea) leaf extract, azulene, diazolidynyl urea, iodopropynyl butylcarbamate, essential oils of lavandula angustifolia (lavender) and citrus aurantium amara (neroli).

Directions: After cleansing, apply an even layer over face and neck and relax for 10 –15 minutes. Rinse well with cool water and follow with toner and appropriate protective treatment. Use once a week. Keep refrigrated for an extracooling sensation also a wonderful treatment product post-laser, glycolic or microdermabrasion.

Supplied by: Sharona Silva Inc.

Replenishing Toner

Ingredients: De-ionized water/aqua, glycerine, *Tamarindus indica* seed extract (tamarind), *Cucumis sativus* (cucumber) fruit extract, *Chamomile recutita* (matricaria) flower extract (chamomile), *Aloe barbadensis* (*Aloe vera*) leaf juice, *Camellia sinensis* (japanese green tea) leaf extract, quaternized wheat protein, saccharomyces/sea salt ferment (biomin), allantoin, polysorbate 20, disodium EDTA. *Citrus grandis* (grapefruit seed extract) (0.3%), diazolidinyl urea (0.1%), iodopropynyl

butylcarbamate (trace 0.001%), rosehips, rose petals essential oil, marshmallow, white willow.

Action: Remove last traces of make-up and dirt particles, restore pH-balance, provide an extra boost of hydration, enhance skin tone and texture, facilitate quicker and more effective absorption of skin moisturizers and serums.

Pumpkin Peel Neutralizer

This is alcohol free neutralizer. It hydrates as it neutralizes the 30% propumpkin peel.

Ingredients: Water, glycerine, *Tamarindus indica* seed extract, cucumber fruit extract, *Chamomile recutita* flower extract, *Aloe vera* leaf juice, Japanese green tea leaf extract, Wheat protein, biomin, allantoin, polysorbate 20, grapefruit seed extract, phenoxyethanol, optiphen as natural preservative.

Action: This neutralizer acts as skin moisturizer.

Supplied by: NCN Por Skin Care, USA.

Melt Away Stress 24-Hour Moisturizing Wash

This moisturizer is enriched with chamomile and an exclusive AromaSoothe™ fragrance with hints of lavender that lasts even after rinsing, this unique moisturizing wash transforms into a creamy, comforting lather that helps to feel relaxed like after a warm bath. And this soothing wash is clinically shown to moisturize for 24 hours without leaving behind any greasy afterfeel.

Ingredients: Water, Glycerin, coco-glucoside, sodium lauroamphoacetate, sodium laureth sulfate, hydroxypropyl starch phosphate, acrylates/beheneth-25 methacrylate co-polymer, fragrance, phenoxyethanol, PEG-150 distearate, styrene/acrylates copolymer, methylparaben, polysorbate 20, tetrasodium EDTA, propylparaben, ethylparaben, mineral oil, polyquaternium-7, *Chamomilla recutita* (matricaria) flower extract, red 33, blue 1, and citric acid.

Directions: Help to feel relaxed by breathing deeply and pouring the body wash on a wet washcloth, pouf, or hand. Lather with this moisturizing wash to cleanse and calm.

Supplied by: Johnson and Johnson Consumer Companies, Inc.

Storage of Products

Store in cool and dry place. Keep away from direct strong light and heat.

Dosage

Chamomile is formulated in many dosage forms. It comes as a topical cream, topical ointment, topical lotion, oral inhalation, powder, tea, and volatile oil known as essential oil.

Usual adult dosage

A. **Dermatitis:** For treatment of dermatitis apply cream or lotion topically 4 times daily.

B. **Tension:** For treatment of tension associated with disorder of nervous system use:

   1. Liquid extract (1:1 in 45–70% ethanol), 1 to 4 ml orally 3 times daily.

   2. Tea powder, orally 1–4 cups of tea daily or as needed.

C. **GI-discomfort:** The average daily dose for treatment of any GI-discomfort is 2–8 g, 3 times a day of fluid-extract (1:1 in 45% ethanol). Dose 1–4 ml orally three times daily.

Usual pediatric dosage

Dermatitis: Cream, topical: Apply 4 times daily.

Tension: Tea, and oral: 1 to 4 times daily, amount dependent on age.

GI-doses: 2 g, 3 times daily or fluid extract of 0.6–2 ml as single dose. Internal use is only in patients older than 3 years.

Administration: Chamomile products can be administrated in many forms. The common ones are: Oral, topical, inhalation, solution for bath, and infusion.

Patents

An herbal formulation comprises: Slippery elm bark; German chamomile flower; fenugreek seed; fennel seed; skullcap herb; cranberry fruit; peppermint leaf; a mixture of Chinese herbs, comprising atractylodes root, capillary artemisia herb, codonopsis root, Job's tears seed, schisandra fruit, agastache whole plant, Chinese licorice root, Chinese thoroughwax, ginger root, Korean ash branches bark, magnolia bark, phellodendron bark, poria cocos root, psyllium seed, Chinese goldthread, Chinese white peony root, costus root, silver root, tangerine peel, and angelica root; and methylsulfonylmethane (Watson and Watson, 2003).

A toothpaste composition for inducing sleep while simultaneously promoting intraoral cleanliness, which includes toothpaste base ingredients and at least one sleep-inducing natural herb or hormone. The sleep-inducing natural herbs and hormone are selected from the group consisting of chamomile, lemon balm, passion flower, and valerian, and the hormone melatonin. The sleep-inducing natural herbs are in a range of 0.25% to 18% by weight of the composition (Arthur, 2004).

The present invention relates to deodorants and other body care products comprising a CO_2 extract of the hops plant having bacteriocide/bacteriostat properties wherein the CO_2 extract has a very low level of essential hops oils (Bergeron et al., 2009).

The present invention is directed to a composition for the treatment of the hair roots in order to prevent or at least to reduce the loss of hairs, to restore the growth of the hairs and to prevent or at least to reduce the formation of white hairs. This composition contains an extract of at least one *Chamomilla* plant and/or of at least one *Achillea* plant (Jamel et al., 2010).

A kit and method for a two-step acne treatment regimen. The kit comprises a wash composition and a cream treatment composition. The method comprising the steps of: Using a wash composition to clean acne affected skin of an individual; and applying a cream treatment composition to the acne affected skin (Saha and Sher, 2011).

BIBLIOGRAPHY

1. *Arthur Z. Sleep inducing toothpaste made with natural herbs and a natural hormone.* United States Patent Application 20040185014. Application number: 10/391004. Publication date: 09/23/2004. http://www.freepatentsonline.com/y2004/0185014.html

2. Bergeron C, Gafner S, Lafrance JL. HOPS based deodorant. United States Patent Application 20090098075. Application number: 12/243329. Publication date: 04/16/2009. http://www. freepatentsonline. com/y2009/0098075.html

3. Chandrashekhar VM, Halagali KS, Nidavani RB, Shalavadi MH, Biradar BS, Biswas D, Muchchandi IS. Anti-allergic activity of German chamomile (*Matricaria recutita* L.) in mast cell mediated allergy model. J Ethanopharmacol. 2011, (In press).

4. Fonseca CM, Quirino MR, Patrocinio MC, Anbinder AL. Effects of *Chamomilla recutita* (L.) on oral wound healing in rats. Med Oral Patol Oral Cir Bucal. 2011; (In press).

5. Ganzera M, Schneider P, Stuppner H. Inhibitory effects of the essential oil of chamomile (*Matricaria recutita* L.) and its major constituents on human cytochrome P450 enzymes. Life Sciences.2006; 78(8): 856–861.

6. http://ncnskincare.com/exfoliate-c-2/pumpkin-enzyme-kit-special-offer-p-26

7. http://ucdenver.edu/academics/colleges/pharmacy/Resources/CurrentStudents/ExperientialProgram/Documents/nutr_monographs/Monograph-chamomile.pdf

8. http://www.911balsam.com/?p=245&lang=en

9. http://www.answers.com/topic/chemical-composition-of-bergamot-oil

10. http://www.johnsonsforyou.com/products_ mas.jsp?id=2

11. http://www.nutriherb.net/chamomile_herb_plant.html

12. Jamel YA, Ahmed AD, Rafida M, Hamed A, Sari KM. Composition for the treatment of the hair roots. United States Patent Application 20100124577. Application number: 12/325867. Publi-cation date: 05/20/2010. http://www.freepatentsonline.com/y2010/0124577. html

13. Paulsen E, Otkjaer A, Andersen KE. The coumarin herniarin as a sensitizer in German chamomile [*Chamomilla recutita* (L.) Rauschert, Compositae]. Contact Dermatitis. 2010; 62 (6): 338–342.

14. Presibella MM, Bôas LDBV, Belletti KMDS, Santos CADM, Santos AMW. Comparison of Chemical Constituents of *Chamomilla recutita* (L.) Rauschert essential oil and its anti-chemotactic activity. Brazilian Archives of Biology and Technology. 2006; 49 (5): 717–724.

15. Saha A, Sher S. Acne treatment. United States Patent Application 20110129552. Application number: 12/629033. Publication date: 06/02/2011. http://www.freepatentsonline.com/y2011/0129552.html.

16. Shikov AN, Pozharitskaya ON, Makarov VG, Kvetnaya AS. Antibacterial activity of *Chamomilla recutita* oil extract against *Helicobacter pylori*. Phytother Res. 2008; 22(2): 252–253.

17. Silva MR, Ferreira AF, Bibas R, Carneiro S. Clinical evaluation of fluid extract of *Chamomilla recutita* for oral aphthae. J of Drugs in dermatology. 2006; Availablr from: http://findarticles.com/p/articles/mi_m0PDG/is_7_5/ai_n26952918/

18. Watson TS, Watson BF. Herbal formulation. United States Patent Application 20030129260. Application number: 10/041314. Publication date: 07/10/2003. http://www.freepatentsonline.com/y2003/0129260.html

19. WHO Monograph on selected Medicinal plants: Http://www.who.int/medicines/library/trm/medicinalplants/monographs.shtml.

Cinchona succirubra

Plant profile

Kingdom: *Plantae* – Plants
Superdivision: *Spermatophyta* – Seed plants
Class: *Magnoliopsida* – Dicotyledons
Order: *Rubiales*
Genus: *Cinchona* L. – Cinchona

Subkingdom: *Tracheobionta* – Vascular plants
Division: *Magnoliophyta* – Flowering plants
Subclass: *Asteridae*
Family: *Rubiaceae* – Madder family
Species: *Cinchona succirubra*

Common name: Peruvian Bark (Red Cinchona)

Vernacular name

San: Sinkona, Kunayanah, **Eng:** Quinine Bark, Peruvian Bark, **Hin:** Quinine, **Kan:** Sinkona, Barkino, **Mal:** Sinkona, Koyina, **Tam:** Cinkona, Konia, **Tel:** Cinkona, Jaddapatta

Ayurvedic preparations

Tvak Churna: 1–2 gm, **Stava:** 125–500 mg

Ayurvedic properties

Guna: Laghu, Ruksha, **Rasa:** Tikt, **Vipak:** Katu, **Veerya:** Ushan, **Prabhav:** Ishamjvar-ghan.

INTRODUCTION

The plant known as the cinchona is a tall evergreen tree that often reaches between fifty to a hundred feet in height when fully mature. The leaves of the cinchona are flat and broad, marked off by large veins running in the lamina which has a shiny green surface. Cinchona flowers are white, pink or red in color and are often elongated; they are also covered all over thickly with silky hairs, giving them a very distinct appearance.

Distribution

A perennial tree native to the jungles of the west coast of South America, it is now cultivated for medicines in India, East Africa, and the East Indies. This tree was at the time cultivated largely on the island of Java in the Indonesian archipelago. In India it is cultivated in Annamalai hills in Coimbatore (above 1600 m), Nilgiri hills in Tamil Nadu (1800–2100 m) and Darjiling hills in West Bengal (above 1700 m).

Chemical Constituents

Its main alkaloid is cinchonidine. The total Alkaloid in the root, stem and bark is 7.6, 5.5 and 3.3 per cent respectively. Out of these amounts, quinine constitutes 0.76–1.42, 1.1–1.74 and 0.8–1.76% respectively in the root, stem and bark. The other active ingredients are quinoline alkaloids, including quinidine, and cinchonine; glycosides; tannins; quinic acid.

- Cinchonine and cinchonidine (stereo-isomers with R = vinyl, R′ = hydrogen)
- Quinine and quinidine (Stereoisomers with R = vinyl, R′ = methoxy)
- Dihydroquinidine and dihydroquinine (stereoisomers with R = ethyl, R′ = methoxy).

Medicinal uses

The bark is used to prevent and treat malaria. Also given for liver conditions associated with an enlarged spleen, anorexia, indigestion, hyperchlorhydria (excessive stomach acid production), cramps, myalgia (muscle pain), and fevers with excessive temperature. It has also been used to help prevent flu. The active ingredient quinine is prescribed as an antimalarial and for muscle cramps, and is in several over-the-counter painkilling and cold remedies. Quinidine is given for certain types of cardiac arrhythmias (irregular heart beats). Bark is used for febrifuge, tonic and astringent; valuable for influenza, neuralgia and debility. The liquid extract is useful as a cure for drunkenness. Extracts are ingredients of herbal formulas used to stimulate saliva and gastric secretions in the treatment of loss of appetite and dyspeptic discomfort. An infusion made from the bark of the cinchona is effectively used as a gargle for treating sore, infected throats and other oral problems. Herbal remedies made from the cinchona plant are used in the treatment of muscular cramps, particularly the muscular cramps that come in the night. Remedies made from the cinchona are also helpful in bringing relief from chronic arthritis and related problems. Cinchona is used in the Indian Ayruvedic system of medicine for the treatment of problems such as sciatica and dysentery, in addition to problems connected with kapha and other disorders.

Cosmetic uses

The powdered bark is often used in tooth-powders, owing to its astringency. Cinchona in decoction is a useful gargle and a good throat astringent. Besides being used as an anti-malarial, quinine has been used for treating various conditions, including hemorrhoids and varicose veins (as hardening agent), and in eye lotions for its astringent, bactericidal, and anesthetic effects. In cosmetics, extracts of cinchona are primarily used in hair tonics, reportedly for stimulating hair growth and controlling oiliness.

Pharmacological Action

This herb is combined with camomile, lemon balm, marshmallow root, angelica root and hops to treat anorexia nervosa. With yarrow, mint, and pleurisy root used in influenza and with crampbark and prickly ash bark for cramps. Quinine, quinidine, cinchonine, cinchonidine, and cinchonicinol have shown *in vitro* inhibitory activity against the cytotoxicity of polymorphonuclear leucocytes. Potent monoamine oxidase inhibiting activity *in vitro* was found from quinine, cinchonicinol, and cinchonaminone derived from *C. succirubra*. Acts as astringent; reduces or prevents fever; stimulates digestion; reduces muscle tension and spasm. It is an antiseptic, antiperiodic, antiphlogistic, antipyretic and antimiasmatic, a diminisher of reflex action, a protoplasmic poison, and a direct emmenagogue and an oxytocic. A recent use for quinine drugs has been for the treatment of muscle spasms and leg cramps.

Toxicological Studies

Quinine has been used at doses of 325 mg to 1 g as the sulfate salt. Classical doses of the crude bark were approximately 1 g.

Clinical Investigation

A meta-analysis of eight (four published and four unpublished) randomized, double-blind, placebo-controlled trials, seven of which had a crossover design to determined the effect of quinine for the treatment of nocturnal leg cramps. When individual patient data from all crossover studies were pooled, persons had 3.60 (95% confidence interval [CI] 2.15, 5.05) fewer cramps in a 4-week period when taking quinine compared with placebo. This compared with an estimate of 8.83 fewer cramps (95% CI 4.16, 13.49) from pooling published studies alone. The corresponding relative risk reductions were 21% (95% CI 12%, 30%) and 43% (95% CI 21%, 65%), respectively. Compared with placebo, the use of quinine was associated with an increased incidence of side effects, particularly tinnitus. This study confirms that quinine is efficacious in the prevention of nocturnal leg cramps (Hing et al., 1998).

In 2002, a double-blind placebo study was undertaken in which 98 people with nocturnal leg cramps were given 400 mg of quinine daily for 2 weeks. The results stated that quinine administered at this dose effectively reduced the frequency, intensity, and pain of leg cramps without relevant side-effects. This use has fueled the natural product market and more people are looking for natural quinine bark as an alternative to the synthesized prescription drugs for this purpose.

Market Formulations

Quinine Bark (Red Cinchona) Cream

Ingredients: Water, purified, palm ester stearates, emulsifying wax, jojoba oil, green tea extract (antioxidant), glucose peroxide, lactoperoxide, methylparaben (0.1%), quinine.

Action: It is recommended for hemorrhoids, varicose veins, leg cramps and as an insect repellant, as well as for its astringent, bactericidal, and anesthetic actions in various other conditions.

Quinine Bark (Red Cinchona) Salve/ Ointment

Ingredients: Beeswax, benzoin (preservative), Jojoba oil, quinine.

Action: It is recommended for varicose veins, leg cramps and anesthetic actions in various other conditions.

Directions: Apply this cream morning and evenings, or as directed by a health care practitioner. On a moist cotton wool pad or with the fingertips, apply to the desired area of the body. Massage onto thoroughly cleansed skin with a gentle circular motion.

Manufacturer: Bianca Rosa.

Quinine Bark (Red Cinchona)-325 mg

Ingredients: Gelatin, water, Quinine bark-325 mg.

Action: Acts against night leg cramps.

Directions: Take 1 capsule, 3 times daily, with meals.

Manufacturer: TerraVita.

Quinine Bark (Red Cinchona) Powder

Ingredients: Quinine.

Action: Appropriately treats anxiousness, anxiety, fever, digestion, affliction, and lots more.

Direction: Stir 1/4 of a teaspoon into a glass of water and consume 3 times daily, with meals.

Manufacturer: TerraVita.

Storage of Products

Store in cool and dry place. Keep away from direct strong light and heat.

Dosage

Use three times daily.

Decoction: Use 0.3–1g of dried bark.

Liquid extract: Use 0.3–1 ml of standardized preparations.

Tincture: Use 2–4 ml of standardized preparations.

Recommended daily dosages in Germany: 1–3 g herb.

Patents

The present invention relates to a composition comprising, in a physiologically acceptable medium, at least one copolymer of a styrene monomer and of an ethylenically unsaturated dicarboxylic acid and at least one anti-seborrheic active agent. Composition according to claim 1, in which the anti-seborrheic active agent is chosen from retinoic acid; benzoyl peroxide; sulfur; vitamin B_6; selenium chloride; sea fennel; mixtures of extract of cinnamon, of tea and of octanoylglycine; the mixture of cinnamon, sarcosine and octanoylglycine; zinc salts; copper derivatives; extracts of plants of the species *Arnica montana, Cinchona succirubra, Eugenia caryophyllata, Humulus lupulus, Hypericum perforatum, Mentha piperita,* *Rosmarinus officinalis, Salvia officinalis* and *Thymus vulgaris* and others (Baldo F, 2008).

BIBLIOGRAPHY

1. Baldo F. Composition for combating the localized hyperpigmentation of dark skin. United States Patent 20080119527. Application No: 11/898426. Publication date: 05/22/2008. http://www. freepat entsonline.com/y2008/0119527.html
2. Blumenthal, M (Ed.): The Complete German Commission E Monographs: Therapeutic Guide to Herbal Medicines. American Botanical Council. Austin, TX. 1998.
3. Hing MMS, Wells G, Lau A. Quinine for Nocturnal Leg Cramps. J Gen Intern Med. 1998; 13(9): 600–606.
4. http://www.answers.com/topic/pharmac ology-biological-activities-of-boneset
5. http://www.herbs2000.com/herbs/herbs_ cinchona.htm
6. http://www.medical-explorer.com/medi cinal-ingredients-c/cinchona-succirubra_ 1.html
7. Wealth of India. Vol-III. p. 563.

Citrus aurantium

Plant profile

Kingdom: *Plantae* – Plants
Superdivision: *Spermatophyta* – Seed plants
Class: *Magnoliopsida* – Dicotyledons
Order: *Sapindales*
Genus: *Citrus* L. – Citrus

Subkingdom: *Tracheobionta* – Vascular plants
Division: *Magnoliophyta* – Flowering plants
Subclass: *Rosidae*
Family: *Rutaceae* – Rue family
Species: *Citrus aurantium* – Sour orange

Vernacular name

Beng: Komla Lebu, **Eng:** Orange, **Hind:** Santra, Narangi, **Guj** and **Punjabi:** Santra, **Kan:** Kithilai **Marathi:** Madhura naranga, **Ori:** Kamala, **Tam:** Kichili Pazham, **Tel:** Kamala Pandu

Ayurvedic properties

Gunna: Guru, Tikshan, **Rasa and Vipaka:** Amal, **Veerya:** Ushan.

INTRODUCTION

The tree ranges in height from less than 10 ft (3 m) to 30′ (9 m). The tree is more erect and has a more compact crown than the sweet orange. It has smooth, brown bark, green twigs, angular when young, and flexible, not very sharp, thorns from 1 into 3 1/8 in (2.5–8 cm) long. The evergreen leaves are aromatic, alternate, on broad-winged petioles much longer than those of the sweet orange, usually ovate with a short point at the apex; 2 1/2 to 5 1/2 in (6.5–13.75 cm) long, 1 1/2 to 4 in (3.75–10 cm) wide; minutely toothed; dark-green above, pale beneath, and dotted with tiny oil glands. The highly fragrant flowers, borne singly or in small clusters in the leaf axils, are about 1 1/2 in (3.75 cm) wide, with 5 white, slender, straplike, recurved, widely-separated petals surrounding a tuft of up to 24 yellow stamens. The fruit is round, oblate or oblong-oval, 2 3/4 to 3 1/8 in (7–8 cm) wide, rough-surfaced, with a fairly thick, aromatic, bitter peel becoming bright reddish-orange on maturity and having minute, sunken oil glands. There are 10 to 12 segments with bitter walls containing strongly acid pulp and from a few to numerous seeds. The center becomes hollow when the fruit is full-grown.

Distribution

The sour orange is native to Southeastern Asia. The plant is natives of the South Sea islands, especially Fiji, Samoa, and Guam. It is grown in orchards or groves only in the orient and the various other parts of the world where its special products are of commercial importance, including Southern Europe and offshore islands, North Africa, the Middle East, Madras, India, West Tropical Africa, Haiti, the Dominican Republic, Brazil and Paraguay.

Chemical Constituents

The main constituents of orange peel are the volatile oil and an amorphous, bitter glucoside called aurantiamarin. Other constituents include hesperidin, a colorless, tasteless, crystalline glucoside occurring mainly in the white zest of the peel, isohesperidin, hesperic acid, aurantiamaric acid, and a bitter acrid resin. In the peel of immature fruits, the chief constituents are naringin and hesperidin, while in the fruit flesh it is umbelliferone. p-Octopamine and p-synephrine, both adrenergic agonists, are the most frequently mentioned biogenic amines found in bitter orange peel.

p-Octopamine   p-Synephrine

Medicinal uses

Bitter orange is also employed in herbal medicine as a stimulant and appetite suppressant. It is also used as antibacterial, antiemetic, antifungal, antispasmodic, antitussive, aromatherapy, carminative, contraceptive, diaphoretic, digestive, sedative; stimulant; stomachic; tonic. The fruit is antiemetic, antitussive, carminative, diaphoretic, digestive and expectorant. They are used in the treatment of dyspepsia, constipation, abdominal distension, prolapse of the uterus, rectum and stomach. The seed and the pericarp are used in the treatment of anorexia, chest pains, colds, coughs, etc. The flowers, prepared as a sirup, act as a sedative in nervous disorders and induce sleep. An infusion of the bitter bark is taken as a tonic, stimulant, febrifuge and vermifuge. Tea prepared from fruit is used to relieve headache. Fruit peel powder is used in face pack against sunstroke and skin blemishes. It is also used to dissolve kidney stones. It purifies blood and improves immunity. "Bitter orange oil", expressed from the peel, is in demand for flavoring candy, ice cream, baked goods, gelatins and puddings, chewing gum, soft drinks, liqueurs and pharmaceutical products.

Cosmetic uses

The essential oil is used in aromatherapy. It is used in treating depression, tension and skin problems. A semi-drying oil obtained from the seed is used in soap making. Essential oils obtained from the peel, petals and leaves are used as a food flavoring and also in perfumery and medicines. The oil from the flowers is called 'Neroli oil'-yields are very low from this species. Neroli oil, mixed with vaseline, is used in India as a preventative against leeches. It has also reported used in cosmetic related products such like sunless tanning, conditioner, shampoo, make-up remover, exfoliant/scrub, blush, mask, facial cleanser. As from related research and application suggestions, *Citrus aurantium* can be used to prevent skin fragility and perks up skin tone and metabolism boosting.

Soap substitute: Throughout the pacific island, the crushed fruit and the macerated leaves, both of which make lather in water, are used as soap for washing clothes and shampooing the hair.

Perfumery: All parts of the sour orange are more aromatic than those of the sweet orange. The flowers are indispensable to the perfume industry and are famous not only for the distilled Neroli oil but also for "orange flower absolute" obtained by fat or solvent extraction. During favorable weather in Southern France, 1,000 kg of flowers will yield 1,000–1,500 g of oil.

Petitgrain oil is distilled from the leaves, twigs and immature fruits, especially from the Bergamot orange. Both Petitgrain and the oil of the ripe peel are of great importance in formulating scents for perfumes and cosmetics. Petitgrain oil is indispensable in fancy Eau de Cologne. The seed oil is employed in soaps. The epicarp is a very versatile compound and is used for its astringent quality, as well as the fact that it helps to prevent skin fragility and perks up skin tone. The aforementioned properties make this material good for purpose of cellulite treatment. Normally, combined with many other herbs such like butchers broom, horsetail, rosemary leaf, *Centella asiatica*, black pepper seed extracts, etc.

Pharmacological Action

Synephrine-rich *Citrus aurantium* extracts have antidepressant effects (Kim et al., 2001) and whole *C. aurantium* peel contains citral, limonene, and several citrus bioflavonoids, including hesperidin, neohesperidin, naringin, and rutin. Weak evidence hints that these substances might have antiviral effect (Kim et al., 2001). The essential oil of *C. aurantium* contains linalool and the fragrant substance limonene has antianxiety and sedative effects (Carvalho-Freitas et al., 2002).

Antiobesity effect of *C. aurantium* contains synephrine which is a stimulant with similar properties as caffeine and ephedrine. It claims to have similar effects by increasing energy expenditure, increasing metabolism and suppressing appetite (Haaz et al., 2006).

In this study, creams were formulated based on the antioxidant potential of herbal extracts and evaluated. Different types of herbal creams were formulated from the ethanolic extracts of *Glycyrrhiza glabra* (root and stolons), *Phyllanthus emblica* (fruit), *Lycopersicon esculentum* (fruit), *Curcuma longa* (rhizomes), *Aloe vera* (leaf) and *Citrus aurantium* (outer peel) namely F1, F2, F3, F4. All the formulations showed good spreadability, good consistency and no evidence of phase separation. From the present study it can be concluded that it is possible to develop creams containing herbal extracts having antioxidant property and can be used as the provision of a barrier to protect skin (Nair et al., 2009).

Toxicological Studies

Acute exposure: In male mice treated by gavage with essential oil from *C. aurantium* peel (0.5 or 1.0 g/kg), the latency period of tonic seizures was increased. In addition, treatment with the higher dose significantly increased hypnotic activity and anxiolytic activity. Sprague-Dawley rats orally administered a single dose of TJ-41 (up to 10 g/kg) or TJ-43 (2 or 8 mg/kg), herbal drug mixtures containing ~10% of *C. aurantium* peel, showed no toxic signs. Additionally, no deaths were reported.

Subcutaneous (s.c.) injection of p-synephrine (1500 mg/kg [8.971 mmol/kg]) caused effects on the sense organs, convulsions, and respiratory stimulation in mice. Administration of (R)-(-)-p-synephrine (700 mg/kg [4.19 mmol/kg]) and the S-enantiomer (1500 mg/kg [8.971 mmol/kg]) by s.c. injection also caused convulsions, as well as dyspnea and cyanosis in mice; oral administration of the compounds (1 mg/kg [6 µmol/kg] R and 0.3 mg/kg [2 µmol/kg] S) increased body temperature.

Daily oral administration (gavage) of *C. aurantium* fruit extracts standardized to 4 or 6% synephrine to male Sprague-Dawley rats for 15 days caused a significant dose-dependent decrease in food intake and body weight gain. Some deaths occurred in all treatment groups. No marked changes were seen in blood pressure; however, ventricular arrhythmias with enlargement of the QRS complex were observed.

The purpose of this study was to determine if relatively pure synephrine or synephrine present as a constituent of a bitter orange extract produced developmental toxicity in rats. Sprague-Dawley rats were dosed daily by gavage with one of several different doses of synephrine from one of two different extracts. Caffeine was added to some doses. Animals were sacrificed and fetuses were examined for the presence of various developmental toxic endpoints. At doses up to 100 mg synephrine/kg body weight, there were no adverse effects on embryolethality, fetal weight, or incidences of gross, visceral, or skeletal abnormalities. There was a decrease in maternal weight at 50 mg synephrine/kg body weight when given as the 6% synephrine extract with 25 mg caffeine/kg body weight; there was also a decrease in maternal weight in the caffeine only group. This decrease in body weight may have been due to decreased food consumption which was also observed in these two groups. Overall, doses of up to 100 mg synephrine/kg body weight did not produce developmental toxicity in Sprague-Dawley rats (Hansen et al., 2011).

Clinical Investigation

Many studies have been conducted regarding the effects of products whose ingredients include both ephedrine and bitter orange. Some report no adverse effects, while others report that the use of bitter orange with stimulants such as ephedra causes cardiotoxicity. When used in products for weight loss, which may be combined with ephedra, adverse effects included sensitivity to light, an increase in blood pressure, and heightened anxiousness.

Seville orange (*Citrus aurantium*) extracts are being marketed as a safe alternative to ephedra in herbal weight-loss products. *C. aurantium* contains synephrine (oxedrine), which is structurally similar to epinephrine. Although no adverse events have been associated with ingestion of *C. aurantium* products thus far, synephrine increases blood pressure in humans and other species, and has the potential to increase cardiovascular events. Additionally, *C. aurantium* contains 6', 7'-dihydroxybergamottin and bergapten, both of which inhibit cytochrome P450-3A, and would be expected to increase serum levels of many drugs. There is little evidence that products containing *C. aurantium* are an effective aid to weight loss. Synephrine has lipolytic effects in human fat cells only at high doses, and octopamine does not have lipolytic effects in human adipocytes (Fugh Berman and Myers, 2004).

Market Formulations

SlimEasy™

Weight loss formula is able to effectively control fat absorption and reduce excessive accumulation of body fats, stabilize and regulate metabolism of blood fats in the body, and remove unnecessary fatty muscles and wrinkles, thus achieving healthy and slim body posture without the need to deliberately change the usual eating and drinking habits.

Ingredients: Lotus leaf extract (*Folium nelumbinis*), bitter orange extract (*Citrus aurantium*), green tea extract (*Camellia sinensis O. ktze*), cassia tora (*Cassia obtusifolia*), Danshen (*Radix Salvia miltiorrhiza*), American Ginseng (*Panax quinquefolium* L.), vitamin E, vitamin C.

Action: Lotus leaf extract and green tea extract help to suppress appetite and promote lipolysis (the process of breaking down fat into free fatty acids and glycerol). Lipolysis must take place before fat can be burned. *Citrus aurantium* (synephrine) stimulates lipolysis in body cells. Ginseng extract and Danshen extract boost energy and buffer the effects of stress. Vitamins E and C nourish and support the body through the rigors of dieting and weight loss.

Direction: Take 1 capsule (300 mg) once a day before meal with plenty of water (drink at least eight glasses of water daily). Do not take the product for more than consecutive 3 months.

Storage: Store SlimEasy in a cool, dry place between 59 and 77°F (15 and 25°C), kept away from heat, moisture, and strong light. Keep SlimEasy out of the reach of children and away from pets.

Supplied by: Botaniex Plc. China.

Skinception™: It is an intensive stretch Mark Therapy, has been scientifically formulated with a series of clinically proven active ingredients that smooth and diminish the appearance of stretch marks and scars from pregnancy, weight loss, weight gain, growth spurts, and surgery.

Ingredients: Water, caprylic capric triglycerides, isopropyl palmitate, pentylene glycol, ethoxydiglycol, beta glucan, butylene glycol, *Nelumbo nucifera* leaf extract, cetyl hydroxyethycellulose, rutin, palmitoyl oligopeptide, palmitoyl tetrapeptide-7, *Phaseolus lunatus* (green bean) extract, *Siegesbeckia orientalis* extract, hydrolyzed soybean fiber, stearyl alcohol, magnesium aluminum silicate, glyceryl stearate, sodium stearoyl lactylate, panthenol, allantoin, *Citrus grandis* (grapefruit) peel oil, *Citrus aurantium Dulcis* (orange) peel oil, cymbopogon schoenanthus oil, sodium lactate, steareth-21, xanthan gum, DMDM hydantoin.

Action: Beta glucans are applied to the skin for anti-wrinkle purposes, but have a much longer history of use in treating eczema, burns, and wounds, and have shown to reduce scarring after surgery. Long used as a moisturizer capable of forming a film over the skin, beta glucan is an anti-irritant that stimulates collagen deposition.

Nelumbo nucifera leaf extract is extracted from lotus leaf, it works to sooth inflammation and preserve the architecture of the extracellular matrix (the non-living tissue that supports the skin cells), which is made largely of proteins. *Nelumbo nucifera* contains a substantial quantity of a protein-repair enzyme called lisoaspartyl methyltransferase. For skin care, it is known to reduce the appearance of skin surface irregularities and works to both hydrate and tighten the skin.

Rutin is a bioflavonoid found in many plants, fruits, and vegetables, including buckwheat, apple peel, citrus fruits and black tea. It has powerful antioxidant properties and also stabilizes vitamin C, making it more potent. Together, rutin and vitamin C (as strong antioxidants) fight free radical damage and thereby help maintain a stronger collagen matrix.

Phaseolus lunatus (green bean) extract is full of antioxidants, tannins, and polyphenols, the purified extract contains a protein that inhibits inflammation and the subsequent damage and scarring. *Phaseolus lunatus* also aids in wound prevention and recovery by contributing to the regeneration of collagen.

Siegesbeckia orientalis extract is widely used as an anti-inflammatory and an emollient for skin repair. It has been used for thousands of years to promote wound healing. Helps restore the collagen matrix and promote natural collagen production.

Citrus grandis (grapefruit) peel oil is an astringent that firms and protects the skin. It also acts as a tonic that restores, invigorates, and tones the skin and tissues, and purifies the skin of toxins.

Citrus aurantium dulcis (orange) peel oil is an essential oil used to tone and restore the skin. Also acts as an antibacterial and high in antioxidants.

Cymbopogon schoenanthus oil is the oil of lemon grass, it is a fresh, citrus-smelling herb used as an antiseptic and astringent to cleanse oily skin and close pores.

Supplied by: Leading Edge Health, USA.

Palmers Eventone Fade Cream

This cream is formulated for oily skin and it works well to moisturize the skin.

Ingredients: Hydroquinone (2%), octyl salicylate (3%), water, glyceryl stearate, mineral oil (paraffinum liquidum), stearyl stearate, propylene glycol, PEG 75 lanolin, cetearyl alcohol, ceteareth 20, PEG 100 stearate, dimethicone, isopropyl myristate, tocopheryl acetate (vitamin E), ascorbyl palmitate (vitamin C) (vitamin C palmitate), citric acid, magnesium aluminum silicate, sodium lauryl sulfate, imidazolidinyl, urea,

bilberry (*Vaccinium myrtillus*) Extract (*Vaccinum myrtilus*), sugar cane (*Saccharum officinarum*) extract, sugar maple extract (*Acer saccharinum*), orange extract (*Citrus aurantium dulcis*), lemon extract (*Citrus medica limonum*), bearberry (*Arctostaphylos uva-ursi*) extract, sodium sulfite, sodium metabisulfite, xanthan gum, methylparaben, disodium EDTA, propylparaben, corn oil (*Zea mays*), glyceryl oleate, BHA, BHT, T butyl hydroquinone, fragrance.

Action: This cream is marketed to help fade the appearance of freckles, moles and sun spots. It will also help to even out the skin tone.

Direction: This **cream** may work well as a moisturizer but can take up to 8 weeks for results to show.

Supplied by: Underarmwhitening.org (on line).

Thieves Foaming Hand Soap

Ingredients: Water, decyl glucoside, cocamidopropyl hydroxysultaine, syzygium aromaticum (clove) flower bud oil, citrus limon (lemon) peel oil, *Cinnamomum verum* (cinnamon) bark oil, *Eucaplyptus radiata* leaf oil, *Rosmarinus officinalis* (rosemary) leaf oil, *citrus aurantium* dulcis (orange) oil, *Aloe barbadensis* leaf juice, tocopheryl acetate, *Ginkgo biloba* leaf extract, retinyl palmitate, *Camellia sinensis* leaf extract, cetyl hydroxyethylcellulose, citric acid, phenoxyethanol, caprylyl glycol, ethylhexylglycerin, and hexylene glycol.

Action: This hand soap will cleanse, defend, and condition the skin with the therapeutic-grade essential oil blend thieves, pure lemon and orange essential oils, aloe, *Ginkgo biloba*, and vitamin E. Dispensed as a rich foam. It contains gentle ingredients so it can be used often without drying or stripping the skin.

Direction: Pump foam onto hands, lather, then rinse thoroughly. Avoid contact with eyes. If contact occurs, flush eyes liberally with clean, cool water for several minutes. If redness or irritation occurs, discontinue use. If symptoms persist contact your physician.

Supplied by: Health Scents, USA.

Storage of Products

Store in cool and dry place. Keep away from direct strong light and heat.

Dosage

Many health professionals recommend 1 to 2 g of dried bitter orange peel simmered for 10 to 15 min in a cup of water. Three cups are usually recommended as a daily dosage. As a tincture, 2 to 3 ml is usually recommended, also to be taken three times a day. The strongly acidic fruit of the bitter oranges stimulates the digestion and relieves flatulence: An infusion of the fruit is thought to soothe headaches, calm palpitations, and lower fevers. The juice helps the body eliminate waste products, and, being rich in vitamin C, which helps the immune system ward off infection.

A typical recommended dosage of such products ranges from 100 to 150 mg two to three times daily.

Patents

An appetite suppressant toothpaste formulations which simultaneously suppresses the users appetite while promoting intraoral cleanliness. The toothpaste composition includes toothpaste base ingredients; and at least one of appetite suppressant and appetite depressant herbs. The toothpaste base ingredients include a combination of known amounts of vegetable glycerin; sorbitol, hydrated silica; purified water; xylitol; carrageenan; sodium lauryl sulfate; and titanium dioxide and a flavoring agent. The appetite suppressing and depressing herbs include at least one of *Garcinia cambogia*; *Gymnema sylvestre*; kola nut; *Citrus aurantium*; yerba mate; and *Griffonia simplicifolia* and comprise a range of substantially 5.5% to substantially 22% by weight of the composition. The appetite suppressing and depressing herbs may further include at least one of guarana, green tea, myrrh, guggul lipid and black current seed oil. Alternatively, the toothpaste composition may be in the form of a dental cream or mouthspray (Zuckerman, 2002).

The present invention consists of and all natural deodorant/antiperspirant which contains olive oil, almond oil, wheat germ oil as gentle soothing emolients, *Salvia officinalis* and *Cupressus semperiens* as natural astringents, *Lavandula angustifolio* and *Citrus aurantium* as natural, mild antiseptic compounds. Beeswax or hydrogenated castor oil may be added as a gentle thickening agent where the inventive formulation is to be used in cream or solid form. In such a manner, skin irritation, swelling, occlusions and inflammations are prevented (Gale, 2003).

A cosmetic composition comprising basalt suitable for use as a body scrub to remove dead skin from the human body such as, for example, from the hands, feet, elbows, and knees and its method of its preparation (Gitomer and Dalton, 2004).

An appetite suppressant mouth spray formulation which simultaneously suppresses the user's appetite while promoting intraoral health. The mouth spray composition includes mouth spray base ingredients; and a plurality of appetite suppressant and appetite depressant herbs. The mouth spray base ingredients include a combination of water; glycerin; xylitol; propylene glycol; sodium benzoate, wintergreen oil and a flavoring agent. A plurality of appetite suppressing and depressing herbs selected from the group consisting of *Griffonia simplicofolia*; *Garcina cambogia*; kola nut; guarana; yerba mate; myrrh gum; *Citrus aurantium*; *Gymnema sylvestre* and green tea (Zuckerman, 2006).

The use of a compound having formula (I) is described as an agent for the browning of skin or hair, formula (I) wherein: R1 and R2 are mutually independently H, OH, C1-C10-alkyl, C1-C10-O-alkyl or O-prenyl, R3 is H, OH, O-glucose or O-rhamnose and R4 is a monosaccharide radical or an oligosaccharide radical having 2, 3, 4 or 5-carbohydrate units, with the proviso that the compound having formula (I) is not used in the form of a preparation based on *Citrus auranitium dulcis* (Vielhaber et al., 2008).

The disclosure herein is directed to organic skin-based products that comprise an excess of 70% organic ingredients. The skin care products described herein are composed of a plurality of organic juices and other organic materials (Jochim and Behnke, 2008).

A composition and method of use, therefore, that allows for permanent straightening or curling of human hair that is not irritating to the skin, that allows for the immediate shampooing of hair thereafter and that provides for re-treatment of the hair without additional damage. The composition primarily uses dimethyl sulfone (MSM) and high temperatures to break and reform disulfide bonds as a means to allow for the penetration of low molecular weight proteins into the shaft of the hair. When curling the hair, it is understood that devices that will stretch hairs into curls at the temperatures expressed in the scope of the patent will produce a permanent curled hair. A test curl will indicate the heat and temperature required to permanently curl the hair being treated (Saute and Saute, 2010).

BIBLIOGRAPHY

1. Carvalho-Freitas MI, Costa M. Anxiolytic and sedative effects of extracts and essential oil from *Citrus aurantium* L. Biol. Pharm. Bull., 2002; 25: 1629–1633.

2. Fugh Berman A, Myers A. *Citrus aurantium*, an ingredient of dietary supplements marketed for weight loss: current status of clinical and basic research. Exp Bio Med. 2004; 229(8): 698–704.

3. Gale C. All natural gentle deodorant and antiperspirant. United States Patent Application 20030206973. Application number: 10/138852. Publication date: 11/06/2003. http://www.freepatentsonline.com/y2003/0206973. html

4. Gitomer TJ, Dalton PE. Body scrub cos-metic composition. United States Patent Application 20040028630. Application number: 10/238321. Publication date: 02/12/2004.

http://www.freepatentsonline. com/y2004/0028630.html

5. Haaz S, Fontaine KR, Cutter G, Limdi N, Perumean-Chaney S, Allison DB. *Citrus aurantium* and synephrine alkaloids in the treatment of overweight and obesity: An update. Obes. Rev., 2006; 7(1): 79–88.

6. Hansen DK, Juliar BE, White GE, Pellicore LS. Developmental toxicity of *Citrus aurantium* in rats. Birth Defects Research Part B: Developmental and Reproductive Toxicology, 2011; 92: n/a. doi: 10.1002/bdrb. 20308

7. http://ntp.niehs.nih.gov/ntp/htdocs Chem_Background/ExSumPdf/Bittero range.pdf

8. http://www.hort.purdue.edu/newcrop/morton/sour_orange.html

9. http://www.mdidea.com/products/proper/proper09005.html

10. http://www.pfaf.org/user/Plant.aspx?LatinName=Citrus+aurantium

11. Jochim M, Behnke K. Compositions for Juice-Based Skin Cleansers. United States Patent Application 20080171011. Application number: 11/933425. Publi-cation date: 07/17/2008. http://www. freepatentson line.com/y2008/0171011. html

12. Kim KW, Kim HD, Jung JS, et al. Characterization of antidepressant-like effects of p-synephrine stereoisomers. Naunyn Schmiedebergs Arch Pharmacol. 2001;364: 21–26.

13. Nair SS, Majeed S, Sankar S, Jeejamol, Mathew M. Formulation of some antioxidant herbal creams. Hygeia. 2009, 1(1): 44–45.

14. Saute R, Saute S. Composition and method for hair straightening and curling. United States Patent Application 20100307526. Application number: 12/802035. Publication date: 12/09/2010. http://www.freepaten tsonline.com/y2010/0307526.html

15. Vielhaber G, Schaper K, Herrmann M. Use of Glycosylated Flavanones for the Browning of Skin or Hair. United States Patent Application 20080305054. Appli-cation number: 11/577846. Publi-cation date: 12/11/2008. http://www. freepatentsonline. com/y2008/0305054. html

16. Zuckerman A. Appetite suppressant mouth spray. United States Patent Appli-cation 20060193795. Application number: 11/066876. Publication date: 08/31/2006. http://www.freepatentsonline.com/y2006/0193795.html

17. Zuckerman A. Appetite suppressant tooth-paste. United States Patent Application 6475810. Application number: 09/749021. Publication date: 11/26/2002. http://www.freepatentsonline.com/6485710.html

Citrus lemon

Plant profile

Kingdom: *Plantae* – Plants

Superdivision: *Spermatophyta* – Seed plants

Class: *Magnoliopsida* – Dicotyledons

Order: *Sapindales*

Genus: *Citrus* L. – Citrus

Subkingdom: *Tracheobionta* – Vascular plants

Division: *Magnoliophyta* – Flowering plants

Subclass: *Rosidae*

Family: *Rutaceae* – Rue family

Species: *Citrus lemon* – Lemon

Vernacular name

San: Nimbaka, Jambira, **Beng:** Pati Lebu, **Eng:** Lemon, **Guj:** Motu Limbu, Goddiya, **Hind:** Nimbu, **Kan:** Nimbuhannu, **Mal:** Cherunarakam, Cherunaranga, **Tam:** Elimichapazham, **Tel:** Nimma kaya

Ayurvedic properties

Gunna: Guru, Tikshan, **Rasa and Vipaka:** Amal, **Veerya:** Ushan

INTRODUCTION

These plants are large shrubs or small trees, reaching 5–15 meter tall, with spiny shoots and alternately arranged evergreen leaves with an entire margin. The leaves are green having shine and are oval shape. Buds are reddish. The flowers are solitary or in small corymbs, each flower 2–4 cm diameter, with five (rarely four) white petals and numerous stamens; they are often very strongly scented. The fruit is a hesperidium, a specialised berry, globose to elongated, 4–30 cm long and

4–20 cm diameter, with a leathery rind surrounding segments or "liths" filled with pulp vesicles. Pulp is pale-yellow, in 8 to 10 segments, juicy, acid. Some fruits are seedless, most have a few seeds, elliptic or ovate, pointed, smooth, 3/8 in (9.5 mm) long, white inside.

Distribution

It is supposed to have been introduced into Southern Italy in 200 AD and to have been cultivated in Iraq and Egypt by 700 AD. It reached Sicily before 1000 and China between 760 and 1297 AD. Some are believed that citrus is to have originated in the part of Southeast Asia bordered by Northeastern India, Myanmar (Burma) and the Yunnan province of China. Major commercial citrus growing areas include Southern China, the Mediterranean Basin (including Southern Spain), South Africa, Australia, the Southernmost USA and parts of South America. In the United States, Florida, California, Arizona and Texas are major producers, while smaller plantings are present in other Sun Belt states.

According to UN 2007 data, Brazil, China, the US, Mexico, India, Spain and Pakistan are the world's largest citrus-producing countries.

Chemical Constituents

The fruit juice mainly contains sugars and fruit acids, mainly citric acid (8%). Lemon peel consists of two layers: The outermost layer (pericarp, zest) contains an essential oil (6%), that is mostly composed of limonene (90%) and citral (5%) plus traces of citronellal, α-terpineol, linalyl and geranyl acetate. The inner layer (mesocarp), on the other hand,

contains no essential oil but a variety of bitter flavone glycosides and coumarin derivatives.

It also content Cineole, trihydroxy-flavonone, methoxyphenyl, hexen, acetic acid, adenosine, alpha-bergamotene, alpha-terpinene, and alpha-humulene.

Medicinal uses

It is used to lower the body heat and to give tone and vitality to the body. It is best for treatment of colds, coughs, throat diseases, headache, stomach ache, indigestion and rheumatism. It is acts as a disinfectant. Lemon juice helps to control blood pressure, purifies blood, reduces swollen spleen, and strengthens immune system as it has vitamin C, B, B_2, calcium and iron. It protects the body against germs and bacteria. Lemon juice cures menorrhagia, nose-bleeding, hepatitis, gastric ulcer if taken several times daily. Lemons have been used as a household remedy for sore throat, gastric and liver troubles, heartburn, biliousness, asthma, vomiting, travel sickness, etc. Lemon juice mixed with glycerine is used for chapped lips or chilblains. For constipation, the juice of a lemon as taken in a glass of hot water one-half hour before breakfast.

Cosmetic uses

Lemon juice is valued in the home as a stain remover, and a slice of lemon dipped in salt can be used to clean copper-bottomed cooking pots. Lemon juice has been used for bleaching freckles and is incorporated into some facial cleansing creams. **Lemon peel oil** is much used in furniture polishes,

Citric acid   Limonene   α-terpineol   Citral

detergents, soaps and shampoos. It is important in perfume blending and especially in colognes. Externally, it is used for treatment of corns, warts, burns, dandruff can be taken care of by applying the juice externally on the site of the ailment.

Pharmacological Action

Cytotoxic effects of Citrus peel extracts were evaluated in the root meristem cells of *Allium cepa*. Ethanolic (80%) extracts were tested for antimitotic and genotoxic effects. Each test concentration of *Citrus lemon* and *Citrus sinensis* extracts significantly reduced the mitotic index. *Citrus lemon* extracts were more effective from *Citrus sinensis* extracts. Furthermore, the extracts shows genotoxic effects such as breaks, bridges, stickiness, pole deviation and micronuclei on root meristem cells but not in very high levels. The effect was directly related to the concentrations of the fruit extracts. These observation shows that Citrus fruit extracts can be harmful for organisms (Ali and Celik, 2007).

The current study was designed to investigate the effect of *Citrus lemon*. L. Burm. (Rutaceae) fruits in experimental liver damage. The ethanol extract of *Citrus lemon*. fruits was evaluated for its effects on experimental liver damage induced by carbon tetrachloride, and the ethyl acetate soluble fraction of the extract was evaluated on HepG2 cell line. The ethanol extract normalized the levels of aspartate aminotransferase (ASAT), alanine aminotransferase (ALAT), alkaline phosphatase (ALP), and total and direct bilirubin, which were altered due to carbon tetrachloride intoxication in rats. Three doses of ethanol extract (i.e. 150, 300, and 500 mg/kg) were evaluated (Bhavsar et al., 2007).

The present study has been specifically designed to investigate the hypolipidemic effects of *Citrus lemon* juice in rabbits after high cholesterol diet for four weeks. The *Citrus lemon* juice (1 ml/kg/day) revealed a significant reduction in serum cholesterol, triglycerides; low density lipoprotein levels and resulted in an increase in high density lipoprotein. These results suggest that the hypocholesterolemic effects of *Citrus lemon* juice may be due to its antioxidant effect (Khan et al., 2010).

The antioxidant and antinociceptive activities of *Citrus lemon* essential oil (EO) were assessed inmice or *in vitro* tests. EO possesses a strong antioxidant potential according to the scavenging assays. Moreover, it presented scavenger activity against all *in vitro* tests. Orally, EO (50, 100, and 150 mg/kg) significantly reduced the number of writhes, and, at highest doses, it reduced the number of paw licks. Whereas naloxone antagonized the antinociceptive action of EO (highest doses), this suggested, at least, the participation of the opioid system. Further studies currently in progress will enable us to understand the action mechanisms of EO (Campelo et al., 2011).

The objective of the present work was to evaluate the *in vitro* anthelmintic potency of the ethanolic extract of *Citrus lemon* L. peels using Indian earthworms (*Pheretima posthumad*). The various concentrations (25–100 mg/ml) of the ethanolic extract were tested *in vitro* for anthelmintic potency by determination of time of paralysis and time of death of worm. Piperazine citrate (15 mg/ml) used as standard. The result of present study indicates that the *Citrus lemon* L. potentiate to paralyze earthworm and also caused its death after some time. Thus, the present study demonstrate that the *Citrus lemon* L. as an potent anthelmintic has been confirm as the ethanolic extracts of peels displayed activity against the earthworm used in study (Bairagi et al., 2011).

Toxicological Studies

Currently no toxicity information is available for **Citrus lemon.** Lemon juice is very low in pH, about 2.0. Its sour taste sometimes causes burning sensation if encounters with mouth, tongue, and lip ulcers. In addition, if taken large amounts may exacerbate acid—peptic disease and stomach ulcer conditions.

Clinical Investigation

The study was a prospective, open, non-comparative, phase III clinical trial. A total of 35 patients, who were diagnosed as suffering from dandruff, and who were willing to give informed written consent were included in the study. All the patients were advised to apply the "Anti-dandruff shampoo", twice a week for a period of 6 weeks. The predefined primary efficacy endpoints were reduction in dandruff lesions, reduction in overall scalp inflammation and healing of existing lesions. The predefined secondary safety endpoints measures were incidence of adverse events and overall patient compliance to the drug treatment. All the adverse events either reported or observed by the patients were recorded with information about severity, date of onset, duration and action taken regarding the study drug. This study observed a significant reduction in the mean scores of itching and white scales of dandruff. The subjective evaluation revealed remarkable symptomatic and clinical improvement in 2 weeks period. The excellent anti-dandruff action of "Anti-dandruff shampoo" might have been due to the synergistic antifungal, anti-inflammatory and local immunostimulatory actions of its ingredients. Therefore, it may be concluded that, "Anti-Dandruff Shampoo" is effective and safe in the management of dandruff (Ravichandran et al., 2004).

Clinical experiences *in vitro* and clinical studies have demonstrated the curative potency and safety of Citrus/Cydonia compositum in seasonal allergic rhinitis treatment. Hence, comparative efficacy and safety study has conducted of two routes of administration (nasal spray versus subcutaneous injections) by a national, randomised, comparative clinical trial with two parallel groups. 23 patients fulfilled the study requirements. Immunologic outcome assessments were blinded to group assignment. 23 patients were randomized and from 22/23 patients (11 in each group) blood samples were analyzed before and after treatment. Both routes of administration demonstrate immunological and clinical effects, with larger inflammatory and innate immunological effects of the nasal spray route and larger allergen-specific clinical effects of the subcutaneous route, and are safe (Baars et al., 2011).

Market Formulations

Himalaya Herbals Anti-wrinkle Cream

Himalaya Herbals anti-wrinkle cream is a herbal ayurvedic formulation containing 14 natural botanical ingredients. Himalaya's anti-wrinkle cream is packed with anti-oxidants. It nourishes, moisturizes and emollients and thus keeps skin fresh and soft. It prevents oxidative skin damage from free radicals, appearance of wrinkles and skin laxity. It delays wrinkles and smoothens fine lines.

Ingredients: Aloe vera, Papaver rhoeas (red poppy, rakta-posta), *Vitis vinifera* (grapes, draksha), *Citrus limon* (Lemon, nimbaka), *Solanum lycopersicum* (Tomato, raktamaci), *Santalum album* (Sandalwood tree, chandana).

Action: Aloe vera, one of the key ingredients in the cream, are helpful in relaxing muscles, enhancing the healing process and treating inflammatory skin disorders. The anthraquinones also scavenge the H_2O_2 on the skin surface, thereby preventing free radical damage to the skin surface. Its collagen contractile enhancement is also a potent factor in the treatment of wrinkles. *Rubia cordifolia* also provides antioxidant activity. This maintains the glutathione content in the physiological system, contributing to the management of wrinkles. *Vitis vinifera* or grapes provide the free-radical scavenging activity thus increasing smoothness and reducing fine wrinkles. Flavonoids also provide additional hydration to the skin. *Citrus lemon* enhances the potential of bio-skin surface, adding to the anti-wrinkle activity. Ferrulic acid present in *Solanum lycopersicum* provides acyl-biosynthesis in the endothelial skin, aiding a role in antioxidant activity. Curcumnoids from *Curcuma longa* shows a marked reduction in alpha-5-beta-1; softening the structure, preventing from the age-related wrinkles. Glabridin from *Glycyrriza glabra* reduces inflammation, stretch marks and aids in the prevention of wrinkles.

Direction: Massage gently over cleansed face twice a day. Use regularly. This is a herbal cream that suits all skin types and has no side effects.

Supplied by: Himalaya Herbals, Bangalore, India.

Anti Spot™

It is used for mild acne vulgaris, especially in adolescence, oily skin, spots, pimples, blackheads.

Ingredients: A specific mixture of three herbs: *Inula helenium, Saponaria officinalis, Citrus limonum.*

Action: Anti Spot™ extract derived highly anti-inflammatory effects. In the other hand, lemon peel oil is widely known as antiseptic and anti-inflammatory agent. The mixture of the three extracts have a positive synergetic effect in treating acne. This is combined with mildly antiseptic and antiinflammatory activity. Indeed some of the herbal constituents add to skin moisture, softness and texture.

Total Body Detox

Ingredients: Citrus limonum 1X, *Taraxacum officinale* 2X, *Uva-ursi* 2X, *Berberis vulgaris* 3X, *Capsicum annuum* 3X, *Galium aparine* 3X, *Lobelia inflata* 3X, *Nux vomica* 3X, Schisandra 6X, Tylophora 6X, Brain 8X, *Cor suis* 8X, *Glandula suprarenalis suis* 8X, *Hepar suis* 8X, *Hepar sulphuris calcareum* 8X, Intestine 8X, *Kidney* 8X, *Lacticum acidum* 8X, *Lung* 8X, *Lymph* 8X, *Magnesia phosphorica* 8X, *Spleen* 8X, Thyroidinum 8X. Demineralized water, 10% ethanol, 8% glycerin, contains no starch, salt, preservatives, wheat, yeast or milk derivatives.

Recommended dosage: Adults: 15 drops under the tongue 3 times a day or add 45 drops to 8–16 ounces of water and sip throughout the day. Take 15 minutes away from food or drink. Children 4 to 12: one-half adult dose. Consult a physician for use in children under 12 years of age.

Warnings: As with any drug, if pregnant or breastfeeding; or under age 18; ask a health professional before use. Do not use if tamper evident seal around neck of bottle is broken or missing. Keep out of reach of children. In case of overdose, get medical help or contact a Poison Control Center right away.

Storage: Store in a cool, dry place after opening.

Distributed by: HCG Weightloss System, St. Petersburg.

Citrus Limonum Hand Wash

A sulfate-free hand wash formulated with organic sicilian lemon oil known for its antibacterial properties and its fresh, uplifting fragrance. Gentle cleansing agents derived from corn, coconut and oat respect the skin and help prevent dryness.

Ingredients: Citrus aurantium bergamia (bergamot) leaf extract, decyl glucoside, glycerin, cocamidopropyl betaine, *Citrus medica limonum* (lemon) peel oil, limonene, citral, xanthan gum, polysorbate 20, sodium lauroyl oat amino acids, phenoxyethanol, sodium hydroxymethylglycinate, panthenol, lactic acid, ethylhexylglycerin.

Action: Lemon oil from cold pressed sicilian lemons refreshes and helps to prevent dryness, Oat amino acids gently cleanse the hands.

Method of Preparation

1. *Mouth ulcer and throat infection:* Mix 1 part of lemon juice with 2 parts of water and gargle it for few minutes. This will cure mouth ulcer and throat infection.

2. *As hair rinse:* Take about half cup of fresh lemon juice and mix well in about one quart of water. Use as a final rinse after routine shampooing. The acidic nature of lemon juice will lighten the hairs and cause the cuticles to tighten, creating shine. The lightening effect is particularly outstanding in blonde hairs.

3. Homemade shampoo

- 1 whole egg
- 1 tsp olive oil
- 1 tsp lemon juice
- A few bits of your favorite soap
- 1/2 cup water

Soak these soap pieces into warm water for few hours till these are soft and slimy. Now mix well all ingredients into a bowl with the help of a fork or whisk wire. This could be your first homemade hair product and store it in a glass bottle. You can use it for 2–3 days if kept outside the refrigerator. However, if a longer storage is required put it in the refrigerator.

4. Homemade hair conditioner

- 3–4 tbsp henna powder
- 1 tablespoon hair oil of your choice
- 1 egg
- juice of 1/2 lemon

First beat the egg with a whisk beater till it is frothy. Add henna powder in sections and continue beating. Add some water and lemon juice to get reasonable consistency. Leave the mixture to set for one hour. Apply gently and thoroughly to hair and scalp. Let it stay for one hour before you rinse and shampoo. Further no conditioner is required.

5. Hair packs

- 1 Egg
- 2 tablespoon honey
- 1 tablespoon lemon juice
- 2 tablespoon olive oil (any massage oil can be used like, almond oil, mustard oil, coconut oil or sesame oil)
- Few drops of essential oil

Mix all the constituents well in a bowl and apply over the entire face. Leave it for twenty to forty minutes, followed by shampooing with luke warm water. End with a cold water rinse. This is an effective hair treatment for all hair types.

6. Hair spray: Lemon hair spray, this is a quick recipe, which works well for all types of hairs. All you need;

- 2 to 4 lemons
- 4 cups of water
- Drops of essential oil

Cut lemons in slices and place in a sauce pan along with water. Let it simmer for fifteen minutes till half of the water has evaporated. Strain the liquid and add essential oil. Pour this in a spray bottle and keep refrigerated. Here it stays well for a week.

Storage of Products

Store in cool and dry place. Keep away from direct strong light and heat.

Dosage

Clinical information is limited. To increase citrate levels, 120 ml of lemon juice, containing citric acid 5.9 g, was diluted and consumed daily (Kang et al., 2007).

Patents

The two basic main ingredients are castor oil and a special lemon extract. The special lemon extract is made from fresh lemon peel. The peel, including the bioflavonoids membrane, is blended with purified water until it is liquidified. Then the mixture is filtered through a sanitized cloth. Other ingredients that have been found helpful include: inositol, choline (from bitartate), niacinamide or nicotinic acid, manganese in chelated form, bioflavonoids, and folic acid. Finally, perfume and sodium benzoate (as a preservative) can be added (Olguin, 2000).

A hair conditioner-scalp stimulator solution made of liquid extracts from coconut, lemon and lime fruit (John, 2002).

The invention concerns a hair lotion stimulating hair growth. It consists of a mixture comprising at least the following constituents: iodised salt, lemon juice, fresh onion, citric acid and alcohol, a fragrance designed to neutralise the onion smell being preferably added to said mixture. Said lotion concerns the industrial and commercial field of production and commercial distribution of hair care products, and is suitable for both women and men, independently of their age (Ahmed, 2004).

A non-toxic mucosal disinfectant for topical application in the nose has a composition of 91% isopropyl alcohol of at least 50% by weight; sesame oil not exceeding 45% by weight, lemon oil of about 2% by weight, aloe of about 5–10% by weight, and optional components of chlorhexidine gluconate and grapefruit seed extract. All of the components are mixed homogeneously, with the sesame oil supplementing and neutralizing the dehydrating effect of the alcohol (Manuel and Shanley, 2005).

A topical composition and method of preparation is disclosed wherein olive oil, bees wax, lemon juice and boric acid is combined to yield a cream for application to burns. The composition makes the patient more comfortable and has been shown to promote more rapid healing with less scarring than many other products (Soubhie, 2005).

Disclosed are eye drops to which are added a clove perfume and a peppermint perfume, or further at least one additional perfume selected from the group consisting of a eucalyptus perfume and a lemon perfume, for improved sensation upon instillation and suppression of peculiar chemical smells of eye drops (Nakayama and Nemoto, 2008).

BIBLIOGRAPHY

1. Ahmed K. Hair lotion stimulating hair growth. United States Patent 20040170594. Application number: 10/481885, Publication date: 09/02/2004.

2. Ali O, Celik TA. Cytotoxic effects of peel extracts from *Citrus limon* and *Citrus sinensis*. Caryologia. 2007; 60 (1-2): 48-51.

3. Baars EW, Jong M, Nierop AFM, Boers E, Savelkoul HFJ. Citrus/Cydonia Compositum Subcutaneous Injections versus Nasal Spray for Seasonal Allergic Rhinitis: A Randomized Controlled Trial on Efficacy and Safety. International Scholarly Research Network. Volume 2011, Article ID 836051, 1–11.

4. Bairagi GB, Kabra AO, Mandade RJ. Anthelmintic activity of *Citrus limon* L. (Burm) peels extract in Indian adult earthworm. International Journal of Pharmacy and Technology. 2011; 3(1): 1913–1919.

5. Bhavsar SK, Joshi P, Shah MB, Santani DD. Investigation into Hepatoprotective Activity of *Citrus limon*. Pharmaceutical Biology. 2007; 45 (4): 303-311.

6. Campelo LML, de almeida AAC, de Freitas RLM, Cerqueira GS, de Sousa GF, Saldanha GB et al., Antioxidant and Antinociceptive Effects of Citrus limon Essential Oil in Mice. Journal of Biomedicine and Biotechnology. 2011; 1–8. Article ID 678673.

7. http://ayurvedicmedicinalplants.com/plants/2445.html

8. http://en.wikipedia.org/wiki/Citrus

9. http://ezinearticles.com/?Natural-Hair-Care-With-Lemon&id=1398444

10. http://factoidz.com/antidandruff-and-disinfectant-lemon-medicinal-properties/

11. http://www.ayushveda.com/herbs/citrus-limon.htm

12. http://www.hort.purdue.edu/newcrop/morton/lemon.html

13. http://www.naturalremediesathome.com/lemons-medicinal-properties

14. http://www.online-family-doctor.com/fruits/lemon.html

15. John L. Hair conditioner and scalp treatment solution. United States Patent Application 20020168326. Application number: 09/850071, Publication date: 11/14/2002. http://www.freepatentsonline.com/y2002/0168326.html

16. Kang DE, Sur RL, Haleblian GE, Fitzsimons NJ, Borawski KM, Preminger GM. Long-term lemonade based dietary manipulation in patients with hypocitraturic nephrolithiasis. J Urol. 2007; 177(4):1358–1362.

17. Khan Y, Khan RA, Afroz A, Siddiq A. Evaluation of hypolipidemic effect of *Citrus lemon*. Journal of Basic and Applied Sciences. 2010; 6(1): 39–43.

18. Manuel V. Jr, Shanley LM. Nontoxic mucosal disinfectant containing isopropyl alcohol, sesame oil, aloe, and lemon oil. United States Patent 6869623. Application number: 10/623816, Publication date: 03/22/2005. http://www.freepatentsonline.com/6869623.html

19. NakayamaH, Nemoto F. Eye drops. United States Patent 20080003310. Application number: 11/854171, Publication date: 01/03/2008. http://www.freepatentsonline.com/y2008/0003310.html.

20. Olguin ME. Hair growth stimulant. United States Patent Application 6159475. Application number: 09/179732, Publi-cation date: 12/12/2000. http://www. freepatentsonline.com/6159475.html

21. Ravichandran G, Bharadwaj VS, Kolhapure SA. Evaluation of the clinical efficacy and safety of "Anti-Dandruff Shampoo" in the treatment of dandruff. The Antiseptic. 2004; 201(1): 5–8.

22. Soubhie E. Composition for the treatment of burns, sunburns, abrasions, ulcers and cutaneous irritation. United States Patent 20050112208. Application number: 10/717914, Publication date: 05/26/2005. http://www.freepatentsonline.com/y2005/0112208.html

Cocos nucifera

Plant profile

Kingdom: *Plantae* – Plants
Subkingdom: *Tracheobionta* – Vascular plants
Superdivision: *Spermatophyta* – Seed plants
Division: *Magnoliophyta* – Flowering plants
Class: *Liliopsida* – Monocotyledons
Subclass: *Arecidae*
Order: *Arecales*
Family: *Arecaceae* – Palm family
Genus: *Cocos* L. – Coconut palm
Species: *Cocos nucifera* – Coconut palm

Vernacular name

San: Narekela; Trinadruma, **Beng:** Narkel, **Eng:** Coconut palm **Guj:** Nariyal, **Hind:** Nariyal, **Mar:** Naral, **Kashmiri:** Narjeel, **Kan:** Thengini kai, Tengu, **Mal:** Thenga, **Ori:** Nadia, **Tam:** Thenga, **Tel:** Kobbari, **Punjabi:** Gola

Ayurvedic properties

Rasa: Madhura, Kashaya, **Guna:** Guru, Snigdha, **Veerya:** Seeta, **Vipaka:** Madhura.

INTRODUCTION

Cocos nucifera trees have a smooth, columnar, light grey-brown trunk, with a mean diameter of 30–40 cm at breast height, and topped with a terminal crown of leaves. A tall, unbranched palm grows up to 35 meters in height. Trunk slender and slightly swollen at the base, usually erect but may be leaning or curved. Leaves pinnate, feather shaped, 4–7 m long and 1–1.5 m wide at the broadest part. Leaf stalks 1–2 cm in length and thornless. Flowers are small; light yellow, in clusters that emerge from canoe-shaped sheaths among the leaves. Male flowers are small and more numerous. Female flowers fewer and occasionally completely absent; larger, spherical structures, about 25 mm in diameter. Fruit roughly ovoid, up to 5 cm long and 3 cm wide, composed of a thick, fibrous husk surrounding a somewhat spherical nut with a hard, brittle, hairy shell. The nut is 2–2.5 cm in diameter and 3–4 cm long. Inside the shell is a thin, white, fleshy layer known as the 'meat'. The interior of the nut is hollow but partially filled with a watery liquid called 'coconut milk'. The meat is soft and jelly-like when immature but becomes firm with maturity. Coconut milk is abundant in unripe fruit but is gradually absorbed as ripening proceeds. The fruits are green at first, turning brownish as they mature; yellow varieties go from yellow to brown. Within the shell is a single seed.

Distribution

Coconut is believed to have its origins in the Indo-Malayan region, from whence it spread throughout the tropics. It is native to coastal areas (the littoral zone) of Southeast Asia (Malaysia, Indonesia, Philippines) and Melanesia. In prehistoric times wild forms (niu kafa) are believed to have been carried eastward on ocean currents to the tropical pacific islands (Melanesia, Polynesia, and Micronesia) and westward to coastal India, Sri Lanka, East Africa, and tropical islands (e.g. Seychelles, Andaman, Mauritius) in the Indian ocean. In these regions, the palms were able to establish themselves on sandy and coralline coasts. Coconut is either an introduction or possibly native to the pacific coast of Central America. Its natural habitat was the narrow sandy coast, but it is now found on soils ranging from pure sand to clays and from moderately acidic to alkaline.

Chemical Constituents

It contains fixed oil, 57.5–71%; volatile oil, wax containing the myricyl ester of cerotic acid. Meat: potein, 6.3%; vitamins A, B, and C; nonyl alcohol; methyl heptyl ketone; methyl undecyl ketone; capronic, decylic, caprylic, lauric and myristic acids; lecithin; stigmasterin, phytosterin; choline; globulin; galactoaraban; galactomannan. It also contains water, 93%; protein, 0.5%; ash, 1%; saccharose; oxidase; catalase, diastase. Khopra (desiccated coconut) contains about 60 to 70% coconut oil. Coconut oil is composed, as might be inferred by its high melting point of over 32°C, mostly of triglycerides of saturated fatty acids. Lauric (dodecanoic acid; 40 to 55%) and myristic acid (tetradecanoic acid; 15 to 20%) dominate, but several other fatty acids are found at concentrations of 5 to 10%. The two short-chain acids caprylic (octanoic) and capric (decanoic) acid (which are responsible for the smell of overaged coconut oil), the long-chain palmitic acid (hexadecanoic acid) and oleic acid, which is the only unsaturated fatty acid found at significant amounts. Furthermore, a great number of alkyl pyrazines were identified (pyrazine, methyl pyrazine, dimethyl pyrazines, vinyl pyrazine, isopropyl pyrazine).

Lauric acid

Myristic acid

Medicinal uses

Coconuts are used in folk remedies for tumors. It is also used as anthelmintic, anti-dotal, antiseptic, aperient, aphrodisiac, astringent, bactericidal, depurative, diuretic, hemostat, pediculicide, purgative, refrigerant, stomachic, styptic, suppurative, and vermifuge. It is a folk remedy for abscesses, alopecia, amenorrhea, asthma, blenorrhagia, bronchitis, bruises, burns, cachexia, calculus, colds, con-stipation, cough, debility, dropsy, dysentery, dysmenorrhea, earache, erysipelas, fever, flu, gingivitis, gonorrhea, hematemesis, hemo-ptysis, jaundice, menorrhagia, nausea, phthisis, pregnancy, rash, scabies, scurvy, sorethroat, stomach, swelling, syphylis, toothache, tuberculosis, tumors, typhoid, venereal diseases, and wounds. The liquid extracted from the stem is used in treating weakness after childbirth. Juice from the midrib at the lower base of the leaf is used in treating maternal postpartum illness. Coconut milk is used to treat fish poisoning. The root is also employed in treating stomachache and blood in the urine. The Oil from the kernel is rubbed onto stiff joints. The oil is also used to treat rheumatism and back pains.

Cosmetic uses

Coconut oil is excellent as a skin moisturizer and softener. Coconut oil massage works wonders for the skin, moisturizing it and keeping it supple. The fat in the oil helps in reducing the appearance of wrinkles, without causing any kind of irritation. Skin problems, such as psoriasis, dermatitis and eczema, can also be treated with coconut oil application. Fractionated coconut oil is often used in the manufacture of essences, massage oils and cosmetics. The oil is also used in expensive skin care products. The oil will aid in removing the outer layer of dead skin cells, making the skin smoother. The skin will become more evenly textured with a healthy "shine". Pure virgin coconut oil is the best natural ingredient for skin lotion available. It prevents destructive free-radical formation and provides protection against them. It helps

to keep connective tissues strong and supple so that the skin does not sag and wrinkle. In some cases it may even restore damaged or diseased skin.

Traditionally, coconut oil has long been used for manufacturing soaps, detergents and shampoos and detergents. The oil is considered to produce a rich lather that is superior to other oils, in any kind of water, right from mineral-rich hard water to seawater. The antibacterial, antiviral and antifungal properties present in the oil make it an effective disinfectant as well.

Coconut oil has immense healing pro-perties associated with it. When applied on scrapes and cuts, it forms a thin, chemical layer, protecting the wound from outside dust, bacteria and virus. Also, it speeds up the healing process of bruises, by repairing damaged tissues.

One of the most popular usages of coconut oil is that of being a hair care product. It helps to remove dandruff problems. Coconut oil also contains effective proteins that nourish hair better than any over-the-counter con-ditioners and hair care products. Massaging head with coconut oil, using circular motions, can be extremely helpful in relieving stress and mental fatigue. Apart from its soothing nature, the aroma of the oil also helps to lower stress level.

Pharmacological Action

Analgesic/Antioxidant: Antinociceptive and free radical scavenging activities of *Cocos nucifera* L. (Palmae) husk fiber aqueous extract: The study demonstrated the analgesic and radical scavenging properties of CN aqueous extract from the husk fiber. Topical treatment of rabbits with the extract did not induce significant dermic or ocular irritation.

Antioxidant: In vitro evaluation of anti-oxidant properties of *Cocos nucifera* Linn. water: The antioxidant activity as most significant in fresh samples of coconut water, diminishing with heat. Maturity also drastically decreased the scavenging ability. The scavenging ability may be partly attributed to the ascorbic acid, an important constituent of coconut water.

Hypertension: The control of hypertension by use of coconut water and mauby: two tropical food drinks provided significant decreases, approximately double the largest values seen with single interventions.

Antineoplastic: The husk fiber of *Cocos nucifera* L. (Palmae) is a source of anti-neoplastic activity. Study results on the *in vitro* antitumoral activities of aqueous extracts of the husk showed antitumoral activity against a leukmia cell line. Study suggests a very inexpensive source of new antineoplastic and anti-multidrug resistant drugs.

Burn wound healing property: Study concluded that the oil of *Cocos nucifera* is an effective burn wound healing agent. There was significant improvement in burn wound contraction in the group treated with the combination of CN and silver sulphadiazine. It suggests *C. nucifera* can be a cheap and effective adjuvant to other topical agents.

Antiulcerogenic: A study of warm water crude extract of coconut milk and a coconut water dispersion showed that coconut milk and water had protective effects on ulcerated gastric mucosa. The coconut milk provided stronger protection on indomethacin-induced ulceration than coconut water in rats.

Antihelmintic: A study of the liquid extracted from the bark of the green coconut and butanol extract on mice showed that the *Cocos nucifera* extracts may be useful in the control of intestinal nematodes.

Protein content: Study showed native coconut proteins consisted of four major polypeptides. The proteins had a relatively high level of glutamic acid, arginine and aspartic acid.

Antineoplastic activity: Study of aqueous extracts of *Cocos nucifera* showed antitumoral activity against leukemia cell line K562 and suggests a potential for an inexpensive source of new antineoplastic and anti-multidrug resistant drugs.

Toxicological Studies

There is no known toxicity for coconut oil. The FDA includes it in its generally recommended as safe (GRAS) list. An easy supplement would be to use it as cooking oil. It tolerates moderately high-cooking temperatures, but best to keep it below smoking point of 350 degrees. As in any other cooking oil, avoid overheating because of toxic by-products. When available, the best is the "virgin" coconut oil, made from fresh coconuts, extracted by boiling, fermentation, refrigeration, mechanical press or centrifuge, not subjected to high temperatures or chemical solvents. Also available as refined, bleached, and deodorized (RBD) coconut oil, usually made from dried coconut, copra, that might have undergone sun-drying, smoking or kiln processing, using higher temperatures and chemical solvents.

Another research showed the effect of coconut water with paracetamol toxicity on some liver enzyme markers {aspartate aminotransferase (AST), alanine amino-transferase (ALT) and alkaline phosphatase (ALP)}, and total protein, urea total bilirubin and conjugated bilirubin were studied. Group 1 represented negative control fed normal rat diet and water *ad libitum*. Group 2 represented positive control which was administered (i.p.) single dose (400 mg/kg b.w.) and a single concentration (25 mg/kg) of the toxicant (paracetamol). While groups 3, 4 and 5 were administered varying doses (2000 mg/kg, 3000 mg/kg and 4000 mg/kg b.w.) of coconut water. Groups 6, 7 and 8 were administered same range of doses after exposure to the toxicant (paracetamol). The activities of the liver enzyme markers in group 2 (positive control) were found to significantly increased (p is lesser than 0.05) when compared with the negative control (group 1) and other test groups. The test groups administered 3000 and 4000 mg/kg respectively after exposure to paracetamol significantly decreased (p is lesser than 0.05) the activities of the liver enzyme markers as compared with the positive control. The groups (6, 7 and 8) treated with combined therapy had significant increase (p is lesser than 0.05) in levels of total protein and urea and significantly decreased (p is lesser than 0.05) in the levels of total (unconjugated) and conjugated bilirubin (Nwodo et al., 2010).

Cocos nucifera (coconut) oil, oil from the dried coconut fruit, is composed of 90% saturated triglycerides. It may function as a fragrance ingredient, hair-conditioning agent, or skin-conditioning agent and is reported in 626 cosmetics at concentrations from 0.0001 to 70%. The related ingredients covered in this assessment are fatty acids, and their hydrogenated forms, corresponding fatty alcohols, simple esters, and inorganic and sulfated salts of coconut oil. The salts and esters are expected to have similar toxicological profiles as the oil, its hydrogenated forms, and its constituent fatty acids. Coconut oil and related ingredients are safe as cosmetic ingredients in the practices of use and concentration described in this safety assessment (Burnett et al., 2011).

Clinical Investigation

This study aimed to determine the effectivity and safety of virgin coconut oil compared with mineral oil as a therapeutic moisturizer for mild-to-moderate xerosis. A randomized double-blind controlled clinical trial was conducted on mild-to-moderate xerosis in 34 patients with negative patch-test reactions to the test products. These patients were randomized to apply either coconut oil or mineral oil on the legs twice a day for 2 weeks. Patients and the investigator separately evaluated, at baseline and at each weekly visit, skin symptoms of dryness, scaling, roughness, and pruritus by using a visual analogue scale and grading of xerosis. Coconut oil and mineral oil have comparable effects. Both oils showed effectivity through significant improvement in skin hydration and increase in skin surface lipid levels. Safety was demonstrated through no significant difference in TEWL and skin pH. Safety for both was further demonstrated by negative

patch-test results prior to the study and by the absence of adverse reactions during the study (Agero and Verallo-Rowell, 2004).

Coconut oil obtained from the nuts of *Cocos nucifera* was formulated into creams in order to standardize its use and present it in an elegant form. Using the fusion method, oil in water (o/w) creams were formulated in concentrations of 5 to 40% w/w of oil. The release of active ingredients from creams was investigated using cream challenge and skin inoculation tests, whereby creams were exposed to various spots on skin inoculated with *P. aeruginosa* ATCC 7853, *E. coli* ATCC 9637, *P. vulgaris* (clinical isolate), *B. subtilis* ATCC 607 and *C. albicans* ATCC 10231. The stability of creams was also evaluated using a standard method. The results showed that active ingredients of the coconut oil were released from the creams; this was shown from the good antimicrobial activity of the cream confirming that all formulation ingredients were compatible and did not interfere with activity of the oil. The creams were also found to be stable, as a result of their ability to withstand shock and maintain their physical characteristics (Oyi et al., 2010).

An open-label pilot study on four weeks of virgin coconut oil (VCO) has carried out to investigate its efficacy in weight reduction and its safety of use in 20 obese but healthy Malay volunteers. Efficacy was assessed by measuring weight and associated anthropometric parameters and lipid profile one week before and one week after VCO intake. Safety was assessed by comparing organ function tests one week before and one week after intake of VCO. Paired t-test was used to analyse any differences in all the measurable variables. *Results revealed* only waist circumference (WC) was significantly reduced with a mean reduction of 2.86 cm or 0.97% from initial measurement (p = .02). WC reduction was only seen in males (p <.05). There was no change in the lipid profile. There was a small reduction in creatinine and alanine transferase levels. Thus results concluded that VCO is efficacious for WC reduction especially in males and it is safe for use in humans (Liau et al., 2011).

Market Formulations

Psorvate Ointment

Superior formulation with a unique combination of natural extracts and oils for long-lasting relief from itching, irritation, redness, scaling and flaking commonly associated with Psoriasis. It reduces hyperkeratosis, scaling, dermal vessel dilation and tortuosity. It offers immediate relief and long-lasting remission.

Ingredients: Natural extracts and/or essential oils from *Wrightia tinctoria*, *Cocos nucifera*, beeswax.

Dosage and application: Apply minimum quantity of ointment on the treatment site and rub in for a few seconds to spread to a thin film on the surface. Avoid contact with eyes. If contact occurs, rinse thoroughly with water. Stop use and ask a doctor if condition worsens or does not improve after regular use. Keep out of the reach of children.

Supplied by: Elder Health Care Ltd., Mumbai.

Nude Lip Liner Pencil

It creates the appearance of fuller looking lips. Prevents lip color products from bleeding into fine lines around mouth and creates natural looking lip definition. It helps lip colour stay on longer when used as a base underneath other lip color products.

Ingredients: Titanium dioxide (CI 77891), hydrogenated palm kernal glycerides, hydrogenated coconut oil (*Cocos nucifera*), microcrystalline wax (*Cera microcristallina*), hydrogenated palm glycerides, mica, myristyl myristate, rhus succedanea fruit wax, behenyl alcohol, hydrogenated coco-glycerides, cetyl palmitate, copernicia cerifera (carnuaba) wax, hydrogenated castor oil sorbitan palmitate, iron oxides (CI 77492, CI 77491, CI 77499), stearalkonium hectorite, propylene carbonate (nude end) hydrogenated palm kernal glycerides, titanium

dioxide (CI 77891), hydrogenated coconut oil (*Cocos nucifera*), microcrystalline wax (*Cera microcristallina*), hydrogenated palm glycerides, mica, myristyl myristate, iron oxides (CI 77491, CI 77492), rhus succedanea fruit wax, behenyl alcohol, hydrogenated coco-glycerides, cetyl palmitate, copernicia cerifera (carnauba) wax, hydrogenated castor oil, sorbitan palmitate, Stearalkonium hectorite, propylene carbonate.

Supplied by: Becca, USA.

Radiance Body Lotion With Royal Jelly

This body lotion is formulated with Royal jelly, to significantly enhance skin radiance. Sun flower oil nourishes and moisturizes, while mica a light reflecting mineral, enhances skin's natural glow.

Ingredients: Water, sunflower oil, vegetable glycerin, coconut oil, stearic acid, beeswax, mica, royal jelly, glucose, aloe extract, sage extract, nettle extract, rosemary extract, chamomile extract, xanthan gum, orange oil, lemon oil, ylang ylang oil, sucrose distearate, sodium borate, tocopherol, glucose oxidase and lactoperoxidase.

Direction: Apply evenly to arms, legs, shoulders and neck for a natural radiant glow. It can be used as a daily facial moisturizer too.

Nourishing Milk and Honey Body Lotion

It soothing milk meets honey in this moisturizing lotion. Honey acts as natural humectant, while aloe and sunflower oil soothe and moisturize. It is perfect for dry and sensitive skin.

Ingredients: Water, sunflower oil, vegetable glycerin, coconut oil, stearic acid, bees wax, fragrance, orange wax, tocopheryl acetate and tocopherol, rosemary extract, *Aloe barbadensis*, milk powder, honey, sodium borate, xanthan gum, sucrose stearate, glucose and glucose oxidase, peru balsam, orange oil, beta carotene.

Direction: Apply the lotion all over the body and hands as often as needed.

Supplied by: Herbal Remedies.

Parachute Advanced Body Lotion

It is 100% natural moisturiser for soft and silky skin with milk of coconut fruit.

Ingredients: Water, *Cocos nucifera* oil, glycerin, coconut milk, steareth-21, steareth-2, fragrance, carbomer.

Action: Steareth-21: The polyethelyne glycol ethers of stearyl alcohol. Used as emulsifier. Steareth-2: derived from palm oil. Used as mild cleanser and emulsifier. Carbomer: It is an excellent thickening agent. Hypoallergenic, does not support bacterial growth. It also has a nice "skin feel" producing solutions and gels that feel rich and luxurious to the touch.

Parachute Advanced Coconut Hair Oil

Parachute advanced is coconut hair oil with a rich and deep fragrance. It is also lighter than other coconut oils, which makes it ideal for the young, appearance conscious consumer. With regular use of parachute advanced, hair strength is known to increase by up to 16%. What's more, when applied prewash, Parachute Advanced reduces protein loss up to 28% and restores the health of hair.

Ingredients: Pure coconut oil, ylang ylang oil.

Action: Parachute advanced contains essential oils of ylang ylang to give it a rich and deep fragrance. It penetrates the hair roots better than other hair oils, strengthening the hair from within.

Direction: Daily before bath.

Supplied by: Marico Ltd., Mumbai.

Storage of Products

Well closed wide mouth bottle, away from sunlight.

Patents

The invention relates to a cosmetic composition with a moisture-regulating effect based on plant active ingredients with a long-lasting moisturizing effect of up to 24 hours. The cosmetic comprises in each case a watery plant milk of the fruit of *Elaeis guinensis*, of the leaves of *Phoenix canariensis*, of the rice husks of *Oryza sativa*, of the fruit of *Cocos nucifera* and a honey complex consisting of acacia honey, eucalyptus honey, pine honey and lavender honey (Golz-Berner and Zastrow, 2007).

Herbal compositions that, when used in an effective amount, are suitable for the regression of chronic inflammatory skin disorders, including eczema, psoriasis and seborrheic dermatitis. The compositions, formulated as ointments, oils and shampoos, are comprised a non-aqueous extract of *Wrightia tinctoria*, an extract of *Tragia involucrata* L., extracts of *Salix* L., *Cocos nucifera* and pharmaceutically or cosmetically acceptable excipients suitable for topical use on humans (Reddy et al., 2009).

The invention relates to a cosmetic composition with a moisture-regulating effect based on plant active ingredients with a long-lasting moisturizing effect of up to 24 hours. The cosmetic comprises in each case a watery plant milk of the fruit of *Elaeis guinensis*, of the leaves of *Phoenix canariensis*, of the rice husks of *Oryza sativa*, of the fruit of *Cocos nucifera*, and a honey complex consisting of acacia honey, eucalyptus honey, pine honey and lavender honey (Karin and Leonhard, 2011).

A method of treating skin comprising topically applying to skin in need thereof a compositing comprising *Nymphaea gigantea*, *Syzygium moorei*, *Cupaniopsis anacardioides*, *Archidendron hendersonii*, *Tristaniopsis laurina*, *Brachychiton acerifolius*, *Stenocarpus sinuatus*, *Alphitonia excelsa*, *Eucalyptus coolabah*, *Plumeria alba*, *Cocos nucifera* or *Tamarindus indica* extract (Florence et al., 2011).

BIBLIOGRAPHY

1. Agero AL, Verallo-Rowell VM. A randomized doubleblind controlled trial comparing extra virgin coconut oil with mineral oil as a moisturizer for mild to moderate xerosis. Dermatitis. 2004; 15(3): 109–116.

2. Burnett CL, Bergfeld WF, Belsito DV, Klaassen COD, Marks Jr JG, Shank RC, Slaga TJ, Snyder PW, Andersen FA. Final report on the safety assessment of *Cocos nucifera* oil

and related ingredients. Int j Toxicol. 2011; 30(3): 5S-16S.

3. Florence T, Gan D, Hines M. Topical skin formulations comprising botanical extracts. United States Patent Application 20110052737. Application number: 12/855391. Publication date: 03/03/11. http://www.freepatentsonline.com/y2011/0052737.html

4. Golz-berner K, Zastrow L. Moisture regulating cosmetic. United States Patent Application 20070269400. Application number: 11/739294, Published in: 11/22/07. http://www.freepatentsonline. com/y2007/0269400.html

5. http://ayurvedicmedicinalplants.com/plants/2465.html

6. http://lifestyle.iloveindia.com/lounge/uses-of-coconut-oil-5690.html

7. http://www.agripinoy.net/medicinal-plant-niyog-coconut-cocos-nucifera.html

8. http://www.uni-graz.at/~katzer/engl/Coco_nuc.html

9. http://www.volcanicearth.com/coconutoil.html

10. http://www.worldagroforestry.org/treedb2/AFTPDFS/Cocos_nucifera.pdf

11. Karin GB, Leonhard Z. Moisture-regulating cosmetics. Patent number: 7906158. Application number: 11/739, 294, Published in: 03/11/2011. http://www.patentgenius.com/patent/7906158.html

12. Liau KM, Lee YY, Chen CK, Rasool AHG. An Open-Label Pilot Study to Assess the Efficacy and Safety of Virgin Coconut Oil in Reducing Visceral Adiposity. ISRN Pharmacology. 2011; 2011, Article ID 949686, 1-7. doi:10.5402/2011/949686

13. Nwodo OFC, Joshua PE, Nwachukwu O, Igbozulike O, Ozoemena N. Effect of Coconut (*Cocos nucifera*) Water on Some Serum Liver Enzyme Markers and Kidney Parameters in Paracetamol-Induced Rats. J of Pharmacy Research. 2010; 3(2): Available at: http://jpronline. info/article/view/1647.

14. Oyi AR, Onaolapo JA, Obi RC. Formilation and antimicrobial studies of Coconut (*Cocos nucifera*) oil. Research Journal of Applied Sciences, Engineering and Technology. 2010; 2(2): 133–137.

15. Reddy BNB, Reddy VNK, Torgalkar A, Murugan NR. Herbal compositions for the regression of chronic inflammatory skin disorders. US Patent number: 20090142422. Application number: 11/999057, Published in: 06/04/09. http://www.freepat entson line.com/y2009/0142422.html

Crataegus laevigata

Plant profile

Kingdom: *Plantae* – Plants

Superdivision: *Spermatophyta* – Seed plants

Subkingdom: *Tracheobionta* – Vascular plants

Division: *Magnoliophyta* – Flowering plants

Class: *Magnoliopsida* – Dicotyledons
Order: *Rosales*
Genus: *Crataegus* L. – Hawthron

Subclass: *Rosidae*
Family: *Rosaceae* – Rose family
Species: *Crataegus laevigata* D.C– smooth hawthron

Vernacular name

Eng: Hawthorn

Other common names

Hagedorn, haw, hazels, gazels, ladies meat, bread and cheese tree, mayblossom, hedgethorn, maybush, mayflower, whitethorn

Ayurvedic properties

Rasa: Sour, **Veerya:** Heating, **Vipaka:** sour, **Doshas:** Balancing for vata, may increase pitta and kapha in excess.

INTRODUCTION

It is a large shrub or small tree growing to 8 m (rarely to 12 m) tall, with a dense crown. The leaves are 2–6 cm long and 2–5 cm broad, with 2–3 shallow, forward-pointing lobes on each side of the leaf. The hermaphrodite flowers are produced in corymbs of 6–12, each flower with five white or pale pink petals and two or three styles, and are pollinated by midges. The fruit is a dark red pome 6–10 mm diameter, slightly broader than long, containing 2–3 nutlets. It is distinguished from the related common Hawthorn, *C. monogyna*, in the leaves being only shallowly lobed, with forward-pointing lobes, and in the flowers having more than one styles.

Distribution

This woody tree grows in temperate regions in the Northern hemisphere. Hawthorn is

Amygdalin

Vitexin

Epicatechin

Chlorogenic acid

native to Western and Central Europe, from Great Britain (where it is uncommon, and largely confined to the Midlands) and Spain east to the Czech Republic and Hungary.

Chemical Constituents

Active ingredients found in hawthorn include tannins, flavonoids (such as vitexin, rutin, quercetin and hyperoside), oligomeric pro-anthocyanidins (OPCs, such as epicatechin, procyanidini, and particularly procyanidin B-2), flavone C, tripene acids (such ursolic acid, oleanolic acid, and crataegolic acid), and phenolic acids (such as caffeic acid, chlorogenic acid and related phenolcarboxylic acids). Standardization of hawthorn products is based on content of flavonoids (2.2%) and OPCs (18.75%). It also contains amygdalin. The bark contains the alkaloid crataegin, isolated in greyish-white crystals, bitter in taste, soluble in water. The red fruit is rich in bioflavonoids and is the most common part employed in herbal medicine.

Medicinal uses

• Hawthorn is considered a potent tonic for the heart. This herb directly benefits the functioning of the heart. It can dilate blood vessels, increase the heart's energy supply and improve its pumping ability. These powerful cardiac effects can probably be traced to its abundant supply of plant compounds called flavonoids, especially procyanidolic oligomers (PCOs)—which act as potent antioxidants.

• Hawthorn also seems to block enzymes that weaken the heart muscle, thereby strengthening its pumping power. This property is especially useful for people with mild congestive heart failure, who do not require strong heart medications. It can also correct irregular heartbeat (cardiac arrhythmia). Moreover, the antioxidant properties of hawthorn may help to protect against damage associated with the build up of plaque in

the coronary arteries. Clinical trials have also found benefits in the elderly with no heart conditions but simply deteriorating heart function with age.

• Hawthorn is also useful in the recovery period after a heart attack by strengthening the heart muscle, and improving blood flow and oxygen to the heart.

• Hawthorn has also a calming effect, and is an effective sleeping aid for some people who suffer from insomnia.

• Hawthorn can also preserve collagen—the protein that forms connective tissue—which is damaged in such diseases as arthritis.

• Hawthorn is reported to be beneficial for heart complaints during menopause, anxiety and nervous disorders.

The fruit contains bioflavonoids, are also strongly antioxidant, helping to prevent or reduce degeneration of the blood vessels. The fruit is antispasmodic, cardiac, diuretic, sedative, tonic and vasodilator. Both the fruits and flowers of hawthorns are well-known in herbal folk medicine as a heart tonic and modern research has borne out this use. The fruits and flowers have a hypotensive effect as well as acting as a direct and mild heart tonic. They are especially indicated in the treatment of weak heart combined with high blood pressure, they are also used to treat a heart muscle weakened by age, for inflammation of the heart muscle, arteriosclerosis and for nervous heart problems. Prolonged use is necessary for the treatment to be efficacious. It is normally used either as a tea or a tincture. The bark is astringent and has been used in the treatment of malaria and other fevers. The roots are said to stimulate the arteries of the heart.

Cosmetic uses

The bioflavonoids in hawthorn are potent antioxidants. Hawthorn has a normalising and tonifying action on the heart and circulatory system. It is used by sports people to sustain the heart. In a cosmetic preparation, it is acceptable vehicle, diluent or carrier. It is used with other active ingredients and the blend could be useful for the photoprotection of the skin and/or the hair, whether before, during and/or after exposure to UV radiation, and to the use of same for preventing and/or attenuating the damage caused by such UV irradiation. It is also use to prevent skin irritation with other mixed ingredients.

Pharmacological Action

A placebo-controlled, randomised, parallel group, multicentre trial conducted and shown the efficacy and safety of a standardised extract of fresh berries of *Crataegus oxyacantha* L. and Monogyna Jacq (Crataegisan) in patients with cardiac failure NYHA class II (Total 143 patients for 8 weeks treatment). Subjective cardiac symptoms at rest and at higher levels of exertion were assessed by the patient on a categorical rating scale. An overall assessment of efficacy at the final visit was provided by the patient and the investigator. The result is confirmed in the PP population (p = 0.047). The subjective assessment of cardiac symptoms at rest and at higher levels of exertion did not change significantly and the patient and investigator overall assessment of efficacy were similar for the two groups. The significant improvement, due to the fact that dyspnoea and fatigue do not occur until a significantly higher wattage has been reached in the bicycle exercise testing allows the conclusion that the recruited NYHA II patients may expect an improvement in their heart failure condition under long-term therapy with the standardised extract of fresh *Crataegus berries* (Degenring et al., 2003).

The hypoglycemic effect of an aqueous extract of hawthorn leaves (*Crataegus oxyacantha*) was investigated in normal and streptozotocin (STZ) diabetic rats. After a single dose or 9 daily doses, oral administration of the aqueous hawthorn extract produced a significant and dose-dependent decrease on blood glucose levels in STZ diabetic rats, but had no effect on blood sugar levels in normal rats. No changes were observed in basal plasma insulin concentrations after treatment in normal or STZ diabetic rats. In addition, the acute toxicity study of the extract was investigated in mice. The results obtained showed that the aqueous hawthorn extract had a high LD_{50} value (13.5 g/kg) in mice. Results conclude that an aqueous extract of hawthorn leaves exhibits a potent antihyperglycemic activity in STZ rats, but not in normal rats, without affecting basal plasma insulin concentrations (Jouad et al., 2003).

The clinical study was performed to investigate the effects of hawthorn for hypertension in patients with type 2 diabetes taking prescribed drugs by randomized controlled trial. Patients with type 2 diabetes ($n = 79$) were randomized to daily 1200 mg hawthorn extract ($n = 39$) or placebo ($n = 40$) for 16 weeks. There was a significant group difference in mean diastolic blood pressure reductions ($p = 0.035$): the hawthorn group showed greater reductions than the placebo group. There was no group difference in systolic blood pressure reduction from baseline. No herb–drug interaction was found and minor health complaints were reduced from baseline in both groups. This is the first randomized controlled trial to demonstrate a hypotensive effect of hawthorn in patients with diabetes taking medication (Walker et al., 2006).

The present study was carried out to test free-radical-scavenging, anti-inflammatory, gastroprotective, and antimicrobial activities of ethanolic extract of hawthorn berries. Phenolic compounds represented 3.54%, expressed as gallic acid equivalents. Determination of total flavonoid aglycones content yielded 0.18%. The percentage of hyperoside, as the main flavonol component, was 0.14%. With respect to procyanidins content, the obtained value was 0.44%. DPPH radical-scavenging capacity of the extract was concentration-dependent, with EC_{50} value of 52.04 µg/ml (calculation based on the total phenolic compounds content in the extract). Oral administration of investigated extract caused dose-dependent anti-inflammatory effect in a model of carrageenan-induced rat paw edema. In comparison to indomethacin, given in a dose producing 50% reduction of rat paw edema, the extract given in the highest tested dose (200 mg/kg) showed 72.4% of its activity. Gastroprotective activity of the extract was investigated using an ethanol-induced acute stress ulcer in rats with ranitidine as a reference drug. Hawthorn extract produced dose-dependent gastro-protective activity, with the efficacy comparable to that of the reference drug. Anti-microbial testing of the extract revealed its moderate bactericidal activity, especially against gram-positive bacteria *Micrococcus flavus*, *Bacillus subtilis*, and *Lysteria monocytogenes*, with no effect on *Candida albicans*. All active components identified in the extract might be responsible for activities observed (Tadic et al., 2008).

A double-blind placebocontrolled study was aimed at finding beneficial effects of *C. laevigata* on biomarkers of coronary heart disease (CHD). The study included 49 diabetic subjects with chronic CHD who were randomly assigned to the treatment for 6 months with either a micronized flower and leaf preparation of *C. laevigata* (400 mg three times a day) or a matching placebo. Blood cell count, lipid profile, C-reactive protein, neutrophil elastase (NE) and malondialdehyde were analyzed in plasma at baseline, at one month and six months. Differences between groups did not reach statistical significance at 6 months. No significant changes were observed in the rest of parameters. In conclusion, *C. laevigata* decreased NE and showed a trend to lower LDL-C compared to placebo as add-on-treatment for diabetic subjects with chronic CHD (Dalli et al., 2011).

Toxicological Studies

No toxicity has been directly related to Crataegus preparations. The acute parenteral toxicity (LD_{50}), tested in different animals, was found in a range of 18 ± 4 ml/kg, with that of individual constituents ranging from 50 to 2,600 mg/kg. The acute oral toxicity was

reported to be in a range of 18.5 ± 3.8 ml/kg and 6 g/kg, respectively. In humans, therapeutic doses of hawthorn did not have adverse effects.

Single-dose toxicity studies have demonstrated that rats and mice tolerate 3g/kg body weight, by gastric lavage, of a standardized hydroalcoholic extract of the leaves with flowers (containing 18.75% oligomeric procyanidins) without any clinical symptoms of toxicity. The intraperitoneal median lethal dose (LD_{50}) was 1.17 g/kg body weight in rats and 750 mg/kg body weight in mice. No toxic effects were observed in a repeat-dose toxicity study in which rats and dogs were given a standardized extract (containing 18.75% oligomeric procyanidins) at doses of 30, 90 and 300 mg/kg body weight daily by the intragastric route for 26 weeks (Schlegelmilch and Heywood, 1994).

Hawthorn has shown promise in the treatment of New York Heart Association functional class II congestive heart failure (CHF) in both uncontrolled and controlled clinical trials. There are also suggestions of a beneficial effect on blood lipids. Trials to establish an anti-arrhythmic effect in humans have not been conducted. The recommended daily dose of hawthorn is 160–900 mg of a native water-ethanol extract of the leaves or flowers (equivalent to 30–169 mg of epicatechin or 3.5–19.8 mg of flavonoids) administered in two or three doses. At therapeutic dosages, hawthorn may cause a mild rash, headache, sweating, dizziness, palpitations, sleepiness, agitation, and gastrointestinal symptoms. Hawthorn may interact with vasodilating medications and may potentiate or inhibit the actions of drugs used for heart failure, hypertension, angina, and arrhythmias. The limited data about hawthorn suggest that it may be useful in the treatment of NYHA functional class II CHF (Rigelsky and Sweet, 2002).

Clinical Investigations

A 2008 update of an earlier Cochrane systematic review, which appraised 10 trials involving

855 patients, concluded that hawthorn extract has significant benefits as an adjunctive treatment for patients with chronic heart failure. Most of the trials concerned reported improvements in exercise tolerance, maximal workload, and symptoms such as shortness of breath and fatigue (Pittler et al., 2008).

A more recent randomised, double-blind, placebocontrolled multicentre study involving 2681 patients who took 900 mg daily of a standardised hawthorn leaf and flower extract (equivalent to 3.6–5.9 g original dried herb) alongside their existing cardiac medications for 24 months, identified a trend for cardiac mortality reduction, but this failed to reach statistical significance. However, in the subgroup of patients with less compromised left ventricular function, a significant reduction in sudden cardiac death was measured (Holubarsch et al., 2009).

Another trial involving 900 mg per day of the same hawthorn preparation added to conventional heart failure therapy in a group of 120 patients with chronic heart failure for a six-month period, also failed to produce significant benefits, although a modest difference in left ventricular ejection fraction was measured (Zick et al., 2009).

However, the clinical evidences on *C. laevigata* as cosmetic formulations are still lacking.

Market Formulations

Although hawthorn flowers and berries have been used primarily as tonics for the heart and circulatory system, they have also been used as mild diuretics and for their astringent quality in the relief of sore throats.

Hawthorn Berries (Crataegus laevigata) 450 mg—100 Capsules

Active ingredient: Each 'preservative free' capsule contains 450 mg of *Crataegus laevigata* fruit powder.

Therapeutic use: Traditionally, used in Europe to help maintain a healthy cardiovascular system. It is recommended this

product be taken over a period of at least two months for its full nutritional benefit to be realised.

Dosage: For adults: take two capsules with a meal twice daily.

Supplied by: Natural Health Direct, Australia.

Gaia Skin Naturals Hawthorn Berry A/F 1 Ounces (Gaia Herbs)

Ingredients: Extract—Hawthorn berry, leaf, and flower (*Crataegus* sp.) Whole plant standardization process—oligomeric procyanidins 20 mg other ingredients: 60% pure vegetable glycerin, water.

Direction: Take 30–40 drops of extract in a small amount of water 3–4 times daily between meals. Not to be used during pregnancy or lactation. If you have a medical condition or take pharmaceutical drugs, please consult with your doctor before use. Use only as directed on label. Keep away from children.

Vitalia-Tincture

Ingredients: SF/01 *Sutherlandia frutescens-* 5 ml/50 ml alcohol based solution; CL/01 *Crataegus laevigata*—5 ml/50 ml alcohol based solution; SM/01 *Silybum marianum*— 5 ml/50 ml alcohol based solution, ES/03 *Eleuttherococcus senticosus*—15/50 ml alcohol based solution, UD/01 *Urtica dioica*—5/50 ml alcohol based solution, PG/03 *Panax ginseng*— 15/50 ml alcohol based solution.

Action: *Sutherlandia frutescens* may assist the immune system as immune modulator which is often suppressed in cases of fatigue. *Crataegus laevigata* may strengthen the heart to provide more blood and oxygen to the brain and body cells. *Silybum marianum* is an excellent remedy for detoxification—when the liver is aided in cleaning the body, vitality returns. *Eleuttherococcus senticosus* is known to raise energy levels and support adrenals. It also support the immune system to overcome adverse effects. *Urtica dioica* is known for its antiallergic effects often implied in fatigue. *Panax ginseng* is an antioxidant—

it also stimulates the immune system and can balance the release of stress hormones.

Storage of Products

Well closed container and away from sunlight.

Dosage

Infusion: Steep 1 tsp flowers in 1/2 cup water. Take 1 to 1 1/2 cups per day, a mouthful at a time. Sweeten with honey if desired.

Decoction: Use 1 tsp crushed fruit with 1/2 cup cold water. Let stand for 7–8 hours, then bring quickly to a boil and strain. Take 1 to 1 1/2 cups per day, a mouthful at a time, sweetened with honey if desired.

Also, use 1/2 oz hawthorn berries simmered in 1 pint of water for 20 minutes, along with 1 tsp of cinnamon and taken 3 times a day after meals, sweetened with honey as a heart tonic.

Tincture: use concentrated preparations under medical direction.

The recommended daily dose of hawthorn berries is 160–900 mg of a native water-ethanol extract of the leaves or flowers (equivalent to 30–169 mg of epicatechin or 3.5–19.8 mg of flavonoids) administered in two or three doses.

Patents

The present invention relates to the use of formulations from *Crataegus* for the preparation of pharmaceutical preparations or food supplements for the prophylaxis and/or treatment of tumour diseases, the formulations substantially comprising constituents of *Crataegus* which are soluble in polar solvents. The invention furthermore relates to the use of *Crataegus* plant parts for the preparation or infusions. Compositions for the prophylaxis and/or treatment of tumour diseases are furthermore included (Schumacher, 2002).

A preparation for oral administration, in which plant material is comminuted

and treated by mixing thoroughly with an extractant. After mixing, porous inorganic particles are added to the resulting suspension, thoroughly mixed with mixture and dried. The preparation provides a homogeneous distribution of the active ingredient as well as long stability and good bioavailability of the active ingredient (Frater-Schroder and Frater, 2006).

Cosmetics containing the following components (A) an oil-absorbing powder having a squalane-absorbing ability of 1 ml/g or more; and (B) a compound having an affinity for component; (A) lower than that of sebum, and having an ability to absorb the sebum and release component; (B) in exchange for the sebum when the cosmetic is applied onto the skin. These cosmetics can adequately absorb the sebum and thus exhibit excellent effects of relieving or preventing oiliness of the skin and worsening in makeup. By selecting an appropriate compound to be released in exchange for the absorbed sebum, a substance capable of improving the skin qualities such as moistness, tightness, and tension can be allowed to penetrate into the skin. Moreover, a bodily feeling of the removal of the sebum can be imparted to users thereby (Kume et al., 2008).

The invention relates to microcapsules, and a continuous microencapsulation water-in-oil-in-water microencapsulation process through *in situ* and interfacial polymerization of the emulsion. The formulation comprises a continuous water phase having a dispersion of microcapsules which contain oil drops and wherein the inside of each oil phase drop — containing optionally oil-soluble materials — there is a dispersion of water, or aqueous extract or water dispersible material or water soluble material. The oil drops are encapsulated with a polymerisable material of natural origin. Such microcapsules are appropriated for spray-dry processes, to be used as dry powder, lyophilised, self-emulsifiable powder, gel, cream and any liquid form. The active compounds included in the microcapsules are beneficial to the health and other biological purposes. Such formulations are appropriate to be incorporated in any class of food, especially for the production of nutraceuticals, as well as cosmetic products (such as rejuvenescence creams, anti-wrinkle creams, gels, bath and shower consumable products and sprays). The preparations are adequate to stabilise compounds added to the food, media for cultivating microbes and nutraceuticals, especially those which are easily degradable or oxidizable (Moser et al., 2011).

BIBLIOGRAPHY

1. Dalli EE, Colomer MC, Tormos J, Cosin-Sales J, Milara E, Esteban G, Saez. *Crataegus laevigata* decreases neutrophil elastase and has hypolipidemic effect: a randomized, double-blind, placebo-controlled trial.(Report). Phytomedicine: International Journal of Phytotherapy and Phytopharmacology. Urban and Fischer Verlag. 2011. *HighBeam Research.* 12 Sep. 2011.

2. Degenring FH, Suter A, Weber M, Saller R. A randomised double blind placebo controlled clinical trial of a standardised extract of fresh Crataegus berries (Crataegisan) in the treatment of patients with congestive heart failure NYHA II. Phytomedicine 2003; 10(5):363–369.

3. Frater-schroder M, Frater G. Preparation for oral administration. United States Patent 7135198. Application number: 10/259910, Publication date: 11/14/2006. http://www.freepatentsonline.com/7135198.html

4. Holubarsch CJF, Colucci WS, Meinertz T, Gaus W, Tendera M. The efficacy and safety of *Crataegus* extract WS 1442 in patients with heart failure: The SPICE trial. Eur J Heart Fail. 2009;10(12):1255–1263.

5. http://len7288.hubpages.com/hub/Health-Benefits-of-Hawthorn

6. http://www.india-herbs.com/herbal_glossary/id/1831

7. Jouad H, Lemhadri A, Maghrani M, Burcelin R, Eddouks M. Hawthorn evokes a potent anti-hyperglycemic capacity in streptozotocin-induced diabetic rats. J Herb Pharmcother. 2003; 3(2):19–29.

8. Kume T, Kawada H, Jisai Y, Sano T. Cosmetics. United States Patent 7427407. Application number: 10/468393, Publication date: 09/23/2008. http://www.free patentsonline.com/7427407.html

9. Moser M, Casaña GV, Gimeno Sierra B, Gimeno Sierra M. Continuous multi-microencapsulation process for improving the stability and storage life of biologically active ingredients. United States Patent 20110064778. Application number: 12/885165, Publication date: 03/17/2011.

10. Pittler MH, Guo R, Ernst E. Hawthorn extract for treating chronic heart failure. Cochrane Database of Systematic Reviews, 2008;(1): CD005312.

11. Rigelsky JM, Sweet BV. Hawthorn: pharmacology and therapeutic uses. Am J Health Syst Pharm. 2002; 59(5): 417–422.

12. Schlegelmilch R , Heywood R. Toxicity of Crataegus (Hawthorn) Extract (WS 1442). International Journal of Toxicology. 1994; 13(2): 103–111.

13. Schumacher KU. Use of Crataegus formulations for prophylaxis and treatment of neoplastic diseases. United States Patent 6440451. Application number: 09/700174, Publication date: 08/27/2002. http://www. freepatentsonline. com/6440451. html

14. Tadic VM, Dobric S, Markovic GM, Dordevic SM, Arsic IA, Menkovic NR, Stevic T. Anti-inflammatory, Gastroprotective, Free-Radical-Scavenging, and Antimicrobial Activities of Hawthorn Berries Ethanol Extract. *J. Agric. Food Chem.*, 2008; 56 (17): 7700–7709.

15. Walker AF, Marakis G, Simpson E, Hope JL, Robinson PA, Hassanein M, Simpson HCR. Hypotensive effects of hawthorn for patients with diabetes taking prescription drugs: A randamised controlled trial. British J of General Practice. 2006; 56: 437–443.

16. Zick SM, Vautaw BM, Gillespie B, Aaronson KD. Hawthorn extract randomised blinded chronic heart failure (HERB CHF) trial. Eur J Heart Fail. 2009;11(10): 990–999.

Cucurbita pepo

Plant profile

Kingdom: *Plantae* – Plants

Subkingdom: *Tracheobionta* – Vascular plants

Superdivision: *Spermatophyta* – Seed plants

Division: *Magnoliophyta* – Flowering plants

Class: *Magnoliopsida* – Dicotyledons

Subclass: *Dilleniidae*

Order: *Violales*

Family: *Cucurbitaceae* – Cucumber family

Genus: *Cucurbita* L. – Gourd

Species: *Cucurbita pepo* L.– Field pumpkin

Vernacular name

San: Karkaru, Kurkaru, Kurlaru, Kushmanda, **Beng:** Kumra, **Marathi:** Kohala, **Hindi:** Safed Kaddu, **Eng:** Pumpkin, **Kan:** Bude-Kumbala-Kayi, Bileegumbala, Boodugumbala, **Mal:** Kumpalam, Kumpalanna, **Tam:** Suraikayi, Parangi, **Tel:** Budadegummadi, Budide-Gummadi

Ayurvedic properties

Rasa: Madhura, Kashaya, **Guna:** Guru, Snigda, **Veerya:** Seeta.

INTRODUCTION

Pumpkin is a creeping plant which is creeping or semi-shrubby, annual, velvet-hairy. Broadly ovate-heartshaped to triangular-heartshaped leaves, 20–30 cm long, 20–35 cm broad, are often with three to five deep lobes, and with toothed margins. Tendrils have two to six branchlets, or are simple and little developed tendrils in the semi-shrubby types. The plant has solitary, flowers borne in leaf axils. The male flowers have stalks 7–20 cm long, a bell-shaped sepal cup of 9–12 mm, linear sepals 1.2–2.5 cm long. Flowers are tubular/bell-shaped, 5–10 cm long, which are divided into five petals for up to one-third or more of its length. Flowers have three stamens. The female flowers have sturdy stalks, 2–5 cm long. The ovary is round, ovoid, cylindrical, smooth, and the sepal cup is very small. The fruit is very variable in size and shape. It is smooth to heavily ribbed, with a rigid skin varying in color from light to dark green, plain to minutely speckled with cream or green contrasting with yellow, orange or two-coloured. The flesh is cream to yellowish or pale orange. Fruit a large, globose to ovoid, obovoid, cushion-shaped or cylindrical berry, weighing up to 50 kg when mature, with a wide range of colours, with small, raised, wartlike spots or smooth, sometimes deeply grooved; flesh whitish to yellow or orange, many-seeded; fruit stalk pentagonal in section, not enlarged at apex. Seeds obovoid, flattened, 1–1.5 cm × 0.5–1 cm, usually white or tawny, surface smooth to somewhat rough, margin prominent.

Distribution

The centre of origin of *Cucurbita pepo* is Mexico, where it was domesticated at least 5000 years ago. It is sometimes believed that it was domesticated on separate occasions in Mexico and the United States as data from archaeological research and molecular studies suggest that two lineages of domesticated taxa exist in *Cucurbita pepo*. It was introduced in Europe with other *Cucurbita* species during the 16th century. *Cucurbita pepo* is less heat resistant than *Cucurbita moschata* Duchesne, and for that

reason less appropriate for tropical Africa, yet it is grown on a large-scale in all countries.

Chemical Constituents

Pumpkin seeds contain sizable amounts of protein (35%) and approximately 50% fatty oil, whose fatty acid profile is dominated by unsaturated fatty acids, namely linoleic and oleic acid. There are many trace constituents like tocopherols (0.1%) and phytosterols (total 0.1 to 0.5%); of the latter group, many are specific for the family of even the species. HPLC analysis of the powerful pigments is found in pumpkin seed oil reveal a number of carotenoids—the main components being beta carotene and lutein. In addition other carotenoids present include-violaxanthin,

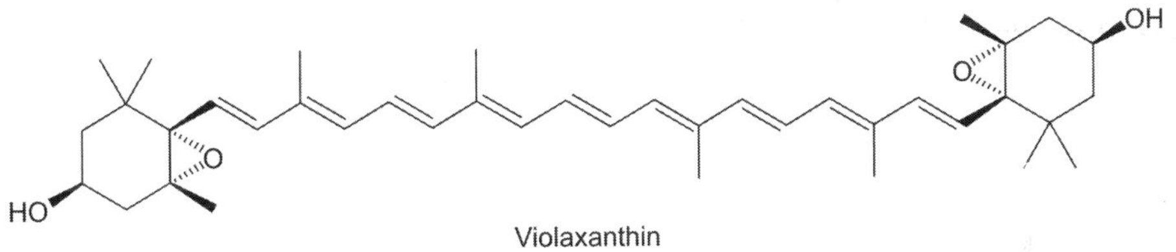

Violaxanthin

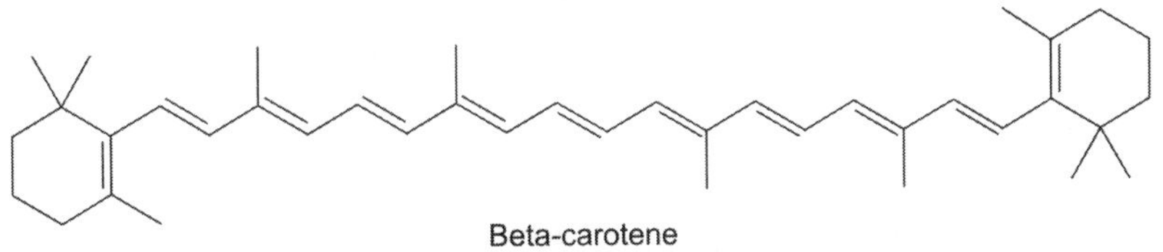

Beta-carotene

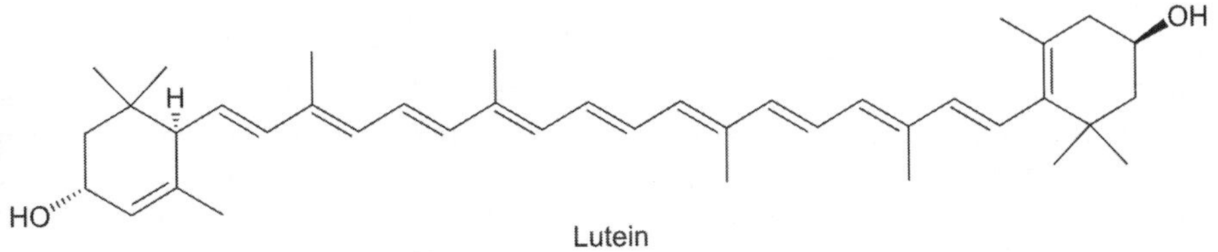

Lutein

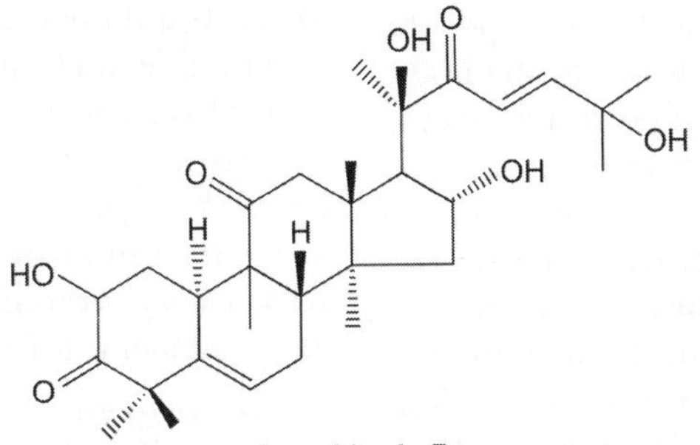

Cucurbitacin D

luteoxanthin, auroxanthin ($C_{40}H_{56}O_4$), flavoxanthin, chrysanthemaxanthin, alpha-cryptoxanthin, beta-cryptoxanthin and alpha-carotene. The plant also contains steroids like delta-5, delta-7 phytosterol. The seed contains 34–54% of a semi-drying oil. Used for lighting. Cucurbitin (1.66–6.63%) has been identified as (–)3-amino-3-carboxypyrrolidine which is active for antihelmintic effect of the seed. The pumpkin flower is a source of protein. Glutamic and aspartic acid, leucine, valine, phenylalanine, and tryptophan are among the amino acids identified.

Medicinal uses

Traditionally in folk medicine, the remedies made from the pumpkin were used in the treatment of disorders such as kidney inflammation and to eliminate intestinal parasites. It is use as taenicide, diuretic, in strangury and urinary affections, in gastritis, enteritis and febrile diseases. The expressed oil of the pumpkin seeds, in doses of 6 to 12 drops, several times a day, is said to be a most certain and efficient diuretic, giving quick relief in scalding of urine, spasmodic affections of the urinary passages, and has cured gonorrhea.

It is a gentle and safe remedy for a number of complaints, especially as an effective tapeworm remover for children and pregnant women for whom stronger acting and toxic remedies are unsuitable. The seeds are mildly diuretic and vermifuge. The complete seed, together with the husk, is used to remove tapeworms. The seed is ground into a fine flour, then made into an emulsion with water and eaten. It is then necessary to take a purgative afterwards in order to expel the tapeworms or other parasites from the body. The seed is used to treat hypertrophy of the prostate. The seed is high in zinc and has been used successfully in the early stages of prostate problems. The diuretic action has been used in the treatment of nephritis and other problems of the urinary system. The leaves are applied externally to burns. The sap of the plant and the pulp of the fruit can also be used. The fruit pulp is used as a decoction to relieve intestinal inflammation. Pumpkin is also indicated in cases of hormonal disorders or adolescent behavior, menopause disorder, intestinal parasites or sexual hyper-excitability.

Cosmetic uses

Pumpkin seed oil is highly nourishing and lubricating oil, and is useful for all skin types. It is especially good if used to combat fine lines and superficial dryness and to prevent moisture loss. *Cucurbita pepo* seed oil is used as an occlusive skin-conditioning agent and is known as pumpkin seed oil. In external use, pumpkin is recommended for treating burns, inflammations and abscesses. It softens the skin and diminishes the inflammatory processes of mucous. A number of hydrating and anti-wrinkle creams contain pumpkin. Pumpkin facials make use of a pumpkin enzyme mask that can soothe and protect skin; the essential nutrients can reach deep within skin pores and absorb quickly. Pumpkin peels can also help with breakouts and acne problems. The nutrients can help soothe and reduce inflammation naturally, helping skin recover from acne-related damage. Pumpkin works as an accelerator for other ingredients, helping the skin absorb vital nutrients and vitamins quickly.

Pharmacological Action

The study focuses on the anti-inflammatory and analgesic activity of methanolic extract of powdered ripe fruit flesh of *Cucurbita pepo* Linn in three different doses (100, 200 and 300 mg/kg b.w.) on Carrageenan and formalin induced inflamed rats against standard drug indomethacin (10 mg/kg b.w.). Results were reported as mean ± SD. It was concluded from these studies that the extract prepared from *Cucurbita pepo* Linn possess potential anti-inflammatory and analgesic activity which was comparable with standard drug indomethacin (Karpagam et al., 2011).

The present study was performed to evaluate the effect of aqueous and alcoholic extracts of *Cucurbita pepo* on methyl isobutyl ketone induced rats. Alcoholic and aqueous extracts were orally administered in rats for 30 days after the induction of depression. The antidepressant activity was examined using forced swim Test (FST) in rats. The results showed that the extracts decreased immobility time with the increase swimming time (or improved behavioral activity) and significantly increased the enzymic and nonenzymic antioxidant status in brain and serum and also HDL and LDL levels in serum. Finally, concluded that *C. pepo* seed extract possess significant antioxidant and antidepression activity (Umadevi et al., 2011).

Present study was taken up to evaluate antimutagenic activity of aqueous extract of *Cucurbita pepo* by bone marrow micronucleus test (MNT) in mice. Mitomycin C (4 mg/kg, i.p.) was used as a genotoxic challenge and bone marrow of control and mitomycin C treated mice was collected. In MNT, the bone marrow smears were stained with May-Grunwald's followed by Giemsa stain. Polychromatic and Normochromatic erythrocytes were counted and P/N ratio was calculated. Hence, *Cucurbita pepo* has significant antimutagenic activity (Reddy and Sudhakar, 2011).

Toxicological Studies

For toxicity studies the extracts of *Cucurbita pepo* Linn in the dosage of 300 and 500 mg/kg b.w were administered in two groups of rats respectively n = 4. The mortality rates were observed after 72 hours. Test for acute toxicity was found to be non-toxic at the dosage of 300 and 500 mg/kg b.w. *Cucurbita pepo* Linn and did not cause death of the animals tested (Karpagam et al., 2011).

Acute toxicity study was done according to organization for economic co-operation and development (OECD) guideline 423. Toxicity study was performed up to the dose level of 2000 mg/kg, no mouse died. So the LD_{50} was 2000 mg/kg and the ED_{50} was 1/10 the of LD_{50} was 2000/10 = 200 mg/kg (Prasad, et al, 2011).

Clinical Investigation

In a multicenter controlled study involving more than two thousand subjects, a product containing pumpkin seeds was evaluated for the treatment of benign prostate hyperplasia (BPH). The results indicated that, not only were pumpkin seeds effective in reducing symptoms associated with BPH, especially in its early stages, but also no side effects of note were reported by the patients involved in the trial (Friederich et al., 2000).

Market Formulations

Pumpkin makes an excellent face mask ingredient for all skin types, especially environmentally damaged or sensitive skin. High in vitamin A (skin healing), C (anti-oxidant) and zinc, the pumpkin soothes, moisturizes and acts as a carrier, assisting the other mask ingredients to absorb deeper into the skin and intensifying the results.

Enzymatic Therapy Better Bladder for Women

Ingredients: Total carbohydrate < 1 g, sodium 10 mg, pumpkin (*Cucurbita pepo*) 500 mg, cellulose, modified cellulose gum, silicon dioxide, modified cellulose, magnesium stearate, carnauba wax, and soylecithin.

Action: Pumpkin seed provides equally significant support for bladder health in women.

Direction: One tablet daily.

Supplied by: Rite Care Pharmacy, Inc. Encino, CA.

Method of Preparation
Pumpkin Honey Mask

Ingredients

4 tablespoons of peeled pumpkin

4 teaspoons of corn flour

2 teaspoons of *Aloe vera* gel

2 teaspoons of mashed pineapple

1 teaspoon of green tea

½ teaspoon of sunflower oil.

Preparation: Prepare green tea and let it rest a bit. While that is cooling down, combine pineapple, honey, pumpkin, and aloe gel. Add to that sunflower oil and then green tea, mix well. Apply the mix on the face and neck and leave on for 15–20 minutes. Rinse off with warm water and pad dry. This amount is enough for 4 applications. Keep it in a glass jar, tightly closed and refrigerated. It is not to be used after 2 weeks.

Pumpkin Pie Face Mask

Ingredients: 2 teaspoons cooked or canned pumpkin, pureed (*see* above for benefits), one-half teaspoon honey (humectant, regenerative) one-quarter teaspoon milk (or soymilk)(alpha hydroxyl acid, enzymes digest skin cells).

Optional ingredients: For dry skin: One-quarter teaspoon heavy whipping cream (moisturizing; alpha-hydroxyl acid) or one-half teaspoon brown sugar (exfoliates, moisturizes, alpha-hydroxyl acid).

For oily skin: One-quarter teaspoon apple cider (tonic action promotes skin circulation; alpha hydroxyl acid; regulates pH) or one-quarter teaspoon cranberry juice (high in antioxidants critically important to the utilization of essential fatty acids to maintain balanced, nourished skin.

Directions

Combine the ingredients for your facemask. Mix gently and apply to face avoiding the eye area. Rest and relax for 10–15 minutes while your pumpkin pie facemask gently exfoliates, nourishes and conditions the face. Rinse with warm water and apply the appropriate moisturizer for skin type.

Storage of Products

Well closed container and away from sunlight.

Dosage

The herbal pumpkin remedy is prepared from a seed mixture by beating about 2 ounces or 60 grams of the dried seeds in about the same amount of sugar mixed in some milk or water. The prepared mixture is added up to a pint and drunk during a fasting stage. The dosage regimen consists of three essential doses, each dose is drunk once every two hours in the day, the last and third dose of the day is followed several hours later, by a drink of castor oil to complete the dosage regimen.

Limited high-quality clinical trials exist to support therapeutic dosing. Fresh pumpkin juice 4 ml/kg of body weight was administered in a study conducted among patients with type 2 diabetes. However, it would take 100 g of crushed fruit to equal 75 ml of juice. Pumpkin seed 30 g provide approximately 4 mg of iron. When administered to non-pregnant adults for 4 weeks, iron status improved. Pumpkin seed 23 g per 100 ml was used in a study of anthelmintic action.

Patents

The present invention relates to cosmetic compositions having retarding action on the regrowth of superfluous hair, more particularly to cosmetic compositions containing fatty acids and antiandrogenic sterols from Serenoa (*Serenoa repens*) and/or from cucurbita seeds (*Cucurbita pepo*) (Di Pierro, 2001).

Cosmetic composition for the oxidative treatment of hair or skin, prepared by mixing of at least two components, in which dehydroascorbic acid or a dehydroascorbic acid salt or a dehydroascorbic acid derivative is generated from ascorbic acid, ascorbic acid derivative and ascorbic acid salt prior to application by an enzyme that catalyzes the enzymatical oxidation of said of ascorbic acid as well as a process for carrying out the oxidative treatment of keratin, particularly for the oxidative post-treatment of reduced hair in the process of permanent deformation of hair using said composition (Cassier et al., 2007).

A pharmaceutical formulation of a calcium salt and a dry plant extract in the form of a coated tablet, in which the formulation has a core of at least one dry plant extract, enveloped by at least one coating of at least one calcium salt. The plant extracts for the core may be selected from: *Vitex agnus castus* (chaste tree); *Belamcanda chinensis* (leopard lily); *Cimicifuga racemosa* (black cohosh); *Trifolium pratense L.* (purple trefoil); *Oenothera biennis hom.* (primrose); *Glycine soja* (soybean); *Serenoa repens* (saw-palmetto); *Urtica dioica* (stinging nettle), in particular its root; *Cucurbita pepo* (pumpkin), in particular its seed; *Pygeum africanum*; as well as suitable mixtures of these. Methods for the use of the formulation in treating osteoporosis and for manufacturing the formulation are provided (Popp, 2009).

BIBLIOGRAPHY

1. Cassier T, Lauscher D, Schreiber B, Kilgore J, Allwohn J. Composition for the oxidative treatment of hair or skin fixative compostion and method for permanent deformation of hair. United States Patent Application 20070092471. Application no: 10/572790, publication date: 04/26/2007.

2. Di Pierro F. Cosmetic compositions having retarding action on the regrowth of superfluous hair. United States Patent Application 20010033849. Application no: 09/781301, publication date: 10/25/2001.

3. Friederich M, Theurer C, Schiebel-Schlosser G. Prosta Fink Forte capsules in the treatment of benign prostatic hyperplasia. Multicentric surveillance study in 2245 patients [Article in German]. Forsch Komplementarmed Klass Naturheilkd. 2000; 7(4): 20020–4.

4. http://herbalnation.blogspot.com/2008/09/pumpkin-seed-cucurbita-pepomedicinal. html

5. http://theherbaldigest.com/tag/cucurbitin/

6. http://www.drugs.com/npp/pumpkin.html

7. http://www.flowersofindia.in/catalog/slides/Pumpkin.html

8. http://www.herbs2000.com/herbs/herbs_pumpkin.htm

9. http://www.liveandfeel.com/medicinalplants/pumpkin.html

10. http://www.organicfoodee.com/herbs/pumpkinseedoil/

11. http://www.pfaf.org/user/plant.aspx ? LatinName=Cucurbita+pepo

12. http://www.prota4u.org/protav8.asp? &g=pe&p=Cucurbita+pepo+L.

13. http://www.uni-graz.at/~katzer/engl/Cucu_pep.html

14. Karpagam T, Varalakshmi B, Bai JS, Gomathi S. Effect of different doses of *Cucurbita pepo* Linn extract as an antiinflammatory and analgesic nutr-aceautical agent on inflamed rats. Int J of Pharm Res and Development. 2011; 3(3): 184–192.

15. Popp MA. Pharmaceutical formulation consisting of a plant dry extract with a calcium coating. United States Patent Application 7629005. Application no: 10/480078, publication date: 12/08/2009.

16. Prasad KS, Anandan R, Kumar KM, Reddy KR, Sudhakar Y. Evaluation of anti-mutagenic activity of *Cucurbita pepo linn* by bone marrow micro nucleus test in swiss albino mice against mito-mycin-c induced mutations. Journal of Pharmacy Research. 2011; 4(4): 1041–1043.

17. Reddy KR, Sudhakar Y. Evaluation of antimutagenic activity of Cucurbita pepo linn by bone marrow micro nucleus test in swiss albino mice against mito-mycin-c induced mutations. J Pharm Res. 2011; 4(4): Available at: http://jpronline.info/article/view/6857.

18. Umadevi P, Murugan S, Jennifer suganthi S, Subakanmani. S. Evaluation of Anti-depressant like activity of *Cucurbita pepo* seed extracts in rats. International Journal of Current Pharmaceutical Research. 2011; 3(1): 108–113.

Cucumis sativus

Plant profile

Kingdom: *Plantae* – Plants
Superdivision: *Spermatophyta* – Seed plants
Class: *Magnoliopsida* – Dicotyledons
Order: *Violales*
Genus: *Cucumis* L. – Melon

Subkingdom: *Tracheobionta* – Vascular plants
Division: *Magnoliophyta* – Flowering plants
Subclass: *Dilleniidae*
Family: *Cucurbitaceae* – Cucumber family
Species: *Cucumis sativus* – Garden cucumber

Vernacular name

San: Trapusha, Sukasa, **Beng:** Sasa, **Eng:** Cucumber, **Hind:** Khira, **Mar:** Kankri, **Kan:** Soutekay, **Guj:** Tansali, **Mal:** Vellari, **Ori:** Kaknai, **Tam:** Vellarikkay, **Tel:** Dosakaya

Ayurvedic properties

Gunna: Ruksh (dry) and laghu (light), **Rasa:** madhur (sweet); **Veerya:** Seeta (cold).

INTRODUCTION

It is an annual creeper plant, diclinous. Stalk length reaches 1.5 m, orbicular, orbicular-cut or cut, prostrate, branching, with simple short tendrils, with varying degrees of pubescence. Bushes with stalk's length of 20–30 cm are formed. The root is a rod and ramified, its great bulk is in an arable layer of soil. Leaves are alternate, five-bladed (blades sharp), less often integral (oval). Length can be short (it is less than 12 cm), average (12–15 cm), or long (more than 15 cm). Flowers are auxiliary, diclinous, it is rarely bisexual. Male flowers are single or in 3–5–7 floral cluster on the main stalk and axils of the first order, female flowers are solitary or in pairs on branches of the second and the subsequent orders. The corolla yellow, funnel-shaped, part is deeper than half, teeth sharp; 4 stamens have grown

together in pairs and one is free. Ovary is long, pubescence. Fruits are various in form, size and coloring. Fruits in technical ripeness are dark green, green or light green, equally-colored or with white spots and strips. Their surface may be smooth, fine-tubercular, large-tubercular, furrowed, with thorns of different coloring. Length is from 5 up to 30 cm and more. Weight from 50 to 400 g. Transverse incision is orbicular or orbuscularly-three-edged. Pulp of the fruit is ochroleucous or pale-green. Weight of fruits when ripe reaches from 50 g up to 3–4 kg, length from 5–10 up to 80–100 cm, the form from spherical up to lengthened-cylindrical, coloring from almost white to dark brown. There may be 20–25 fruits on a plant. Seeds are white, lengthened-elliptic, length from 9 to 10 mm up to 16–18 mm.

Distribution

India is considered the native land of cucumbers. It is cultivated in open ground up to the 65th degree of Northern latitude in the European part of Russia and until the 61st degree of Northern latitude in the Asian part of Russia (Central regions of Yakutia). The basic crops are concentrated in the central areas of Russia, the Volga region, on Northern Caucasus, Ukraine, in Kazakhstan, Western Siberia. It is also found in Africa, Indonesia, Thailand, Australia.

Chemical Constituents

Its seed contains 42% protein and 43% fats. Seed pulp contains yellow colored oil. The oil in the cucumber contains 22.3% linoleic acid, 58.5% oleic acid, 6.8% palmitic acid and 3.7% stearic acid. The fresh cucumber is a very good source of vitamin C, vitamin K, and potassium. It also contains vitamin A, vitamin B_6, thiamin, folate, pantothenic acid, magnesium, phosphorus, potassium, copper, and manganese. The serial parts of the cucumber plant contain a 14 α-methyl-phytosterol, α- and β-amyrin, multiflorenol, isomultiflorenol, 24-methylenecycloartenol, cycloartenol, triucallol. Presence of a cytokinin-binding protein, isopentenyl adenosine trialcohol is also reported in the cotyledons.

to treat dyspepsia in children. The seed is cooling, diuretic, tonic and vermifuge. The emulsion made by bruising cucumber seeds and rubbing them up with water is much used in catarrhal infections and diseases of the bowels and urinary passages. The fruit is depurative, diuretic, emollient, purgative and resolvent. The fresh fruit is used internally in the treatment of blemished skin, heat rash, etc, and also used externally as a medicine for burns, sores. A decoction of the root is diuretic.

Cosmetic uses

The fruit extract is well-known for its moisture-binding, soothing, tightening and anti-inflammatory properties and provides nourishment and a gentle softening of the skin.

It is reduces the sweeling and dark circles around the eyes. The puffiness and tiredness of eyes can reduce by using this fruit as eye pad.

Cucumber including cucumber seeds and cucumber fruit is excellent for rubbing over the skin to keep it soft and white. Cucumber fruit is cooling, healing and soothing to an irritated skin, whether caused by sun, or the effects of a cutaneous eruption, and cucumber plant juice is in great demand in various forms as a cooling and beautifying

Linoleic acid

Oleic acid

Medicinal uses

Plant pacifies vitiated pitta, edema, kidney diseases, fever, insomnia, headache, burning sensation, jaundice, hemorrhages, urinary retention constipation, calculus and general debility. The leaf juice is emetic; it is used

agent for the skin. Cucumber fruit has soothing effects on the skin and improves moisture retention. Cucumber soap is used by many women, and a cucumber wash applied to the skin after exposure to keen winds is extremely beneficial. Emollient ointments prepared from the cucumber plant were formerly considerably employed in irritated states of the skin, but they have been largely superseded by non-fatty cosmetics. The most frequently used preparation of cucumber plant and fruits at the present time in the cosmetic preparation known as cucumber jelly, which is used as a soothing application in roughness of the skin, etc. Cucumber consists of a jelly of tragacanth, quince seeds or some similar mucilaginous drug, flavoured with cucumber juice, which imparts to the preparation a characteristic odor. Seed extracts are ingredients made from the cucumber, *Cucumis sativus*. These ingredients are used in a wide variety of products including bath products, soaps and detergents, lotions, cleansing products, nail care products, make-up and eye make-up, as well as hair care products. Seed extract function as skin conditioning agents–skin conditioning agents.

Soothing, cooling, toning facial: Simmer half a peck of quince blossoms covered with water for an hour; cut 2 large cukes of cucumber into very thin slices and mince; put in pan with blossoms and boil for 5 minutes; when cold, pour into bottles. Use by smearing on face and leave for 10 minutes before washing.

Pharmacological Action

The aqueous fruit extract of *Cucumis sativus* L. was screened for free radical scavenging and analgesic activities. The extract was subjected to *in vitro* antioxidant studies at 250 and 500 µg/ml and analgesic study at the doses 250 and 500 mg/kg, respectively. The free radical scavenging was compared with ascorbic acid, butylated hydroxyl anisole (BHA), whereas, the analgesic effect was compared with diclofenac sodium (50 mg/kg). The *C. sativus* fruit extract showed maximum antioxidant and analgesic effect at 500 µg/ml and 500 mg/kg, respectively. The presence of flavonoids and tannins in the extract as evidenced by preliminary phytochemical screening suggests that these compounds might be responsible for free radical scavenging and analgesic effects (Kumar et al., 2010).

Three antimicrobial sphingolipids were separated by bioassay-guided isolation from the chloroform fraction of the crude methanol extract of cucumber (*Cucumis sativus* L.) stems. They were evaluated to show antifungal and antibacterial activity on test microorganisms including four fungal and three bacterial species. The results indicated that sphingolipids could be the main antimicrobial compounds in the crude methanol extract of cucumber stems (Tang et al., 2010).

The effect of hydroalcoholic and butanolic extract obtained from *C. sativus* seeds in a model of streptozotocin (STZ)-induced diabetic (type I) rats was studied. Normal and diabetic male Wistar rats (STZ, 60 mg/kg, intraperitoneal) were treated daily with vehicle (5 ml/kg), hydroalcoholic (0.2, 0.4, 0.8 g/kg) and butanolic extract (0.2, 0.4, 0.8 g/kg) and glibenclamide (1 and 3 mg/kg) separately and treatment was continued for 9 days. both hydroalcoholic (22.5–33.8%) and butanolic (26.6–45.0%) extracts were effective on diminishing blood glucose level and controling the loss of body weight in diabetic rats compared to controls after 9 days of continued daily therapy. It is concluded that *C. sativus* seeds extracts (hydroalcoholic and butanolic) had a role in diabetes control probably through a mechanism similar to euglycemic agents (Minaiyan et al., 2011).

Toxicological Studies

No toxicity has been directly related to this plant.

Clinical Investigation

This study was designed to develop a topical skincare cream water in oil (w/o) emulsion of 3% cucumber extracts versus its vehicle (Base) as control and evaluates its effects on skinmelanin, skin erythema, skin moisture, skin sebum and transepidermal water loss (TEWL). Hydroalcoholic cucumber (*Cucumis sativus) fruit* extract was entrapped in the inner aqueous phase of w/o emulsion. Base containing no extract and a formulation containing 3% concentrated extract of *C. sativus* was formulated. Both the base and formulation were applied to the cheeks of 21 healthy human volunteers for a period of 4 weeks. The expected pharmaceutical stability of creams was achieved from 4 weeks *in vitro* study period. Odor disappeared with passage of time due to volatilization of lemon oil. The base showed insignificant (p > 0.05) effects on all skin parameters except sebum that was not significant, whereas the formulation showed statistically significant (p < 0.05) effects on skin sebum secretion. TEWL and erythema was increased while skin melanin and skin hydration level was decreased by formulation. The results showed a good stability over 4 weeks of observation period of both base and formulation and the formulation has anti sebum secretion, bleaching and moisturizing effects (Akhtar et al., 2011).

Polyherbal fairness cream was formulated and clinically evaluated its safety and efficacy with daily appled during 30 days. Total 50 numbers of subjects were participated and finally concluded that the formulation was safe for human use (Soni et al., 2011).

Market formulations: Cucumis sativus (cucumber) fruit extract is a skin-conditioning agent and it is used in cosmetic industry in variety of products. It may be found in hair care products, make-up, bath products, cleansing products, and lotions.

Sensitive Skin Cleanser

- Gently cleanses while providing hydration for sensitive, post-procedure skin
- Specially formulated with anti-irritants to protect highly sensitive skin
- Thoroughly removes environmental pollutants, excess oil and make-up

Ingredients: Water, *Carthamus tinctorius* (safflower) seed oil, cetyl ethylhexanoate, glyceryl stearate, PEG-100 stearate, glycerin, cetearyl alcohol, tocopheryl acetate, *Chamomilla recutita* (matricaria) flower extract, *Calendula officinalis* flower extract, *Cucumis sativus* (cucumber) fruit extract,

dimethicone, ceteareth-20, polysorbate 20, aminomethyl propanol, carbomer, disodium EDTA, phenoxyethanol, methylparaben, propylparaben, isobutylparaben, butylparaben, ethylparaben.

Action: Chamomilla recutita (matricaria) flower extract is an anti-inflammatory with skin repairing properties that help neutralize skin irritants. *Cucumis sativus* (cucumber) fruit extract shown to reduce melanin production for natural skin lightening effect. *Calendula officinalis* (pot marigold) flower extract has skin healing, soothing, antiseptic, and anti-inflammatory properties.

Direction: Splash face with warm water. Gently massage cleanser over face. Rinse thoroughly. As with any skin care product, avoid getting in eyes. If contact occurs, rinse eyes thoroughly with water.

Supplied by: Skin medica.

GreatSkin® Firming Night Cream

Firming night cream is a heavy moisturizer that deeply penetrates the skin and will prevent excess dryness. This therapeutic vitamin and moisturizing treatment increases the skin's vitality and seals in the moisture, augmenting skin's resilience. It contains collagen which restores elasticity, and vitamin E which helps in healing and nourishes the skin.

Ingredients: Purified water, carbomer 940, octyl sterate, sorbitol, collagen, glyceryl stearate, stearic acid, sesame oil, olive oil, apricot kernal oil, jojoba oil, avocado oil, squalene, kakui nut oil, tocopherol acetate (vitamin E acetate), cetyl alcohol, elastin, *Aloe vera*, triethanolamine, extracts of lemon balm, comfrey, sage, marigold, marsh mallow, teaberry, jasmine, cucumber, panthenol, sodium laurel sulfate, allantoin, retinyl plamitate (vitamin A), ergocalciferol (vitamin D), PPG-20 methyl glucose ether, benzalkonium chloride, tetrasodium EDTA, biotin, RNA sodium, imiazolidinyl urea, methylparaben, propylparaben.

Action: It works by increasing your skin's vitality and sealing in moisture. With collagen it restores elasticity and vitamin E for healing. Deeply penetrates to prevent excess dryness. With sesame oil, olive oil, apricot kernel oil, jojoba oil, avocado oil, squalene, kakui nut oil, *Aloe vera*, vitamin E, lemon balm, comfrey, sage, marigold, teaberry, jasmine, cucumber, vitamin A, and vitmin D. Apply before bedtime after cleansing and toning.

Direction: Before bedtime, apply evenly over face and throat after cleansing and toning.

Supplied by: Great Skin Pvt Ltd.

Apricot Almond Anti-wrinkle Cream

This anti-wrinkle cream enriched with natural extracts of apricot and almond known for their moisturizing and regenerative properties. This cream fortified with vitamin E, a powerful antioxidant. Cucumber, Fennel and lemon help clarifying and skin toning. Sesame oil helps brighten the complexion and could diminish the appearance of fine lines.

Ingredients: Almond oil, apricot kernel oil, lemon oil, fennel oil, cucumber oil, mash extract, sesame oil, tea, perfume, water, LLP, beeswax, GMS, stearic acid, glycerine, cetyl alcohol, borax, emulsifying wax, sodium propyl paraben, sodium methyl paraben, 2 phenoxy ethanol.

Action: The apricot (*Prunus armeniaca*) is a species of *Prunus* and it possesses the highest levels of carotenoids which are antioxidants. The apricot oil moisturizes and helps to regenerate the skin. The almond (*Prunus dulcis*) oil moisturizes and helps regenerate the skin. The cucumber has Thiamine, riboflavin, niacin, B_5, B_6 vitamins and more. Cucumber oil helps clarify and tone the skin. Fennel oil helps clarify and tone the skin. Lemon oil helps clarify the skin and sesame oil helps brighten the complexion.

Direction: Apply to freshly cleansed face. Repeat whenever required. For best results, use morning and night after cleansing.

Supplied by: Herbal Destination, USA.

Himalaya Herbals Deep Cleansing Milk

Ingredients: Cucumbur (*Cucumis sativus*) (100 mg) shikimate dehydrogenase is extracted from cucumber pulp. It helps to keep the facial skin soft, has healing and soothing effect on the damaged skin and exerts a natural sunscreen. It acts as a toner and lightens the facial skin. Soapnut (*Sapindus Mukorossi*) (20 mg) contains Saponins. It cleanses the skin of oily secretion and is even used as a cleanser for washing hair and a hair tonic, and forms a rich, natural lather.

Action: Its important ingredients are soapnut and cucumber. They deep clean the skin gently and effectively by dissolving dirt, oil and make-up. Skin is left feeling refreshingly clean and yet soft and moisturized. While soapnut penetrates deep into pores and removes dirt and excess oily secretions, cucumber soothes and softens the skin.

Direction: Gently massage with fingertips in a circular motion on face and neck. After one minute wipe-off with a moist cotton pad. Rinse with water. Regular use makes skin refreshingly clean, healthy and radiant.

Supplied by: Himalaya Drug Company, India.

Method of Preparations
Cucumber Foner

- 1/2 cucumber with peel, chopped
- 3 tablespoons witch hazel
- 2 tablespoons distilled water

Put all of the ingredients in a blender or food processor and blend until smooth. Pour the mixture through a fine-mesh sieve to remove all of the solids, then pour the toner into a clean bottle with a tight-fitting lid. Store this toner in the refrigerator for a longer shelf-life — it should last for several weeks.

To use, apply the toner to face using a clean cotton ball.

Soothe Puffy Eyes

The high water content helps to hydrate tender skin in the eye region, while the chill of a refrigerated cucumber helps contract blood vessels in the area—both effects combine to reduce swelling.

To use cucumbers as an eye treatment, grab a cold cucumber from the refrigerator and cut two thick slices. Find a comfortable place to relax and set the cucumbers over closed eyes for about 10 to 15 minutes.

Cucumber-avocado Facial Mask

- 1/2 cup chopped cucumber
- 1/2 cup chopped avocado
- 1 egg white
- 2 teaspoons powdered milk

Blend all the ingredients together until they form a smooth, paste-like consistency. The mask can either be used immediately or left in the fridge for half an hour first.

Massage 2 tbsp of the mask onto face and neck using circular upward motions. Relax for 30 minutes, or until the mask is dry. Rinse off with warm water, then follow with a cold water rinse, pat dry.

Cucumber with Honey Toner

Take 1 medium cucumber, peeled and cut up into pieces 2 teaspoon. Puree cucumber in a blender. Line a sieve with cheesecloth and set the sieve over a glass bowl or measuring cup. Pour the cucumber puree through the sieve and let it stand for 15 minutes for the juices to drip into the bowl. Pour the clear juice into a clean bottle and add honey.

To use, shake the bottle and saturate a cotton pad with the lotion. Sweep over face, neck, and chest morning and night, and let it air dry (about 3 to 4 minutes). Store toner covered in the refrigerator for up to 1 week, makes about 1/2 cup.

Cucumber Lotion for Sunburn

Chop up a cucumber and squeeze out the juice with a lemon-squeezer. Mix this with a quantity of glycerine and rose water mixed together in equal parts.

Cucumber juice is used in the preparation of glycerine and cucumber creams. After expression and clarification, it is treated with alcohol, benzoin or salicylic acids being added as preservatives.

Emollient ointments prepared from the cucumber were formerly considerably employed in irritated states of the skin, but they have been largely superseded by non-fatty cosmetics. The most frequently used preparation of cucumber at the present time is the cosmetic preparation known as *cucumber jelly*, which is used as a soothing application in roughness of the skin, etc. It consists of a jelly of tragacanth, quince seeds or some similar mucilaginous drug, flavoured with cucumber juice, which imparts to the preparation a characteristic odor.

Cucumber Ointment

Incorporate 1 part of distilled spirit of cucumbers with 7 parts of benzoinated lard. The spirit is made by distilling a mixture of 1 part of grated cucumbers with 3 parts of diluted alcohol, returning the first 2 parts or distillates which come over. This spirit is permanent and ointment or cream made from it keeps well.

Cucumber milk: Is made of the following ingredients: 1 oz soap, 1 oz olive oil, 1 oz wax, 1 oz spermaceti, 1 lb almonds, 4 1/2 pints freshly expressed cucumber juice, 1 pint extract of cucumber, 2 lb alcohol.

Storage of Products

Store in cool place and away from sunlight.

Patents

The present invention relates to a synergistic herbal formulation comprising an active fraction from *Azadirachta indica* designated as fraction A and a fraction from *Citrullus colocynthis* designated as fraction B, along with a fraction C containing an antioxidant from *Cucumis sativus* extract; and pharmaceutically accepted a carrier wherein the ratio of the components ranging between

2 and 5.5% of fraction A; about 0.5 to 2.5% of fraction B; about 0.1–0.4% of extract of *Cucumis sativus* and about 82–97% of carrier or additive (Behl et al., 2004).

A skin treatment regimen comprising applying to facial skin, in order, at least once per day, effective amounts of five distinct products, including a daily cleanser, an exfoliating masque with thermal infusion, a daily balancer, a daily eye cream and a daily moisturizing cream (Bottiglieri et al., 2007).

The invention was directed to compositions for antiacne sunscreen. The sunscreen composition has the unique ability to treat and prevent acne in addition to screen both UV-A and UV-B radiation. In particular, the sunscreen composition includes a sunscreen base, at least one UV-A deactivator, at least one UV-B deactivator, and at least one antiacne agent. The UV-A deactivator may be avobenzone and the UV-B deactivator may be selected from one of the following oxybenzone, octisalate, octyl methoxycinnamate, or a mixture thereof (Kunin, 2009).

BIBLIOGRAPHY

1. Akhtar N, Mehmood A, Khan BA, Mahmood T, Khan HMS, Saeed T. Exploring cucumber extract for skin rejuvenation. African Journal of Bio-technology. 2011; 10(7): 1206–1216.

2. Behl HM, Sidhu OP, Mehrotra S, Pushpangadan P, Singh SC. Nontoxic dental care herbal formulation for preventing dental plaque and gingivitis. United States Patent Application 20040191337. Application number: 10/397967, Publication date: 09/30/2004. http://www.freepatentsonline.com/y2004/0191337.html

3. Bottiglieri P, Saute RE, Keefe CR, Abbruzzese A. Skin care method and products. United States Patent Application 20070243155. Application number: 11/406745, Publication date: 10/18/2007. http://www.freepatentsonline.com/y2007/0243155.html

4. http://ayurvedicmedicinalplants.com/plants/3322.html

5. http://botanical.com/botanical/mgmh/c/cucum123.html

6. http://www.agroatlas.ru/cultural/Cucumis_sativus_K_en.htm

7. http://www.ayushveda.com/herbs/cucumis-sativus.htm

8. http://www.cosmeticsinfo.org/ingredient_details.php?ingredient_ id=1690

9. http://www.disabled-world.com/artman/publish/cucumber_benefits. shtml

10. http://www.flowersofindia.in/catalog/slides/Cucumber.html

11. http://www.naturalcosmeticsupplies.com/cucumber.html

12. http://www.spicesmedicinalherbs. com/cucumber-cucmis-sativus-veggie. html

13. Kumar D, Kumar S, Singh J, Narender, Rashmi, Vashistha BD, Singh N. Free Radical Scavenging and Analgesic Activities of *Cucumis sativus* L. Fruit Extract. J Young Pharm. 2010; 2(4): 365–368.

14. Kunin A. Anti-Acne Sunscreen Composition. United States Patent Application 20090246156. Application number: 12/059761, Publication date: 10/01/2009. http://www.freepatentsonline.com/y2009/0246156.html

15. Minaiyan M, Zolfaghari B, Kamal A. Effect of Hydroalcoholic and Buthanolic Extract of Cucumis sativus Seeds on Blood Glucose Level of Normal and Streptozotocin-Induced Diabetic Rats. Iranian Journal of Basic Medical Sciences. 2011; 14 (5): 436–442.

16. Soni HA, Gandhi VJ, Bhatt SB, Shah UD, Shah VN. Evaluation of safety and efficacy of polyherbal fairness cream. International Research j of Pharmacy. 2011; 2(1): 99–103.

17. Tang J, Meng X, Liu H, Zhao J, Zhou L, Qiu M, Zhang X, Yu Z, Yang F. Antimicrobial activity of sphingolipids isolated from the stems of cucumber (*Cucumis sativus* L.). Molecules. 2010;15(12): 9288–97.

Curcuma longa

Plant profile

Kingdom: *Plantae* – Plants
Superdivision: *Spermatophyta* – Seed plants
Class: *Liliopsida* – Monocotyledons
Order: *Zingiberales*
Genus: *Curcima L.* – Curcuma

Subkingdom: *Tracheobionta* – Vascular plants
Division: *Magnoliophyta* – Flowering plants
Subclass: *Zingiberidae*
Family: *Zingiberaceae* – Ginger family
Species: *Curcuma longa* – Turmeric

Vernacular name

San: Haridra; Rajani, **Beng:** Halud, **Eng:** Turmeric, Indian Saffron, **Guj:** Haldarvo, **Hindi:** Haldi, Haridra; **Marathi:** Heddi, **Kan:** Yettega, **Mal:** Manja-kadamva, **Tam:** Manja-kadamba, Manja-kadambe, Manjal, **Tel:** Pasupu-kadamba

Ayurvedic properties

Gana: Kusthaghna, Haridradi, Shirovirechana, **Guna:** Laghu (light) and Ruksha (rough), **Rasa:** Katu (pungent) and Tikta (bitter), **Veerya:** Ushna, **Vipaka:** Katu (pungent), **Dosha:** Tridoshic at normal dosages;
Prabhava: Purifies the skin and complexion, **Manas Tri Guna:** Sattva.

INTRODUCTION

Turmeric is a very important spice in India, which produces nearly the whole world's crop and uses 80% of it. Turmeric usage dates back nearly 4000 years, to the Vedic culture in India, when turmeric was the principal spice and also of religious significance. Turmeric is a perennial plant that grows 5–6 feet high in the tropical regions of Southern Asia, with trumpet-shaped, dull yellow flowers. Its roots are bulbs that also produce rhizomes, which then produce stems and roots for new plants. Turmeric is fragrant and has a bitter, some what sharp taste.

Distribution

It is native to tropical South Asia and needs temperatures between 20°C and 30°C and a considerable amount of annual rainfall to thrive. In India it is distributed especially in Andhra Pradesh and Tamil Nadu. Cultivated extensively in India within tropical climate.

Chemical Constituents

Turmeric contains an essential oil (max. 5%), which contains a variety of sesquiterpenes, many of which are specific for the species. Most important for the aroma are turmerone (max. 30%), ar-turmerone (25%) and zingiberene (25%). Conjugated Diarylheptanoids (1,7-diaryl-hepta-1,6-diene-3,5-diones, e.g. curcumin) are responsible for the orange color and probably also for the pungent taste (3 to 4%). Curcumin, a polyphenol compound, is responsible for the yellow color of turmeric and is thought to be the most active pharmacological agent. Natural curcumin, isolated from *Curcuma longa*, contains curcumin I (diferuloyl methane as the major constituent), as well as curcumin II (6%) and III (0.3%). Turmeric may be standardized to contain approximately 95% curcuminoids per dose. The dried root of turmeric reportedly contains 4–8% curcumin, of which curcumin I is the most abundant, but may not be the most biologically active. Curcumin is insoluble in water and ether, but is soluble in ethanol, dimethylsulfoxide, and other organic solvents.

Medicinal uses

Turmeric is used extensively in foods for both its flavor and color. Turmeric has a long tradition of use in the Chinese and Ayurvedic systems of medicine. The rhizome (root) of turmeric has long been used in traditional Asian medicine to treat gastrointestinal upset, arthritic pain. Turmeric has long been used as an anti-inflammatory, to treat digestive disorders, osteoarthritis, atherosclerosis, cancer and liver problems and for the treatment of skin diseases and wound healing. It is believed to strengthen the overall energy of the body, relieve gas, dispel worms, regulate menstruation, dissolve gallstones, and relieve arthritis. The active components of turmeric reduce the destructive activity of parasites or roundworms. It has also demonstrated in animals a protective effect on the liver, anti-tumor action and ability to reduce inflammation and fight certain infections.

Curcumin 95%, a potent antioxidant is believed to be the most bioactive and soothing portion of the herb turmeric. Curcumin is posses the properties like antioxidant, anti-inflammatory, anti-platelet, cholesterol-lowering antibacterial and anti-fungal effects. It contains a mixture of powerful antioxidant phytonutrients known as Curcuminoids. Curcumin 95% inhibits cancer at initiation, promotion and progression stages of tumor development. It is a strong anti-oxidant, which supports colon health, exerts neuroprotective activity and helps maintain a healthy cardiovascular system.

Cosmetic uses

It is also used as a key ingredient in some cosmetic products like cream, soaps and cleansers. Turmeric paste is traditionally used by Indian women to keep them free of superfluous hair and as an antimicrobial. Turmeric paste, as part of both home remedies and Ayurveda, is also said to improve the skin and is touted as an anti-aging agent. Staining oneself with turmeric is believed to improve the skin tone and tan. Turmeric is currently used in the formulation of some sunscreens. Isolated tetrahydrocurcuminoids (THC) from turmeric are colorless compounds that might have antioxidant and skin-lightening properties, and might be used to treat skin inflammations, making these compounds useful in cosmetics formulations. Turmeric have been formulated to heal and prevent dry skin, treat skin conditions such as

Curcumin

Ar-turmerone

Zingiberene

eczema and acne, and retard the aging process. Washing in turmeric improves skin complexion and also reduces hair growth on body. Nowadays there are lots of herbal products in the market in which main herb used is turmeric as natural ingredient. it brings a healthy glow to the skin and makes them beautiful. They also help to restore or maintain youth by controlling wrinkle and crease formation on the surface of the skin. Turmeric can also benefit skin conditions including: eczema, psoriasis and acne. Effectual healing properties of turmeric have made it accepted after ingredient in cosmetics and drugs, as the leaf oil of turmeric and extract can also be used as bio-pesticides and sunscreens. Turmeric powder is valuable for its aromatic, stimulatory and carminative properties as well as for curing many ailments both minor such as sore throat and coryza as well as major afflictions. It is eminent for accelerated healing of both septic and non-septic wounds. Turmeric is also very effective tonic and a blood purifier. It is also skin-friendly and constitutes an important ingredient of many creams and lotions.

Pharmacological Action

Turmeric is used for heartburn (dyspepsia), stomach pain, diarrhea, stomach bloating, loss of appetite, jaundice, liver problems and gallbladder disorders. It is also used for headaches, bronchitis, colds, lung infections, fibromyalgia, leprosy, fever, menstrual problems. Other uses include depression, Alzheimer's disease, water retention, worms, and kidney problems.

The purpose of the present study was to elucidate the protective efficacy of curcumin against cadmium chloride induced colon toxicity in Swiss albino mice. Four batches of Swiss-albino mice, each batch comprising six mice, were treated with cadmium chloride and curcumin. Comparative observations of colon in cadmium-treated groups clearly showed induction of histopathological damage as evidenced by emptied goblet cells, lacerated and suppressed mucosa, destruction

of surface epithelium, marked decline in mucin production and hemorrhagic colon, which were not seen in control and curcumin treated group. In animals pretreated with curcumin and subsequently treated with cadmium chloride, the histopathology of colon clearly delineates the protection accorded by curcumin, as the induced damage by cadmium chloride was negligible and the number of empty goblet cells, hemorrhagic areas, suppression of mucosa was significantly reduced (Singh et al., 2011).

Toxicological Studies

No significant toxicity has been reported following either acute or chronic administration of turmeric extracts at standard doses. At very high doses (100 mg/kg body weight), curcumin may be ulcerogenic in animals, as evidenced by one rat study (Ammon and Wahl, 1991).

Turmeric powder (*Curcuma longa* Linn) produced no acute toxicity in mice when given at the dose of 10 g/kg body weight. LD_{50} of ethanolic extract in mice was more than 15 g/kg either by p.o., s.c. or i.p. administration. Subchronic toxicity of turmeric powder at 0.03, 2.5 and 5.0 g/kg/day was investigated for 6 months in 96 wistar rats. No evidence of other abnormalities has been observed (Sittisomwong et al., http://webdb.dmsc.moph.go.th/ifc_herbal/research_poison/pdf_research/i_2345 749 44683422. pdf).

Clinical Investigation

Animal study results indicate that topical application of curcuminoids, even at low dosages, inhibits skin tumor cells, probably through antioxidant mechanisms. Topical pretreatment of skin in animal studies also significantly inhibited the oxidative effects of ultraviolet radiation (Thiele and Elsner, 2001).

Volunteers were randomly assigned to apply 2 mg of a curcuminoids-loaded solid lipid nanoparticles facial cream to one half of their face and 2 mg of a placebo cream base to the other half for an 8-week period. Results on skin wrinkles, hydration, melanin content, biological elasticity, and viscoelasticity were measured with sophisticated testing equipment. Skin irritation was measured by medical observation and also by testing water loss and skin pH. After the 3rd week, test results of all aspects on the treatment side were significantly better than the control side, with no sign of skin irritation (Plianbangchang et al, 2007).

Women between the ages of 40 and 60 from two ethnic groups (Caucasian and Asian) were randomly assigned to use the turmeric cream or a control placebo cream twice daily on the face for eight weeks. Photos taken at baseline, four, and eight weeks were evaluated by experts for changes in fine lines, wrinkles, and age spots. Their findings determined an approximately 15% positive difference between the test and control group (Brauser, 2010).

Market Formulations

Vicco Turmeric Cream

This vanishing cream prevents the penetration of ultraviolet rays of the sun into the skin. It keeps pimples and acne at bay, giving skin a blemish-free complexion. It rejuvenates and revitalizes the skin from within, leaving it soft, supple and young-looking.

Ingredients: Turmeric (16.0 % w/w) and sandalwood (0.5% w/w) and other excipients.

Action: Turmeric prevents the penetration of ultraviolet rays of the sun into the skin and thus maintains the original color of the pigment of the skin, and sandalwood oil is supposed to be the most cooling element.

Method of prepartion: The extracts of Turmeric, sandalwood oil and other ingredients fall into the vessels with gravitational force untouched by hand. As soon as the production is complete, the cream automatically falls into the storage vessels underneath, from where it is transferred through pipe line to the automatic tubefilling machine.

Direction: It can be used in face and body with rubbing.

Supplied by: Vicco Lab, Goa.

Turmeric Curcumin Herbal Soap Bar Acne Whitening Skin

It improves skin complexion, reduces hair growth on body. Maintain youth by controlling wrinkle and crease formation on the surface of the skin.

Ingredients: Turmeric, tamarind, vitamin E, palm kernel oil, sodium cocoate (coconut oil), sweet basil, glycerin, water.

Action: Tamarind deeply exfoliates and whitens, while curcumin and honey bring out the natural radiance of the skin. They stimulate circulation and invigorate to create a healthy glow. Turmeric acts as a natural anti-bacterial agent which helps prevent acne and skin rashes. Vitamin E provides moisture and revitalizes the skin.

Direction: For face and body wash twice daily.

Supplid by: Super Resources Co., Ltd., Thailand.

Himalaya Herbals Purifying Neem Face Wash for All Skin Types with Neem and Turmeric

It is a soap-free herbal formulation. It gently removes impurities and prevents pimples.

Ingredients: Each gram contains: Ext *Nimba* 50 mg, *Haridra* 5 mg.

Action: Neem is kown for its anti-bacterial property that kills problem-causing bacteria, and turmeric, which effectively controls acne and pimples.

Direction: Moisten face, apply a small quantity of purifying neem face wash and gently work up lather with a circular motion. Wash off and pat dry. Use twice daily.

Supplied by: Himalaya Drugs Pvt. LTD., India.

Shadew Ayurvedic Lemon Turmeric Cream

The formulation contains turmeric, which helps to make the skin appear soft, smooth, clear and bright.

Ingredients: The formulation contains turmeric, loemon and other excipients.

Action: Turmeric helps to make the skin appear soft, smooth, clear and bright. Lemon and turmetic also provides emollients and hydration to dry skin, protecting it from the effects of soap, chlorine and exposure to the elements.

Supplied by: Shahnaz Ayurveda Pvt Ltd, Delhi, India.

Method of Preparation

Basic Turmeric Face Mask (For Dry Skin)

A typical basic turmeric face mask, the milk soothes and eases your skin, while the turmeric cleans and makes it glow.

Ingredients

- 1 tablespoon of milk powder
- 1 teaspoon turmeric paste

Method

Mix the milk powder and turmeric paste until it is a smooth paste. Spread the paste gently and equally with fingertips on clean face and neck. Keep the eye area clear then lie down and leave the mask on for 15–20 minutes, skin should start to feel tight and dry. Then wash it off with cold water; pat the skin dry with a clean towel.

Facial Mask (For Acne Skin)

It is a simple turmeric facial mask for acne prone skin.

Ingredients

- 2 tablespoons curd
- 1/4 teaspoon turmeric powder
- 1/2 teaspoon sandalwood powder

Method

Mix the curd, turmeric and sandalwood until have a smooth and uniform paste. Pat the paste gently and equally with fingertips on clean face and neck. Keep the eye area clear. Leave the mask on for 15–20 minutes then wash it off with warm water and a soft warm wash cloth, end with a splash of cold. Pat the skin dry with a clean towel. Finally, apply a moisturizer.

Facial Mask (For Oily Skin)

Ingredients:

- Coconut oil
- 1/4 teaspoon turmeric powder

Method

Mix the coconut oil and turmeric until have a smooth and uniform paste. Smooth the paste gently and equally with fingertips on clean face and neck. Keep the eye area clear then leave the mask on for 20 minutes. Wash it off with warm water and a soft warm wash cloth, end with a splash of cold. Pat skin dry with a clean towel. Finally, apply a moisturizer.

Storage of Products

It should be stored only in air-tight container free from light, since it will lose its flavor. Turmeric should be handled carefully because it will stain clothes and fingers.

Dosage

Doses of 500–8,000 mg of powdered turmeric per day have been used in human studies. Standardized extracts are typically used in lower amounts, in the 250–2,000 mg range.

Adult: The following are doses recommended for adults:

- Cut root: 1.5–3 g per day
- Dried, powdered root: 1–3 g per day
- Standardized powder (curcumin): 400–600 mg, 3 times per day
- Fluid extract (1:1) 30–90 drops a day
- Tincture (1:2): 15–30 drops, 4 times per day

Patents

This invention provides compositions and methods to treat medical disorders, such as skin conditions, using a combination of turmeric extracts, and hydroxy acids, and/or alpha-1-antitrypsin. Skin conditions such as psoriasis, severe dryness, itchiness, stretch marks, acne, and microbial infection are treated by topical application of the compositions (Phan, 2003).

Herbal compound formulations using herbs *Curcuma longa*, *ocimum sanctum* (holy basil) and *Emblica officinalis* (Amla) in their natural forms and a method of packaging the same for dietary supplement as tablets, wafers, capsules and as teabags where the packaging of individual dosages and the outer packaging may use visual display of color that represents the proportion and presence of different herbs in the herbal compound (Singhal, 2009).

A topical skin care product is provided that contains therapeutic concentrations of turmeric and other constituents which is colorless upon application to the skin. The product can be used for cosmetic, protective, and healing purposes without staining skin or clothing yellow (Bommarito, 2010).

Disclosed are natural compositions including at least five core ingredients that provide pain relief without the side effects of synthetic pain relievers. Further provided are methods of making the disclosed herbal formulations or compositions, and methods of treating mammals that include administering the herbal compositions to a mammal in need thereof (Lomax and Pohlman, 2011).

A nutritional supplement useful for promoting good overall health and relieving minor aches and pains associated with inflammation and muscle soreness includes a plurality of active herbal ingredients (turmeric extract, white willow extract, passion flower extract, valerian root extract, and/or licorice root extract), a plurality of non-herbal active ingredients (magnesium glycinate, vitamin B_1 (thiamine HCI), vitamin B_2 (riboflavin), vitamin B_3 (niacinamide), vitamin B_6 (pyridoxine HCl), folic acid, vitamin B_{12}, and/or beta carotene). The nutritional supplement may be in the form of an effer-vescent tablet that is dissolvable in a potable liquid and can include one or more sweeteners and flavorings (Tuchinsky, 2011).

BIBLIOGRAPHY

1. Ammon HPT, Wahl MA. Pharmacology of *Curcuma longa*. Planta Medica 1991; 57:1–7.

2. Bommarito AA. Topical turmeric skin care products. United States Patent 7763289. Application no: 12/148242, Publication date: 07/27/2010. http://www.freepatentsonline.com/7763289. html

3. Brauser D. Turmeric Cream Decreases Signs of Aging. Medscape Medical News. [Online] March 16, 2010. http://www. medscape.com/viewarticle/ 718563.

4. http://curcumalonga.com/

5. http://www.advance-health.com/curcumin.html

6. http://www.motherherbs.com/curcuma-longa.html

7. http://www.natural-homeremedies-for-life.com/turmeric-facial-mask.html

8. http://www.turmeric.co.in/turmeric_cosmetic_use.htm

9. http://www.umm.edu/altmed/articles/turmeric-000277.htm

10. http://www.uni-graz.at/~katzer/engl/Curc_lon.html

11. Lomax L, Pohlman K. Herbal pain killer compositions. United States Patent 20110076327. Application no: 12/895200, Publication date: 03/31/2011. http://www.freepatentson line. com/y2011/0076327.html

12. Phan D. Methods of treatment for skin disorders using turmeric extract and a hydroxy acid. United States Patent 20030113388. Application no: 10/313912, Publication date: 06/19/2003. http://www.freepatentson line.com/y2003/0113388.html

13. Plianbangchang P, Tungpradit W, Tiya-boonchai W. Efficacy and Safety of Curcuminoids Loaded Solid Lipid Nanoparticles Facial Cream as an Anti-Aging Agent. *Naresuan University Journal* 15(2).

[Online] 2007. http://office.nu. ac.th/nu_journal/pdf/journal/15(2)73-81.pdf.

14. Singh P, Mogra P, Sankhla V, Deora K. Protective effects of curcumin on cadmium chloride induced colon toxicity in Swiss albino mice. Journal of Cell and Molecular Biology 2011; 9(1): 31–36.

15. Singhal TC. *Curcuma longa* (turmeric) based herbal compound formulations as dietary supplements. United States Patent 20090252818. Application no: 12/383175, Publication date: 10/08/2009. http://www.freepatentsonline.com/y2009/0252818.html

16. Sittisomwong N, Leelasangaluk V , Chivapat S, Wangmad A, Ragsaman P, Chuntarachaya C. Acute and Subchronic Toxicity of Turmeric. 2533; 32 (3): 101–111. http://webdb. dmsc. moph.go.th/ifc_herbal/research_poison/pdf_research/i_23457494 4683422. pdf

17. Thiele J, Elsner P. Oxidants and antioxidants in cutaneous biology. s.l.: Karger Publishers, 2001. ISBN 3805571321, 9783805571326.

18. Tuchinsky DB. Nutritional supplement. United States Patent 20110142968. Application no: 13/030267, Publication date: 06/16/2011. http://www. freepatentsonline.com/y2011/0142968.html

Daucus carota

Plant profile

Kingdom: *Plantae* – Plants

Superdivision: *Spermatophyta* – Seed plants

Class: *Magnoliopsida* – Dicotyledons

Order: *Apiales*

Genus: *Daucus* L. – Wild carrot

Subkingdom: *Tracheobionta* – Vascular plants

Division: *Magnoliophyta* – Flowering plants

Subclass: *Rosidae*

Family: *Apiaceae* – Carrot family

Species: *Daucus carota* L.– Queen Anne's lace

Vernacular name

San: Shikha-mula,Garijara, **Beng:** Gajor, **Eng:** Carrot, **Mar:** Gajar, **Guj:** Gajar, **Hind:** Gajjar, **Kan:** Gajjari, manjal mulangi, **Mal:** Carrot, Manjamullanki, **Ori:** Gajara, **Tam:** carrot, karttukkilangu, **Tel:** gajjara-gedda

Ayurvedic properties

Rasa: Madhura, **Guna:** Guru, Snigdha, **Veerya:** Seeta.

INTRODUCTION

Daucus carota is a variable biennial plant, usually growing up to 1 m tall and flowering from June to August. The umbels are claret-coloured or pale pink before they open, then bright white and rounded when in full flower, measuring 3–7 cm wide with a festoon of bracts beneath; finally, as they turn to seed, they contract and become concave like a bird's nest. The flowering stem grows to about 1 metre (3 ft) tall, that produce a fruit called a mericarp by botanists, which is a type of schizocarp. The dried umbels detach from the plant, becoming tumbleweeds.

Distribution

The carrot appears to have originated in Central Asia, in the hills of Punjab and Kashmir in India, with a secondary centre of distribution in Asia, Europe and North Africa around the Mediterranean. It is cultivated in India, Malaysia, Indonesia, Philippines, Central, East and West Africa, South America, the Caribbean and Australia. The variety of carrot found in north India is rare everywhere except in Central Asia and other contiguous regions, and is now growing in popularity in larger cosmopolitan cities in South India. The North Indian carrot is pink-red comparable to plum or raspberry or deep red apple in color (without a touch of yellow or blue) while most other carrot varieties in world are from orange to yellow in colour.

Chemical Constituents

The carrot is an extremely rich source of vitamin A. The name carotene, which is a form of pro vitamin, has been derived from carrot. The carotene is converted into vitamin A by the liver and it is also stored in our body. Carrots are rich in sodium, sulfur, chlorine and contain traces of iodine. The main chemical constituents of carrot seed oil include α-pinene, camphene, β-pinene, sabinene, myrcene, γ-terpinene, limonene, β-bisabolene, geranyl acetate and carotol.

Medicinal uses

Carrots have been known to have medicinal properties that allow it to be used as an astringent and antiseptic substance. Carrot has been used as a cleansing medicine, because of its activities which basically helps the kidney in the removal of waste in the body. It also aids in the stimulation of one's menstruation because of the same properties. For children, raw carrot roots are a safe and recommended treatment for threadworms. In the same manner, regular intake of cultivated carrot roots allows the stimulation of urine flow. It manifests activities that help the skin, by virtue of its anti-oxidant contents. Regular intake of carrot roots, in any manner, though recommended to be taken in juice form, improves eyesight. Particular substances that are contained in the roots have been discovered to be anti-cancer. An infusion of carrot seeds (1 teaspoon per cup of boiling water) is believed to be diuretic, to stimulate the appetite, reduce colic, aid fluid retention and help alleviate menstrual cramps. The dried flowers are also used as a tea as a remedy for dropsy. Chewing a carrot immediately after food kills all the harmful germs in the mouth. It cleans the teeth, removes the food particles lodged in the crevices and prevents bleeding of the gums and tooth decay. Carrot soup is supposed to relieve diarrhea and help with tonsilitus.

Cosmetic uses

Carrot has been regarded by the ancient healers as the 'herbal healer' of skin diseases. Carrot root pulp appears to have several

Beta-carotene

beneficial attributes for skin care. Dermatologists agree that the most important step for preventing wrinkles is to stop the sun's ultraviolet rays from reaching the skin. In the past it was used as a poultice to treat skin ulcers and cancerous sores. The pulp can be used as a face mask to help clear blemishes and to soften rough patches, leaving the skin silky smooth. Carrot seed oil relieves fluid retention and can be beneficial in cases of anorexia, while it revitalizes and tones the skin, helping in cases of dermatitis, eczema and rashes. Dry skin, with impurities, acne, difficulty in tanning, sunburns, eritema, premature appearance of wrinkles—all these things can depend largely on an insufficient intake of this vitamin. Anti-inflammatory, revitalizes and tones the skin. It is used to treat dermatitis, eczema, rashes, as well as wrinkles. It also promotes the healing of cuts, abrasions and stubborn sores. Carrot seed diluted in wheatgerm oil is recommended as a suntan lotion. It is indeed used in some suncare lotions on the market. Eye care—carrot brings relief of eyestrain and inflammation of the eyes. Nourishing—carrot promotes healthy skin and is beneficial to areas of the skin that are regularly exposed to the sun and tend to blister and peel. Moisturising—carrot aids in repair of skin tissue and helps in the treatment of dry, chapped and scaling skin conditions. The fresh root, finely chopped, can be used as a beauty mask for the face. Carrot oil has vitamins A and E which are very useful in conditioning the scalp.

Pharmacological Action

Antioxidant properties of water-soluble anthocyanin obtained from the carrot (*Daucus carota*) callus cultures were evaluated by the estimation of the amount of hydroperoxide formed by the autoxidation of linolenic acid at pH 2, 4 and 7. The results indicate that anthocyanin obtained from carrot showed strong antioxidant activity at pH 2, 4 and 7 by the linolenic acid auto-oxidation system (Ravindra and Narayan, 2003).

The aim of this study was to investigate the influence of carrot seed extract (CSE) on spermatogenesis, number and motility of sperms in cauda epididyme in male rats. After 4 weeks treatment, fasting serum samples were obtained for the sex hormone analysis. Under anesthesia, testis, cauda epididymides and sperm ducts were dissected and sperm count, motility and cauda epididymis sperm reserves (CESR) were determined. The extract could also protect testis from the gentamicin-induced necrosis. The CSE administration caused about 3.5-times increase in the LH levels even in spite of receiving 5 mg/kg/day gentamicin with no significant effect on FSH levels. It concluded that the extract has opposite effects on male and female reproductive systems (Mohammad et al., 2009).

The present study was undertaken to determine the gastroprotective potential of the fresh juice extract of the roots of *Daucus carota*. The juice extract of the roots of *Daucus carota* (DCE) was tested orally at the dose of 200 and 400 mg/kg body weight, on gastric ulceration experimentally induced by pylorus ligation, aspirin and ethanol induced. The DCE at the dose of 200 and 400 mg/kg, significantly decreased gastric volume, free acidity, total acidity and ulcer index, while it increased the pH and the mucus content as compared with control. The DCE at a dose of 400 mg/kg produced 60.45, 56.80 and 43.51% significant inhibition when gastric ulceration were induced by pylorus ligation, aspirin and ethanol, respectively. The DCE possesses gastroprotective property and the results supported traditional uses of the roots of this plant in the treatment of gastric ulcer and acidity (Khatib et al., 2010).

Toxicological Studies

Daucus carota seed extract was administered orally at a dose of 5 mg/kg initially and mortality was observed for 3 days followed by toxicity was studied with the doses of 50, 300 and 2000 mg/kg. All the doses were found nontoxic. The extract did not produce any mortality even at 2000 mg/kg dose administered. Three submaximal doses (100, 200 and 400 mg/kg, p.o.) were also found to be safe in mice (Vasudevan et al., 2006).

Market Formulations

Problem Skin Night Lotion

It will help reduce oiliness, refine the skin, reduce pore size, sort out the causes of the problems also balance the moisture and hydration levels of the skin.

Ingredients: Bacocalmine™, ACNET®, *Croton lechleri*, Evermat®, borage oil, horse chestnut, *Centella asiatica* extract, chamomile, vitamin E, rosehip, green tea, carrot seed.

Action: Bacocalmine™ calmed irritated skin in 100% of the test volunteers and has an anti-oxidizing effect, while fighting fungal infection normally present in acne skin and furthermore, clears and neutralizes toxins on the skin. ACNET® fights hyperseborrhea (excessive oiliness) in the skin, cutting down excess production of sebum by 54.1%, and reduces inflammation in the skin by up to 72%. The resin is obtained from the Croton Lechleri tree and it has on preventing and healing inflammatory skin conditions. Evermat® is help control the production of sebum or oil in the skin and to reduce pore size. Borage oil is helps to soften and soothe the skin, while reducing skin redness, especially in sensitive skins. Horse chestnut extract protects and strengthens the veins and capillaries, prevents cellular filtration (leaking), promotes better and healthier collagen, fights inflammation. *Centella asiatica* extract helps in the healing of any lesions, and prevents excess scar tissue from forming on the healing wounds. *Centella asiatica* helps to repair the walls of the veins and capillaries in the skin, thereby promoting healthy circulation. Chamomile is of great benefit to problem or acne skin, as it helps cleanse the pores of impurities and reduces puffiness. Tocopherol (the cosmetic name for vitamin E) is an extremely effective anti-oxidant, helping

with wound healing and the penetration of other active ingredients into the skin. Rosehip is high in bio-available vitamin C, which boosts the regeneration of the skin, with a calming function as well. Green tea is proven to reduce inflammation, which is present when suffering from skin problems such as acne and pimples—it is the inflammatory condition that causes the skin redness. Carrot seed oil improves the complexion and helps to remove toxic build up from the skin. Even skin prone to acne, pimples, spots and zits has to repair itself while you sleep and carrot seed oil is an excellent ingredient in this regard, as it helps to rejuvenate the skin, accelerating the regeneration of healthy skin cells, while toning and firming the skin.

Direction: Place a small amount of the problem skin night lotion on tips of fingers, rub together, and then gently massage with upward and outward strokes to your face and neck.

Supplied by: Dermaxime products, South Africa.

Ladymax™

It is an anti-acne gel. It is effective for oily and acne-prone skin, fighting against the four parameters responsible in the acne process: hyperseborrhea, hyperkeratosis, inflammation and bacterial growth.

Ingredients: Tea tree oil, *Psidium guajava* extract, *Terminalia chebula* extract, *Aloe vera*, *Daucus carota* extract, oleanolic acid, nordihydroguaiaretic, butylene glycol, PEG-60 almond glyceride, caprylyl glycol, glycerine, carbomer, diazolidinyl urea, chlorphenesin, methylisothiazolinone, AHA, triethanolamine, deionised water.

Action: **Tea tree oil** works as a topical antiseptic agent which helps to promote faster healing with less scarring. It also helps to remove excess oil. *Psidium guajava* extract contributes to clear the skin. Its astringent properties make it an excellent healing agent. It also decreases the chance of bacterial growth. *Terminalia chebula* extract inhibits the production of skin oil. Its usage leads to a shrink of the sebaceous glands. *Aloe vera* possesses healing activity and minimizes scarring. *Daucus carota* extract helps to boost the strength of the epithelial layer and blood circulation to the skin prompting better waste removal and healing to affected areas. Oleanolic acid inhibits 5-α reductase to fight hyperseborrhea.

Direction: Twice daily on the acne area after cleansing the skin. It is important to apply regularly for at least 6 weeks. In some instances, it should be treated in combination with Completec® , Fame Chlorella® and Ladymax® (or) Rejuvir® oral therapy.

Storage condition: Store below 30°C.

Supplied by: Fame Pharmaceuticals.

Storage of Products
It should be stored in air-tight container free from light and in cool place.

Dosage
Dose is depends on the type of formulation and it should not be more than 500 mg. The oil of carrot is used not more than 10–20 ml in the cosmetic formulation.

Patents
The present invention provides for plant extracts and dermatological formulations comprising one or more plant extracts that are capable of inhibiting one or more extracellular proteases selected from the group of: matrix metalloprotease-1 (MMP-1), matrix metalloprotease-2 (MMP-2), matrix metalloprotease-3 (MMP-3), matrix metalloprotease-9 (MMP-9) and human leukocyte elastase (HLE). The present invention further provides for a rapid method for screening plant extracts to identify those having the above activity that are suitable for incorporation into the dermatological formulations of the invention. The invention also provides for the use of the plant extracts as dermatological agents suitable for the treatment or prevention of various dermatological conditions, including wrinkling or sagging of the skin, irradiation induced skin and/or hair damage, deepening of skin lines, elastotic changes in the skin, as well as for the routine care of the skin, hair and/or nails (Behr et al., 2007).

The disclosure herein is directed to organic skin-based products that comprise an excess of 70% organic ingredients. The skin care products described herein are composed of a plurality of organic juices and other organic materials (Jochim and Behnke, 2008).

BIBLIOGRAPHY
1. Behr S, Duret P, Gendron N, Guay J, Lavallee B, Page B. Plant extracts and dermatological uses thereof. United States Patent Application 20070122492. Application no: 10/533025. Publication date:05/31/2007. http://www. freepatentsonline.com/y2007/0122492. html

2. http://ayurvedicmedicinalplants.com/ plants/2104.html

3. http://mdidea.com/products/new/ new06901.html

4. http://www.agripinoy.net/medicinal-plant-carrot-daucus-carota-l.html

5. http://www.anagen.net/carrot.htm

6. http://www.best-home-remedies.com/ herbal_medicine/vegetables/carrot. htm

7. http://www.carrotmuseum.co.uk/nutrition3. html

8. http://www.danielbaileyonline.com/2011/ 04/cosmetic-uses-carrots.html

9. http://www.essentialoils.co.za/essential-oils/carrot-seed.htm

10. http://www.flowersofindia.in/catalog/ slides/Carrot.html

11. http://www.indiaparenting.com/ homeremedies/data/008.shtml

12. Jochim M, Behnke K. Compositions for Juice-Based Moisturizers. United States Patent Application 20080171031. Application no: 11/933428. Publi-cation date: 07/17/2008. http://www. freepatentsonline. com/ y2008/0171031. html

13. Khatib N, Angel G, Nayna H, Kumar JR. Gastroprotective activity of the aqueous extract from the roots of *Daucus carota* L. in rats. International Journal of Research in Ayurveda and Pharmacy, 2010; 1(1): 112–119.

14. Mohammad N, Khaki A, Fatemeh FA, Mohammad RR. The Protective Effects of Carrot Seed Extract on Spermatogenesis and Cauda Epididymal Sperm Res-erves in Gentamicin Treated Rats. Yakhteh Medical Journal, 2009; 11(3); 327–333.

15. Ravindra PV, Narayan MS. Antioxidant activity of the anthocyanin from carrot (*Daucus carota*) callus culture. Int J food Sci Nutr. 2003; 54(5): 349–355.

16. Vasudevan M, Gunnam KK, Parle M. Antinociceptive and anti-inflammatory properties of *Daucus carota* seeds extracts. J of Health Science. 2006; 52(5): 598–606.

Echinaceae angustifolia

Plant profile

Kingdom: *Plantae* – Plants

Subkingdom: *Tracheobionta* – Vascular plants

Superdivision: *Spermatophyta* – Seed plants

Division: *Magnoliophyta* – Flowering plants

Class: *Magnoliopsida* – Dicotyledons

Subclass: *Asteraidae*

Order: *Asterales*

Family: *Asteraceae* – Aster family

Genus: *Echinacea* Moench – Purple coneflower

Species: *Echinacea angustifolia* DC.–Black Sampson Echinacea

Other common names

Narrow-Leaf Purple Coneflower, Narrow-Leaf Echinacea, Black Susan, Hedgehog, Missouri Snake Root, Coneflower, Kansas Snake Root, Snake Root, Black Sampson, Comb Flower, Indian Head

INTRODUCTION

Echinacea angustifolia is an herbaceous plant, the thick, black, pungent root of which sends up from year to year, a slender, sometimes somewhat stout stem, bristling with hairs, and from 2 to 3 feet high. The leaves are 3-veined and hispid-pubescent, vary in shape from broad lanceolate to lance-linear. At the base they become slender, and the lowermost have short petioles. The involucre consists of about 2 rows of lanceolate, scaly bracts. As the flower develops the disk is at first concave, but as the growth progresses it becomes ovoid, the receptacle taking on a sharply conical form. The linear-lanceolate, chaff-like bracts of the receptacle are firm, remain permanently attached, are boat-shaped and concave, and become narrowed into a stiff, spine-like crisp, extending beyond the disc-flowers. The ray-flowers are narrow, and from 1 to 2 inches long. They are rose or purple, and drooping or pendant, and, while withering, are yet persistent. The root of Echinacea varies in thickness, from that of an ordinary lead pencil to that of the little finger. The deep-brown or reddish-brown epiderm is shrunken and wrinkled longitudinally, and is often disposed in spiral folds upon the subdermal portion of the root. The woody portion, as seen upon transverse section (*see* illustration of root), is composed of medullary rays, separated by a greenish, pulp-like substance. When broken the dried root exhibits a grainy and apparently rotten aspect.

Distribution

Echinacea angustifolia, De Candolle, is an entirely different plant, found only in prairie regions, and not occurring East of the prairie regions of Illinois, and has never been used under the name Black Sampson. *Echinacea angustifolia* is an indigenous plant of the composite order, growing chiefly in the Western states, from Illinois to Nebraska, and Southward through Missouri to Texas, It thrives best in rich prairie soil. It has also been stated that it grows in rocky and sandy soil. The plant, however, which is abundant in Kansas, Nebraska and neighboring localities.

Chemical Constituents

Essential oil (0.2–2%) and echinacoside (0.4–1.7%) are present in *E. angustifolia* root.

Quantitative analysis confirms the presence of echinacoside, cynarine, chicoric acid, chlorogenic acid derivatives and other biologically active ingredients including a volatile oil, alkamides, polyalkenes, polyalkynes, caffeic acid derivatives, and polysaccharides. The volatile oil contains, among other compounds, pentadeca-(1,8-Z)-diene (44%), 1-pentadecene, ketoalkynes and ketoalkenes. More than 20 alkamides, mostly isobutylamides of C_{11}–C_{16} straight-chain fatty acids with olefinic or acetylenic bonds, or both, are found in the roots. The main alkamide is a mixture of isomeric dodeca-2,4,8,10-tetraenoic acid isobutylamides. Caffeic acid ester derivatives present include echinacoside, cynarin, and chicoric acid. Cynarin is present only in *E. angustifolia*, thus distinguishing it from the closely related *E. pallida*.

Medicinal uses

Echinacea was traditionally used by the North American Comanche tribe used as an herbal remedy for toothaches and sore throats. While the Sioux, another major North American tribe, used it for the treatment of rabies, any snakebite, and for all septic conditions arising from injury. The ability of the herb to enhance and boost the performance of the immune system is well known traditionally and has been extensively documented down the years. In addition, the herbal remedies made from the Echinacea possess an antibiotic action and also act like interferons in their anti-viral action

Echinacoside

Cynarine

within the body. The herbal remedy is also known to have an amazing fungicidal effect and it is extensively used as an anti-allergenic herbal remedy. Some studies have also recorded the Echinacea as possessing an anti-tumor activity. It was used in the treatment of burns on the skin. Nowadays, the herbal remedies derived from the Echinacea herb also function as a blood cleansing remedy for the holistic treatment of various skin problems including boils and all types of topical and internal abscesses, and urticaria, it is increasingly used in the treatment of different types of infections such as tonsillitis, the common colds, the flu, and various other chest infections. The herbal remedy is also used in the treatment of asthma and other viral diseases like the glandular fever, it is also used in the treatment of candidiasis and to suppress post-viral fatigue syndrome in patients. Research is still being conducted on the possible beneficial effects of Echinacea treatment on symptoms of HIV and AIDS in patients. Respiratory problems and chilblains are also treated using the herbal remedies prepared from the Echinacea herb. The circulation of blood in the body is benefited by the stimulating effect of the Echinacea herb, this is especially true when the herb is taken in the form of a hot infusion, the hot herbal infusion also helps by stimulating the process of sweating and this rapidly brings the fevers down, while at the same time enhancing the natural defenses of the body and enabling the immune system to shake off the infection, which has induced the condition of fever initially. Echinacea also possesses a strong anti-inflammatory effect on the body, this ability is particularly important in the treatment of long-term arthritis and in the treatment of gout. This property of the herb is also good for all types of inflammatory conditions affecting the reproductive system in women.

Cosmetic uses

Echinacea may further work to cleanse and detoxify the body of wastes through the skin.

Chicoric acid, cynarine and other compounds from *E. angustifolia* have antihyaluronidase activity, which may reduce inflammatory changes in damaged tissues.

Several Echinacea constituents have protected collagen from degradation during exposure to free radicals, leading to suggestions that echinacea may be helpful in protecting against sun damage to skin. Echinacoside, chlorogenic acid, cynarine, 3,5-dicaffeoylquinic acid are acts as an antioxidant, radical scavenging and anti-inflammatory activity. Phytosterols, amino acids, polysaccharides: moisturizing the skin and provide nutrients to the skin. It improving dull and tired skin and prevents collagen loss to the skin. The extract of this plant also can help in the treatment of acne, eczema, burn wounds, gum inflammations, etc. When applied on the skin, echinacea 15% pressed herb (non-root) juice semisolid preparation has been used daily for wounds and skin ulcers.

Some of the root preparations of this herb for beneficial uses as, decoction—the herb root can be used in this decoction form and patients affected by infections, can consume 10 ml doses, once every 1–2 hours during the entire acute stage of infection and the treatment can be used as and when necessary. Wash—the herbal root decoction or the diluted herbal tincture can be used to wash infected wounds on the body. Carefully bathe the affected area as frequently as possible using the tincture or the decoction. This will not only result in reducing the infection but will also heal the wound in no time. Gargle—the Echinacea can also be used as a gargle, and 10 ml of the root tincture in a glass of warm water can be used to gargle when dealing with sore throats in different patients. Powder—the root can be powdered down and the herbal dust is often used for the treatment of various infected skin conditions including boils—when treatment should be combined with other herbs such as the marshmallow, during the treatment of eczema infections on the skin.

Pharmacological Action

Echinacea's fine plant-based antiseptic, antiviral and antifungal qualities support the body's own natural efforts to stimulate resistance to infection, diseases, fever, blood poisoning, common colds and flu. *Echinacea angustifolia* may be effective against vaginal yeast infections, such as Candida, and other fungal infections. This species of Echinacea is specifically thought to be helpful in cases of trichomoniasis, the vaginal infection caused by *Trichomonas vaginalis*, a single-cell protozoan parasite. Currently, it is being used as a possible chemopreventative agent.

The effect of chemically defined extracts from *Echinacea angustifolia* roots on rhinovirus infection was evaluated. A total of 437 volunteers were randomly assigned to receive either prophylaxis (beginning seven days before the virus challenge) or treatment (beginning at the time of the challenge) either with one of these preparations or with placebo. The results for 399 volunteers who were challenged with rhinovirus type 39 and observed in a sequestered setting for five days were included in the data analysis. Evaluation of blinding revealed that the proportion of volunteers who believed they were taking the active medication during the treatment phase of the study ranged from 21 of 51 (41 percent) to 24 of 48 (50 percent) in the active-treatment groups; in the placebo group, 37 of 103 volunteers (36 percent) believed they were receiving the active medication. Further result showed prophylaxis with the echinacea preparations had no significant effect on rhinovirus infection and there was no effect on the infection rate in the groups that received echinacea only in the treatment phase of the study. The quantitative virus titer was also not affected by either prophylaxis or treatment with these echinacea preparations. Finally the results demonstrate that, as tested, the putative active constituents of *E. angustifolia* do not have clinically significant effects on rhinovirus infection or illness (Ronald et al., 2005).

The study aimed to determine interference with doxorubicin chemotherapy, and if fractions and compounds from *Echinacea angustifolia* roots protected the cells. Cervical and breast cancer cells were treated with the Echinacea samples and doxorubicin. At 0.05 and 0.5 microM doxorubicin concentration, cynarine increased HeLa cell growth by 48–125% and 29–101%, respectively ($p < 0.01$). At 0.05 microM doxorubicin concentration, chicoric acid increased cell growth by 23–100% ($p < 0.01$). Results indicate that phenolic compounds are responsible for proliferative activity. Studies with individual compounds show that chicoric acid and cynarine interfered with cells treated with 0.5 microM doxorubicin. The results of this study show that Echinacea herbal medicines affect cell proliferation despite cancer treatment, and that herbal medicines require further study with respect to anticancer drugs (Huntimer et al., 2006).

Toxicological Studies

Polinacea™ is a new immonomodulating *E. angustifolia* standardized extract. Through direct action on T cells, it increases the immune functions when compared to very well-known Echinacea based specialties sold as drug in Europe (Morazzoni et al., 2005). The studies conducted on Polinacea™ (oral acute and subacute toxicity) indicate that the product has no toxic effect at all being the $LD_{50} > 2000$ mg/kg. No signs of any clear toxicological effect were seen at any of the dose levels investigated (the high dose 1g/kg/day may be considered the "no observed adverse effect level"- NOAEL in the study). This specific characteristic entitles Polinacea™ to be considered a high profile candidate as a "functional ingredient" in the enhanced food category.

Laboratory analysis of blood, urine, and organ specimens from animals treated with Echinacea products provides some additional evidence of safety. Experiments using rats and mice fed up to 8 g/kg/day over several weeks failed to demonstrate measurable adverse effects.

Clinical Investigation

For URI treatment, numerous human trials have found Echinacea to reduce duration and severity, particularly when initiated at the earliest onset of symptoms. However, the majority of trials, largely conducted in Europe, have been small or of weak design. Negative results exist of a U.S. trial in adults, which used a whole-plant Echinacea preparation containing *Echinacea angustifolia*. Another clinical trial reported in July 2005 also did not demonstrate any clinical benefit. However, a 2006 meta-analysis investigating the efficacy of Echinacea found that the likelihood of experiencing a clinical cold was 55% higher with placebo than with Echinacea (based on three trials). The sum of the current evidence is conflicting and further well-designed studies are needed before a definitive conclusion can be drawn. Lack of benefit in children ages 2–11 has also been reported. For URI prevention (prophylaxis), daily Echinacea has not been shown effective in human trials. Preliminary studies of echinacea taken by mouth for genital herpes and radiation-associated toxicity remain inconclusive. Topical Echinacea juice has been suggested for skin and oral wound healing, and oral/injectable Echinacea for vaginal *Candida albicans* infections, but evidence is lacking in these areas (Web article: The Healthline Site).

Market Formulations

Echinacea angustifolia STEMS G™

It is an anti-stress boost for dull and tired skin. It stimulates the synthesis of new collagen on *in vitro* cultures of human fibroblasts.

Ingredients: Preparation of 20% of *Echinacea angustifolia* stem cells in plant glycerine (80%), xantham gum 0.3%, without preservatives. The formulation is compatible with O/W emulsions, serum, and pH d" 6. Mixing was carried out during the cooling phase (< 50°C).

Direction: 20 minutes application of a face mask with 1.5% of *Echinacea angustifolia*

stems G™. It increases hydration by 28% and elasticity by 11%.

Storage: Store the product in the original, well-closed container, in a cool, dry area and protected from light.

Actifirm Recovery Foner (Glycerin) Cream

This combination of cosmetic is used for the basic skin care and gives a benefit for skin firming, lifting and wrinkle care.

Ingredients: Glycerin, water, dipropylene glycol, betaine, sodium hyaluronate, *Betula alba* juice, glycereth-26, trehalose, ethoxydiglycol, *Althaea rosea* flower extract, *Aloe barbadensis* leaf extract, soluble collagen, saccharide isomerate, royal jelly extract, butylene glycol, *Phellinus linteus* extract, PEG-60 hydrogenated castor oil, bis-PEG-18 methyl ether dimethyl silane, xylitol, *Aspalathus Linearis* leaf extract, *Chamomilla recutita* (matricaria) flower extract, methylparaben, arginine, ammonium acryloyldimethyltaurate/VP copolymer, carbomer, alcohol, *Centella asiatica* extract, *Echinacea angustifolia* extract, *Rosmarinus officinalis* (rosemary) leaf extract, fragrance, adenosine, disodium EDTA.

Indication and usage: After cleansing, wet the cotton pad and apply gently on face from center to outwards following the facial contour.

Dosage and administration: Take appropriate amount according to your skin condition and apply every morning and night.

Supplied by: Gowoonsesang Cosmetics Co., Ltd.

Storage

Keep container tightly closed. Keep container in a cool, well-ventilated area. Do not store above 24°C (75.2°F). All the formulated cosmetic products also stored in cool place and away from sunlight.

Dosage

There is no proven effective medicinal dose for Echinacea. Echinacea is commercially available as capsules, expressed juice, extract, tincture and tea. A common dosing range studied in trials is 500 to 1,000 milligrams of Echinacea in capsule form taken by mouth three times daily for five to seven days. As an extract, 300 to 800 milligrams of Echinacea has been taken by mouth two to three times daily for up to six months.

E. angustifolia Root

Unless otherwise prescribed, hot water (about 150 ml) is poured over about 0.5 teaspoon (about 1 g) of powdered plant material, allowed to steep for 10 minutes, passed through a strainer, and taken orally three times a day between meals.

Liquid Extract

(1 : 5, 45% ethanol), 0.5–1 ml three times daily.

Tincture

(1 : 5, 45% ethanol), 2–5 ml three times daily.

Patents

A novel throat spray composition is provided that is useful for topical application to sore throats. The throat spray composition contains *Piper methysticum* (Kava Kava) as its main active ingredient, and is a suitable alternative to phenol-based over-the-counter throat sprays. *Piper methysticum* is used for its little known or utilized analgesic property, which provides a soothing and numbing effect to the throat. Additional ingredients include *Echinacea angustifolia, Eucalyptus globulus, Thymus vulgaris, Lycopodium clavatum, Phytolacca decandra, Capsicum annum, Mentha piperita*, and phosphorus, which together with *Piper methysticum*, offer temporary relief of sore throat pain, irritation, difficulty swallowing, and symptoms of hoarseness or laryngitis (Watkins and Coyne, 2000).

The present invention relates to topical compositions containing: (a) *Ginkgo biloba terpenes*; (b) floroglucinols pure or in mixture thereof, extracted from *Humulus lupulus, Hypericum sp* and *Mirtus sp*; (c) *Zanthoxylum bungeanum* or *Echinacea angustifolia* lipophilic

extract; for the treatment of atopic dermatitis, skin allergic conditions and acne (Bombardelli, 2009).

The invention provides formulations containing sanguinarine, chelerythrine or chelidonine or their salts or extracts containing them, mixed with suitable vehicles and/or excipients, for the treatment of common skin warts and verrucas, anal and vulvar warts and psoriatic plaques (Bombardelli et al., 2011).

A unique, synergistic formulation for the treatment of first-degree skin burns is provided for topical use and alleviates the full spectrum of symptoms and concerns associated with epidermal burns, including pain, blistering, redness, swelling, compromised skin integrity, risk of infection and scarring. The formulation contains natural homeopathic extracts combined with a pharmaceutically acceptable carrier suitable for topical administration. The formulation may include three or more of the following homeopathic extracts: *Cantharis vesicatoria, Echinacea angustifolia, Calendula officinalis* and *Hypericum perforatum* (Manickam, 2011).

BIBLIOGRAPHY

1. Bombardelli E, Fontana G, Morazzoni P, Riva A, Ronchi M. Formulations with sanguinarine, chelerythrine or chelidonine for the treatment of warts, verrucas and psoriatic plaques. United States Patent Application 20110014306. Application number: 12/921476, Publication date: 01/20/2011. http://www.freepatentsonline. com/y2011/0014306.html

2. Bombardelli E. Compositions for the treatment of atopic dermatitis, skin allergic conditions and acne. United States Patent Application 7597915. Application number: 10/579966, Publication date: 10/06/2009. http://www.freepatentsonline.com/597915.html

3. http://herbalinformation.awardspace.com/?cm=e&fn=echinacea_angustifolia

4. http://naturalspedia.com/herbsremedies/echinacea/toxicity.htm

5. http://www.chiro.org/nutrition/ABST RACTSEchinacea_Monograph. shtml

6. http://www.healthline.com/natstan dardcontent/echinacea

7. http://www.henriettesherbal.com/ eclectic/kings/echinacea.html

8. http://www.herbalextractsplus.com/ echinacea-angustifolia.cfm

9. http://www.herbs2000.com/herbs/ herbs_echinacea.htm

10. http://www.irbtech.com/fileadmin/ user_upload/PDF_Prodotti/Brochure_ Echinacea.pdf

11. http://www.mdidea.com/products/ herbextract/echinacea/data14.html

12. Huntimer ED, Halaweish FT, Chase CC. Proliferative activity of *Echinacea angustifolia* root extracts on cancer cells: Interference with doxorubicin cytotoxicity. Chem Biodivers. 2006; 3(6):695–703.

13. Manickam SS. Antipyrotic Formulation for the Treatment of Epidermal Burns. United States Patent Application 20110206774. Application number: 13/098795, Publication date: 08/25/2011. http://www. free patentsonline. com/y2011/0206774. html

14. Morazzoni P, Cristoni A, Di Pierro F, Avanzini C, Ravarino D, Stornello S, Zucca M, Musso T. *In vitro* and *in vivo* immune stimulating effects of a new standardized *Echinacea angustifolia* root extract (Polinacea TM). Fitoterapia. 76; 2005: 401–411.

15. Ronald BT, Rudolf B, Karin W, Thomas CH, David JG. An Evaluation of *Echinacea angustifolia* in Experimental Rhinovirus Infections. N Engl J Med 2005; 353: 341–348.

16. Watkins MB, Coyne JS. Sore throat spray. United States Patent Application 6159473. Application number: 09/103965, Publication date: 12/12/2000. http://www.freepate ntsonline.com/6159473.html

Eclipta alba

Plant profile

Kingdom: *Plantae* – Plants

Superdivision: *Spermatophyta* – Seed plants

Class: *Magnoliopsida* – Dicotyledons

Order: *Asterales*

Genus: *Eclipta*

Subkingdom: *Tracheobionta* – Vascular plants

Division: *Magnoliophyta* – Flowering plants

Subclass: *Asteridae*

Family: *Asteracear* – Aster family

Species: *Eclipta alba*

Common name

False daisy

Vernacular name

San: Bhringaraj, maarkava, kesharanjana, **Beng:** Kesuriya, **Eng:** Bhringaraj, **Guj:** Bhangaro, **Hind:** Bhangara, **Mar:** Maakaa, **Kan:** Garujalu, **Mal:** Kayyonni, **Tam:** Karisalankanni, **Tel:** Galagara, Guntakalagara.

Ayurvedic properties

Guna (properties): ruksha (dry), laghu (light), **Rasa (taste):** katu (pungent) , tikta (bitter), **Veerya:** ushna, **Vipaka:** katu, **Karma (Action):** Pacifies vata and kapha, increases pitta. Heals wounds, works as anti-inflammatory, improves appetite, digestion, stimulates bile secretion, reduces pain in joints, digests ama. Seeds are aphrodisiac, reduces fever, makes hair black and lusturous.

Ayurvedic preparations

- Bhrngaraja taila
- Sadbindu taila
- Bhrngarajadi curna
- Bhrngaraja ghrta, etc.

INTRODUCTION

This tropical annual is a creeping and moisture-loving herb; it has a short, flat or round, brown stem and small white flowers on a long stalk. It grows 3″ tall; the leaves are opposite and lance-shaped and grow 11–13 cm long. Flowers—Have both ray flowers and disk flowers in inflorescense. Floral heads 6–8 mm in diameter, solitary, white, achene compressed and narrowly winged. Flowering takes place in July–October. Stems are erect and grow to an average height of 45–50 cm. The stem is single from base and has many branches. The fruits are many seeded and the seeds are black, resemble cumin seeds.

Distribution

Eclipta grows abundantly in the tropics and it is native of the USA. It is widely distributed throughout India, China, Thailand and Brazil.

Chemical Constituents

Eclipta alba (L.) contains wide range of active principles which include coumestans, alkaloids, flavonoids, glycosides, poly-acetylenes, triterpenoids. The leaves contain

Nicotinic acid

Glutamic acid

Stigmasterol

Wedelolactone

stigmasterol, a-terthienylmethanol, wedelolactone, demethylwedelolactone and demethylwedelolactone-7-glucoside. The roots give hentriacontanol and heptacosanol. The roots contain polyacetylene substituted thiophenes. The aerial part is reported to contain a phytosterol, P-amyrin in the n-hexane extract and luteolin-7-glucoside, P-glucoside of phytosterol, a glucoside of a triterpenic acid and wedelolactone in polar solvent extract. The polypeptides isolated from the plant yield cystine, glutamic acid, phenyl alanine, tyrosine and methionine on hydrolysis. It also contains nicotine and nicotinic acid. *Eclipta alba* oil has various chemical compounds that include saponin, tannin, vitamin A and ecliptine.

Medicinal uses

Internally, bhrngaraja is useful in many diseases. It is keen stimulant to digestive system. It augments the appetite and improves digestion. It is an effective cholegouge, hence benevolent in hepatosplenomegaly as well as hepatitis. The fresh juice of the plant (15–30 ml) is given along with castor oil curbs, the tendency for getting repeated worm infestations. Bhrngaraja as an adjunct relieves mucous and facilitates free breathing in asthma and cough. It is one of the best blood purifiers, stimulates the liver and alleviates the general oedema all over the body. It also augments the hemoglobin percentage in anemia. The fresh juice of plant with marica (pipernigrum) fruit powder is recommended in the treatment of anasarca and anemia. Bhrngaraja augments the acuity of vision and is salutary in night blindness. The seeds are commonly used as aphrodisiac in sexual debility. As bhrngaraja has diuretic property, it is beneficial for dysuria and in various skin disorders like urticaria. It also works well as a brain tonic. For that purpose, the powder of its dried leaves 1 gm, honey 3 gm, rock candy 5 gm and ghee 6 gm mixture is daily recommended. Commonly, the fresh juice of bhrngaraja is used. As it looses its properties on boiling, its decoction is not prepared.

Cosmetic uses

In glandular swellings and filariasis, it is used to mitigate swelling and pain, by applying the paste. The chronic and infected wounds get cleansed and heal better with application of its paste. It also alleviates the dispigmentation of the skin and renders normal colour to the skin. The fresh juice of the plant is instilled in the ears and eyes in otitis and conjunctivitis to ameliorate pain. The massage with its fresh juice is effective to alleviate headache. The nasal drops of its juice, mixed with milk, are beneficial in migraine. The medicated oils of bhrngaraja are widely used as hair tonic and to prevent hairfall and premature graying of the hair. Bhrngaraja is the most common ingredient incorporated in numerous market preparations of various hair oils. The oil is useful in skin disorders, headache, insomnia and mental disorders. It has restorative properties that can help assist in the treatment of premature greying of the hair, alopecia and balding. It works as a general antiseptic and astringent oral mouthwash for infected gums and loose teeth. It helps in maintaining and rejuvenating teeth, bones, hearing, memory and sites. A black dye obtained from *Eclipta alba* is also for dyeing hair and tattooing.

The *Eclipta alba* (bhringraj) leaf juice boiled with sesame or coconut oil for anointing the head to render the hair black and luxuriant has been practiced since ancient times. Oil prepared out of amla (Indian gooseberry) and Eclipta and sometimes with brahmi is very popular formula as hair oil and is used to blacken the hair. An herbal poultice is made with sesame oil and used over glandular swellings and various skin conditions. The leaf juice is also effective when applied externally to treat minor cuts, abrasions, and burns. *Eclipta alba* (bhringraj) is one of the key ingredients of IHT 9—Herbal hair shampoo. This shampoo arrests hair loss and promotes hair growth. Some useful cosmetic applications are:

- Reduces inflammation.
- Prevents eruption of acne.

- Removes scars and marks of acne.
- Restores and rejuvenates affected skin.
- Expels body toxins which cause acne.
- Enhances the moistness of skin.
- A best complete natural acne remedy.
- Makes smooth and flawless skin.

Pharmacological Action

The present study was deals with the investigation of standardized and phytochemically evaluated aqueous and hydro-alcoholic extracts of the plant *Eclipta alba* for sedative, muscle-relaxant, anxiolytic, nootropic and anti-stress activities. The hydrolyzed fraction of the aqueous extract was also subjected to similar studies in rats. The results point towards the potential neuropharmacological activity of the plant *Eclipta alba* as a nootropic and also having the property of attenuating stress induced alterations (Thakur and Mengi, 2005).

The present study investigated the ability of 100 and 200 mg/kg of aqueous extract of *Eclipta alba* to circumvent aggression. Foot shock induced aggression and water competition test were utilized as models for screening of antiaggressive activity. *Eclipta alba* significantly minimized dominance (p < 0.05) which is correlated to the level of aggression particularly with 200 mg/kg in the water competition test. A tangible behavioral submission was observed with 100 and 200 mg/kg and of *Eclipta alba* in the foot shock induced test (Otilia et al., 2008).

The antianaphylactic activity of alcoholic extract of *Eclipta alba* with two different doses of 250 and 500 mg/kg was studied by using different animal models such as effect on mast cell degranulation using rat mesentery, passive cutaneous anaphylaxis using rat, by measuring leakage of Evans blue dye in skin, passive paw anaphylaxis using rat by measuring the paw volume by plethysmometer. *Eclipta alba* showed beneficial effect on paw anaphylaxis induced by antiserum and also on infiltration of various inflammatory cells as well as on

histamine release from lungs (Patel et al., 2009).

The study was aimed to investigate the efficacy of methanol extract of *Eclipta alba* as hair growth promoter. Pigmented C57/BL6 mice, preselected for their telogen phase of hair growth were used. In these species, the truncal epidermis lacks melanin-producing melanocytes and melanin production is strictly coupled to anagen phase of hair growth. The extract was applied topically to assess telogen to anagen transition. The results revealed the methanol extract of whole plant when tested for hair growth promoting potential, exhibited dose dependent activity in C57/BL6 mice. The activity was assessed by studied the melanogenesis in resected skin, follicle count in the subcutis, skin thickness and surrogate markers in vehicle control and extract treated animals. Finally concluded that the methanol extract of *Eclipta alba* may have potential as a hair growth promoter (Datta et al., 2009).

The present study confirmed the antibacterial potential of aerial parts extracts of *Eclipta alba* in solvents like acetone, ethanol, methanol, aqueous and hexane against selected gram positive and gram negative bacterial species. The antibacterial studies were done by agar well diffusion methods. The MIC and MBC methods were also used. Hexane extract of *Eclipta alba* showed high antibacterial activity against *S. aureus*, *B. cereus*, *E. coli*, *S. typhi*, *K. pneumoniae*, *S. pyogenes* and *P. aeruginosa*, whereas acetone, ethanol, methanol and aqueous extracts showed intermediate activity against *S. aureus*, *B. cereus*, *E. coli*, *S. typhi*, *K. pneumoniae*, *P. aeruginosa*, *P. mirabilis* and *S. pyogenes*. The inhibitory activities of all the extracts reported were compared with standard antibiotics (Ciprofloxacin 25 µg/ml) (Pandey et al., 2011).

Eclipta alba has diverse pharmacological and antioxidative properties which aroused interest to obtain insight into the radio-protective effect of its aqueous leaf extract against gamma radiation induced haematological alterations in Swiss albino mice. The results indicated that *E. alba* aqueous extract modulates radiation induced hematological alterations which may be probably due to its anti-oxidative activity (Singh, et al. 2011).

The study was determined the median lethal dose (LD$_{50}$) of aqueous extract of *E. alba* (L.) Hassk (asteraceae) on male Swiss albino mice. The experimental groups additionally received orally for seven successive days the aqueous leaf extract of *E. alba* at 500 – 3000 mg/kg b.wt. The median acute toxicity (LD$_{50}$) of the compound was determined to be 2413.407 mg/kg b.wt. Histopathological manifestations include hepatocyte apoptosis and changes in liver architecture of groups of mice (group-V and group-VI) receiving higher experimental doses of aqueous extract of *E. alba* (Tanuja et al., 2011).

Toxicological Studies

Acute toxicity study was carried out with the experimental mice groups received in addition orally, the different concentrations doses of 500 mg/kg, 1750 mg/kg, 2000 mg/kg, 2500 mg/kg and 3000 mg/kg b.wt of aqueous leaf extract of *E. alba,* prepared by dissolving 500 – 3000 mg of dried powder of *E. alba.* leaves in 10 ml of distilled water. The treatment volume of aqueous extract was determined based on body weight. The toxicological effects were observed in terms of mortality expressed as LD$_{50}$. Results concluded that aqueous extract of *E. alba* did not produce any toxic symptoms or mortality upto the dose of 1750 mg/kg b.wt. in Swiss albino mice. Based on the experimental observations the acute oral LD$_{50}$ of the extract was calculated as 2413.407 mg/kg b.wt (Singh et al., 2011).

Clinical Investigation

The study was an open, non-comparative, non-randomized, phase III clinical trial, conducted as per the ethical guidelines of Declaration of Helsinki and was approved by Institutional Ethics Committee. Twenty-five patients of both sexes, from the age group of 20–45 years, who were clinically diagnosed as suffering from mild to moderate dandruff. All patients were advised to apply 10 ml of "Anti-Dandruff Hair Oil", twice daily for a period of 2 weeks with gentle massage to the entire scalp and were advised to leave the "Anti-Dandruff Hair Oil" on the scalp for a contact period of minimum 3–4 hours after application. All patients were followed for a period of 2 weeks. Clinical assessment of scalp lesions was done objectively and also subjectively. Thorough scalp examination was done after the completion of 1 week and at the end of the study. The severity of the dandruff symptom was recorded on a score scale and patients, with help of a linear analogue scale, did the subjective assessment. There was a highly significant reduction (p < 0.001) in the mean score for itching and white scales at the end of 2 weeks. There were no clinically significant adverse reactions, either reported or observed, during the entire study period and overall compliance to the treatment was excellent. Therefore, it may be concluded that, "Anti-Dandruff Hair Oil" is effective and safe in the management of dandruff (Vyjayanthi et al., 2004).

The ointment formulations containing extracts of the *Curcuma longa, Tridax procumbens* and *Eclipta alba* herbs formulated and their wound healing activity was evaluated on the experimentally induced open wound in albino rats. Formulations containing different concentration of herbal extracts were prepared and applied topically and their results were studied on the 4th day, 6th, 8th, 12th and 16th day postoperatively and compared with the controls. It was found that treated wound showed the faster rate of wound contraction than the control and wound contractions increase with the concentration of herbal extracts (Srivastava and Mishra, 2009).

Hair formulation of *Eclipta alba* Hassk [Asteraceae] 10% w/v, *Hibiscus rosa sinensis* Linn [Malvaceae] 10 % w/v, *Nardostachys jatamansi* [Valerianceae] 5 % w/v concentration in the form of herbal oil were prepared. Three months evaluation of a herbal hair oil vs coconut oil was conducted on human volunteers with hair fall problem. Combining assay was performed to evaluate the efficacy of the herbal hair oil. The findings of the study show that the test oil was effective in reducing the hair fall (Thorat et al., 2009).

In the study, it was envisaged to prepare herbal formulations containing petroleum ether extracts of *Eclipta alba*, methanolic extract of *Eclipta alba* and various fixed oil like mustard oil, coconut oil and sesame oil for evaluating the formulations for hair growth promoting potential activity. Inhibition of alopecia by administration of *Eclipta alba* extract gel was evaluated by hair growth ignition and completion, area responded, anagen/telogen ratio and histopathological skin section done for hair growth promoting potential activity. The earliest hair growth was seen in methanolic extract gel. All the experimental groups have shown a highly significant increase in density of hair, area responded and hair bulb present as compared to control group. Finally it was concluded that methanolic extract of *Eclipta alba* definitely promotes hair growth by inducing anagen in telogen (resting) phase hair follicles (Agrawal and Dewangan, 2011).

Market Formulations

Keshbeam Hair Oil

Keshbeam hair oil is a 100% safe, proprietary selection of special herbs renowned for their special properties and blended in a powerful effective combination. It is offered to people who suffer from hair fall, dandruff, itching scalp, premature graying and bald scalp conditions. This hair oil acts as a best conditioner. Encourage new growth by allowing oxygen to hair follicles. The ingredients of the hair oil stimulate and nourish the scalp. It is a complete hair care treatment at home. It makes the hair soft and shiny and initiates hair growth.

Ingredients: Bringa (*Eclipta alba*); brahmi (*Centella asiatica*); nimba (*Azadirachta indica*); amla (*Emblica officinalis*), coconut oil (q.s).

Action: *Eclipta alba* is having good antifungal activity hence removes dandruff and also prevents its formation. The plant shows anti-inflammatory properties and hence soothes scalp skin. A mixture of Eclipta and coconut oil keeps the hair dark and lustrous. Brahmi has a positive effect on the circulatory system. It improves the flow of blood while strengthening the veins and capillaries. Relieves stress and depression, energize, flagging mental powers, fight sterility, ward off a nervous breakdown, and improve reflexes. It induces good sleep. Neem is having anti-inflammatory, antifungal and antibacterial activity and it prevents scalp from infections and prevents hair fall. Amla prevents premature graying of hair and makes them strong and free from dandruff.

Direction: Hairs and scalp should be properly dried before application. Use the oil according to the scalp requirement only. Start with a few drops of hair oil in the palm of hand and rub into the scalp with fingertips. Massage the scalp well. Then brush the hair, gently and thoroughly, from scalp to ends. Shake well before each use.

Supplied by: Ayurhelp, Mysore.

CleanComb® Anti-hairloss Lotion

This lotion is a unique preparation clinically proven to prevent hairloss in men and women. It is made from the extracts of carefully selected vital natural ingredients, which treat and nourish the hair roots effectively.

Ingredients: Extracts of *Hibiscus rosa sinensis*, *Lawsonia alba*, *Eclipta alba*, *Emblica officinalis*, and *Aloe barbadensis* in a water miscible gel base.

Action: The lotion cleans the scalp and opens up hair follicles. It penetrates all layers of the scalp to reach the hair roots and then stimulates blood vessels, causing vasodilation. It also provides nourishment, strengthens hair root and shaft and holds hair roots firmly, thereby arresting hair loss.

Direction: As directed by the skin specialists.

Supplied by: Brihans Natural Products Pvt. Ltd., Maharashtra.

Pigmento Tablet

Pigmento is very effective in treating problems like vitiligo, white patches, leucoderma, unevenly shady skin, skin patches, dark spots, etc. as this herb is powered by various herbs, this herb is supposed to cast no side effects on the body and is also 100% chemical free.

Ingredients: Abhrak 0.225 mg, loh 0.225 mg, tamra 0.225 mg, *Emblica officinalis* 0.375 mg, *Terminalia belerica*-0.375 mg, *Terminalia chebula* 0.375 mg, *Hingul shuddha* 0.75 mg, purified asphaltum shilajit shuddha 0.9 mg, kajjali 1.125 mg, *Caesalpinia bonducella* 1.875 mg, shankha 3.75 mg, *Plumbago zeylanica* 3.75 mg, *Sapindus mukorossi* 3.75 mg, *Acacia catechu* 3.75 mg, *Nardostachys jatamansi* 3.75 mg, sonageru 6.5 mg, navayas loh 6.5 mg, *Myristica fragrans* 7.5 mg, *Caryophyllus aromaticus* 7.5 mg, *Elettaria cardamomum* 7.5 mg, purified *Ferula foetida* 7.5 mg, *Cuminum cyminum* 7.5 mg, *Cinnamomum iners* 7.5 mg, saindhav 7.5 mg, *Piper nigrum* 7.5 mg, yavakshar 7.5 mg, *Cyperus rotundus* 7.5 mg, *Cinnamomum cassia* 7.5 mg, *Carum copticum* 7.5 mg, *Zingiber officinale* 7.5 mg, *Lawsonia alba* 11.25 mg, *Swertia chirata* 11.25 mg, yashad 15 mg, *Piper longum* 15 mg, *Picrorrhiza kurrooa*18.5 mg, *Acorus calamus* 30 mg, *Panchamrit parpati* 37.5 mg, *Melia azadirachta* 3.75 mg, *Aloe barbadensis* 11.1 mg, *Ipomoea turpethum* 11.25 mg, *Embelia ribes* 11.25 mg, *Butea frondosa* 11.25 mg, *Lawsonia alba* 16.8 mg, *Amoora rohituka* 22.5 mg, *Tinospora cordifolia* 22.5 mg, *Boerhaavia diffusa* 22.5 mg, *Tephrosia purpurea* 22.5 mg, *Solanum nigrum* 22.5 mg, *Eclipta alba* 33.6 mg, *Andrographis paniculata* 45.0 mg, *Psoralea corylifolia* 93.75 mg.

Action: It helps in correcting the liver due to presence of kutki, sarpankha, kalmegh, etc. It straps up the direct rays from sun which helps in stimulating the melanocytes activity by bakuchi and various other herbs present in it. It also helps in reducing the pigmentation and dark spots to facilitate good fair skin.

Direction: The dosage recommended to patients are two tablets thrice daily with water for adults and one tablet thrice daily for kids below 12 years of age.

Supplied by: Herbal Cure India.

Storage of Products

Store the products in cool place and away from sunlight.

Dosage

For extract, recommended dose: 0.05% to 0.2%.

Tinctures: 0.5–3 ml daily.

Infusion: 1– 2 cups daily.

Patents

The present invention provides a herbal black dye from natural materials comprising of *Juglans regia, Indigofera tinctoria, Terminalia chebula, Acacia accocina, Lawsonia inermis, Trigonella foenum-graecum, Sapindus mukorossi, Eclipta alba, Embelica officinalis, Acacia catechu* and *Piper betle,* the dye derived being safe, non-toxic, anti-allergic, anti-dandruff and free from toxic symptoms like itching (Palpu et al., 2006).

A method and composition for improving human hair, having a composition of neutraceuticals that include *Eclipta alba,* chlorophyll, and an absorption enhancer, in which the absorption enhancer is selected from the group consisting of bioperine and phosphatidyl choline and *Eclipta alba* is included in an amount between 0.1 and 60 parts admixed by weight of the composition. The method and composition can be administered: (a) sublingually, having a solute, preferably ethanol, and preferably in the range between 0.1% and 30% by volume; (b) orally by encapsulation in a gelatinous capsule; and/or (c) by chewable mixture having a gum material and/or a flavor (McCarthy, 2008).

A topical hair care formulation and method of making includes combining generally equal amounts of *Eclipta alba* herbal powder and

Emblica officinalis herbal powder to form a herbal powder mixture. The herbal powder mixture is steeped in an amount of base oil for a period of time to form a brown solution. A portion of an amount of olive oil is combined with an amount of biotin and an amount of methylsulfonylmethane to form a natural mixture. A remainder of the amount of olive oil is combined with the brown solution. An amount of Jojoba oil is combined with the brown solution. An amount of rosemary oil is combined with the brown solution. An amount of lavender oil is combined with the brown solution. The natural mixture is combined with the brown solution. The brown solution is settled for a period of time to form the topical hair care formulation (Anderson, 2010).

BIBLIOGRAPHY

1. Agrawal SS, Dewangan G. Evaluation of *Eclipta alba* gel for its hair growth promoting activity. Ethnopharmacology. 2011; Article ID: Inventi:pep/542/11.

2. Anderson LD. Topical Hair Care Formulation and Method of Making. United States Patent Application 20100092582. Application no: 12/578410, publication date: 04/15/2010. http://www.freepatentsonline.com/y2010/0092582.html

3. Datta K, Singh AT, Mukherjee A, Bhat B, Ramesh B, Burman AC. *Eclipta alba* extract with potential for hair growth promoting activity. J Ethanopharmacol. 2009. 124 (3): 450–456.

4. http://ayurhelp.com/plants/eclipta-alba.htm

5. http://en.wikipedia.org/wiki/Eclipta_ alba

6. http://www.ayurhelp.com/articles/Bhringaraj.htm.

7. http://www.herbalcureindia.com/herbs/bhrngaraja.htm

8. http://www.naturalcosmeticsupplies.com/eclipta-alba-oil.html.

9. http://www.tropilab.com/eclipta. html

10. McCarthy R. Method and composition for improving hair growth. United States Patent Application 20080220058. Application no: 11/716224, publication date: 09/11/2008. http://www. freepatentsonline. com/y2008/0220058. html

11. Otilia JF Lobo, David B, Annamalai AR, Manavalan R. Evaluation of antiaggressive activity of *Eclipta alba* in experimental animals. Pak. J. Pharm. Sci., 2008; 21 (2): 195–199.

12. Palpu P, Pal M, Dixit BH, Banerji R, Rao CV. Herbal dye and process of preparation thereof. United States Patent Application 20060143838. Application no: 11/027575, publication date: 07/06/2006. http://www.freepatentsonline. com/y2006/0143838. html

13. Pandey MK, Singh GN, Sharma RK, Lata S. Antibacterial activity of *Eclipta alba* (L.) Hassk. Journal of Applied Pharmaceutical Science. 2011; 01 (07): 104–107.

14. Patel MB, Panchal SJ, Patel JA. Anti-anaphylactic activity of alcoholic extract of *Eclipta alba*. Pharmacology, 2009; 1(3): 244–250.

15. Singh A, Kumar R, Nivedita, Singh JK, Tanuja. Radioprotective effect of *Eclipta alba* (L.) against radiation induced haematological changes in Swiss albino mice. Journal of Natural Products. 2011; 4: 177–183.

16. Srivastava S, Mishra N. Evaluation of polyherbal formulation for wound healing activity. Der Pharmacia Lettre. 2009; 1(1): 157–161.

17. Tanuja, Singh A, Nivedita, Kumar R. Evaluation of acute toxicity of aqueous extract of *Eclipta alba* and its effects on liver of male Swiss albino mice. Journal of Herbal Medicine and Toxicology. 2011; 5 (2): 89–95.

18. Thakur VD, Mengi SA. Neuropharmacological profile of *Eclipta alba* (Linn.) Hassk. J Ethanopharmacol. 2005; 102(1): 23–31.

19. Thorat R, Jadhav V, Kadam V, Sathe N, Save A, Ghorpade V. Evaluation of a herbal hair oil in reducing hair fall in human volunteers. Int J Pharm Res Development. 2009; 6: ISSN 0974–9446.

20. Vyjayanthi G, Kulkarni C, Abraham A, Kolhapure SA. Evaluation of anti-dandruff activity and safety of polyherbal hair oil: An open pilot clinical trial. The Antiseptic. 2004; 101(9): 368–372.

Emblica officinalis

Plant profile

Kingdom: *Plantae* – Plants

Subkingdom: *Tracheobionta* – Vascular plants

Superdivision: *Spermatophyta* – Seed plants

Division: *Magnoliophyta* – Flowering plants

Class: *Magnoliopsida* – Dicotyledons

Subclass: *Rosidae*

Order: *Euphorbiales*

Family: *Euphorbiaceae* – Spurge family

Genus: *Emblica L.* – leaf flower

Species: *Emblica officinalis* – Emblic

Vernacular name

San: Dhatri, Amlaka, Adiphala, **Beng:** Amloki, Amla, **Eng:** Gooseberry, Emblic Myrobalam, **Guj:** Amali, Ambala, **Hind**: Amla, Amlika, **Kan:** Amalak, Bettadanelli, **Mal**: Nellika, **Tam:** Nellikkai, Malanelli, **Tel:** Amalakkamu, Usirikai, **Ory:** Aanla, **Punjabi:** Olay

Ayurvedic properties

Rasa (taste): Sour and astringent are the most dominant, but the fruit has five tastes, including sweet, bitter, and pungent, **Veerya (nature):** cooling, **Vipaka (taste developed through digestion):** Sweet, **Guna (qualities):** Light, dry, **Doshas (effect on humors):** It is especially effective for pitta.

INTRODUCTION

The tree is small to medium in size, reaching 8 to 18 m in height, with a crooked trunk and spreading branches. Branchlets are glabrous or finely pubescent, 10–20 cm long, usually deciduous; leaves are simple, sub-sessile and closely set along branchlets, light green, resembling pinnate leaves. The flowers are greenish-yellow. Fruits are nearly spherical, light greenish yellow, quite smooth and hard on appearance, with six vertical stripes or furrows.

Distribution

The crop has wide adaptability and can be found in diverse climatic and soil conditions ranging from the Western and Eastern Himalayas, Arawali ranges, Vindhya and Western Ghats. It is common all over tropical and sub-tropical India and also found in Burma. Tt is abundant in deciduous forests of Madhya Pradesh and in the semi-arid regions of Maharashtra, Gujarat, Rajasthan, Andhra Pradesh, Karnataka, Tamil Nadu and the Arawali ranges in Haryana, Kandi area in

Punjab, Himachal Pradesh extending to Ghar area of Uttar Pradesh. It also grows in tropical and subtropical parts of Sri Lanka, Malay Peninsula and China. In Sri Lanka, it is very common in exposed places on Patna land in the moist regions up to 4000 feet altitude.

Chemical Constituents

The fruit is a very rich source of vitamin C. The seeds contain a fixed oil, phosphatides, and an essential oil. The fruit bark, and the leaves are rich in tannin. The root contains ellagic acid and lupeol and bark contains leucodelphinidin. The seeds yield a fixed oil (16%) which is brownish-yellow in colour. It has the following fatty acids: linolenic (8.8%), linoleic (44.0%), oleic (28.4%), stearic (2.15%), palmitic (3.0%) and myristic (1.0%).

quantity of amla oil to the head before bathing removes diseases of the eyes, night blindness and bilious giddiness. Amla confection is used in syphilis, flatulence, bronchitis, asthma and consumption. Amla has most anti-diabetic property. To prevent or cure diabetes take a fresh one-fourth cup of amla or a teaspoon of amla powder with a teaspoon of turmeric powder everyday.

Cosmetic uses

Amla fruit is one of the strongest rejuvenators. Amla rebuilds new tissues and increases the red blood cell count. It cleanses the mouth, strengthens the teeth, stops the bleeding of gums and improves eyesight. It nourishes the bones and promotes the growth of healthy, lustrous hair and strong nails. The fruits are used in the preparation of writing inks and

Ellagic acid   Lupeol   Leucodelphinidin

Medicinal uses

Fresh fruit of **amla** (*Emblica officinalis*) is refrigerant, diuretic and laxative. Green fruit is exceedingly acid. Fruit is also carminative and stomachic. Dried fruit is sour and astringent. Flowers are cooling and aperient. Bark is astringent. The herb is also aphrodisiac, hemostatic, nutritive tonic, rejuvenative. It increases red blood cell count. It removes excessive salivation, nausea, vomiting, giddiness, spermatorrhoea, internal body heat and menstrual disorders. Because it is also cooling, it increases sattwa, and is an excellent liver tonic. The dried amla fruit is astringent and useful in cases of diarrhoea and dysentery. It is also a very important ingredient in the famous Chyavanaprash, and a constituent of Triphala powder. The application of a small

hair dyes. The dried fruit is used as shampoo for the head. An essential oil, distilled from the leaves, is used in perfumery.

Skin sores and wounds: The milky juice of the leaves is a good application to sores. Grind the bark of *Emblica officinalis* (10 g) into a paste and apply to the cut or wound area once daily for 2 to 3 days. Alternatively, squeeze *Emblica officinalis* leaves and extract the juice to the cut once daily for 3 to 4 days. Healing occurs when the dynamic harmony of the *doshas* is restored.

Skin whitening: Skin lightening agents have been widely used to either lighten and depigment the skin in the Asia, Far East and Middle East countries, whereas in the European market products tend to be employed for age spots and freckles.

Sore eyes: Infusion of the leaves is applied to sore eyes. The dried fruit immersed in water in a new earthen vessel for a whole night yields a decoction which is used as a *collyrium* (a medical lotion applied to the eye as an eyewash) in ophthalmia. It may be applied cold or warm. In another treatment an infusion of the seeds is also used as a collyrium and applied with benefit to recent inflammations of the conjunctive and other eye complaints. The exudate collected from incisions made on the fruit is applied externally on inflammation of the eye.

Pharmacological Action

The aim of the present study was to investigate the efficacy of *Emblica officinalis* fruit (EO) to inhibit UV-B-induced photoaging in human skin fibroblasts. Mitochondrial activity of human skin fibroblasts was measured by MTT-assay. Quantifications of pro-collagen 1 and matrix metalloproteinase 1 (MMP-1) release were performed by immunoassay techniques. Results revealed EO stimulated, otherwise UV-B-inhibited cellular proliferation and protected pro-collagen 1 against UV-B-induced depletion via inhibition of UV-B-induced MMP-1 in skin fibroblasts (10–40 µg/ml, p > 0.001). The results of the present study suggests that EO effectively inhibits UV-B-induced photoaging in human skin fibroblast via its strong ROS scavenging ability and its therapeutic and cosmetic applications remain to be explored (Adil et al., 2010).

Present study was investigated an antimicrobial potential of aqueous (infusions, decoctions) and methanlic extracts (1:2 and 1:5 concentrations) of *Emblica officinalis* (amla) against seven pathogenic bacteria, namely *Staphylococcus aureus, Staphylococcus saprophyticus, Escherichia coli, Enterococcus faecalis, Enterococcus cloacae, Proteus vulgaris* and *Klebsiella pneumoniae* by well diffusion method. The results concluded that *Emblica officinalis* definitely possess potent antimicrobial activities and this can serve as an important platform for the development of inexpensive, safe and effective medicines (Kumar et al., 2011).

The present study was revealed the effect of standardized hydroalcoholic extract of *E. officinalis* fruit (HAEEO) against kainic acid (KA)-induced seizures, cognitive deficits and on markers of oxidative stress. Rats were administered KA (10 mg/kg, i.p.) and observed for behavioral changes, incidence, and latency of convulsions over 4 hours. Results indicated that pretreated with HAEEO (500 and 700 mg/kg, i.p.) significantly (p < 0.001) increased the latency of seizures as compared with the vehicle-treated KA group and significantly prevented the increase in TBARS levels and ameliorated the fall in GSH. Thus it was concluded that these neuroprotective effects may be due to the antioxidant and anti-inflammatory effects of HAEEO (Golechha et al., 2011).

The aim of the present study was to investigate the protective role of *E. officinalis* against isoproterenol (ISP)-induced cardiotoxicity in rats and elucidate the possible mechanism involved. Rats were administered *E. officinalis* (100, 250 and 500 mg/kg, p.o.) or vehicle (normal saline) for 30 days, with concurrent subcutaneous injections of ISP (85 mg/kg, at 24 h interval) on 29th day and 30th day. Further the results of the present study was demonstrated cardioprotective potential of *E. officinalis* attributed to its potent antioxidant and free radical scavenging activity as evidenced by favorable improvement in hemodynamic, contractile function and tissue antioxidant status (Ojha et al., 2011).

Toxicological Studies

Hydroalcoholic extract of *E. officinalis* was orally administered in rats for determination of acute, chronic toxicity and LD_{50}. The doses like 250, 500 and 1000 mg/kg b.wt. were selected and the mortality in the rats occurred within 48 hours was noted. For chronic toxicity, the rats were given the dose for 3 weeks and the gross effects were observed. The results revealed there was no mortality observed after 48 hours of drug administration. Hence the LD_{50} of the extract is more than 1000 mg/kg b.wt. Further during 3 weeks study, *E. officinalis* did not cause any untoward effect. No gross observational effects were observed at any of the doses of the plant extract during acute and chronic toxicity studies (Pandey and Pandey, 2011).

Clinical Investigation

The *in vitro* and *in vivo* activity of various topical antioxidants and nutritional supplements was assessed in a randomized double-blind placebo-controlled study conducted over 8 weeks on 30 female volunteers (aged 48 to 59 years), with moderate dry skin and photodamage. All subjects twice daily applied an anocolloidal gel and/or daily took two capsules of an oral diet supplement. The test formulations were antioxidant-enriched, containing ascorbic acid, tocopherol, alpha-lipoic acid, melatonin, and *emblica*. Oral and topical administration of the antioxidant compounds resulted in significant reduction in oxidative stress and lipid peroxidation. Free radicals recovered in blood serum and on skin, *in vivo*, and ROS induced by *in vitro* irradiation of leucocytes with UV-B, were also decreased in antioxidant-treated patients. The results concluded that all of the compounds, including *emblica*, showed therapeutic potential, in topical or systemic form, as photoprotectants and agents intended to lower the oxidative stress of photoaging-affected individuals (Morganti et al., 2002).

Recent investigations with *emblica* also have dermatologic implications or suggestions of potential cutaneous benefits. A two-stage skin carcinogenesis study was performed in Swiss albino mice to ascertain the chemopreventive action of *E. officinalis* fruit extract. A single application of 7, 12-dimethyabenz(a) anthrecene (DMBA) was used to induce skin cancer, followed two weeks later by thrice weekly applications of croton oil (through the end of the 16-week experiment) to promote tumors. A statistically significant difference in tumor incidence, yield, and burden as well as cumulative number of papillomas favored the mice treated with *E. officinalis* as opposed to the control group. The results concluded that *E. officinalis* fruit extract exhibited chemopreventive potential against DMBA-induced skin cancer in Swiss albino mice (Sancheti et al., 2005).

The effects of amla extract on human skin fibroblasts was studied with a focus on *in vitro* production of procollagen and matrix metalloproteinases (MMPs). The WST-8 assay was employed to assess human skin fibroblast mitochondrial activity and an immunoassay to measure procollagen, MMPs and tissue inhibitor of metalloproteinase-1 (TIMP-1) released from the fibroblasts. The results revealed that amla extract controls collagen metabolism in therapeutic and cosmetic applications in a concentration-dependent fashion. It also promoted fibroblast proliferation and in a concentration- and time-dependent manner stimulated procollagen synthesis. Finally the results concluded the potential of amla have mollifying and therapeutic as well as cosmetic benefits (Fuji et al., 2008).

Market Formulations

Natural amla oil: Natural amla oil enriches hair growth and pigmentation. It prevents premature graying of hair, dandruff, increases the strength of hair follicles (and thus preventing hairfall). Customarily, a small amount of natural amla oil is applied to the hair after washing. This not only brings forth a rich, natural shine and soft texture to the hair, but also helps rejuvenate hairs that are dull and damaged. It also prevents split hair ends.

Ingredients: Amalaki fruit (*Emblica officinalis*), brahmi leaf (*Bacopa monnieri*), ela seed (*Elettaria cardamomum*), gulab petal (*Rosa centifolia*), kapoor kachri rhizome (*Hedichium spicatum*), bhringraj whole plant (*Eclipta alba*), musta rhizome (*Cyperus rotundus*), natural sesame oil as base.

Action: Application of natural amla oil over the scalp has a cooling effect. As such, it keeps the mind cool and promotes sound sleep. Natural amla oil along with various other formulations such as drugs like brahmi, shikakai and reetha are very beneficial for mind, hair and body. Natural amla oil is useful for increasing memory.

Direction: Part the hair and apply oil all over the scalp. Massage the scalp gently with fingers in a circular motion, so that the oil gets absorbed into the scalp gradually, leave for an hour or more before washing with an all-natural shampoo.

Supplied by: Ganga Prasad Puneet Kumar, Faridabad, Haryana.

Henna Amla Hair Oil

Amla oil has a long tradition is being used for improving the health of hair and scalp. It is one of the oldest natural hair conditioners. This oil is light and gets absorbed easily within the hair producing no sticky impact. It nourishes hair and makes them stop falling.

Ingredients: Mainly amla, henna and other herbs.

Action: Strengthen roots of hair and nourishes hair-shafts. Promoting new hair growth and strengthening current hair. It soften and condition the hair and increasing hair luster and vibrancy. It protects hair from damaging effects of the sun and harsh weather by improving resiliency over time and also replenishing and rejuvenating for dry, damaged hair, and helps prevent excessive brittleness and split-ends.

Direction: Massage the scalp gently with fingers in a circular motion, so that the oil gets absorbed into the scalp gradually, leave for an hour.

Supplied by: Izuk Impex, New Delhi.

Dabur Amla Hair Oil

Dabur amla hair oil is a non-sticky hair oil and moisturizing treatment for long hair. It can be used as an overnight revitalizing treatment or as an intensive, weekly hair treatment.

Dabur amla works to strengthen the hair as well as promotes hair growth.

Ingredients: Emblica officinals extract, petroleum based emollient paraffinum liquidum, canola oil, oleic acid, t-butyl hydroquinone, yellow no. 10, green no. 6 and red no. 17, and fragrance.

Action: Amla extract works to strengthen the hair and prevent graying. Petroleum based emollient paraffinum liquidumand canola oil acts as emollient ingredients which help to soften the hair by coating and protecting the hair shaft, preventing moisture from escaping. Oleic acid acts as emulsifiers that prevent oil and water based ingredients from separating. In addition, these ingredients work to thicken and give the hair oil a smooth texture. t-butyl hydroquinone is the anti-oxidant ingredient and it helps to reduce damage caused by overexposure to the sun. Yellow no. 10, green no. 6 and red no. 17 are used as dyes and fragrances.

Method of preparation: Amla can be boiled with coconut oil and used.

Supplied by: Dabur India Pvt., Ltd., Ghaziabad, UP.

Storage of Products

It can be easily stored at room temperature for a period of 3 years. Keep away from sunlight.

Dosage

Amla juice: 3 teaspoonful twice daily with water.

Amla capsule: 1–2 capsules twice daily with plain water.

Patents

The present invention relating to a herbal antiallergic composition which comprises a synergistic mixture of extracts from the fruits of *Terminalia chebula,* bark of *Albizia lebbeck, Terminalia bellerica* and *E. officinalis.* The present invention also contains the fruits of *Piper longum, Piper nigrum* and of rhizomes of *Zingiber officinale* and thoroughly mixed to get the final composition which has potent antiallergic activity. The invention also relates to a process for the preparation of such composition. The composition is particularly useful for the treatment of allergic conditions (Agarwal and Agarwal, 2004).

The present invention concerns cosmetic compositions and methods comprising an extract of E. officinalis and at least one ingredient chosen from dihydroxy acetone, a dibenzoyl methane derivative, ultrafine particles of zinc oxide, ultrafine particles of titanium oxide, astaxanthin, retinoids, alpha-hydroxy acids, beta-hydroxy acids, polyhydroxy acids, hydroquinone, compounds useful for the treatment of dandruff, hair colorants, hair pigments, and hair dyes. The addition of E. officinalis extract to these compositions has the advantage of increasing their stability and therefore increasing or prolonging their effectiveness in such cosmetic compositions (Hansenne et al., 2004).

A method to retard, prevent and reverse the sign of skin photo-damage, comprising administering, to a human subject in need thereof, a therapeutically effective dose of a formulation comprising an extract of Emblica officinalis as the active ingredient (Chaudhuri, 2005).

The present invention provides a herbal black dye from natural materials comprising of Juglans regia, Indigofera tinctoria, Terminalia chebula, Acacia accocina, Lawsonia inermis, Trigonella foenum-graecum, Sapindus mukorossi, Eclipta alba, Emblica officinalis, Acacia catechu and Piper betle, the dye derived being safe, non-toxic, anti-allergic, anti-dandruff and free from toxic symptoms like itching (Palpu et al., 2007).

An extract of Emblica officinalis (amla): Transdermal formulation having an extract of Emblica officinalis having exhibiting greater migration of vitamin C across a skin surface as compared to a transdermal formulation having vitamin C without the extract. Ext-

ract of Emblica officinalis exhibiting greater migration of H⁺ ions across a skin surface as compared to a transdermal formulation having vitamin C alone. Method of preparing an extract of Emblica officinalis (Antony, 2009).

Disclosed a novel herbal composition for maintaining/caring the skin around the eye, the composition comprising extracts of Saxifraga ligulata syn. Bergenia ligulata, Cipadessa baccifera and Emblica officinalis and cosmeceutical excipients thereof. Also disclosed a method of extraction, delivery system comprising the same and use thereof (Mitra et al., 2011).

The present invention is directed to a method of reducing the appearance of skin changes associated with intrinsic and/or extrinsic aging, involving the steps of: (a) providing a first composition containing: (i) from about 20 to about 50% by weight of at least one hydroxy acid; (ii) optionally, from about 1 to about 15% by weight of at least one weak organic acid; and (iii) remainder, to 100%, water; (b) providing a second composition containing: (iv) from about 20 to about 50% by weight of L-ascorbic acid powder; (v) optionally, from about 1 to about 10% by weight of an extract of *emblica*; (vi) from about 10 to about 60% by weight of a water-soluble film-forming polymer; and (vii) remainder, to 100%, of a water-absorbing thickening agent; (c) mixing (a) and (b) to form a finished composition; (d) applying the finished composition onto the skin, just after mixing, in order to form a masque on the skin; and (e) removing the masque from the skin after a predetermined period of time (Feng and Rousseau, 2011).

BIBLIOGRAPHY

1. Adil MD, Kaiser P, Satti NK, Zargar AM, Vishwakarma RA, Tasduq SA. Effect of *Emblica officinalis* (fruit) against UVB-induced photoaging in human skin fibroblasts. J Ethanopharmacol. 2010; 132 (1): 109–114.

2. Agarwal RK, Agarwal A. Herbal composition having antiallergic properties and a process for the preparation there of. United States Patent 6730332. Application No.10/019389, Publication date: 05/04/2004. http://www. freepatentsonline.com/6361804.html

3. Antony B. Amla extract for transdermal application. United States Patent Application 20090238780. Application No.12/382602, Publication date:09/24/2009. http://www.freepatentsonline.com/y2009/0238780.html

4. Chaudhuri RK. Method for protection of skin against sun-induced damage by oral administration of an extract of *Emblica officinalis* (syn. *Phyllanthus emblica*). United States Patent 20050089590. Application No.10/691959, Publication date: 04/28/2005. http://www. freepatentsonline. com/y2005/0089590.html

5. Feng S, Rousseau G. Novel skin peel composition in masque form. United States Patent Application 20110086000. Application No.12/895125, Publication date: 04/14/2011. http://www. freepatentsonline. com/y2011/0086000.html.

6. Fuji T, Wakaizumi M, Ikami T, Saito M. Amla (*Emblica officinalis* Gaertn.) extract promotes procollagen production and inhibits matrix metalloproteinase-1 in human skin fibroblasts. J Ethno-pharmacol. 2008;119 (1):53–57.

7. Golechha M, Bhatia J, Ojha S, Arya DS. Hydroalcoholic extract of *Emblica officinalis* protects against kainic acid-induced status epilepticus in rats: Evidence for an anti-oxidant, anti-inflammatory, and neuro-protective intervention. Pharm Biol. 2011; 49 (11): 1128–1136.

8. Hansenne I, Galdi A, Fares H, Foltis SP. Cosmetic composition comprising an extract of *Emblica officinalis* and methods of using same. United States Patent 20040028642. Application No.10/424111, Publication date: 02/12/2004. http://www. freepatentsonline.com/y2004/0028642.html

9. http://en.wikipedia.org/wiki/Phyllanthus_emblica.

10. http://www.dweckdata.com/published_papers/emblica_officinalis.pdf.

11. http://www.gits4u.com/agri/agri5e. htm.

12. http://www.tnsmpb.tn.gov.in/images/amla.pdf

13. Kumar A, Tantry BA, Rahiman S, Gupta U. Comparative study of antimicrobial activity and phytochemical analysis of methanolic and aqueous extracts of the fruit of *Emblica officinalis* against pathogenic bacteria. J Tradit Chin Med. 2011; 31(3):246–50.

14. Mitra SK, Saxena E, Babu UV. Herbal composition for maintaining/caring the skin around the eye, methods of preparing the same and uses there of. United States Patent Application 8007837. Application No.12/839734, Publication date: 08/30/2011. http://www. freepatentsonline.com/8007837. html

15. Morganti P, Bruno C, Guarneri F, Cardillo A, Del Ciotto P, Valenzano F. Role of topical and nutritional supplement to modify the oxidative stress. Int J Cosmet Sci. 2002; 24(6): 331–339.

16. Ojha S, Golechha M, Kumari S, Arya DS. Protective effect of *Emblica officinalis* (amla) on isoproterenol-induced cardiotoxicity in rats. Toxicol Ind Health. 2011 (Under publication).

17. Palpu P, Pal M, Dixit BS, Banerji R, Rao CV. Herbal dye and process of preparation thereof. United States Patent 7186279. Application No.11/027575, Publication date: 03/06/2007. http://www.freepatentsonline. com/7186279.html

18. Pandey G, Pandey SP. Phytochemical and toxicity study of *Emblica officinalis* (amla). Int Res J Pharmacy. 2011; 2 (3): 270–272.

19. Sancheti G, Jindal A, Kumari R, Goyal PK. Chemopreventive action of *Emblica officinalis* on skin carcinogenesis in mice. Asian Pac J Cancer Prev. 2005; 6(2):197–201.

Eucalyptus globules

Plant profile
Kingdom: *Plantae* – Plants

Superdivision: *Spermatophyta* – Seed plants

Class: *Magnoliopsida* – Dicotyledons

Order: *Myrtales*

Genus: *Eucalyptus L' Her* – Gum

Subkingdom: *Tracheobionta* – Vascular plants

Division: *Magnoliophyta* – Flowering plants

Subclass: *Rosidae*

Family: *Myrtaceae* – Myrtle family

Species: *Eucalyptus globulus*– Tasmanian bluegum

Common name: Blue gum.

Vernacular name

San: Tailapatra, Sugandhapatra, **Eng:** Eucalyptus

Other names: Tailaprana and nilgiri taila are the other names used for the Eucalyptus.

Ayurvedic properties

- *Gunna (properties):* Snigdh (slimy) and Laghu (light)
- *Rasa (taste):* Tickt (bitter), kashaya (astringent) and katu (pungent)
- *Viraya (potency):* Ushan (hot)

INTRODUCTION

The bark sheds often peeling in large strips. The broad juvenile leaves are borne in opposite pairs on square stems. They are about 6 to 15 cm long and covered with a blue-grey, waxy bloom, which is the origin of the common name "blue gum". The mature leaves are narrow, sickle-shaped and dark shining green. They are arranged alternately on rounded stems and range from 15 to 35 cm in length. The adult leaves are alternate, lanceolate and 6–12 inches long and 1–2 inches broad. Its flowers are cream in colour. The buds are top-shaped, ribbed and warty and have a flattened operculum (cap on the flower bud) bearing a central knob. The cream-colored flowers are borne singly in the leaf axils and produce copious nectar that yields a strongly flavored honey. The fruits are woody and range from 1.5 to 2.5 cm in diameter. Numerous small seeds are shed through valves (numbering between 3 and 6 per fruit) which open on the top of the fruit. It produces roots throughout the soil profile, rooting several feet deep in some soils. They do not form taproots. The appearance of its bark varies with the age of the tree. Its bark consists of long fibers and can be pulled off in long pieces. Stems of the seedlings and coppice shoots are quadrangular. Flowers are in cymose panicles. The fruit is a capsule.

Distribution

It is an evergreen tree, one of the most widely cultivated trees native to Australia. They typically grow from 30 to 55 m (98 to 180 ft) tall. The tallest currently known specimen in Tasmania is 90.7 m tall. There are historical claims of even taller trees, the tallest being 101 m (330 ft). The natural distribution of the species includes Tasmania and Southern Victoria (particularly the Otway ranges and Southern Gippsland). There are also isolated occurrences on King Island and Flinders Island in Bass Strait and on the summit of the You Yangs near Geelong. There are naturalized non-native occurrences in southern Europe (Galicia, Akamas, Cyprus and Portugal), Southern Africa, New Zealand, Western United States (California), Hawaii and Macaronesia. It is widely grown in Tamil Nadu, Andhra Pradesh, Gujarat, Haryana, Mysore, Kerala and in the Nilgiri Hill. It grows well in deep, fertile, well drained loamy soil with adequate moisture. It is also found in Nagarhole National Park and Bandipur National Park in India.

Chemical Constituents

The essential oil of Eucalyptus used in medicine is obtained by aqueous distillation of the fresh leaves. It is a colorless or straw-colored fluid when properly prepared, with a characteristic odour and taste, soluble in its own weight of alcohol. Oil yield ranges from 1.0–2.4% (fresh weight), with cineole being the major isolate. The most important constituent is Eucalyptol, present in *E. globulus* up to 70 per cent of its volume. It consists chiefly of a terpene and a cymene. Eucalyptus oil contains also, after exposure to the air, a crystallizable resin, derived from Eucalyptol. The concentration of α-terpeneol was estimated to be 28%. The leaves are rich in tannins and ellagitannins, and also contain 2–4% triterpenes (ursolic acid derivatives), a series of phloroglucinol-sesquiterpene-coupled derivatives (macrocarpals B, C, D, E, H, I and J) and flavonoids (rutin, quercetin, quercitrin and hyperoside).

Medicinal uses

The medicinal Eucalyptus oil is probably the most powerful antiseptic of its class, especially when it is old, as ozone is formed in it on exposure to the air. It has decided disinfectant action, destroying the lower forms of life. Internally, it has the typical actions of a volatile oil in a marked degree. Eucalyptus oil is used as a stimulant and antiseptic gargle. Locally applied, it impairs sensibility. It increases cardiac action. An emulsion made by shaking up equal parts of the oil and powdered gum-arabic with water has been used as a urethral injection, and has also been given internally in drachm doses in

Eucalyptol   Cymene   Cineole

pulmonary tuberculosis and other microbic diseases of the lungs and bronchitis. In croup and spasmodic throat troubles, the oil may be freely applied externally. The oil is an ingredient of 'catheder oil,' used for sterilizing and lubricating urethral catheters. In large doses, it acts as an irritant to the kidneys, by which it is largely excreted, and as a marked nervous depressant ultimately arresting respiration by its action on the medullary centre. For some years Eucalyptus-chloroform was employed as one of the remedies in the tropics for hookworm, but it has now been almost universally abandoned as an inefficient anthelmintic, Chenopodium oil having become the recognized remedy. In veterinary practice, Eucalyptus oil is administered to horses in influenza, to dogs in distemper, to all animals in septicaemia. It lowers the blood sugar. It brings relief to the patients of Asthma and bronchitis. It is the excellent topical remedy for aching joints and rheumatism. It helps in improving the blood circulation. Cineole controls airway mucus hypersecretion and asthma via anti-inflammatory cytokine inhibition. Eucalyptus oil also stimulates immune system response by effects on the phagocytic ability of human monocyte derived macrophages.

Cosmetic uses

Eucalyptus ointment is a good ointment for the skin, containing antiseptic and healing properties. Eucalyptus oil is also found in creams and ointments used to relieve muscle and joint pain, and in some mouthwashes. The mouthwashes have been shown to help prevent plaque and gingivitis. It produces very satisfactory results in scurf, chapped hands, chafes, dandruff, tender feet and enlargement of the glands, spots on the chest, arms, back and legs. Apply a piece of clean cotton or lint to wounds after all dirt is washed away. For aches and pains rub the part affected well and then cover with lint. It is also used for parasitic skin affections. Externally it has slightly anesthetic, anti-bacterial, and warming properties and hence

it is a valuable resource treatment of burns, sores, ulcers, scrapes, boils, and wounds. Applied topically as an oil or ointment, it also helps relieve the aching, stiffness, and neuralgia. For outdoor enthusiasts, Eucalyptus rubbed into the skin seems to work well as an insect repellant, especially for mosquitoes and fleas. Eucalyptus oil is also used in personal hygiene products for antimicrobial properties in dental care and soaps. It can also be applied to prevent infection, making detergents.

Pharmacological Action

The efficiency of antioxidant properties of supercritical fluid extract (SCF), methanolic and water extracts of eucalyptus stem bark in a concentration range of 0.10–0.90 mg/ml, was evaluated on two different free-radical species: 2,2-diphenyl-1-picrylhydrazyl free radical (DPPH), β-carotene lineolate radical using electron spin resonance (ESR) spectroscopy. These extracts scavenged all types of investigated radicals in dependence on their applied concentrations. Generally, methanolic extracts possessed better scavenging and antioxidant activity than water extracts, while SCFE extracts exhibited lower activities than methanolic extracts (Vankar et al., 2006).

The antibacterial activities of essential oils from leaves of two Eucalyptus species (globulus and camaldulensis) was determined against *Staphylococcus aureus* Gram (+) and *Escherichia coli* Gram (–) bacteria. The inhibiting activity was evaluated by three methods: aromatogramme, microatmosphere and germs in suspension. Results demonstrated of the leaf essential oils of the two species showed an excellent inhibitory effect on *S. aureus* than that of *E. coli*. These data would indicate the potential usefulness of the two Eucalyptus species as a microbiostatic, antiseptic or as disinfectant agent (Ghalem and Mohamed, 2008).

The antiplasmodial activity was assessed using Lactate Dehydrogenase Assay (pLDH) and the cytotoxicity estimated on LLC-MK2 monkey kidney epithelial cells. Seven extracts from five different plants were significantly active, with very weak or no cytotoxicity. The *Dacryodes edulis* leaves showed the highest activity (IC_{50} of 6.45 µg/ml on 3D7 and 8.2 µg/ml on DD2) followed by the leaves of *Vernonia amygdalina* (IC_{50} of 8.72 and 11.27 µg/ml on 3D7 and DD2 resp.) and roots of *V. amygdalina* (IC_{50} of 8.72 µg/ml on 3D7), *Coula edulis* leaves (IC_{50} of 13.80 µg/ml and 5.79 µg/ml on 3D7 and DD2 resp.), *Eucalyptus globulus* leaves (IC_{50} of 16.80 µg/ml and 26.45 µg/ml on 3D7 and DD2) and *Cuviera longiflora* stem bark (IC_{50} of 20.24 µg/ml and 13.91 µg/ml on 3D7 and DD2) (Zofou et al., 2011).

The aim of the study was to assess the development of polyherbal formulation, i.e. F1, F2 and F3 of Hydroalcoholic extract *Eucalyptus gloubulus* in combination with eugenol and piperine will be evaluated first time for pharmacological (cyclooxygenase inhibitory activity) aspect. The polyherbal formulations were prepared in various doses and anti-inflammatory activity studied on polyherbal formulations. The formulations F1, F2 and F3 reduced the inflammation induced by carrageenan by 49.18%, 48.72% and 47.58 % on oral administration respectively as compared to the control treated group (Gomase et al., 2011).

The antidiabetic potency of the aqueous extract of the powder mixture of dried fruits of *Eucalyptus globules* and rhizomes of *Curcuma zedoaria* in a ratio of 10:1 was investigated using streptozotocin as the diabetogenic agent. The extract produced a significant antihyperglycemic activity in dose dependant manner. The highest oral dose tested (600 mg/kg) produced significant antihyperglycemic activity when compared with that of standard Glibenclamide (180 µg/kg). Results concluded that the extract prepared from the above mentioned formulation possesses potential anti-hyperglycemic activity (Sadiq et al., 2011).

Toxicological Studies

The probable lethal dose of pure Eucalyptus oil for an adult is in the range of 0.05 ml to 0.5 ml/per kg of body weight (Hindle, 1994). If consumed internally at low dosage as a flavoring component or in pharmaceutical products at the recommended rate, cineole based 'oil of Eucalyptus' is safe for adults. However, systemic toxicity can result from ingestion or topical application at higher than recommended doses (Darben et al., 1998). In other research, the oral acute toxicity study in mice demonstrated that the LD_{50} value for the *Eucalyptus globulus* leaf tissue was 4.5 g/kg (Jouad et al., 2004).

Clinical Investigation

The dermal toxicity of canisep, a herbal cream containing oils of *Ocimum sanctum* [*O. tenuiflorum*], *Acorus calamus*, *Eucalyptus globulus*, *Linum usitatissimum*, and *Cinnamomum camphora*, was tested on adult Winstar rats. Under acute toxicity test, rats were exposed to Canisep at 2000 mg/kg body weight for 24 h by applying the cream on shaved areas. Under sub-acute toxicity test, rats were exposed to canisep at 500 and 1000 mg/kg body weight for 28 days. Canisep at all dosage was non-toxic, as it did not significantly affect feed consumption and body weight, and did not cause skin abnormality in treated rats (Sachan et al., 2002).

Market Formulations

Topical use or inhalation of eucalyptus oil at low concentrations may be safe, although significant and potentially lethal toxicity has been consistently reported with oral use and may occur with inhalation use as well. All routes of administration should be avoided in children.

Hustagif® Balsam

This balm is for children and adults. Soft skin care balm to rub gently on the skin of chest and back.

Ingredients: Eucalyptus globulus leaf oil, *Lavandula angustifolia* (Lavender) oil, *Thymus vulgaris* (Thyme) oil.

Application: Caring balm for relaxed breathing.

Supplied by: Dentinox, Berlin.

Noxzema

Noxzema is a skin cleanser designed to remove dirt, oil and makeup from facial skin. It has a distinct fragrance and a characteristic tingle.

Ingredients: Water, stearic acid, linseed oil, glycine soja oil, camphor, menthol, ammonium hydroxide, *Eucalyptus globulus* leaf oil, propylene glycol, gelatin, calcium hydroxide.

Action: The primary function of water is as a solvent and emulsifier. It helps to combine the water and oil ingredients to create the resulting cream. Stearic acid acts as both a cleansing agent and an emulsifying agent, helping to lift dirt from the skin as well as combine the oil and water ingredients of the product. Linseed oil can also soften skin, reduce skin irritations, and even aid in wound healing by decreasing redness and swelling. Glycine soja oil functions as an antioxidant, an emollient to help the water and oil products mix, and a skin-conditioning agent. The main function of ammonium hydroxide in cosmetic products is to help adjust the pH or acidity of the final product. Camphor is a white waxy crystalline solid with a very distinct odor. It is often used in cosmetic and personal care products to soften synthetic polymers and relieve pain. Menthol is primarily as a fragrant ingredient. *Eucalyptus globulus* leaf oil is used as a fragrance as well as a skin conditioning agent. Propylene glycol is one of the most widely used organic alcohols in personal care products. It attracts water and therefore acts as a moisturizer, reducing dry skin flaking. It also helps stabilize the final product, increasing its shelf-life. Gelatin is a natural protein. When used in face masks, gelatin binds to dead skin cells and oil. When the mask is removed, the dirt is removed with it, leaving smoother, softer skin. Calcium hydroxide is used to control the pH balance of the product. Because calcium hydroxide is a base, it can neutralize excess acid, which could be irritating to the skin. Noxzema contains acids such as stearic acid, and therefore, may need pH adjustment by calcium hydroxide, to achieve a more neutral, less irritating end product.

Supplied by: The Lance Armstrong Foundation (LIVESTRONG.COM), Santa Monica, CA.

Canisep

This cream is used for abrasions, cuts, lacerations, burns, scalds and surgical wounds, pruritus, pyoderma, dermatitis, callosites, dry/moist eczema, ringworm, sinuses and fistulae. Canisep acts as antimicrobial agent. It inhibits the growth of micro-organisms. It acts as an antifungal and antimaggot agent. It controls maggot infestation.

Ingredients: **Oils** of atasi (*Linum usitatissimum*) 50 mg, tailaparna (*Eucalyptus globulus*) 30 mg, karpura (*Cinnamomum camphora*) 25 mg, tulasi (*Ocimum sanctum*) 12 mg, vacha (*Acorus calamus*) 8 mg Pdr. Tankana (shuddha) 18 mg.

Storage: In room temperature.

Supplied by: The Himalaya Drug Company, India.

Bathroom and Toilet Cleaner

This gel will leave the bathroom and toilet spotlessly clean. Use on the sinks, bath, toilet and shower. It can also use as a general cleaning gel on surfaces in the kitchen.

Ingredients: Water, lauryl glucoside, pinus sylvestris (pine) leaf oil, guar gum, xanthan gum, citric acid, *Eucalyptus globulus* leaf oil, potassium sorbate, sodium benzoate, glycerin, acetic acid, ascorbic acid, caprylic acid, *Cinnamomum zeylanicum* (cinnamon) leaf oil, lactic acid, ethanol, *Citrus aurantium* amara (bitter orange) fruit extract, *Yucca schidigera* stem extract.

Supplied by: **Better** Earth, South Africa.

Anti Comedo Lotion

Alcoholic-watery lotion with acid pH of 4,8 for follow-up treatment of face after take out of pimples, optimal as well for treatment of other skin impurities.

Ingredients: Water, alcohol, isopropyl alcohol, aluminium lactate, sodium lactate, *Eucalyptus globulus* oil.

Supplied by: Dr. Baumann Cosmetic GmBH, CH-8807 Freienbach.

Moroccan Beldi Black Soap

It was used as a product of dermatology and later became a real beauty tool for the body suitable for every type of skin. It prepares the skin for exfoliation. The skin will be softer and ready for a scrub. Combined with the action of the Kessa glove will remove impurities and dead skin.

Ingredients: Water, olive oil, *Eucalyptus globulus* essential oil, potassium hydroxide.

Action: It removes dead skin (the rough and dry outer surface). Deep cleans the skin by removing toxins and dead skin cells, making skin softer. The exfoliation increases the blood and lymph circulation. Ideal to reduce in-growing hair. Eucalyptus helps relax the body and adds to the soap some antiseptic, antibacterial and anti-inflammatory properties.

Direction: After 15 minutes in hot bath, hot shower, steam room or hammam, using only plain water, apply soap on a warm and wet body. Avoid the face and eyes. Leave on for up to 10 minutes, then rinse with warm water.

Supplied by: Argan Oil Tree, UK.

Storage of Products

Store in a well-filled, tightly closed container, protected from heat and light.

Dosage

Eucalyptol, U.S.P.: Dose, 5 drops. Ointment, B.P.

Daily dosage: Several drops or 30 ml essential oil in 500 ml lukewarm water rubbed into the skin for local application; 5–20% essential oil in liquid and semisolid preparations; 5–10% in hydroalcoholic preparations. Mouthwash: 20 ml of a 0.91 mg/ml solution, gargled twice daily. Inhalation: 12 drops/150 ml boiling water.

Eucalyptus oil (for topical application): Add ½ –1 ml (15–30 drops) of oil to 1/2 cup of carrier oil (sesame, almond, olive, etc.). For inhalation, add 5–10 drops of oil to 2 cups boiling water. Place towel over head and inhale steam.

Patents

The invention as presently conceived discloses a chemical formula for the composition of a liquid topical analgesic. It is produced from a unique blend of herbal essential oils and plant- and/or vegetable-derived oils. Although the precise proportions and method of mixture are held in confidence, it comprises at least olive oil, castor oil, grapeseed oil, almond oil, apricot kernel oil, vitamin E, various herbal essential oils, and a stabilizer. Such essential oils may include *Eucalyptus* (*Eucalyptus globulus*, among others), lavender (*Lavandula angustifolia*), tea tree (*Melaleuca alternifloria*), ginger (*Zingiber officinale*), lemongrass (*Cymbopogon citratus*, among others), red thyme (*Thymus vulgaris*), sweet birch, (*Betula lenta*) and ylang ylang (*Cananga odorata*). Alternate and additional essential oils or plant- and/or vegetable-derived oils are anticipated to be used as well. When produced, the mixture would be bottled and promoted to the public to be used as a topical rubbing compound for the skin or for use in a soaking bath. It will reduce pain from sore muscles, bruised ligaments and tendons, lower back pain, arthritis and other similar ailments associated with common aches and pains (Anderson, 2008).

A liquid carrier for topical administration is provided that contains a mineral oil, turpentine and, optionally, camphor, and optionally other terpenoid components containing pinene, and its use to provide an antifungal composition containing a mixture of antifungal essential oils and the carrier system, wherein the mixture of antifungal essential oils includes an effective amount of each of *Eucalyptus globulus*, peppermint, cedarwood, and Manuka; along with the use of this composition in the topical treatment of fungal infection, particularly of the nail (Selner, 2009).

The invention relates to the use of at least davanone and 1,8-cineol for the manufacture of an antimicrobial and/or antiinflammatory composition as well as a method of producing said composition (Faergemann et al., 2010).

BIBLIOGRAPHY

1. Anderson LA. Herbal healing oil. United States Patent 7368135. Application no. 11/491581, Publication date: 05/06/2008. http://www.freepatentsonline.com/7368135. html

2. Darben T, Cominos L, Lee CT. Topical Eucalyptus Oil Poisoning. *Australasian Journal of Dermatology*, 1998, Vol.39, pp.265–267.

3. Faergemann J, Hedner T, Sterner O, Björk L. Antimicrobial and anti-inflammatory composition. United States Patent Application 20100331401. Application no. 12/873711, Publication date: 12/30/2010. http://www.free-patentsonline. com/y2010/0331401. html

4. Ghalem BR, Mohamed B. Antibacterial activity of leaf essential oils of *Eucalyptus globulus* and *Eucalyptus camaldulensis*. *Afr. J. Pharm. Pharmacol.* 2008; 2 (10): 211–215.

5. Gomase PV, Shire PS, Nazim S. Phytochemical evaluation and anti-inflam-matory studies of polyherbal form-ulation. IJPI's Journal of Pharmacognosy and Herbal Formulations. 2011; 1 (4): 63–72.

6. Hindle, R.C., Eucalyptus oil ingestion, *New Zealand Medical Journal*, 1994, pp 185–186.

7. http://ayurveda.ygoy.com/2010/11/23/eucalyptus-benefits-in-ayurveda/

8. http://en.wikipedia.org/wiki/Eucalyptus_globulus

9. http://herbalinformation. awardspace.com/?cm=e&fn=eucalyptus_globulus

10. http://www.botanical.com/botanical/mgmh/e/eucaly14.html

11. http://www.ecoindia.com/flora/trees/eucalyptus-tree.html

12. http://www.gardensablaze.com/Herb EucalyptusMed.htm

13. http://www.umm.edu/altmed/articles/eucalyptus-000241.htm

14. Jouad H, Maghrani M, Hassani RAEI, Eddouks M. Hypoglycemic activity of aqueous extract of *Eucalyptus globulus* in normal and streptozotocin-induced diabetic rats. 2004; 10 (4): 19–28.

15. Sachan A, Jayakumar K, Mitra SK, Ramesh N, Udupa V, Udupa V. Dermal toxicity studies of a herbal formulation "Canisep" cream in rats. Journal of Applied Zoological Researches. 2002; 9(1):7–10.

16. Sadiq MJ, Bheemachari, Shiv Kumar, Vigneshwaran E, Balaji K. A study on antidiabetic potency of mixture of powders of dried fruits of *Eucalyptus globules* and rhizomes of *Curcuma zedoaria*. 2011; 2 (3): 326–332.

17. Selner M. Penetrating carrier, anti-fungal composition using the same and method for treatment of dermatophytes. United States Patent 7601371. Application no. 11/684869, Publication date: 10/13/2009. http://www.freepatentsonline.com/7601371. html

18. Vankar PS, Tiwari V, Srivastava J. Extracts of stem bark of *Eucalyptus globulus* as food dye with high antioxidant properties. Electron. J. Environ. Agric. Food Chem. 2006; 5 (6): 1664–1669.

19. Zofou D, Tene M, Ngemenya MN, Tane P, Titanji VPK. *In vitro* antiplasmodial activity and cytotoxicity of extracts of selected medicinal plants used by traditional healers of western cameroon. Malaria Research and Treatment. 2011; doi:10.4061/2011/561342, Article ID 561342, 6 pages.

Ficus carica

Plant profile

Kingdom: *Plantae* – Plants
Superdivision: *Spermatophyta* – Seed plants
Class: *Magnoliopsida* – Dicotyledons
Order: *Urticales*
Genus: *Ficus* L. – Fig

Subkingdom: *Tracheobionta* – Vascular plants
Division: *Magnoliophyta* – Flowering plants
Subclass: *Hamamelididae*
Family: *Moraceae* – Mulberry family
Species: *Ficus carica* L. – Edible fig

Vernacular name

San: Falgu, Anjeera, Manjula, **Beng:** Dumoor, **Eng:** Fig, **Guj:** Falgu, **Hind**: Anjeer, **Kan:** Anjura, **Mal**: Seema ati, Athipazham, **Ori:** Dimiri, **Tam:** Simaiyatti, **Tel:** Athi pallu.

Ayurvedic properties

Rasa: Madhura, **Guna:** Guru, Snigdha, Sara, **Veerya:** Seeta, **Vipaka:** Madhura

INTRODUCTION

It grows to a height of 6.9–10 m (23–33 ft) tall, with smooth grey bark. *Ficus carica* is well known for its fragrant leaves that are large and lobed. The leaves are 12–25 cm (4.7–9.8 in) long and 10–18 cm (3.9–7.1 in) across, and deeply lobed with three or five lobes. *Ficus carica* have pyriform sicon infructescences, the fleshy fruit fig, with inner unisexual flowers. The fruit is 3–5 cm (1.2–2.0 in) long, with a green skin, sometimes ripening towards purple or brown. *Ficus carica* has milky sap (laticifer). The sap of the fig's green parts is an irritant to human skin.

Distribution

Ficus carica is a monoecious, deciduous tree or a large shrub. It is native to the Middle East. It was later cultivated from Afghanistan to Portugal, and from the 15th century onwards was grown in areas including Northern Europe and the New World. It also distributes in Iran, Turkey, Pakistan, Northern India, Arkansas, Louisiana, California, Georgia, Texas, US, Southwestern British Columbia in Canada, Durango and Coahuila in Northeastern Mexico, as well as areas of Argentina, Australia, Chile, Peru and South Africa.

Chemical Constituents

Stem: Campesterol, hentriacontanol, stigmasterol, euphorbol and its hexacosanate, ingenol and taraxerone. **Leaves:** Moisture, 67.6%; protein, 4.3%; fat, 1.7%; crude fiber, 4.7%; ash, 5.3%; N-free extract, 16.4%; pentosans, 3.6%; carotene, bergapten, stigmasterol, sitosterol, tyrosine, ficusin, taraxasterol, betasitosterol, rutin, sapogenin, calotropenyl acetate, lepeolacetate and oleanolic acid. **Latex:** Caoutchouc (2.4%), resin, albumin, cerin, sugar and malic acid, rennin, proteolytic enzymes (ficin), diastase, esterase, lipase, catalase, and peroxidase. **Seed:** Dried seeds contain 30% of a fixed oil containing the fatty acids: oleic, 18.99%; linoleic, 33.72%; linolenic, 32.95%; palmitic, 5.23%; stearic, 2.1 8%; arachidic, 1.05%.

Medicinal uses

Many medicinal virtues have been ascribed to the fig. It is considered a restorative food which helps in quick recovery after prolonged illness. It removes physical and mental exertion and endows the body with renewed vigor and strength. It is an excellent tonic for the weak people who suffer cracks in lips tongue and mouth. **Constipation:** Taken either fresh or dried, the fig is regarded as a dependable laxative on account of its large cellulose content and its tough skin. The tiny seeds in the fruit possess the property of stimulating peristaltic or wave-like movements of intestines which facilitates easy evacuation of faeces and keeps the alimentary canal clean. **Piles:** Owing to its laxative property, the fig is an excellent remedy for piles. Two or three dried figs should be soaked in cold water in a glass of enamelware in the night after cleaning them thoroughly with hot water. They should be taken next morning. Figs should be taken similarly in the evening. This will remove straining at stools and thus prevent the protrusion of the anus. The piles will be cured with regular use of figs in this manner for three or four weeks. **Asthma:** Figs are considered beneficial in the treatment of asthma. Phlegmatic cases of cough and asthma are treated with success by their use. It gives comfort to the patient by draining of the phlegm. **Sexual weakness:** Figs can be beneficially used in the treatment of sexual debility. They can be supplemented by other dry fruits like almonds and dry dates along with butter. Their use has proved effective in such cases.

Apart from that fresh and dried fruit of *F. carica* is used in cancer, carcinoma, ulcers, hepatomegaly, spleenomegaly. Latex is used in ulcers and gout. Leaves are used in cancer and tumours.

Cosmetic uses

Traditionally, figs have been used to treat eczema, psoriasis (chronic skin disease) and vertigo (white skin patches). Topically, its latex has been used to remove warts and treat skin tumors. The leaves are used for dermatitis. The roasted fruit is emollient and used as a poultice in the treatment of gumboils, dental abscesses, etc. The milky juice of the freshly-broken stalk of a fig has been found to remove warts on the body. When applied, a slightly inflamed area appears round the wart, which then shrivels and falls off. The milky juice of the stems and leaves is very acrid and has been used in some countries for raising blisters. The milky white juice of the leaves is applied on the corns to reduce calluses. It can use for skin sagging. The fresh figs form a nice tonic to weak people who suffer from cracks in lips, tongue and mouth.

Pharmacological Action

The aqueous-ethanolic extract of *Ficus carica* (Fc.Cr) was studied for antispasmodic effect on the isolated rabbit jejunum preparations and for antiplatelet effect using *ex vivo* model of human platelets. The study showed the presence of spasmolytic activity in the ripe dried fruit of *Ficus carica* possibly mediated through the activation of K^+ATP channels along with antiplatelet activity which provides sound pharmacological basis for its medicinal use in the gut motility and inflammatory disorders (Gilani et al., 2008).

Shade dried leaves of *Ficus carica* were extracted using petroleum ether (60–80°) and tested for antihepatotoxic activity on rats treated with 50 mg/kg of rifampicin orally. The parameters assessed were serum levels of glutamic oxaloacetate transaminase, glutamic pyruvic transaminase, bilirubin and histological changes in liver. Liver weights and pentobarbitione sleeping time as functional parameters were also monitored. There was significant reversal of biochemical, histological and functional changes induced by rifampicin treatment in rats by petroleum ether extract treatment, indicating promising hepatoprotective activity (Gond and Khadabadi, 2008).

The study was investigated the antimicrobial activity of methanol (MeOH) extract of figs against oral bacteria. The MeOH extract (MICs, 0.156 to 5 mg/ml; MBCs, 0.313 to 5 mg/ml) showed a strong antibacterial activity against oral bacteria. The combination effects of MeOH extract with ampicillin or gentamicin were synergistic against oral bacteria. Hence it could be employed as a natural antibacterial agent in oral care products (Jeong et al., 2009).

The immunomodulatory effect of ethanolic extract of the leaves of *Ficus carica* (Moraceae) was investigated in mice. The study was carried out by various hematological and serological tests. Administration of extract remarkably ameliorated both cellular and humoral antibody response. It is concluded that the test extract possessed promising immunostimulant properties (Patil et al., 2010).

Five extracts (methanolic, hexanic, ethyl acetate, hexane-ethyl acetate (v/v) and chloroformic) of *Ficus carica* were investigated *in vitro* for their antiviral potential activity against herpes simplex type 1 (HSV-1), echovirus type 11 (ECV-11) and adenovirus (ADV) by the following assays: Adsorption and penetration, intracellular inhibition and virucidal activity. Observation of cytopathic effects was used to determine the antiviral action. The hexanic and hexane-ethyl acetate (v/v) extracts inhibited multiplication of viruses by tested techniques at concentrations of 78 μg ml(–1). These two extracts were possible candidates as herbal medicines for herpes virus, echovirus and adenovirus infectious diseases. All extracts had no cytotoxic effect on Vero cells at all tested concentrations (Lazreg et al., 2011).

The study was aimed to evaluate protective effect of hydroalcohalic extract of *Ficus carica* (HEFC) on Gentamicin (GM)-induced renal proximal tubular damage. The rats were pre-fed experimental diets for 8 days and then received GM (100 mg/kg body weight/day) treatment for 8 days while still on diet. Serum parameters, oxidative stress in rat kidney were analyzed. GM nephrotoxicity was recorded by increased serum creatinine and blood urea nitrogen. GM increased MDA level, whereas decreased catalase, reduced glutathione. In contrast, HEFC alone increased concentration of catalase (CAT) activities, glutathione (GSH) content and decreased malondialdehyde (MDA) level. HEFC supplementation ameliorated GM-induced specific metabolic alterations and oxidative damage due to its intrinsic biochemical/antioxidant properties (Kaksaheb et al., 2011).

Toxicological Studies

Different extracts of *Ficus carica* were prepared and was studied for acute toxicity on mice. Different amounts like 30, 100, 300, 1000, 2000 and 5000 mg/kg of extracts were administered orally to mice. The extracts were given at the doses of 300 and 600 mg/kg/day of body weight. All the animals were found to be safe at dose of 5000 mg/kg. Mice were then observed for incidence of mortality or any sing of toxicity up to 24 hours. Hence the LD_{50} of the extract is more than 5000 mg/kg. Further during 24 hours study, all the extracts did not cause any untoward effect. No gross observational effects were observed at any of the doses of the plant extract during acute toxicity study (Patil and Patil, 2011).

Clinical Investigation

In this study, cytotoxicity of fruit and leaf extracts as well as the latex of *Ficus carica* on human cancerous cell line were evaluated. The viability of the cells was determined by the reduction of 3-(4, 5-dimethylthiazol- 2-yl)-2, 5-diphenyl tetrazolium bromide (MTT) from formazan following 48 h incubation and the absorbance was measured at 540 nm using an ELISA plate reader. The result indicates that the latex and different extracts of *Ficus carica* could reduce the viability of the cancerous cell at concentration as low as 2 mg/ml in a dose dependent manner. The IC_{50} value of the latex was about 17 mg/ml (Khodarahmi et al., 2011).

Market Formulations

Refreshing Fruit Pack

It firms, tightens and rejuvenates facial skin. It removes deeply embedded impurities and used for normal to dry skin.

Ingredients: Each ml contains: Papaya (*Carica papaya*), apple (*Pyrus malus*), cucumber (*Cucumis sativus*), fig (*Ficus carica*).

Action: A rich source of natural alpha hydroxy acids (AHAs) improvees and nourishes skin texture. Fig and cucumber provide cooling effect to the skin. Papaya removes freckles and smoothens the skin. Mineral clay soothes the skin and removes deeply embedded impurities.

Directions: Apply evenly over cleansed face and neck avoiding the area around eyes. Allow it to dry for 10–15 minutes. Remove with wet sponge and wash the skin with cool water. Use once or twice a week. Follow up with face moisturizing lotion.

Supplied by: Himalaya Drug Company, India.

Skin Appetit Detoxifying Nutri-cleanser with Nutrx 8 Complex for Ageless Skin

Ingredients: Water, sodium C14–16 olefin sulfonate, sodium methyl cocoyl taurate, stearic acid, cetyl alcohol, stearyl alcohol, cetearyl alcohol, ceteareth-30, glycol distearate, panthenol-pro vitamin B5, tocopheryl acetate-vitamin E, ascorbyl palmitate-vitamin C, propylene glycol, nutrx 8 complex, blueberry fruit extract- *Vaccinium angustifolium, Cucumis melocantalupensis* fruit extract, grape seed extract—*Vitis vinifera*, hydrolyzed yogurt protein, fig extract—*Ficus carica*, walnut seed extract—*Juglans regia*, cocoa extract—*Theobroma cacao*, honey extract, *Aloe barbadensis* leaf extract, cinnamon bark extract—*Cinnamomum cassia*, hydrolyzed silk, *Macadamia ternifolia* seed oil, tea tree leaf oil—*Melaleuca alternifolia*, methylchloro-isothiazolinone, methylisothiazolinone, fragrance, blue 1, red 33.

Directions: Apply cleanser to face and massage upward gently. Rinse thoroughly with warm water and pat dry. Best when used with other skin appetit products.

Supplied by: Environmental Working Group, USA.

Nevexin Topical Treatment

It is a safe, all natural alternative for effectively removal of warts, genital warts, mosaic warts, plantar foot warts, hand warts, human papilloma virus (HPV), moles on the skin and face, skin tags and syringoma.

Ingredients: Water, octyl palmitate, glyceryl stearate, cetearyl alcohol, PEG-100 stearate, caprylic/cpric triglyceride, sodium hydroxide, triethanolamine, *Ficus Carica*, anacardium occidentale, *Citrus Limon, Chelidonium Majus*, talc, de-ionized water.

Supplied by: Great Wall Herbal, USA, Canada.

Storage of Products

All products are store in cool places and away from sunlight.

Dosage

For adults: There is no proven safe or effective dose for fig. However, as a tea decoction, 1 cup daily of 13 g of *Ficus carica* leaf has been used.

For children: There is no proven safe or effective dose for fig in children, and use is not recommended.

Patents

The invention relates to a cosmetic based on various natural raw materials, which can be used to counterageing processes of the human skin. The inventive cosmetic contains 0.1–5% by weight of an extract from a mixture of fig leaves and fruits, 0.1–3% by weight of an extract from pomegranate fruits, 0.001–0.5% by weight of a ground dry mixture of rosemary stems and leaves, 0.01–3% by weight of liposomes containing an extract from peeled musk melons, 0.1–5% by weight of liposomes containing a plankton extract containing the photolyase enzyme, 0.1–5% by weight of liposomes containing 0.1 to 0.5% by weight, in relation to the liposome weight, of a micrococcus lysate containing the UV-endonuclease enzyme; and up to 100% by weight, other active substances, carrier substances, adjuvants or mixtures thereof (Golz-Berner et al., 2004).

The present invention relates to topical compositions containing a dispersion of composite particles. Each of such composite particles contains one or more core particles encapsulated or entrapped in a polymeric shell. At least some of the core particles are formed of a material capable of causing generation of reactive oxygen species (ROS). Each of the composite particles further contains: (1) a first antioxidant co-encapsulated or co-entrapped with the core particles inside the polymeric shell for quenching or scavenging ROS generated in the vicinity of the core particles, and (2) a second antioxidant coated over the polymeric shell for preventing or reducing oxidative damage to the skin, such as lipid peroxidation. The particle dispersion of the present invention therefore can be readily formulated into topical sunscreen compositions with organic sunscreen agents susceptible to oxidative decomposition or degradation to provide improved protection against skin damage caused by exposure to UV light (Sente et al., 2010).

BIBLIOGRAPHY

1. Gilani AH, Mehmood MH, Janbaz KH, Khan AU, Saeed SA. Ethno-pharmacological studies on anti-spasmodic and antiplatelet activities of *Ficus carica*. J Ethanopharmacol. 2008; 119(1):1–5.

2. Golz-Berner K, Zastrow L, Coty BV. Anti-ageing skin cosmetic. US Patent No: 8,034,385, Application no: 10/520,562, publication date: 01/15/2004. http://patents.com/us-8034385. html

3. Gond NY, Khadabadi SS. Hepato-protective activity of *Ficus carica* leaf extract on rifampicin-induced hepatic damage in rats. Indian J Pharm Sci. 2008; 70:364–366.

4. http://ayurvedicmedicinalplants.com/plants/149.html

5. http://en.wikipedia.org/wiki/Common_fig

6. http://www.ayushveda.com/herbs/ficus-carica.htm

7. http://www.best-home-remedies.com/herbal_medicine/fruits/fig.htm

8. http://www.hort.purdue.edu/newcrop/morton/fig.html

9. http://www.india-herbs.com/herbal_glossary/id/232

10. Jeong MR, Kim HY, Cha JD. Anti-microbial Activity of Methanol Extract from *Ficus carica* Leaves Against Oral Bacteria. Journal of Bacteriology and Virology. 2009; 39 (2): 97–102.

11. Kaksaheb JK, Rajkumar VS, Babasaheb NK, Borade AS. Protective role of hydroalcoholic extract of *Ficus carica* in gentamicin induced nephrotoxicity in rats. Int. J. of Pharm. & Life Sci. (IJPLS). 2011; 2 (8): 978–982.

12. Khodarahmia GA, Ghasemib N, Hassanzadeha F, Safaiea M. Cytotoxic Effects of Different Extracts and Latex of *Ficus carica* L. on HeLa cell Line Iranian Journal of Pharmaceutical Research. 2011; 10 (2): 273–277.

13. Lazreg AH, Gaaliche B, Fekih A, Mars M, Aouni M, Pierre CJ, Said K. *In vitro* cytotoxic and antiviral activities of *Ficus carica* latex extracts. Nat Prod Res. 2011; 25(3):310–319.

14. Patil VV, Bhangale, SC, Patil VR. Studies on immunomodulatory activity of *Ficus carica*. International Journal of Pharmacy and Pharmaceutical Sciences. 2010; 2 (4): 97–99.

15. Patil VV, Patil VR. Evaluation of anti-inflammatory activity of *Ficus carica* Linn. leaves. Indian J of Natural Products and Resources. 2011; 2(2): 151–155.

16. Sente I, Declercq L, Maes DH, Sojka MF, Cummins P, Fthenakis CG et al. Composite particles having an antioxidant-based protective system, and topical compositions comprising the same. United States Patent Application 20100040696. Application no: 12/507145, publication date: 02/18/2010. http://www.freepaten tsonline.com/y2010/0040696.html

Ginkgo biloba

Plant profile

Kingdom: *Plantae* – Plants
Superdivision: *Spermatophyta* – Seed plants
Class: *Ginkgoopsida* – Dicotyledons
Family: *Ginkgoaceae* – Ginkgo family
Species: *Ginkgo biloba* L. – Maidenhair tree

Subkingdom: *Tracheobionta* – Vascular plants
Division: *Ginkgophyta* – Ginkgo
Order: *Ginkgoales*
Genus: *Ginkgo* L – Ginkgo

Vernacular name

Eng: Maidenhair tree, Ginkgo, Fossil tree, **Hin:** Balkuwari, **Urdu:** Pankha Plant, **Kashmiri:** Aziz tree, **Arabic:** Mabad ag.

Other names

Duck foot tree, silver apricot.

INTRODUCTION

It is a large, long lived, tough trees which may live as long as 3000 years. They can be as tall as 165 ft. The tree has long angular branches, with strong deep rooted roots which make it stable in extreme snow and windy conditions. The tree is not evergreen, the leaves turn deep yellow during winters and fall considerably, in winters it is no more than bare branches but in spring and summers it is in full bloom and have the pollination as well. The leaves are fan shaped which have dichotomous venation. The leaves are usually 5–10 cm (2–4 inches), but sometimes up to 15 cm (6 inches) long. Leaves of long shoots are usually notched or lobed, but only from the outer surface, between the veins. They are borne both on the more rapidly-growing branch tips, where they are alternate and spaced out, and also on the short, stubby spur shoots, where they are clustered at the tips. The seed is 1.5–2 cm long. Its fleshy outer layer (the sarcotesta) is light yellow-brown, soft, and fruit-like.

Distribution

The Ginkgo tree is the native of China, Japan and Korea, and is also found in Europe and the U.S. Recently it is also found in India (Kashmir), Gilgat, Iran, Afghanistan and North America and is one of the oldest species of trees in existence today.

Chemical Constituents

The main bioactive components of Ginkgo leaves are flavonoids, biflavonoides, pro-anthocyanidins, and triactonic diterpenes, which include ginkgolide A, ginkgolide B, ginkgolide C, plus ginkgolide J and the sesquiterpene bilobalide; flavonols, including kaempferol, quercetin, and isorhamnetin; flavones, including luteolin and tricetin; biflavones, mainly bilobetin, ginkgetin, isoginkgetin, and sciadopitysin; catechins; proanthocyanidins; sterols and 6-hydroxy-kynurenic acid (6-HKA). Ginkgolide B has

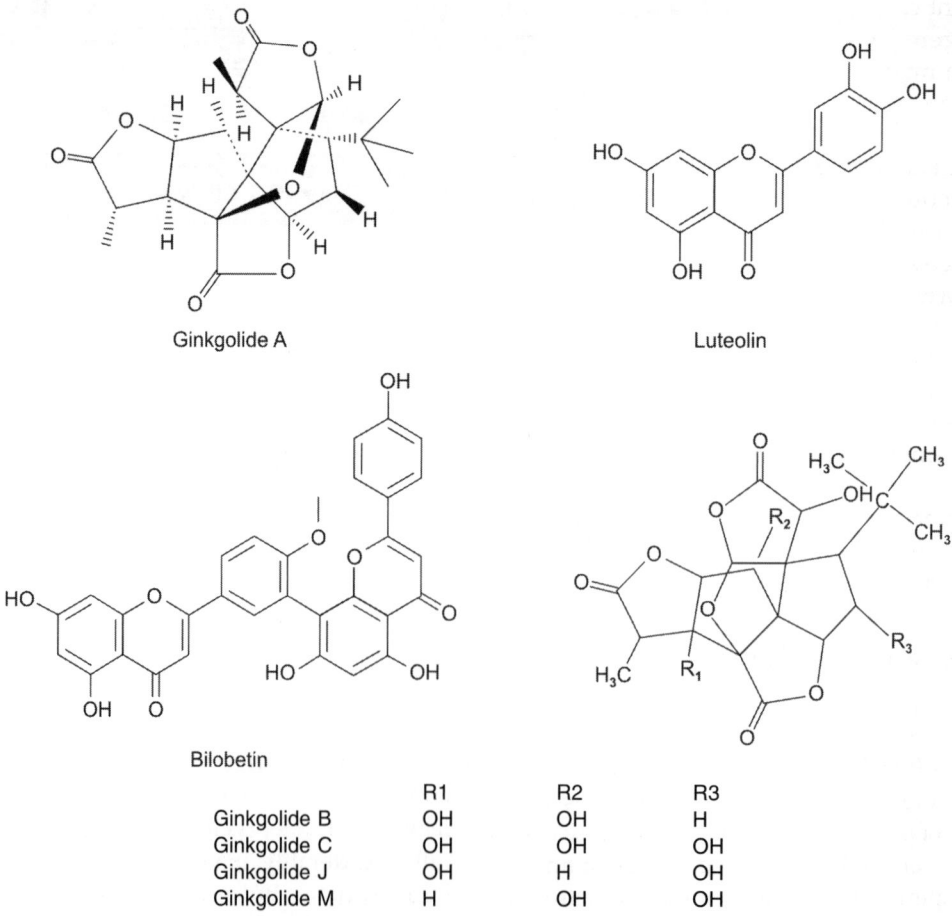

Ginkgolide A

Luteolin

Bilobetin

	R1	R2	R3
Ginkgolide B	OH	OH	H
Ginkgolide C	OH	OH	OH
Ginkgolide J	OH	H	OH
Ginkgolide M	H	OH	OH

been shown to inhibit platelets in the blood from coagulating. The flavonoids in Ginkgo have demonstrated very strong antioxidant effects.

Medicinal uses

Ginkgo has been used for medicinal purposes for almost 5,000 years. In Chinese traditional medicine, it is used to treat asthma, bronchitis, and various brain disorders. In Asia, the seeds of the Ginkgo tree are used to aid digestion and to reduce the intoxicating effects of alcohol. In Europe and North America, Ginkgo extract is used for the treatment of circulatory problems, immune system dysfunction and cognitive disorders, including memory loss.

Cosmetic uses

Gingko biloba helps to increase circulation, prevent capillary fragility and helps to boost collagen formation and create fibroblast, which makes it ideal to use in rejuvenating skin care products. The therapeutic properties of this herb in skin care is that of being a vasodilator, increasing circulation, improving sebaceous secretions, decreasing capillary hyper-permeability, improving tissue irrigation and activating cell metabolism (especially in the cortex, by increasing glucose and oxygen uptake). It also has anti-inflammatory and anti-allergenic properties. It furthermore increases the creation of fibroblast, collagen and extracellular fibronectin, while also exhibiting good anti-oxidant qualities. It is known for anti-aging properties that give a healthier, rosy, younger looking skin.

Creams, emulsions, milks, gels, oils, makeup products, lipsticks and lotions are the other available forms.

Pharmacological Action

The study was investigated the anti-inflammatory mechanism of the polysaccharides of *Ginkgo biloba* leaves (PGBL) by inhibiting leucocyte adhesion. The anti-inflammatory effects of purified PGBL (p-PGBL) were assayed by ear edema induced by xylol and the acute peritonitis model in mice. The effect of p-PGBL on inhibiting the interaction between P-selectin and its ligands was investigated by flow cytometry and flow chamber. Results indicated that p-PGBL can inhibit the inflammatory process through interfering with the interaction between P-selectin and its ligands (Fei et al., 2008).

Effect of administration of the standardized extract of *Ginkgo biloba* leaves (EGb 761) on learning, memory and exploratory behavior was estimated in water maze and hole-board tests. Rats (18-month old) received for three months EGb 761 at doses: 50, 100 and 150 mg/kg b.w. per day. After completion of the behavioral experiment, concentrations of neurotransmitters were estimated in selected brain regions. The increased level of 5-hydroxy-tryptamine (5-HT) in the hippocampus and 5-HIAA (5-HT metabolite) in the prefrontal cortex correlated positively with the retention of spatial memory. Positive correlation between platform crossings in SE during the probe trial and neurotransmitter turnover suggest improvement of spatial memory. Long-term administration of *Ginkgo biloba* extract can improve spatial memory and motivation with significant changes in the content and metabolism of monoamines in several brain regions (Klin et al., 2009).

Free radical scavenging activity of three important plants *Ginkgo biloba*, *Stevia rebaudiana* and *Parthenium hysterophorous* was carried out to evaluate and explore new potential sources of natural antioxidants. The leaves of the three plants were processed. In these experiments the order of the antioxidant activity was, maximum activity shown by methanolic extract of *Ginkgo biloba* followed by *Parthenium hysterophorous* and *Stevia rebaudiana*. Furthermore the ethanolic extract of *Ginkgo biloba* also showed maximum antioxidant activity seconded by *Stevia rebaudiana* and *Parthenium hysterophorous* (Ahmad et al., 2010).

Toxicological Studies

Acute toxicity: In rats and mice, orally administered GBE did not produce evidence of organ damage or impairment of hepatic and renal functions when administered over 27 weeks in doses ranging from 100 to 1,600 mg/kg. Subacute/Subchronic Studies: In rats and mice, orally administered GBE did not produce evidence of organ damage or impairment of hepatic and renal functions when administered over 27 weeks in doses ranging from 100 to 1,600 mg/kg (Salvador, 1995).

The study was established the toxic reaction on compound *Ginkgo biloba* Tablets by ig administration in mice, three times a day. Its acute toxic reactions were observed for 7 days. The results revealed that all the mice were survive and no toxic reaction was shown during the observed period. The average rate of its increasing weight was 24.82 percent. The maximum administration dosage was 3339.6 mg of puerarin given to mice per kilogram, equivalent to men's daily administration dosage by 2360.1 times, hence *Ginkgo biloba* tablets is safe for oral administration (Ping et al., 2007).

Clinical Investigation

Thirty patients with gastric cancer were treated with 2 oral *G. biloba* exocarp polysaccharides (GBEP) capsules twice per day for longer than 1 month. An electron gastroscope was used to measure the area of tumors before and after treatment. Inhibitory and effective rates were then calculated. Results show that GBEP capsules reduced the area of tumors by an effective rate of 73% (Xu et al., 2003).

A cochrane systematic review that included 10 trials with a total of 792 patients and assessed the efficacy of Ginkgo found no convincing evidence to support the use of Ginkgo for recovery after stroke or improvement in neurological deficit at the end of treatment (Zeng et al., 2005).

Market Formulations

Créme Aux Liposomes

Cream with liposomes is a very light textured cream between a gel and an emulsion. It is able to penetrate into the epidermis. All of the active ingredients of this emulsion are extracted from plants.

Ingredients: Mainly sweet almond oil, *Aloe vera*, *Ginkgo biloba*.

Action: Sweet almond oil is rich in fatty acids and proteins to nourish the skin. *Aloe vera* stimulates the healing process, injured and stressed skin is soothed. *Ginkgo biloba* neutralizes free radicals, one of the causes of aging. It also improves capillary blood flow and strengthens tissues.

Direction: Apply in the morning and evening by gently massaging on cleansed face and neck. Due to its rapid absorption, any other cream or make-up can be applied right after.

Supplied by: Make-Up USA, Michigan.

Liposome Gel (Gel Aux Liposomes)

Liposome gel has a very light texture. It is anti-aging gel and suitable for all skin types.

Ingredients: *Ginkgo biloba*, coconut oil.

Action: When a gel or cream containing liposomes is applied to the skin, the liposomes are deposited on the skin and begin to merge with the cellular membranes. In the process, the liposomes release their payload of active materials into the cells. Coconut oil is also used to decrease rough, dry skin.

Direction: Apply in the morning and evening on cleansed face and neck. Due to its rapid absorption, any other cream or make-up can be applied right after.

Supplied by: Make-Up USA, Michigan.

Enzergen Ginkgo Biloba

Enzergen Ginkgo is for centuries the Chinese have taken *Ginkgo biloba* to enhance concentration, memory, cognitive activity and bloodflow. *Ginkgo biloba* extract is a powerful aid to circulatory problems, particularly lack of blood to the brain, which causes memory loss, vertigo, tinnitus, disorientation, headaches and depression for the elders.

Ingredients: Each capsule contains: Capsules contain 40 mg *Ginkgo biloba* extract, equivalent to 2000 mg dry *Ginkgo biloba* herb plus tapioca starch.

Action: It improves concentration and provides healthy blood circulation. It prevents memory associated diseases.

Direction: For adults only, take 1–2 capsules a day with food or as directed by health professionals.

Storage: Store in a cool dry place below 30°C away from sunlight. Keep tightly closed.

Supplied by: L.I.I. Exports Pvt. Ltd., New Delhi, India.

Night Glow Cream *(Brand Name: Bio Valley)*

This cream revitalizes dull tired looking skin, thus helps to maintain healthy youthful glow on the skin.

Ingredients: Rose oil, avocado oil, almond oil, *Ginkgo biloba*, tocopheryl acetate (vitamin E) and evening primrose oil.

Action: Firming, moisturizer, nourishing to the skin.

Direction: Clean the face and apply on the face at night time with gentle rubbing.

Supplied by: Bio Valley, Noida, India.

Storage of Products

Ginkgo must be protected from light and moisture.

Dosage

Ginkgo is available in liquid or solid pharmaceutical forms, for oral intake and parenterally for homeopathic use.

Capsules — 30 mg, 40 mg, 50 mg, 60 mg, 100 mg, 120 mg, 260 mg, 400 mg, 420 mg, 440 mg, 450 mg, 500 mg

Extract — 50:1

Liquid — 40 mg/5 ml

Tablets — 30 mg, 40 mg, 60 mg, 80 mg, 120 mg, 260 mg daily dosage.

Ginkgo biloba extract should be standardized to contain 24% flavone and 6% terpene lactones: 40 to 80 mg three times a day. Studies have demonstrated efficacy with 120 mg daily in 2 to 3 divided doses for dementia, peripheral arterial occlusive disease and for equilibrium disorders like tinnitus or vertigo.

Tincture (1:5): 2 to 4 ml three times a day.

Patents

The present invention relates to a composition which comprises as active ingredients a combination of melatonin, *ginkgo biloba* and biotin. The composition is particularly suitable for producing formulations for topical application in the hair (Schmid, 2006).

Flavanoid components of the *Ginkgo biloba* tree are useful to stimulate the growth of hair and are thus treatment of alopecia or baldness (Gallwitz et al., 2007).

A composition and method of use, therefore, that allows for permanent straightening or curling of human hair that is not irritating to the skin, that allows for the immediate shampooing of hair thereafter and that provides for re-treatment of the hair without additional damage. The composition primarily uses dimethyl sulfone (MSM) and high temperatures to break and reform disulfide bonds as a means to allow for the penetration of low molecular weight proteins into the shaft of the hair. When curling the hair, it is understood that devices that will stretch hairs into curls at the temperatures expressed in the scope of the patent will produce a permanent curled hair. A test curl will indicate the heat and temperature required to permanently curl the hair being treated (Saute and Saute, 2010).

BIBLIOGRAPHY

1. Ahmad N, Fazal H, Abbasi BH, Farooq S. Efficient free radical scavenging activity of *Ginkgo biloba*, *Stevia rebaudiana* and *Parthenium hysterophorous* leaves through DPPH (2, 2-diphenyl-1-picrylhydrazyl). Int J of Phytomedicine. 2010; 2(3): 231–239.

2. Fei R, Fei Y, Zheng S, Gao YG, Sun HX, Zeng XL. Purified polysaccharide from *Ginkgo biloba* leaves inhibits P-selectin-mediated leucocyte adhesion and inflammation. Acta Pharmacol Sin. 2008; 29(4): 499–506.

3. Gallwitz WE, Garrett RI, Gutierrez G. Stimulation of hair growth by *Ginkgo biloba* Flavanoids. United States Patent Application 20070196316. Application no. 10/575285. Publication date: 08/23/2007. http://www. freepatentsonline. com/y2007/0196316. html

4. http://altmedicine.about.com/cs/herbsvitaminsek/a/Ginkgo.htm

5. http://en.wikipedia.org/wiki/Ginkgo_biloba

6. http://jkmpic.blogspot.com/2011/01/ginkgoginkgo-biloba-l-in-hindi-india. html

7. http://medicineherbs.net/medicinal-herbs/ginkgo

8. http://www.rxlist.com/ginkgo-page2/supplements.htm

9. http://www.sigmaaldrich.com/life-science/nutrition-research/learning-center/plant-profiler/ginkgo-biloba.html

10. Klin KB, Piechal A, Joniec I, Pyrzanowska J, Tyszkiewicz EW. Pharmacological and biochemical effects of *Ginkgo biloba* extract on learning, memory consolidation and motor activity in old rats. Acta Neurobiol Exp. 2009; 69: 217–231.

11. Ping PH, Ling WP, Ying C, Xian ZZ, Fen CX, Jing LY, Mei LU. Experimental study on acute toxicity of compound *Ginkgo biloba* tablets in mice. Lishizhen Medicine and Materia Medica Research. 2007; 06: 026.

12. Salvador R.L. Herbal medicine - Ginkgo. CPJ/RPC, 1995; (July-August), 52: 39–41.

13. Saute R, Saute S. Composition and method for hair straightening and curling. United States Patent Application 20100307526. Application no. 12/802035. Publication date: 12/09/2010. http://www.freepatentsonline. com/y2010/0307526.html

14. Schmid HW. Formulations containing melatonin, *ginkgo biloba* and biotin. United States Patent Application 20060099278. Application no. 10/533516. Publication date: 05/11/2006. http://www.freepatentsonline. com/y2006/0099278.html

15. Xu AH, Chen HS, Sun BC, et al. Therapeutic mechanism of *ginkgo biloba* exocarp polysaccharides on gastric cancer. World J Gastroenterol. 2003; 9(11):2424–2427.

16. Zeng X, Liu M, Yang Y, Li Y, Asplund K. *Ginkgo biloba* for acute ischaemic stroke. Cochrane Database Syst Rev. 2005; (4): CD003691.

Glycyrrhiza glabra

Plant profile

Kingdom: *Plantae* – Plants

Superdivision: *Spermatophyta* – Seed plants

Subkingdom: *Tracheobionta* – Vascular plants

Division: *Magnoliophyta* – Flowering plants

Class: *Magnoliopsida* – Dicotyledons

Order: *Fabales*

Genus: *Glycyrrhiza* L. – Licorice

Subclass: *Rosidae*

Family: *Fabaceae* – Pea family

Species: *Glycyrrhiza glabra* L. – Cultivated licorice

Vernacular name

San: Yastimadhu, Madhuka, **Eng:** Licorice, **Hind:** Jothi-madh, Mulhatti, **Ben:** Jashtimadhu, Jaishbomodhu, **Guj:** Jethimadhu, **Kan:** Yastimadhuka, atimaddhura, **Mal:** Iratimadhuram, **Tam:** Atimaduram, **Mar:** Jeshtamadha, **Oriya:** Jatimadhu, **Tel:** Atimadhuranu, Yashtimadhukam

Ayurvedic properties

Rasa: Madhura, **Guna:** Guru, Snigdha, **Veerya:** Seeta.

INTRODUCTION

It is a herbaceous perennial, growing to 1.5 m in height, with pinnate leaves about 7–15 cm (3–6 in) long, with 9–17 leaflets. The flowers are 0.8–1.2 cm ($\frac{1}{2}$ – $\frac{1}{3}$ in) long, purple to pale whitish blue, produced in a loose inflorescence. The fruit is an oblong pod, 2–3 cm (1 in) long, containing several seeds. The roots are stoloniferous. Leaves compound, pinnate, leaf lets 4–7 pairs. The flowers violet in axillary racemes, fruits flattened linear pods, containing flattened seeds. Underground stems woody, pale brown color, fibrous and cream colored, with a peculiar sweetish taste.

Distribution

This plant is native to central and south western Asia and the Mediterranean region. It is cultivated in the Mediterranean basin of Africa, in South Europe and in India. This species is widely distributed in the world from 50 W to 1000 E longitude and 200 to 500 N latitude. It is reported abundant in western China, parts of Asia Minor, Persia, Asian Republics of erstwhile U. S. S. R. and Afghanistan. It is also cultivated in Punjab and sub-Himalayan tracts in India.

Chemical Constituents

The major bio-active constituent of rhizomes is a triterpenoids saponin glycyrrhizin, glycyrrhizinic acid, glabrin A&B, glycyrrhetol, glabrolide, isoglabrolide, isoflavones, coumarins, triterpene sterols, etc.

Medicinal uses

It has anti-inflammatory, antiviral, saponin glycyrrhizin and many important chemicals that are required for good health. It is mainly used in treating cervical cancer, kidney and bladder disorders, HIV, hepatitis B, herpes, inflammation of mucous membrane, food poison, stomachaches, coughs, horse-voice, bronchitis, asthma, ulcers, arthritis, shingles, sunburns, fevers in infants, and also used for insect bites. It is combined with other herbs in treating coughs, colds, sore throat, asthma, stomach and duodenal ulcers, hepatitis, hysteria, food poisoning, hypoglycemia, bronchitis, colitis, diverticulitis, gastritis, some of stress related disorders and nausea. Licorice properties include anodyne, antioxidant, antispasmodic, demulcent, depurative, diuretic, cooling, expectorant, emollient, estrogenic, expectorant, anti-arthritic, tonic, stimulant adrenal, anti-cholesterolemic, antigastritis and anti-allergenic. The licorice glycoside is similar to the natural steroids of the body and has a good effect in treatment of Addison's disease.

Cosmetic uses

Licorice-derived ingredients are used in the formulation of eye makeup, other makeup products, hair care products and skin care products. In cosmetics it works great as a skin brightening and for rashes and viral related dermatological problems, it is also soothing, also good for insect bites. It has skin lightening, anti-inflammatory, anti-irritant, and antioxidant activity. *Glycyrrhiza glabra*

Glycyrrhizinic acid

Glycyrrhetinic acid

Glycyrrhizin

root extract can provide a rose-like scent, improves cell regeneration, tissue repair and skin toning, regulates sebum production to reduce surface shine. Licorice is widely used in asia for its rejuvenating and nutritive properties. It is included in the formulation primarily as an adrenal tonic although it has notable influence on a number of systems including the immune function.

The following functions have been reported for the licorice-derived ingredients: Antioxidants *Glycyrrhiza glabra* (licorice) root extract, *Glycyrrhiza glabra* (licorice) root water.

Cosmetic astringent—*Glycyrrhiza glabra* (licorice) root water.

Flavoring agents—*Glycyrrhiza glabra* (licorice) rhizome/root

Skin conditioning agents—emollients—*Glycyrrhiza glabra* (licorice) root water

Skin conditioning agents—humectant—*Glycyrrhiza glabra* (licorice) root extract

Skin conditioning agents—miscellaneous—*Glycyrrhiza glabra* (licorice) leaf extract, *Glycyrrhiza glabra* (licorice) root, *Glycyrrhiza glabra* (licorice) root extract,

Glycyrrhiza glabra (licorice) root juice, *Glycyrrhiza glabra* (licorice) root powder.

Pharmacological Action

The present study was undertaken to estimate the acetylcholinesterase-inhibiting activity of extracts of *Glycyrrhiza glabra, Myristica fragrans* seeds, and ascorbic acid and compare these values with a standard acetylcholinesterase-inhibiting drug, metrifonate. Aqueous extract of *G. glabra* (150 mg/kg p.o. for 7 successive days), n-hexane extract of *M. fragrans* seeds (5 mg/kg p.o. for 3 successive days), ascorbic acid (60 mg/kg i.p. for 3 successive days), and metrifonate (50 mg/kg i.p.) were administered to young male swiss albino mice. Acetylcholinesterase enzyme was estimated in brains of mice. *G. glabra, M. fragrans*, ascorbic acid, and metrifonate significantly decreased acetylcholinesterase activity as compared with their respective vehicle-treated control groups (Dhingra et al., 2006).

Present work was done to investigate the antiinflammatory and antinociceptive activity of aqueous and ethanolic extracts of *Glycyrrhiza glabra* by using different pain models in swiss albino mice of either sex. Antinociceptive activity was evaluated at 50–200mg/kg i.p. in mice using various experimentally induced pain models like acetic acid induced abdominal constrictions, formalin induced hyperalgesia and tail flick method. From the results it was concluded that both extracts exhibited anti-nociceptive activity by central and peripheral mechanism (Bhandage et al., 2009).

The present study was undertaken to explore the free radical scavenging, anti-microbial and cytotoxic activity of the methanolic extract of *Glycyrrhiza glabra* (Fabaceae) using DPPH, disc diffusion and brine shrimp lethality bioassay methods respectively. In antimicrobial screening, *G. glabra* showed potent antimicrobial activity against almost all the test organisms except *Pseudomonas aeruginosa*. It exhibited highest sensitivity against *Staphylococcus aureus* with the zone of inhibition 22 mm. The extract possessed potent cytotoxic activity having LC_{50} value of 0.771μg/ml. On the other hand, the free radical scavenging activity was found moderate having IC_{50} value of 87.152 μg/ml (Sultana et al., 2010).

A randomized, double-blind, placebo-controlled study was conducted to evaluate the efficacy of GutGard, an extract of *Glycyrrhiza glabra*, in patients with functional dyspepsia. The patients received either placebo or Gut-Gard (75 mg twice daily) for 30 days. Efficacy was evaluated in terms of change in the severity of symptoms (as measured by 7-point Likert scale), the global assessment of efficacy, and the assessment of quality of life using the shortform Nepean Dyspepsia Index. In comparison with placebo, GutGard showed a significant decrease (P d" .05) in total symptom scores on day 15 and day 30, respectively. Similarly, GutGard showed marked improvement in the global assessment of efficacy in comparison to the placebo. GutGard was generally found to be safe and well-tolerated by all patients (Raveendra et al., 2011).

Toxicological Studies

Ethanol (30%) extract of the root, administered orally to mice of both sexes, produced LD_{50} 32.0 ml/kg. Water extract of the dried root (48–58% glycyrrhizin), administered intra peritoneally, orally and subcutaneously to mice and rats, produced LD_{50} 1.5 gm/kg, 16.0 gm/kg, and 4.2 gm/kg, respectively (Nalini Sofia and Thomas, Web article).

Glycyrrhizin (crude extract 48–58%)

LD_{50} values in rats and mice; LD_{50} s.c. 4–4.4 g/kg; LD_{50} i.p. 1.42–1.70 g/kg and LD_{50} oral 14.2–18.0g/kg (Lavekar and Padhi, 2001).

Clinical Investigation

Licorice extract (LE) can also be considered as an effective agent for the treatment of atopic dermatitis. It is expensive and thus used modestly in cosmetics. Creams containing whole licorice (often combined with extract of chamomile) are in wide use as "natural hydrocortisone creams." However, there is only preliminary supporting evidence for this use. In one double-blind, placebo-controlled trial of 30 people, licorice gel at 2% was more effective than placebo or 1% gel for reducing symptoms of eczema (Saeedi et al., 2003).

Liquiritin is another main active ingredient of licorice extract, and it appears to induce skin lightening by dispersing melanin. To see clinical results in melasma, studies demonstrated using 20% liquiritin cream applied at 1 g per day for 4 weeks (Zhu and Gao, 2008).

The current work aimed to formulate a stable w/o cream containing *Glycyrrhiza glabra* extract and studying its effects on skin pigment "melanin". Samples of base and formulation were stored at different accelerated conditions (8°C, 25°C, 40°C, 40°C + 75%RH) for four weeks to predict the stability of creams. Base and formulation were stable at all accelerated conditions regarding color, liquifaction and phase separation. Both base and formulation were applied to the cheeks of human volunteers for four weeks. Different parameters of human skin like melanin, erythema were monitored every week to measure any effect produced by these creams. Significant decrease in skin melanin was produced by formulation, whereas insignificant decrease in this skin pigment was observed by base (Akhtar et al., 2011).

Market Formulations

There are powdered and finely cut root preparations made for teas, tablets, and capsules, as well as liquid extracts. Different ayurvedic preparations are, yashtyadi churna, yashtyadi kwath, swadishta virechan churna, yashtimadhu churna, yashtimadhvadya taila, eladi gutika, kalyanavleha, angamara-daprashamana kashaya, brihat ashwagandha ghrita, brihachchhagaladya ghrita, shatavaryadi ghrita, nasika churna, guduchyadi taila, pippalyaditaila, vyaghri taila, kubjaprasarini taila, vridhihara lepa.

Godiva Licorice Skin Whitening Toner

It cleans and refreshes as it helps even out skin discolorations caused by age, pregnancy, scars or overexposure to the sun.

Ingredients: Glabridin from licorice extract. Also contains: Water, propylene glycol, sodium lactate, lactic acid, PEG-4, polysorbate 20, glycerine, methyl paraben, propyl paraben, fragrance, benzalkonium chloride, ethyl methane carboxamide, sodium erythrobate.

Action: The glabridin in licorice whitens by suppressing the formation of melanin, the pigment responsible for skin discoloration. Over time, the old layer of the pigmented skin is replaced by whiter, creamier skin.

Direction: Wash your face as usual. Moisten cotton pad with toner and stroke over face and neck. Use morning and evening for best results.

Supplied by: Godiva Inc. Philippines.

Rolanjona

It is a female facial mask sheet. It nourishes and repairs the skin of face and neck.

Ingredients: Water, aloe extract, alga extract, almond oil, liquorice extract, lactic acid, xanthic gum, spicery.

Action: It contains almond oil with powerful nourishing function and liquorice extract with whitening and repairing functions, nourishes and repairs skin, replenishes adequate water and nutrition into skin, balances oil secretion, assists to resist oxidation, promotes the formulation of collagen, firms skin, fades spots and keeps skin tender.

Direction: Take out the mask and spread it out on face with gently press with fingers to make it closely cling on face. Relax for 20–30 minutes, then discard the mask and wipe out the remaining liquid on face with soft paper, cotton or wet paper.

Supplied by: Guangzhou Youxi Cosmetics Co., Ltd. , China.

Himalaya Herbals Baby Lotion

A gentle lotion specially formulated to restore the tender skin's natural beauty, leaving the skin soft and supple. It is a prophylactic against skin blemishes due to its moisturizing, nourishing properties on an infant's tender skin.

Ingredients: Country mallow (*Sida cordifolia*); licorice (*Glycyrrhiza glabra*); tinospora gulancha

(*Tinospora cordifolia*); olive oil (*Olea europaea*); almond (*Prunus amygdalus*).

Action: It is enriched with almond oil that restores skin moisture, olive oil, which improves skin lustre, and yashtimadhu and country mallow, which soothes skin.

Direction: Apply all over the body after baby's bath.

Supplied by: Himalaya Drugs Pvt Ltd., India.

Rumalaya Gel

Rumalaya Forte helps to maintain the integrity of joints and muscles and connective tissue. It promotes normal strength and flexibility of the joints and muscles.

Ingredients: Boswellia/shallaki (*Boswellia serrata*) 240 mg; Indian bedellium/Guggul (*Commiphora wightii*) 200 mg; rasna (*Alpinia galanga*) 70 mg; licorice/yashti madhu (*Glycyrrhiza glabra*) 70 mg; small caltrops/gokshura (*Tribulus terrestris*) 60 mg; tinospora gulancha/guduchi (*Tinospora cordifolia*) 60 mg.

Action: Boswellia and Guggulu present in Rumalaya Forte are well known in worldwide medical and scientific communities for their proven efficacy in joint care. Guduchi in Rumalaya Forte is well known for its immunomodulatory activity. Rumalaya Forte also contains *Alpinia galanga, Glycyrrhiza glabra* and *Tribulus terrestris*-herbs that maintain flexibility and support connective tissue.

Direction: 2 capsules twice daily with meals.

Supplied by: Himalaya Drugs Pvt Ltd., India.

Various Preparations

It has various forms like Candy, capsules, teas, tinctures, salves, syrups. Preparations: capsules, teas, balms (salves), tinctures (root). For tea use 1 tbs. powdered licorice root in 1 cup of boiling water (let it step for 5 min). Drink after meals. It can also be used topically for skin infections and herpes/zoster blisters. For ulcers, capsules should be taken before meals. Liver formula (tincture or capsules): licorice root, milk thistle, beet leaf, artichoke leaf. Immune

formula (tincture or capsules): Licorice, garlic, myrrh, oregon grape,goldenseal. Joint and muscle formula (tincture or capsules): cramp bark, licorice, passion flower, turmeric root, valerian, willow bark; Digestive formula (tincture or capsules): licorice, slippery elm, mint; Allergy formula (tincture or capsules): licorice, coleus, Indian ipecac.

Storage of Products

All the products should be stored in well closed containers and away from sunlight.

Dosage

For Internal Application to Adults

Licorice can be taken in the following forms:

- *Dried root:* 1–5 g as an infusion or decoction (boiled), 3 times daily.
- *Licorice 1:5 tincture:* 2–5 ml, 3 times daily.
- *Standardized extract:* 250–500 mg, 3 times daily, standardized to contain 20% glycyrrhizinic acid.

Patents

Disclosed are cosmetic preparations and, more particularly, skin and hair care and cleaning products with an amount of at least one glycyrrhetic antibacterial agent selected from glycyrrhetic acid, glycyrrhetic acid derivatives and extracts of *Glycyrrhiza glabra*, and their salts, effective to inhibit or prevent the growth of bacteria. The cosmetic preparations may include other antibacterial or antibiotic agents. The glycyrrhetic antibacterial agent and, optionally, the antibacterial or antibiotic agents may be provided in microencapsulated form. Also disclosed is the use of such cosmetic preparations for treating acne (Buchwald-Werner, 2006).

There is a composition having at least one of the following active extracts *Butea frondosa, Naringi crenulata, Stenoloma chusana*, or any combinations thereof. There is also a composition having at least one of the following additional extracts *Azadirachta indica*,

Glycyrrhiza glabra, Morinda citrifolia, tomato glycolipid or any combinations thereof in combination with one or more of the active extracts. The compositions and methods of the invention are effective to lighten hair, skin, lips and/or nails (Mahalingam et al., 2007).

Composition for topical treatment of various forms of acne including (a) 20–40% by volume of an extract of a *Glycyrrhiza glabra* plant; (b) 15–30% by volume of an extract of a *Foeniculum officinale* plant; (c) 15–45% by volume of a saturated water solution of at least one mineral salt selected from the group of NaCl, KCl and $MgCl_2$; and (d) 1–3% by volume of an extract of a *Atriplex halimus* plant (Gonen, 2009).

The present invention pertains to a herbal formulation with highly potent wound healing properties, in humans and animals. The composition consists of aqueous extracts of *Azadirachta indica*, in a mixture of natural oils along with herbs viz. *Berberis aristata, Curcuma longa, Glycyrrhiza glabra, Jasminum officinale, Picrorhiza kurrooa, Pongamia pinnata, Rubia folia, Saussurea lappa, Terminalia chebula, Trichosanthes dioica, Capsicum* and *Stellata* wild in well-defined ratios. The invention also includes a process for preparing the formulation by extracting the water-soluble components from bark of *Azadirachta indica* (Saxena, 2010).

A topical composition for anti-aging treatment of the skin includes water soluble extract of *Uncaria* species, *Arabidopsis thaliana* extract, lipoic acid, dimethylethanolamine, tetrahexyldecyl ascorbate, dimethyl sulfoxide, *Glycyrrhiza glabra* (Licorice) root extract, methylsulfonylmethane, phytosterols, D-ribose, tocotrienol, tocopherol, glucosamine hydrochloride, *Pisum sativum* extract; *Bambusa vulgaris* extract, and a dermatologically acceptable liposomal delivery medium. Further disclosed is a method using the topical composition for anti-aging skin treatment utilizing DNA repair in both the nucleus and the mitochondria. The topical composition can be applied on the skin of a person daily, and can also be applied in a dermal infusion treatment, with or without micro-dermal abrasion or skin suction. Topical application of the composition achieves effective reduction of pigmentation spots and degree of winkles, and improvement of skin color tone (Giampapa, 2010).

BIBLIOGRAPHY

1. Akhtar N, Shoaib Khan HM, Iqbal A, Ali Khan B, Bashir S. *Glycyrrhiza glabra* extract cream: Effects on skin pigment "Melanin". International Conference on Bioscience, Biochemistry and Bioinformatics. 2011; IPCBEE Vol.5. 434–439.

2. Bhandage A, Shevkar K, Undale V. Evaluation of antinociceptive activity of roots of *Glycyrrhiza glabra* Linn. Journal of Pharmacy Research 2009; 2(5): 803–807.

3. Buchwald-werner S. Cosmetic preparations with anitacterial properties. United States Patent Application 20060057090. Application No. 10/521722, Publication date: 03/16/2006. http://www.freepatentsonline.com/y2006/0057090.html

4. Dhingra D, Parle M, Kulkarni SK. Comparative brain cholinesterase-inhibiting activity of *Glycyrrhiza glabra, Myristica fragrans,* ascorbic acid, and metrifonate in mice. J med Food. 2006; 9(2):281–283.

5. Giampapa VC. Topical composition for anti-aging skin treatment using dual DNA repair mechanism and method of use. United States Patent Application 20100291190. Application No. 12/781983, Publication date: 11/18/2010. http://www.freepatentsonline.com/y2010/0291190.html

6. Gonen S. Topical treatment of acne with combined herbal extracts and minerals. United States Patent 7498049. Application No. 11/619471, Publication date: 03/03/2009. http://www. freepatentsonline.com/7498049.html

7. http://en.wikipedia.org/wiki/Liquorice

8. http://enchantingkerala.org/ayurveda/ayurvedic-medicinal-plants/irattimadhuram.php

9. http://openmed.nic.in/3195/01/Glycyrrhiza_final.pdf

10. http://www.aminaherbs.com/product.php?id_product=133

11. http://www.cosmeticsinfo.org/ingredient_details.php? ingredient_ id=1935

12. http://www.drchealthandbeauty.com/health/97-herbs-naturopathy-licorice-or-liquorice-glycyrrhiza-glabra.html

13. http://www.indiawebmall.com/medicinal_herbs/mulathi.php

14. Lavekar GS, Padhi MM, Database on medicinal plants used in Ayurveda and Siddha. CCRAS (Central Council for Research in Ayurveda & Siddha) Dept. of Ayush. Govt of India. 2001; Vol 3. p.562–566.

15. Mahalingam H, Jones BC, Mccain N, Limtrakul P. Use of active extracts to lighten skin, lips, hair, and/or nails. United States Patent 7189419. Application No. 10/321706, Publication date: 03/13/2007. http://www.freepatentsonline.com/7189419.html

16. Nalini SH, Walter TM. Review of *Glycyrrhiza glabra*, Linn. http://openmed. nic.in/3195/01/Glycyrrhiza_final.pdf.

17. Raveendra KR, Jayachandra, Srinivasa V, Sushma KR, Allan JJ, Goudar KS et al., An extract of *Glycyrrhiza glabra* (Gut Gard) alleviates symptoms of functional dyspepsia: A randomized, doubleblind, placebo-controlled study. Evidence- Based Complementary and Alternative Medicine. 2011; 2012: Article ID 216970, 9 pages. doi:10.1155/2012/216970.

18. Saeedi M, Morteza-Semnani K, Ghoreishi MR. The treatment of atopic dermatitis with licorice gel. *J DermatologTreat*. 2003;14: 153–157.

19. Saxena M. Herbal formulation for wound healing. United States Patent Application 20100178367. Application No. 12/519137, Publication date: 07/15/2010. http://www.freepatentsonline. com/y2010/0178367.html

20. Sultana S, Haque A, Hamid K, Urmi KF, Roy S. Antimicrobial, cytotoxic and antioxidant activity of methanolic extract of *Glycyrrhiza glabra*. Agric. Biol. J. N. Am., 2010; 1(5): 957–960.

21. Zhu W, Gao J. The use of botanical extracts as topical skin-lightening agents for the improvement of skin pigmentation disorders. J Investig Dermatol Symp Proc 2008; 13: 20–4.

Hibiscus rosa sinensis

Plant profile

Kingdom: *Plantae* – Plants
Superdivision: *Spermatophyta* – Seed plants
Class: *Magnoliopsida* – Dicotyledons
Subkingdom: *Tracheobionta* – Vascular plants
Division: *Magnoliophyta* – Flowering plants
Subclass: *Dilleniidae*

Order: *Malvales*
Genus: *Hibiscus* L. – rosemallow

Family: *Malvaceae* – Mallow family
Species: *Hibiscus rosa sinensis* L. – shoeblackplant

Vernacular name

San: Japa, Ondrapuspi, Ragapuspi, **Eng:** China rose, Hibiscus Tagalog, **Hind:** Gudahal, Jasum, **Ben:** Joba, **Guj:** Jasud **Kan:** Dasaia Gida, **Mal:** Chemparatty, Jasund, **Mar:** Jaswad, **Tel:** Mandara; **Tam:** Chembarathi.

Common names: Chinese Hibiscus, Shoeblackplant, Tropical Hibiscus, Red Hibiscus.

Ayurvedic properties

Rasa: Kashaya, **Guna:** Lakhu, Rooksha, Shlasna, **Veerya:** Seeta.

INTRODUCTION

Hibiscus plants are among the showiest of flowering shrubs, often reaching 30 feet in nature. Roots are cylindrical of 5–15 cm length and 2 cm in diameter, off white in color light brown transverse lenticies. Its fracture is fibrous. Leaves are simple ovate or ovate-lancolate. Leaves are entire at the base and coarsely toothed at the apex. Taste is mucilagenous. Flowers are pedicillate, actinomorphic, pentamerous and complete. Corolla consists of 5 petals, red in colour and about 3 inches in diameter. The fruit (very rarely formed) is a capsule about 3 cm long.

Distribution

Hibiscus rosa sinensis is native to tropical Asia. A native of Southeastern Asia (China), the plant is commonly found throughout the tropics and as a house plant throughout the world. Most ornamental varieties are hybrids. The present wide range of cultivars is considered to be a complex of interspecific hybrids, between 8 or more different species originating from the African East Coast and islands in the Indian and Pacific Ocean.

Chemical Constituents

The main constituents are taraxeryl acetate, beta-sitosterol, campesterol, stigmasterol, cholesterol, erogosterol, lipids, citric, tartaric and oxalic acids, fructose, glucose, sucrose, flavonoids and flavonoid glycosides. hibiscetin, cyanidin and cyanin glucosides, alkanes.

Medicinal uses

Root is demulcent and used for cough. A decoction of root is used for venereal diseases and fevers. Fresh root juice is given for gonorrhoea and powdered root for menorrhagia. Leaves are emollient, aperient, anodyne and laxative. Leaves and stembark are used for abortion. Staminal column is diuretic used for kidney trouble. Flowers are astringent, demulcent, emollient, refrigerant, constipating, hypoglycaemic, aphrodisiac, emmenagogue and used for treating aloparia, burning sensation in the body, diabetes, and menstrual disorders. Buds are used in treatment of vaginal and uterine discharges. Leaves and flowers are good for healing ulcers and for promoting growth and colour of hair.

Cosmetic uses

In ayurvedic medicine, hibiscus petal was used to stimulate thicker hair growth and to prevent premature graying, hair loss and scalp disorders. It acts as a natural emollient hair conditioner and can be used in hair

Hibiscetin

Cyanidin

washes, treatments and vinegar rinses for the hair. Use it in combination with brahmi and amla. It has also been long used as a mild shampoo for babies. In ancient times it was used to produce perfumes and refreshing balms. Hibiscus extract visibly promotes even tone and texture to skin affected by cellulite. Hibiscus extract is used throughout Polynesia, Southeast Asia, and Central and South America for creating an infusion to cleanse, soften, and soothe baby's hair and scalp. Formulated with ultralight kukui oil to detangle and hibiscus extract to calm and seal the hair for maximum sheen. The flower extracts to prevent unwanted pregnancies at an early stage. Traditionally boiled the flowers and leaves of the Hibiscus, then mixed the infusion with herbal oil before applying it to their hair as a stimulant to the growth of luxurious tresses. The Hibiscus flower's juice as an ingredient in black dye for the hair and eyebrows, Indians include hibiscus flower juice in a famous herbal oil and conditioner which is now bottled and sold throughout eastern India under the brand name Jaba Kusam. One reason for the widespread popularity of this oil is its effectiveness against dandruff. This formulation is reputed to be particularly effective against dandruff. Traditional use of the flowers and leaves in India include burning them in ghee to produce a black dye used to blacken eyes and eyebrows.

Pharmacological Action

The assessment of immunomodulatory activity of hydro-alcoholic extract of flowers of *Hibiscus rosa sinensis* Linn. (75, 150 and 300 mg/kg, p.o.) was done by carbon clearance method for non-specific immunity, haemagglutination antibody titre method for humoral immunity and footpad swelling method for cell mediated immunity on wistar albino rats. Results of the studies suggest that the hydro-alcoholic extract of *Hibiscus rosa sinensis* Linn was found to possess significant immunostimulatory action on immune system in dose dependent manner when compared with control group (Gaur et al., 2009).

The present study was undertaken to separate mucilage from leaves of *Hibiscus rosa sinensis* and explore its use as tablet disintegrant. Mucilage extracted from the pods of *Hibiscus rosa sinensis* were subjected to toxicity studies for its safety and preformulation studies for its suitability as a disintegrating agent. Eight formulations were prepared and evaluated for physical parameters such as thickness, hardness, friability, weight variation, drug content, disintegration time and drug dissolution. The formulated tablets had good appearance and better drug release properties. The study revealed that *Hibiscus rosa sinensis* mucilage powder was effective as disintegrant in low concentrations (4%) (Shah and Patel, 2010).

The effect of 200 mg/kg of the aqueous leaf extract of *H. rosa sinensis* on the renal function of hypertensive rats was investigated. The administration of extract shows a significant ($p < 0.05$) increase in the Na^+ level of normotensive rats, thus it may interfere with the normal function of the kidney and hence produce increased salt retention. The results had shown that although *H. rosa sinensis* leave extract reduce blood pressure, the integrity of the kidney may be compromised when this plant is used for the treatment of hypertension (Kate, and Lucky, 2010).

The study was aimed at evaluating the antioxidant activity of the ethanolic extract of *Hibiscus rosa sinensis* flowers at varying concentration *in vitro*. The extract was found to contain large amount of phenolic compounds and flavonoids. Additionally, the reducing power (capacity to reduce Fe^{3+} to Fe^{2+}), the capacity to scavenge hydrogen peroxide, superoxide radicals and nitric oxide were evaluated. The extract exhibited a concentration dependent scavenging activity (Bhaskar et al., 2011).

The antioxidant properties of crude extract of *Hibiscus rosa sinensis* were examined using different *in vitro* analytical methodologies, such as total antioxidant activity determination by ferric thiocyanate, hydrogen peroxide scavenging, 1,1-diphenyl-2-picrylhydrazyl free radical (DPPH) scavenging, 2,22-azino-bis(3-ethylbenzthiazoline-6-sulfonic acid) (ABTS + radical cation) radical cation scavenging activity and superoxide anion radical scavenging by riboflavin-methionine-illuminate system. The crude extract inhibited 94.58% peroxidation of linoleic acid emulsion at 20 µg/ml concentration and finally concluded that the crude extract is an effective natural antioxidant component (Mandade et al., 2011).

The antioxidant activities of *Hibiscus rosa sinensis* Linn. flower petals treatment for 4 weeks were screened in 10% D-glucose feeding in rats. Experimental rats were treated with 10% D-glucose solution to drink orally. The decreased activities of key antioxidant enzymes such as superoxide dismutase, catalase, glutathione peroxidase, and nonenzymatic antioxidants such as glutathione, vitamin E and vitamin C in 10% D-glucose-fed rats were brought back to near-normal range upon *H. rosa sinensis* flower petals treatment. Finally it was concluded that *H. rosa sinensis* exhibited preventive and curative effects in 10% D-glucose-induced oxidative stress in rat heart tissues (Bhuvana et al., 2011).

Toxicological Studies

The acute oral toxicity study of *H. rosa-sinensis* was carried out as per guidelines set by Organisation for Economic Cooperation and Development (OECD). The median lethal dose of the pet-ether (40–60°C), chloroform, alcohol and aqueous was determined by orally administering the extracts in increasing dose levels of 1, 2, 3, 4 and 5 g/kg body weight to healthy adult *Wistar* albino rats of either sex. The animals will be observed continuously for 2 h under behavioural profile, neurological profile and autonomic profiles. After a period of 24 h they will be observed for any lethality or death (% of mortality). All extracts were found to be safe in the doses used and there

was no mortality up to a dose of 200 mg/kg, b.w. Because of this the LD_{50} for pet-ether, chloroform, alcohol and aqueous extracts were considered as 2 g/kg body weight and the ED_{50} (1/10th dose LD_{50}) for these extracts was taken as 200 mg/kg (Birari et al., 2010).

Acute toxicity studies of the hydroalcoholic extract of *Hibiscus rosa sinensis* L. (HRS) in swiss albino mice was performed according to OECD guidelines and shows no signs and symptoms such as restlessness, respiratory distress, diarrhea, convulsions and coma and it was found safe up to 5000 mg/kg (Kandhare et al., 2011).

Clinical Investigation

Petroleum ether extract of leaves and flowers of *Hibiscus rosa sinensis* was evaluated for its potential on hair growth by *in vivo* and *in vitro* methods. *In vivo*, 1% extract of leaves and flowers in liquid paraffin was applied topically over the shaved skin of albino rats and monitored and assessed for 30 days. The length of hair and the different cyclic phases of hair follicles, like anagen and telogen phases, were determined at different time periods. From the study it is concluded that the leaf extract, when compared to flower extract, exhibits more potency on hair growth (Adhirajan et al., 2003).

This study was an open, non-comparative, non-randomized, phase III clinical trial. Twenty-five patients of both sexes, from the age group of 20–45 years, who were clinically diagnosed as suffering from mild to moderate dandruff. All patients were advised to apply 10 ml of "Anti-Dandruff Hair Oil", twice daily for a period of 2 weeks with gentle massage to the entire scalp and were advised to leave the "Anti-Dandruff Hair Oil" on the scalp for a contact period of minimum 3–4 hours after application. All patients were followed for a period of 2 weeks. Clinical assessment of scalp lesions was done objectively and also subjectively. Thorough scalp examination was done after the completion of 1 week and at the end of the study. The severity of the dandruff symptom was recorded on a score scale and patients, with help of a linear analogue scale, did the subjective assessment. The predefined primary efficacy endpoints were rapid clinical improvement and sym-ptomatic control of dandruff. There was a highly significant reduction (p < 0.001) in the mean score for itching and white scales at the end of 2 weeks. In subjective evaluation, majority of patients experienced remarkable overall improvement. There were no clinically significant adverse reactions, either reported or observed, during the entire study period and overall compliance to the treatment was excellent (Vyjayanthi et al., 2004).

Hair formulation of *Eclipta alba* Hassk [Asteraceae] 10% w/v, *Hibiscus rosa sinensis* Linn [Malvaceae] 10 % w/v, *Nardostachys jatamansi* [Valerianceae] 5 % w/v concentration in the form of herbal oil were prepared. Three months evaluation of a herbal hair oil vs coconut oil was conducted on human volunteers with hair fall problem. Combining assay was performed to evaluate the efficacy of the herbal hair oil. The findings of the study show that the test oil was effective in reducing the hair fall (Thorat et al., 2009).

The study was aimed to investigate the efficacy of ethanolic extract of *H. rosa sinensis* flower as hair growth promoter. Female wistar rats were selected for hair growth promotion studies. 2% solutions of vehicle (control), minoxidil (standard) and ethanolic extract of *H. rosa sinensis* flowers respectively were applied on shaved denuded skin of different groups of rats twice a day for thirty days. During this period they were observed visually for pattern of hair growth studies and after treatment period their skin biopiosis were taken for follicular density and cyclic phases of hair growth. On the basis of visual observation of animals and histopathology, ethanolic extract of *H. rosa sinensis* flowers showed shorter hair and take more time for growth and favours telogenic stage of hair follicles as compared to control, thus it showed hair growth retarding activity in spite

of hair growth promoting one (Upadhyay et al., 2011).

Market Formulations

Dabur Vatika Nourishing Hair Conditioner

It helps hair set properly and gives extra shine. It makes hair smooth and untangled for easy combing. Reduces split hair and gives natural colour to the hair.

Ingredients: Lawsonia enermis—excellent conditioner, adds body, shine and colour, hair restorative; *Hibiscus rosa sinensis*—conditioner, promotes hair growth, gives shine; *Acacia concinna*—natural cleanser and conditioner.

Action: After-shampoo care enriched with herbal ingredients like soap fruit, henna and Hibiscus petals, for gentle conditioning and adding life to dull hair. Soap fruit, one of the best natural conditioners, cleans the hair and scalp gently and adds lustre. Hibiscus restores moisture balance, making your hair soft and healthy, while henna adds body and gives natural coloring to your hair.

Direction: Apply over wet hair and scalp. Massage shampoo into scalp with fingertips and work up a rich lather. Rinse thoroughly and repeat, if necessary.

For best results, lather well and leave for almost a minute and then rinse.

Storage: Keep at a cool, dark place.

Supplied by: Dabur India Limited, India.

Herbal Hair Conditioner

It is 100% natural and it nourishes and strengthens for full, shiny, healthy hair. Feed your hair a healthy, balanced diet and strengthen hair fiber, repair damage from salon treatments and restore softness, manageability and shine.

Ingredients: Extracts of Hibiscus (*Hibiscus rosa sinensis*), Gotu kola (*Centella asiatica*), liquid paraffin, fatty alcohol, fatty ester, coco amide propyl betaine, poly 7, silicone emulsion, purified water, preservative, fragrance, color.

Direction: Apply to wet hair and scalp after washing with herbal daily wash shampoo and rinse after 2–3 minutes.

Supplied by: Homeshoppedirect Merchandizing Private Limited, New Delhi, India.

Gentle Baby Shampoo

It is a specially formulated mild shampoo that gently cleanses the hair making it soft, shiny and easy to manage. Does not cause tears in the babies.

Ingredients: Rice (*Oryza sativa*); bengal gram (*Cicer arietinum*); shoe-flower (*Hibiscus rosa sinensis*); vetiver, khas-khas (*Vetiveria zizanioides*).

Action: Hibiscus extract softens, conditions the hair, natural proteins from chickpea give shine and healthy shaft, and khus-khus cools the scalp.

Direction: Wet hair. Apply shampoo. Lather and rinse. Repeat, if required.

Supplied by: Himalaya Drug Company, India.

Storage of Products

All the products should be stored in well closed containers and away from sunlight.

Dosage

Depends on toxicity study, the tropical doses of the leaves and flower extracts in formulation have to be selected.

Patents

Recent patents on *Hibiscus rosa sinensis* as in cosmetic formulations are not available.

BIBLIOGRAPHY

1. Adhirajan N, Kumar TR, Shanmugasundaram N, Babu M. *In vivo* and *in vitro* evaluation of hair growth potential of *Hibiscus rosa sinensis* Linn. Journal of Ethnopharmacology. 2003; 88: 235–239.

2. Bhaskar A, Nithya V, Vidhya VG. Phytochemical screening and *in vitro* anti-oxidant activities of the ethanolic extract of *Hibiscus rosa sinensis* L. Annals of Biological Research., 2011; 2 (5): 653–661.

3. Bhuvana S, Mahesh R, Hazeena begum VM. Antioxidant activities of *Hibiscus rosa sinensis* Linn during 10% D-glucose feeding in rat heart tissues. Journal of Food Biochemistry. 2011; 35 (3): 792–802.

4. Birari RB, Singh A, Giri IC, Saxena N, Shaikh MI, Singh A. Evaluation of anticonvulsant activity of *Hibiscus rosa sinesis* flower extracts. IJPSR. 2010; 1(5): 83–88.

5. Gaur K, Kori ML, Nema RK. Comparative screening of immunomodulatory activity of hydro-alcoholic extract of *Hibiscus rosa sinensis* Linn. and ethanolic extract of *Cleome gynandra* Linn. Global Journal of Pharmacology. 2009: 3 (2): 85–89.

6. http://ayurvedicmedicinalplants.com/plants/1569.html

7. http://www. filipinoherbshealingwonders. filipinovegetarianrecipe.com/gumamela. htm

8. http://www.naturalcosmeticsupplies. com/hibiscus-flower.html

9. http://www.tnsmpb.tn.gov.in/images/HIBISCUS%20ROSA% 20SINENSIS.pdf

10. Kandhare AD, Raygude KS, Ghosh P, Ghule AE, Gosavi TP, Badole SL, Bodhankar SL. Effect of hydroalcoholic extract of *Hibiscus rosa sinensis* Linn. Leaves in experimental colitis in rats. Asian Pacific Journal of Tropical Biomedicine. Accepted 27 September 2011. Available online 28 March 2012. Journal homepage:www.elsevier. com/locate/apjtb.

11. Kate IE, Lucky OO. The effects of aqueous extracts of the leaves of *Hibiscus rosa sinensis* Linn. on renal function in hypertensive rats. African Journal of Biochemistry Research. 2010: 4(2); 43–46.

12. Mandade R, Sreenivas SA, Sakarkar DM, Choudhury A. Radical scavenging and antioxidant activity of *Hibiscus rosa sinensis* extract. African Journal of Pharmacy and Pharmacology. 2011: 5(17); 2027–2034.

13. Shah V, Patel R. Studies on mucilage from *Hibiscus rosa sinensis* Linn as oral disintegrant. International Journal of Applied Pharmaceutics. 2010; 2(1): 18–21.

14. Thorat R, Jadhav V, Kadam V, Sathe N, Save A, Ghorpade V. Evaluation of a herbal hair oil in reducing hair fall in human volunteers. IJPRD/2009/PUB/ARTI/VOL-6/AUG/001.

15. Upadhyay SM, Upadhyay P, Ghosh AK, Singh V, Dixit VK. Effect of ethanolic extract of *Hibiscus rosa sinensis* L., flowers on hair growth in female wistar rats. Der Pharmacia Lettre. 2011; 3 (4) 258–263.

16. Vyjayanthi G, Kulkarni C, Abraham A, Kolhapure SA. Evaluation of anti-dandruff activity and safety of polyherbal hair oil: An open pilot clinical trial. The Antiseptic. 2004; 101(9): 368–372.

Juglans regia

Plant profile

Kingdom: *Plantae* – Plants

Superdivision: *Spermatophyta* – Seed plants

Class: *Magnoliopsida* – Dicotyledons

Order: *Juglandales*

Genus: *Juglans* L. – Walnut

Subkingdom: *Tracheobionta* – Vascular plants

Division: *Magnoliophyta* – Flowering plants

Subclass: *Hamamelididae*

Family: *Juglandaceae* – Walnut family

Species: *Juglans regia* L. – English walnut

Vernacular name

San: Akschota, **Eng:** Walnut, **Hin:** Akhor, Akhrot, **Ben:** Akrot; Bosnian, **Guj:** Akrot, Akharot, **Kan:** Acrota **Mal:** Akrot, Akshotaka vriksham, **Tam:** Akrottu, Akroot Kai, **Mar:** Akrod, **Oriya:** Akhoot, **Tel:** Akroot Kaya, Nattua Krot Vitthu

Common name

Walnut.

Ayurvedic properties

Rasa: Madhura, Kashaya, **Guna:** Guru, **Veerya:** Ushna.

INTRODUCTION

Juglans regia is a large deciduos tree attaining heights of 25–35 m, and a trunk up to 2 m diameter, commonly with a short trunk and broad crown, though taller and narrower in dense forest competition. It is a light-demanding species, requiring full sun to grow well. The bark is smooth, olive-brown when young and silvery-grey on older branches, and features scattered broad fissures with a rougher texture. Like all walnuts, the pith of the twigs contains air spaces. This chambered pith is brownish in color. The leaves are alternately arranged, 25–40 cm long, odd-pinnate with 5–9 leaflets, paired alternately with one terminal leaflet. The largest leaflets are three at the apex, 10–18 cm long and 6–8 cm broad; the basal pair of leaflets are much smaller, 5–8 cm long with the margins of the leaflets entire. The male flowers are in drooping catkins 5–10 cm long, and the female flowers terminal, in clusters of two to five, ripening in the autumn into a fruit with a green, semi-fleshy husk and a brown corrugated nut. The whole fruit, including the husk, falls in autumn; the seed is large, with a relatively thin shell, and edible, with a rich flavor.

Distribution

It is native to the region stretching from the Balkans eastward to the Himalayas and southwest China. The largest forests are in Kyrgyzstan, where trees occur in extensive, nearly pure walnut forests at 1,000–2,000 m altitude. The plant is also extending from Xinjiang province of western China, parts of Uzbekistan and Southern Kirghizia and from lower ranges of mountains in Nepal, Bhutan, Tibet, northern India and Pakistan, through Afghanistan, Turkmenistan and Iran to portions of Azerbaijan, Armenia, Georgia and Eastern Turkey.

Chemical Constituents

Six compounds obtained from the 70% ethanol extract were identified as 1, 2, 3, 4, 6-penta-O-galloyl-3-D-glucose, rugosin C, 1, 2, 3, 6-tetra-O-galloyl-3-D-glugose, tellimagrandin II, casuarictin, 1-degalloylrugosin F. The walnut oils contains oleic acid ranged between 15.9 and 23.7% of total acids, while linoleic acid content ranged from 57.2 to 65.1% and the linolenic acid from 9.1 to 13.6%. Sterols like beta-sitosterol, delta(5)-avenasterol, and campesterol were also found.

Medicinal uses

The cotyledons are used in the treatment of cancer. Walnut has a long history of folk used in the treatment of cancer, some extracts from the plant have shown anticancer activity. The seeds are antilithic, diuretic and stimulant. They are used internally in the treatment of low back pain, frequent urination, weakness of both legs, chronic cough, asthma, constipation due to dryness or anaemia and stones in the urinary tract. Externally, they are made into a paste and applied as a poultice to areas of dermatitis and eczema. The leaves are alterative, anthelmintic, anti-inflammatory, astringent and depurative. They are used internally for the treatment of constipation,

Casuarictin

Oleic acid

chronic coughs, asthma, diarrhoea, dyspepsia, etc. The leaves are also used to treat skin ailments and purify the blood. They are considered to be specific in the treatment of strumous sores. Male inflorescences are made into a broth and used in the treatment of coughs and vertigo. The rind is anodyne and astringent. It is used in the treatment of diarrhoea and anaemia. The oil from the seed is anthelmintic. It is also used in the treatment of menstrual problems and dry skin conditions. The bark and rootbark are anthelmintic, astringent and detergent.

Cosmetic uses

The leaves are also used as a blood purifier and in cure of various skin diseases. The bark is finely powdered and used to prevent bleeding gums and as a mouth rinse. The leaves and bark are antiscorbutic and detergent, and are useful in herpes, eczema, scrofula and syphilis. The walnut shells are ideal as the gritty, rough agent in soap cosmetics and dental cleansers. Walnut scrub contains ground apricot seeds and crushed walnut shells to exfoliate and *aloe vera* to moisturize and soothe. Walnut scrub spreads easily over the skin into a dense lather. Walnut oil is used in the healing of wounds and skin problems. It may also help eliminate warts, and when rubbed on the skin, walnut is reputed to be beneficial for eczema, herpes, psoriasis, and skin parasites. The oil has traditionally been used externally in the treatment of gangrene, leprosy, and wounds. Walnut in a massage blend has great emollient qualities and is a good choice to include when mixing a massage blend or preparing a carrier base. Walnut oil presents good moisturizing, anti-aging, and regenerative and toning properties, and can be used in all anti-wrinkle products and creams for dry, normal, and mature skin, body and hygiene products, massage blends, and lip balms, etc.

Pharmacological Action

The study was revealed the antinociceptive, anti-inflammatory and acute toxicity effects of *Juglans Regia* L. leaves in mice. The acute (intraperitoneally) toxicity was evaluated for 2 days. Antinociceptive activities were done using hot-plate and writhing tests. Anti-inflammatory effects were studied using xylene induced ear edema and cotton pellet tests. The LD_{50} values of *J. regia* aqueous and ethanolic extracts were 5.5 and 3.3 g/kg, respectively. The aqueous (2.87 and 1.64 g/kg) and ethanolic (2.044 and 1.17 g/kg) extracts showed antinociceptive activity in hotplate test. The extract exhibited antinociceptive activity in writhing test, which was not blocked by naloxone. The extracts were also showed anti-inflammatory activity against the chronic inflammation (Hosseinzadeh et al., 2011).

The present study was designed to evaluate the protective effects of *Juglans regia* kernel extract against cigarette smoke extract (CSE)-induced lung toxicities in wistar rats. Lung injury markers lactate dehydrogenase (LDH) activity, total cell count, total protein and reduced glutathione (GSH) in bronchoalveolar lavage fluid (BALF) were evaluated. Glutathione reductase (GR), xanthine oxidase (XO) and catalase activities were measured in lung tissue. *J. regia* extract significantly decreased the levels of LDH, total cell count, total protein and increased the GSH level in BALF, it also significantly restored the levels of GR, catalase and reduced the XO activity in lung tissue. On the basis of these results, protective role of *J. regia* extract against CSE-induced acute lung toxicity in wistar rats is suggested (Qamar and Sultana, 2011).

The study deals with evaluation of the effect of acetone and aqueous extracts of *J. regia* L., a traditional medicinal plant. The efficacy of the plant extracts has been assessed by testing on salivary samples of patients suffering from dental carries. Antimicrobial assay was carried out using disc diffusion method. Acetone extract was found to be more effective as anti-cariogenic medicine. Chlorhexidine was used as a standard (Deshpande et al., 2011).

An anthelmintic activity of different extracts of stem bark of *J. regia* L was developed. Different concentrations of ethyl acetate, acetone, ethanol, methanol and aqueous extracts of the plant material were tested against *Eicinia foetida* as test worms. The bioassay involved determination of the time of paralysis and time of death control. Albendazole was included as standard reference and normal saline as control. The results of this study indicate that the crude acetone and methanolic extracts significantly demonstrated paralysis and also caused death of worms in dose dependent manner (Kale et al., 2011).

The study was undertaken to investigate the effect of root bark of *Juglans regia* (RBJR) organic extracts on cell proliferation, and to determine the molecular mechanism of RBJR-induced cell death by determining the expression of Bcl-2, Bax, caspases, Tp53, Mdm-2 and TNF-α in MDA-MB-231 human breast cancer cells. The results demonstrate that WNRB suppressed proliferation and induced apoptosis in a dose and time dependent manner by modulating expression of key genes. This involved characteristic changes in cytoplasmic and nuclear morphology, DNA fragmentation (TUNEL assay), levels of mRNA and expression of corresponding proteins. As a conclusion the presence of bioactive compound(s) in WNRB capable of killing breast carcinoma cells through induction of apoptosis, and therefore a candidate source of anticancer drugs (Hasan et al., 2011).

Toxicological Studies

The acute (intraperitoneally) toxicity was evaluated for 2 days. Antinociceptive activities were done using hot-plate and writhing tests. Anti-inflammatory effects were studied using xylene induced ear edema and cotton pellet tests. The LD_{50} values of *J. regia* aqueous and ethanolic extracts were 5.5 and 3.3 g/kg, respectively (Hosseinzadeh et al., 2011).

Clinical Investigation

Chronic and acute effects of walnuts on antioxidant capacity and nutritional status in humans were evaluated on selective human. A 19-week, randomized crossover trial was conducted in 21 generally healthy men and postmenopausal women age 50 years to study the dose-response effects of walnut intake on biomarkers of antioxidant activity, oxidative stress, and nutrient status. Results indicated red blood cell (RBC) linoleic acid and plasma pyridoxal phosphate (PLP) were significantly higher after 6 week with 42 g/d walnuts as compared to baseline levels. Overall, changes in plasma total thiols, and other antioxidant biomarkers, were not significant with either walnut dose. Further it concluded that walnut consumption did not significantly change the plasma antioxidant capacity of healthy, well-nourished older adults in this pilot study (McKay et al., 2010).

The bark extracts of *Mimusops elengi* (bakul) and *Juglans regia* (walnut), each in aqueous and acetone solvents were evaluated and compared for antibacterial activity against salivary microflora. The salivary samples were collected from children of 6–12 years of age with moderate caries. Antibacterial assay was carried out using paper disc diffusion method. The antibacterial potential of both the plants was confirmed. The acetone extract of *J. regia* showed highest zones of inhibition and indicated its use as a potent antibacterial agent. Thus, this in vitro study supports its folklore application as a preventive remedy for microbial diseases of hard and soft tissues in the oral cavity (Kulkarni et al., 2011).

Market Formulations
Juglans Regia (Walnut) Shell Powder

Crushed walnut shell is a hard fibrous material ideal as abrasive. Its grit is extremely durable, angular and multi-faceted, yet considered a soft abrasive. Very fine particle size (through 40 mesh, on 100 mesh). Fine light-brown powder, characteristic odour. Insoluble in water.

Ingredient: Prepared from finely crushed walnut shells.

Action: Prepared from finely crushed walnut shells, this material acts as gentle exfoliating agent which assists the surface removal of dead skin cells.

Method of preparation: Crushed walnut shell is a hard fibrous material ideal as abrasive. Its grit is extremely durable, angular and multi-faceted, yet considered a soft abrasive. Very fine particle size (through 40 mesh, on 100 mesh). Fine light-brown powder, characteristic odor. Insoluble in water.

Application: Facial scrubs, skin cleansers, peeling creams and lotions.

Supplied by: Making Cosmetics Inc. Renton WA.

Himalaya Fairness Cream

This cream works throughout the day, making complexion better, each time, every time. It clears dark spots, discolored and pigmented skin and suitable for all skin types.

Ingredients: Rosa centifolia (Persian rose, satapatri), *Citrus reticulata* (mandarin orange, narangi), *Juglans regia* (walnut, akschota), *Aloe vera* (barbados aloe, kumari).

Action: Fairness cream is a 100% natural fairness cream that improves complexion, nourishes and makes the skin soft. A unique formulation with natural ingredients, it ensures ease of application and faster absorption, which provides better nourishment to the skin.

Direction: Clean the face and neck thoroughly with gentle face wash gel/cream. Apply fairness cream all over the face and neck in upward circular motions, twice a day. It is for external use only.

Supplied by: Himalaya Drug Company, India.

Himalaya Gentle Exfoliating Walnut Scrub

Himalaya walnut scrub is a herbal ayurvedic formulation. This gel cleanses, nourishes and moisturizes the skin to make it soft, fresh and radiant.

Ingredients: Walnut, crab apple and wheat germ oil.

Action: Pyrus malus (apple, seva) possess antiseptic, cooling, soothing, nutritive and keratolytic properties. It reduces skin inflammation and helps remove dead skin fragments. *Triticum sativum* (wheat, godhuma) seeds are cooling and oleaginous. Wheat germ oil is a good source of natural vitamins and prevents loss of moisture from the skin. *Juglans regia* (walnut, akschota) its shell acts as exfoliating micro-particles and contains organic acids that are astringent and antiseptic.

Direction: Moisten face and neck and apply, avoiding area around eyes. Massage gently with fingertips in a circular motion for a minute. Wash off and pat dry. Use at least once a week for a glowing, healthy complexion.

Supplied by: Himalaya Drug Company, India.

Various Preparations

Juglans regia (walnut) shell powder is used in the following product types: Exfoliant/scrub; facial cleanser; body wash/cleanser; acne treatment; foot cleansing; sunless tanning; foot odor control; moisturizer; foot moisturizer; anti-itch/rash cream.

The leaves or green walnut shells from walnuts can be combined with henna to create brown hair dye. Walnut dye is juglone, 5-hydroxy-1,4-naphthoquinone. Walnut is often mixed into henna powder for skin. It makes a very fast dark stain.

Storage of Products

All the products should be stored in well closed containers and away from sunlight.

Dosage

It can be administered as decoctions and other galenic preparations for external use. To prepare a decoction, soak 2 teaspoonfuls of drug in 1 cup of water, boil and strain. An infusion is prepared by using 1.5 g of finely cut drug, soaked in cold water, brought to simmer and strained after 3 to 5 minutes. The average daily dose for external use is 3 to 6 g of drug.

Patents

Liquid lamellar cleansing compositions are described that contain synthetic surfactants, hydrophilic emollients and exfoliant particles where 80% or more of the particles have a major axis length of between 100 and 1000 microns. The combination of the mild surfactants, moisturizers, and exfoliants provides the user with simultaneous moisturization and exfoliation in a convenient liquid cleansing product (Massaro et al., 2005).

The present invention provides a herbal black dye from natural materials comprising of *Juglans regia, Indigofera tinctoria, Terminalia chebula, Acacia accocina, Lawsonia inermis, Trigonella foenum-graecum, Sapindus mukorossi, Eclipta alba, Emblica officinalis, Acacia catechu* and *Piper betle*, the dye derived being safe, non-toxic, anti-allergic, anti-dandruff and free from toxic symptoms like itching (Palpu et al., 2006).

The disclosure provides oral compositions having at least two botanical active ingredients derived from plants. The oral composition also includes an orally acceptable vehicle to deliver an effective amount of the at least two active ingredients *in vivo*. The botanical active ingredients provide particularly efficacious antimicrobial (antibacterial, antiviral, and/or antifungal), antioxidant, anti-inflammatory, anti-ageing, and or healing properties to the oral compositions (Cummins, 2009).

BIBLIOGRAPHY

1. Cummins D. Oral compositions containing botanical extracts. United States Patent Application 20090087501. Application no. 12/243669. Publication date: 04/02/2009. http://www.freepatentsonline.com/y2009/0087501.html

2. Deshpande RR, Kale AA, Ruikar AD, Panvalkar PS, Kulkarni AA, Deshpande NR et al. Antimicrobial activity of different extracts of *Juglans regia* L. against oral microflora. International Journal of Pharmacy and Pharmaceutical Sciences. 2011; 3(2): 200–201.

3. Hasan TN, Grace L, Shafi G, Al-Hazzani AA, Alshatwi AA. Anti-proliferative effects of organic extracts from root bark of *Juglans Regia* L. (RBJR) on MDA-MB-231 human breast cancer cells: Role of Bcl-2/Bax, Caspases and Tp53. Asian Pacific Journal of Cancer Prevention. 2011; 12: 525–530.

4. Hosseinzadeh H, Zarei H, Taqhiabadi E. Antinociceptive, anti-inflammatory and acute toxicity effects of *Juglans regia* L. leaves in mice. Iranian Red Crescent Medical Journal, 2011; 13 (1): 27–33.

5. http://ayurveda.ygoy.com/2010/09/11/juglans-regia-medicinal-uses/

6. http://ayurvedicmedicinalplants.com/plants/51.html

7. http://drclarkia.com/juglans_regia. htm

8. http://en.wikipedia.org/wiki/Juglans_regia

9. http://www.naturalcosmeticsupplies.com/walnut-shells-fruit.html

10. http://www.purematters.com/herbs-supplements/w/walnut#dosage

11. Kale AA, Gaikwada SA, Kamblea GS, Deshpandea NR, Salvekara JP. *In vitro* anthelmintic activity of stem bark of *Juglans regia* L. J. Chem. Pharm. Res. 2011; 3(2): 298–302.

12. Kulkarni AA, Deshpande RR, Panvalkar P, Mahajan P, Kale A, Ruikar A, Deshpande NR. Comparative evaluation of antibacterial properties of different extracts of *Mimusops elengi* (bakul) and *Juglans regia* (walnut) against salivary microflora. Research Journal of Pharmaceutical, Biological and Chemical Sciences. 2011; 2(3): 635–642.

13. Massaro M, Goldberg JW, Subramanyan KK, Johnson AW, Slavtcheff CS, Gencarelli SL. Liquid cleansing composition having simultaneous exfoliating and moisturizing properties. United States Patent Application 20050170979. Application no. 11/091636. Publication date: 08/04/2005. http://www.freepatentsonline. com/y2005/0170979. html

14. McKay D, Chen CYO, Yeum Kyung J, Matthan N, Lichtenstein A, Blumberg J. Chronic and acute effects of walnuts on antioxidant capacity and nutritional status in humans: a randomized, crossover pilot study. Nutrition Journal. 2010; 9(1): 21.

15. Palpu P, Pal M, Dixit BS, Banerji R, Rao CV. Herbal dye and process of preparation thereof. United States Patent Application 20060143838. Application no. 11/027575.

Publication date: 07/06/2006. http://www.freepatentsonline.com/y2006/0143838.html

16. Qamar W, Sultana S. Polyphenols from *Juglans regia* L. (walnut) kernel modulate cigarette smoke extract induced acute inflammation, oxidative stress and lung injury in wistar rats. Hum Exp Toxicol. 2011; 30(6):499–506.

Luwsonia inermis

Plant profile

Kingdom: *Plantae* – Plants

Superdivision: *Spermatophyta* – Seed plants

Class: *Magnoliopsida* – Dicotyledons

Order: *Myrtales*

Genus: *Lawsonia* L. – Lawsonia

Subkingdom: *Tracheobionta* – Vascular plants

Division: *Magnoliophyta* – Flowering plants

Subclass: *Rosidae*

Family: *Lythraceae* – Loosestrife family

Species: *Lawsonia_inermis* L. – Henna

Vernacular name

San: Mendika, Rakigarbha, Nakharanjani, **Eng:** Henna, **Hin:** Mehndi, **Ben:** Mehedi, **Kan:** Mayilanchi, **Mal:** Mailanchi, **Tam:** Muruthani, **Mar:** Mendi, **Oriya:** Benjati, **Tel:** Gorintaaku.

Common name

Henna, Al-Khanna, Al-Henna, Jamaica Mignonette, Mehndi, Mendee, Egyptian Privet, Smooth Lawsonia.

Ayurvedic properties

Rasa: Tikta, Kashaya, **Guna:** Lakhu, Rooksha, **Veerya:** Seeta, **Vipaka:** Katu.

INTRODUCTION

Lawsonia inermis is a glabrous branched shrub or small tree (2 to 6 m in height). Leaves are small, opposite, entire margin elliptical to broadly lanceolate, sub-sessile, about 1.5 to 5 cm long, 0.5 to 2 cm wide, greenish brown to dull green, petiole short and glabrous acute or obtuse apex with tapering base. New branches are green in color and quadrangular, turn red with age. Young barks are greyish brown, older plants have spine-tipped branchlets. Inflorescence has large pyramid shaped cyme. Flowers are small, numerous, aromatic, white or red colored with four crumbled petals. Calyx has 0.2 cm tube and 0.3 cm spread lobes. The fruits are small, brown globose capsule, opening irregularly and split into four sections with a permanent style. They are 4–8 mm in diameter, with 32–49 seeds per fruit, and open irregularly into four splits. Seeds have typical, pyramidal, hard and thick seed coat with brownish coloration.

Distribution

It is native to North Africa and South East Asia, and often cultivated as an ornamental plant throughout India, Persia, and along the African coast of the Mediterranean sea. Henna grows better in tropical savannah and tropical arid zones, in latitude between 15° and 25° N and S, produces highest dye content in temperature between 35 and 45°C. Henna is commercially cultivated in Western India, Pakistan, Morocco, Yemen, Iran, Afghanistan, Somalia, Sudan, Libya, Egypt, and Bangladesh. Presently the Pali district of Rajasthan is the most heavily cultivated henna production area in India, with over 100 henna processors operating in Sojat city.

Chemical Constituents

The chemical constituents isolated from *L. inermis* are naphthoquinone derivatives,

phenolic compounds, terpenoids, sterols, aliphatic derivatives, xanthones, coumarin, fatty acids, amino acids and other constituents. The phytochemicals including lawsone, isoplumbagin, lawsoniaside, lalioside, lawsoniaside B, syringinoside, daphneside, daphnorin, agrimonolide 6-O-β-D-glucopyranoside, (+)-syringaresinol O-β-D-glucopyranoside, (+)-pinoresinol di-O-β-D-glucopyranoside, syringaresinol di-O-β-D-glucopyranoside, isoscutellarin3β, hennadiol, (20S)-3β, 30-dihydroxylupane, lawnermis acid, 3-methyl-nonacosan-1-ol, laxanthones I, II, III and lacoumarin, etc. The dye molecule, lawsone is the chief constituents of the plant; its highest concentration is detected in the petioles (0.5–1.5 %).

Medicinal uses

Traditionally the roots are used as bitter, depurative, diuretic, emmenagogue, abortifacient, burning sensation, leprosy, skin diseases, amenorrhoea, dysmenorrhoea and premature greying of hair. The leaves are used as an astringent, acrid, diuretic, emetic, edema, expectorant, anodyne, anti-inflammatory, constipating, depurative, liver tonic, haematinic, styptic, febrifuge, trichogenous, wound, ulcers, strangury, cough, bronchitis, burning sensation, cephalalgia, hemicranias, lumbago, rheumatalgia, inflammations, diarrhoea, dysentery, leprosy, leucoderma,

Lawsoniaside   Lawsone   Isoplumbagin   Lalioside scabies, boils, hepatopathy, splenopathy, anemia, hemorrhages, hemoptysis, fever, ophthalmia, amenorrhoea. Flowers are cardiotonic, refrigerant, soporific, febrifuge, tonic, cephalalgia, burning sensation, cardiopathy, amentia, insomnia, fever. Seeds are used as an antipyretic, intellect promoting, constipating, intermittent fevers, insanity, amentia, diarrhoea, dysentery and gastropathy.

Cosmetic uses

Henna has been used as a cosmetic hair dye for long days. Henna is the only natural hair color with an FDA monograph for safe and effective use as a natural hair color. Henna is cooling and specific for devitalized, fragile, fine hair. This ancient plant has a wonderful conditioning and smoothing effect on the hair. This is an excellent hair conditioner and tonic. The leaves of the plant are astringent and used as a prophylactic against skin diseases. Largely used as a hair dye when mixed with other natural dyes. It is dried and crushed into a fine powder that is then mixed with oils and natural ingredients and applied to the skin. The paste dries and flakes off, leaving a beautiful stain of the design of your choice in your skin that lasts a week and a half of longer depending on skin type and after care. It is harmless to body system and causes no irritation to skin. For an herbal hair bleach, blend henna with orange, lemon and cucumber juice and apply to hair. It controls dandruff, tones and promotes healthy scalp, removes excess oil from the scalp and relieves headaches and scalp tension.

Pharmacological Action

The aim of this study was to investigate whether L. inermis can destroy cancer cells by induction of apoptosis due to decreasing of intracellular H^+ ion level or increasing intracellular free radicals and H_2O_2 levels in cancer cells as a result of oxidative effect or not. At the result of this study the thickness of subcutaneous lipid tissue, diameters of gluteal mass, the pH levels of gluteal mass, the GSH levels at the liver tissue samples and

the MDA levels of the liver tissue samples of these groups were measured. This study showed that L. inermis can be used as a supplementary agent for cancer treatment (Zumrutdal et al., 2008).

The study was investigated some of the effects of its ethanolic root extract of *Lawsonia inermis* on the histology of the ovary of adult female wistar rats. The doses were administered once daily for 21 days, and the rats were sacrificed on the 22nd day. The results showed there was an arrest in the development of the follicles in the ovary of the experimental groups and which was not observed in the control group. Hence it can be concluded that ingestion of *Lawsonia inermis* has some deleterious effects on the ovaries investigated (Towolawi et al., 2010).

The present work was evaluated the *in vitro* anthelmintic potency of the petroleum ether extract of *Lawsonia inermis* leaves using Indian earthworms (*Pheretima posthumad*). The various concentrations (25, 50, and 75 mg/ml) of the petroleum ether extract were tested *in vitro* for anthelmintic potency by determination of time of paralysis and time of death of worm. The result of present study indicates that the *Lawsonia inermis* L. potentiate to paralyze earthworm and also caused its death after some time. Hence the study was demonstrated that the *Lawsonia inermis* L. as an anthelmintic agent (Bairagi et al., 2011).

Toxicological Studies

Acute toxicity evaluation was designed and conducted with three doses of aqueous leaves extract of *Lawsonia inermis* (AELI) in log progression (i.e. 10 mg/kg, 100 mg/kg, 1000 mg/kg) were administered to 4 mice each and 100% death was observed at 1000 mg/kg while no death occurred at the lower doses. Subsequently, 100 female mice were randomly allocated into five groups of 4 mice per dose levels (i.e. 20 mice per dose level) of 100 mg/kg, 200 mg/kg, 400 mg/kg, 800 mg/kg and 1000 mg/kg intraperitoneal (i.p.) AELI respectively. After i.p. administration of AELI, animals were observed every 15 minutes for the first 4 hours, then every 2 hours for the next 8 hours, then daily for 14 days. AELI revealed an intraperitoneal LD50 of 894 mg/kg in mice. No evidence of toxicity was observed at doses up to 800 mg/kg. All animals given 1000 mg/kg became sluggish within 2 hours, fell into sleep within 6 hours and died within 12 hrs of drug administration. No animal had convulsion nor diarrhea (Bello et al., 2010).

An initial study was conducted to determine the dose range of the extract to be used for the study. Five groups of rats were administered i.p. doses of the stock aqueous root extract (0.3 g/ml) volumes corresponding to 200, 400, 800, 1200 and 1600 mg/kg BW. Any sign of change in mental or physical activities were recorded and the number of death in each group (if any) within 24 h was recorded for 10 (ten) days. The result obtained from the study show that the aqueous root extract of L. inermis is slightly toxic; causes delayed toxicity (with no death) which is accompanied by clinical symptoms in the body system of the rats for more than 7 days after i.p. injection. The clinical symptoms include paralysis, total collapse of the body, weakness, laziness and loss of appetite. But spontaneous abortion of the foetus in the pregnant female administered doses ranging from 800 to 1600 mg/kg BW, even though, no death in the pregnant females was observed, and they later regained their composure and normalcy (Mudi et al., 2011).

Clinical Investigation

The objective of the study was to evaluate the efficacy and safety of itch cream as an emmollient and topical antipruritic for symptomatic relief in various xerotic and pruritic dermatological disorders. The cream was formulated with extract of turmeric, saffron, sandalwood, vativer, lata kasturi, mehendi, tulsi, liquorice, turmeric oil, surasar and swarna bhasma. The study was prospective, unicentric, open label, randomized and controlled study that was performed in SSKM hospital, Kolkata.

Children (2–12 years) of either sex with atopic dermatitis *Ichthyosis vulgaris* and impetigo and adults (12–70 years) of both sex with senile pruritus, *Ichthyosis vulgaris* or other xerotic diseases were selected for the study. The results were satisfactory and suggested that itch cream has antipruritic and an emollient effect in patients of atopic dermatitis, *Ichthyosis vulgaris* and other xerotic diseases of mild severity (Chatterjee et al., 2005).

A clinical trial was conducted with 30 patients of eczema with compound formulation of *Olea europea, Lawsonia innermis* and *Nigella sativa* and noted improvement in signs and symptoms of eczema (Nawab et al., 2008).

The antitumour activity of *L. inermis* leaf extract was studied on 7,12-dimethylabenz (a)nthracene (DMBA) induced 2-stage skin carcinogenesis and B16F10 melanoma tumour model using swiss albino mice. Topical application of *L. inermis* leaf extract at a dose level of 1000 mg/kg body weight was found to be effective in reducing the number of the papillomas. The tumor yield was significantly decreased 1.6 respectively as compared with the DMBA treated control group 3.5. In other experiment the effect of cyclophosphamide (CP) alone and in combination with *L. inermis* was studied in B16F10 melanoma tumour bearing mice. The inhibition rate was 25.9% in the CP treated group but these increased to 35.14% with *L. inermis*. The life span time and volume of tumour doubling time were also increased. Results from two models showed that *L. inermis* extract have protective potential against skin tumour (Wasim et al., 2009).

The study is to develop and evaluate hair oil absolutely from herbal origin. Alcoholic extracts of *Citrus limonis, Cuscuta reflexa, Emblica officinalis, Centella asiatica, Allium cepa, Lawsonia inermis, Azadirachta indica, Eclipta alba, Ocimum sanctum* and *Eugenia caryophyllus* were incorporated into the olive oil. Prepared herbal oil was studied for hair growth activity. Above ingredients were selected to formulate herbal hair oil on the basis of their traditional use. The experimental paradigms used were primary skin irritation test and hair growth activity including number of hair, length of hair and hair growth initiation as well as hair growth completion time. Significant increase in hair numbers and length were observed after topical application of the formulation (Sharma et al., 2010).

Market Formulations

Protein Shampoo

It is natural protein nourishment. It enriched with extracts of protein-rich herbs, this new shampoo range nourishes your hair, giving the three essential benefits that hair needs—strength, reduced hair fall and protection from everyday damage.

Ingredients: Chickpea, amla and black myrobalan, haritaki (*Terminalia chebula*), bhringaraja (*Eclipta prostrata*), chanaka (*Cicer arietinum*), yashtimadhu (*Glycyrrhiza glabra*), bibhitaka (*Terminalia bellirica*), madayanti (*Lawsonia inermis*).

Action: Chickpea is a rich source of natural protein. Amla and black myrobalan is natural hair tonics that strengthen hair roots. Licorice helps promote hair growth. *Lawsonia inermis* provides natural color of the hair.

Supplied by: Himalaya Drug Company, India.

Surya Henna Powder

Surya henna powder is a henna based hair color that is 100% natural and vegetal that treats the hair while it colors. It gives the hair a beautiful shiny glow.

Ingredients: Henna (*Lawsonia inermis*), amla *(Phylanthus emblica)*, shikakai *(Acacia conccina)*, indigo *(Indigofera tinctoria)*, aritha *(Terminalia bellerica)*, honey.

Action: Henna *(Lawsonia inermis)*—used as a copperish colorant for the nails, skin and hair, henna has an antimicrobial and antifungal action. Its active principle is the colorant lawsone, which has a great affinity for the hair. Henna forms a protective layer on each strand of hair. Amla *(Phylanthus emblica)*

—very rich in vitamin C, this fruit has tannins that help compose the colour of the powder and is traditionally used as treatment for hair and scalp problems. Shikakai *(Acacia conccina)* —very rich in tannins and low pH, shikakai, the Indian "fruit for the hair", allows for the closing of the hair layers. It is rich in soft saponins that clean the hair without opening the layers. Helps prevent dandruff. Indigo *(Indigofera tinctoria)*—known for its coloring properties in India, indigo gives the hair bluish tonalities. Aritha *(Terminalia bellerica)*—this fruit is rich in saponins that softly wash the hair. Also functions as an astringent and bactericide. Honey is use as an antioxidant and moisturizer.

Direction: Surya henna powder is a complete conditioning and color treatment. The only additional suggestion is to mix in the packet of honey provided and skim milk (or skim yoghurt) if an additional moisturizing effect is desired. Do not mix it with any other substance since this alters the product.

Supplied by: NTP Health Products, Australia.

Surya Henna Cream

Surya Henna Cream is a semi-permanent hair coloring containing hair dyes plus herbs and fruits. It does not damage or change the structure of the hair. It is indicated for all hair types. Color uptake should be monitored in the case of dry, damaged or processed hair.

Ingredients: Deionized water, dipropylene glycol methyl ether, cetearyl alcohol, cetrimonium chloride, babaçu (*Orbignya oleifera*) seed oil, vegetable glycerin, henna (*Lawsonia inermis*), arnica (*Arnica montana*), yarrow (*Achillea millefolium*), chamomile (*Chamomilla recutita*), amla (*Phylanthus embilica*), mallow (*Malva sykvestris*), hazelnut (*Corylus rostrata*), guaraná (*Paulinia cupana*), açaì (*Euterpe oleracea*), carrot (*Daucus carota sativa*), juá (*Zizyphus joaseiro*), aloe vera (*Aloe barbadensis*), acerola (*Malpighia punicifolia*), brazil nut (*Bertholletia exclesa*), hydroxyethylcellulose, Essentials Oils: Rose (*Rosa centifola*), ylang ylang (*Cananga odorata*) jasmine (*Jasminum officinale*) sandalwood

(*Santalum album*), dehydroacetic acid and benzyl alcohol, aminomethyl propanol.

Action: Deionized water acts as transmitter vehicle, dipropylene glycol methyl ether penetrates in the skin and allows undesirable ingredients such as dyes, preservatives and synthetic agents in formulation to enter more easily into the stratum corneum of the skin. Cetearyl alcohol is a very safe natural wax used as a viscose controlling agent. Cetrimonium chloride is used for its conditioning properties. Babaçu (*Orbignya oleifera*) seed oils used for its emollient properties. Vegetable glycerinacts as humectant. Henna (*Lawsonia inermis*) is used for its scalp and skin conditioning properties. Henna forms a protective layer on each hair strand enhancing shine and condition of the cuticle. Arnica (*Arnica montana*) is a capillary stimulant and acts as a vasoprotector. Yarrow (*Achillea millefolium*)—plant that acts as an analgesic and a bactericide. Chamomile (*Chamomilla recutita*)—plant that acts as a dermal protector, decongestant and is a natural tranquilizer. Amla (*Phylanthus embilica*)—is rich in vitamin C. Mallow (*Malva sykvestris*)—plant that acts as an emollient is a natural tranquilizer and protector. Hazelnut (*Corylus rostrata*)—is acts as moisturizes and enriches formula. Guaraná (*Paulinia cupana*)—brazilian fruit that acts as a capillary stimulant enhancing circulation of the scalp. Açaì (*Euterpe oleracea*)—brazilian fruit that moisturizes, acts as an emollient and as a remineralizer. Carrot (*Daucus carota sativa*) softens and protects skin. Juá (*Zizyphus joaseiro*)—Brazilian Jua fruit lends tensoactive, anti-inflammatory and peripheral circulation stimulant properties to the formula. Aloe vera (*Aloe barbadensis*)— plant that is renown for its moisturizing properties, softens and acts as an emollient. Acerola (*Malpighia punicifolia*)—brazilian fruit rich in vitamin C. Brazil Nut (*Bertholletia exclesa*)—moisturizes, revitalizes, conditions and acts as an emollient. Hydroxyethylcellulose: Natural gum of plant origin. Essential Oils: Rose (*Rosa centifola*), ylang ylang (*Cananga odorata*), jasmine (*Jasminum officinale*), sandalwood

(*Santalum album*). Dehydroacetic acid and benzyl alcohol act as preservatives.

Supplied by: NTP Health Products, Australia.

Various Preparations

Henna also imparts excellent conditioning effect as well as thickness to the hair. Henna works more efficiently when the extract of some plants like amla (Indian gooseberry), reetha (soap nuts), shikakai (*Acacia concinna*), etc. are mixed and added to it.

Storage of Products

All the products should be stored in well closed containers and away from sunlight.

Dosage

Topical application in mice more than 1000 mg/kg body weight was investigated as safe for *Lawsonia inermis* leaf extract. It is non-irritant for the rabbit skin and eye. *Lawsonia inermis* 10 % in petrolatum was considered not irritating on human skin after repeated insult patch tests.

Patents

A deodorizing composition comprising as an antimicrobial agent an effective amount of an extract of *Indigofera tinctoria,* or of an antimicrobially active fraction thereof (Rosenberg, 2003).

This invention relates to a cosmetic composition comprising: (1) rosemary (*Rosmarinus officinalis* L. extract)—0.001–10%; (2) henna (*Lawsonia inermis* L. extract)—0.001–10%; (3) melanin—0.000005–5%; and (4) sunflower (*Helianthus annuss* L. extract)—0.0001–10%. Combination of the above naturally-occurring components is suitable for preparing finished cosmetic compositions in association with an acceptable cosmetic base. The compositions are particularly useful for dark hair and exhibit sun protection against UV radiation, protection of hair fiber chemical structure, protection and retention of hair color and appropriate gloss (Masiero, 2004).

The present invention provides a herbal black dye from natural materials comprising of *Juglans regia, Indigofera tinctoria, Terminalia chebula, Acacia accocina, Lawsonia inermis, Trigonella foenum-graecum, Sapindus mukorossi, Eclipta alba, Emblica officinalis, Acacia catechu* and *Piper betle*, the dye derived being safe, non-toxic, anti-allergic, anti-dandruff and free from toxic symptoms like itching (Palpu et al., 2007).

BIBLIOGRAPHY

1. Bairagi GB, Kabra AO and Mandade RJ. Anthelmintic activity of *Lawsonia inermis* L. Leaves in Indian Adult Earthworm. International Journal of Research in Pharmaceutical and Biomedical Sciences. 2011; 2(1): 237–240.

2. Bello SO, Bashar I, Muhammad BY, Onyeyili P. Acute toxicity and uterotonic activity of aqueous extract of *Lawsonia inermis* (Lythraceae). Research Journal of Pharmaceutical, Biological and Chemical Sciences. 2010; 1(3): 790–798.

3. Chatterjee S, Datta RN, Bhattacharya D, Bandopadhyay D. Emmollient and anti-pruritic effect of itch cream in dermatological disorders: A randomized clinical trial. Ind J Pharmacol. 2005; 37 (4): 253–254.

4. Chauhan MG, Pillai APG. Microscopic profile of powdered drug used in Indian system of medicine, Jamnagar, Gujarat, 2007, pp: 84–85.

5. http://ayurvedicmedicinalplants.com/plants/3985.html

6. http://www.4to40.com/ayurveda/index.asp?p=Henna

7. http://www.iaim.edu.in/home%20doctor/burning.html

8. http://www.naturalcosmeticsupplies.com/henna-leaves-powder.html

9. http://www.pharmatutor.org/articles/lawsonia-inermis-henna-traditional-uses-scientific-assessment

10. Masiero S. Cosmetic composition, especially for dark hair. United States Patent Application 20040117921. Application no. 10/714721. Publication date: 06/24/2004. http://www.freepatentsonline.com/y2004/0117921.html.

11. Mudi SY, Ibrahim H, Bala MS. Acute toxicity studies of the aqueous root extract of *Lawsonia inermis* Linn. in rats. Journal of Medicinal Plants Research. 2011; 5(20): 5123–5126.

12. Nawab MD. Mannan A, Siddiqui M. Evaluation of the clinical efficacy of Unani formulations on eczema. Indian J Traditional Know. 2008; 7: 341–344.

13. Palpu P, Pal M, Dixit BS, Banerji R, Rao CV. Herbal dye and process of preparation thereof. United States Patent Application 7186279. Application no. 11/027575. Publication date: 03/06/2007. http://www.freepatentsonline. com/7186279. html.

14. Rosenberg NM. Antibacterial deodorizing compositions. United States Patent 6548052. Application no. 09/519257. Publication date: 04/15/2003. http://www. freepatentsonline.com/6548052. html.

15. Sastri BN. The Wealth of India: Raw Materials, CSIR, New Delhi, 1962, pp: 47–50.

16. Sharma AK, Agarwal V, Kumar R, Kaushik K, Bhardwaj P, Chaurasia H. Development and evaluation of herbal formulation for hair growth. Inter J Curr Trends Sci Tech. 2010; 1(3): 147–151.

17. Towolawi OA, Ashaolu JO, Oyewopo OA, Dare JB, Caxton-Martins EA, Mesole SB. Effect of ethanolic root extract of henna (*Lawsonia inermis*) on the histology of the ovary of adult female wistar rats. Tropical Journal Of Laparo Endoscopy. 2010; 1(1): 39–44.

18. Warrier PK. Indian medicinal plants a compendium of 500 species, Vol. 3, Orient Longman, Chennai, 2004. pp: 303–304.

19. Wasim R, Agrawal RC, Ovais M. Chemopreventive action of *Lawsonia inermis* leaf extract on DMBA-induced skin papilloma and B16F10 Melanoma Tumour. Pharmacologyonline. 2009; 2:1243–1249.

20. Zumrutdal MZ, Ozaslan M, Tuzcu M, Kalender ME, Daglioglu K, Akova1 A et al., Effect of *Lawsonia inermis* treatment on mice with sarcoma. African Journal of Biotechnology. 2008; 7 (16): 2781–2786.

Matricaria chamomilla

Plant profile

Kingdom: *Plantae* – Plants

Subkingdom: *Tracheobionta* – Vascular plants

Superdivision: *Spermatophyta* – Seed plants

Division: *Magnoliophyta* – Flowering plants

Class: *Magnoliopsida* – Dicotyledons

Subclass: *Asteridae*

Order: *Asterales*

Family: *Asteraceae* – Aster family

Genus: *Matricaria* L.– Mayweed

Species: *Matricaria chamomilla*

Vernacular name

San: Babuna, Babunaj, **Eng:** Chamomile, German chamomile, **Hin:** Babuni ki phool, **Ben:** Babunphul **Guj:** Babuna, **Kan:** Shime-shavantige, **Mal:** Shimajevanthi pushpam, **Tam:** Seemaseventhi poo, **Mar:** Babuna, **Tel:** Sinacha mauli pushapamu.

Common name

Amerale, Babunnej, Bayboon, Camomile, Chamomile, Kami-Ture, Manzanilla Dulce, Manzanilla, Papatya.

INTRODUCTION

It is an annual plant found wild along roadsides, in fields, and cultivated in gardens. The leaves are pale green, bipinnate, sharply incised, and sessile. The branched stem is erected, heavily ramified, and grows to a height of 10–80 cm. The long and narrow leaves are bi- to tripinnate. The flower heads are placed separately, they have a diameter of 10–30 mm, and they are pedunculate and heterogamous. The golden yellow tubular florets with 5 teeth are 1.5–2.5 mm long, ending always in a glandulous tube. The 11–27 white plant flowers are 6–11 mm long, 3.5 mm wide, and arranged concentrically. The receptacle is 6–8 mm wide, flat in the beginning and conical, cone-shaped later, hollow—the later being a very important distinctive characteristic of *Matricaria*-and without paleae. The fruit is a yellowish brown achene.

Distribution

The plant is indigenous to Germany and is found in countries like America, Europe,

Germany, Hungary, India. It is a wild growing plant and also cultivated in central European countries. In India it is found in Punjab, Uttar Pradesh, Maharashtra, and Jammu and Kashmir.

Chemical Constituents

Major chemical constituents responsible for physicochemical and therapeutic action of the herb are: Matricaria volatile oil consisting of azulene, sesquiterpenes, sesquiterpene alcohols, paraffin hydrocarbons, methoxycournarin, furfural, and a fatty acid Matricaria contains 0.3–0.5 per cent of a volatile oil consisting of azulene, sesquiterpenes, sesquiterpene alcohols, paraffin hydrocarbons, methoxycoumarin, furfural, and a fatty acid. The volatile oil is a thick fluid freely soluble in alcohol and is blue in colour due to the presence of azulene. Various minor chemical constituents of this drug are salicylic acid glucosides of apigenin, choline and triacontane phytosterol fatty acids.

Azulene   Furfural

Medicinal uses

Chamomile has potential for the treatment of: Blocked tear duct, canker sores (mouth ulcers), colic, diarrhea, eczema, gingivitis (periodontal disease), indigestion and heartburn, insomnia, irritable bowel syndrome, peptic ulcer. Chamomile is a stimulant, bitter, tonic, aromatic, emmenagogue, anodyne, antispasmodic, stomachic. It is used externally to spur wound healing and treat inflammation, and internally for fever, digestive upsets, anxiety, and insomnia. Chamomile was also shown useful for reducing inflammation in arthritis. It also relaxed smooth muscle of the intestine. Chamomile was a popular eye wash for treating conjunctivitis and other reactions. It

had also been found to promote wound healing. Chamomile may help to prevent stomach ulcers and speed their healing. Antispasmodic action of chamomile soothes the menstrual cramps and to lessen the possibility of premature labor. It was also found to stimulate menstruation. Chamomile depresses the central nervous system. It had been shown effective as a tranquilizer from antiquity. Chamomile relieves arthritic joint inflammation in animal studies. The essential oil was found to reduce the time required to heal burns. The herb kills the yeast fungi that cause vaginal infections, as well as certain bacteria. Chamomile impairs the replication of polio virus. Chamomile stimulates the immune system's infection fighting white blood cells (macrophages and B-lymphocytes).

Cosmetic uses

It is used in cosmetics, perfumery and as a flavoring agent. It is good for skin care (most skin-types), acne, allergies, boils burns, eczema, inflamed skin conditions, earache and wounds. It is used commercially in shampoos for fair hair and it can lighten hair color. It makes an excellent wash for sore and weak eyes. The leaves are dried and used in aromatherapy and other herbal medicines. A salve or compress made from chamomile can be used to help treat wounds, burns, cold sores, canker sores, pink eye and other skin irritations, due to its antibacterial and anti-inflammatory properties. It is often added to cosmetics as an anti-allergen.

Pharmacological Action

The antifungal activity of *Matricaria chamomilla* L. flower essential oil was evaluated against *Aspergillus niger* with the emphasis on the plant's mode of action at the electron microscopy level. In the bioassay, *A. niger* was cultured on potato dextrose broth medium in 6-well microplates in the presence of serial two fold concentrations of plant oil (15.62 to 1000 microg/ml) for 96 h at 28 degrees C. Based on the results obtained, *A. niger* growth was inhibited dose dependently with a maximum of approximately 92.50% at the highest oil concentration. These findings indicate the potential of *M. chamomilla* L. essential oil in preventing fungal contamination and subsequent deterioration of stored food and other susceptible materials (Tolouee et al., 2010).

The effect of *Matricaria chamomilla* extract on linear incisional wound healing was studied. Thirty male wistar rats were subjected to a linear 3 cm incision made over the skin of the back. For computing the percentage of wound healing, the area of the wound measured at the beginning of experiments and the next 2, 5, 8, 11, 14, 17, and 20 days. The percentage of wound healing was calculated by Walker formula after measurement of the wound area. Results showed that there were statistically significant differences between treatment and olive oil animals ($p < 0.05$) in most of the days. It concluded that the extract of *M. chamomilla* administered topically has wound healing potential in linear incisional wound model in rats (Jarrahi et al., 2010).

The essential oil and methanol of *Matricaria chamomilla* L. were subjected to screening for their possible antioxidant activity by two complementary test systems, namely 2,2-Diphenylpicrylhydrazyl (DPPH) free radical scavenging and β-carotene-linoleic acid assays. In the DPPH test system, the IC_{50} value of essential oil and methanol extract were respectively 4.18 and 1.83 µg/ml. In the â-carotene-linoleic acid system, oxidation was effectively inhibited by *M. chamomilla*. The essential oil and methanol extract were tested against bacterial and fungal strains using a broth microdilution method. The results suggest that *M. chamomilla*, oil and methanol extract have significant antimicrobial activity (Fatouma et al., 2011).

Toxicological Studies

Acute toxicity studies of aqueous flower extract of *M. chamomilla* were performed on mice. Ten male and 10 female mice were divided into control and experimental groups.

Each group included five males and five females mice. The experimental group received extract and the control group received water by gavage with the aid of a metal gastric needle at a single dose of 5000 mg/kg of the animal weight. The animals were observed carefully every 2 days to record toxic manifestations, and to measure body mass and water and ration consumption. After 14 days, the mice were sacrificed. The livers, kidneys, lungs, and hearts were observed macroscopically and the relative weights (organ/body) were determined. The acute toxicity test after oral administration of 5000 mg/kg of MC extract revealed no toxicity at this dose. There were no significant alterations in water or food consumption, or body weight during the experiment. The body weights, relative weights of the kidneys, livers, lungs and hearts were not statistically different from those of the control group (Saied and Ali, 2009).

Clinical Investigation

In an open, bilateral comparative trial, 161 patients with eczema on their hands, forearms, and lower legs initially treated with 0.1% diflucortolone valerate received one of four treatments: Chamomile cream (Kamillosan), 0.25% hydrocortisone, 0.75% fluocortin butyl ester (a glucocorticoid), or 5.0% bufexamac (a nonsteroidal anti-inflammatory). After 3–4 weeks, the chamomile cream was found to be as effective as hydrocortisone and demonstrated superior activity to bufexamac and fluocortin butyl ester (Aertgeerts et al., 1985).

In a double-blind, randomized, placebo-controlled study, 48 women receiving radiation therapy for breast cancer were treated topically with either chamomile cream or placebo (almond oil) to protect the radiation-treated area. While there were no significant differences between the two groups in objective scores of skin irritation, the patients preferred the chamomilecontaining cream to the placebo for its rapid absorption and stainlessness (Maiche et al., 1991).

A randomized, double-blind study was conducted with 164 cancer patients taking 5-fluorouracil (5-FU) chemotherapy. The patients rinsed three times daily with either a chamomile mouthwash or placebo. After 14 days, no difference was observed between the two groups in the incidence of stomatitis induced by 5-FU (Fidler et al., 1996).

Market Formulations

Chamomile baby cream: This cream is used for treating facial skin and the body, fine children's skin, sensitive adult skin all age. It has comforting effects and softens. This cream is used for fine children's skin of every age, but also sensitive adult skin.

Ingredients: Plant oils—soya bean and almond, German chamomile essential oil, emulsifiers, glycerine, wax, vitamins A, E, pro-vitamins, water.

Direction: Apply the required amount of cream onto the face skin or body and rub gently into it. The skin remains mat after the application.

Shelf time: 6 months.

Supplied by: VenBella, Wellness and Spa, UK.

Baby oil: This oil is used as a daily moisturiser, after bathing and before bed, to moisturize, nourish and soften skin, leaving it looking and feeling fresh and baby soft. The oil is also beneficial for your own skin. Used as a nourishing oil after showering, it is beneficial for those with very sensitive skin. It can also be used as a bath oil.

Ingredients: Chamomile (*Matricaria chamomilla*) oil, violet (*Viola tricolor*) oil, lemon essential oil.

Action: Chamomile flowers contain the volatile oil azuleen that has a very soothing effect on the skin, especially on rashes, skin irritations, and sensitive skin. It is so gentle on skin that it has been nicknamed the "Baby's Herb." Violet has a very delicate flower but is a very hardy plant. It has cleansing saponins and anti-inflammatory properties which are known to help relieve skin problems and is especially beneficial to weeping eczema.

Direction: Add a few drops into the bath before bathing and apply all over as a moisturiser after bathing. Gently massage into the skin until completely absorbed.

Supplied by: Shirley's Herbal Care, Ireland.

Psoriasis ointment: This ointment is designed to help relieve red, dry, flaky skin and soothe the irritation of large patches of psoriasis on the body. Especially it is beneficial for large areas of psoriasis on the elbows, shins and body and it helps to reduce redness, cracking and flaking.

Ingredients: Calendula (*Calendula officinalis*) oil extraction, Comfrey (*Symphytum officinale*) glycerine extraction, Calendula (*Calendula officinalis*) wax extraction, Chamomile (*Matricaria chamomilla*) wax extraction, comfrey (*Symphytum officinale*) wax extraction. Petroleum Jelly (Vaseline®) and mineral oil to the beeswax and sunflower oil in the wax.

Action: Calendula contains natural iodine and resins that is also used when the skin is naturally dry, affected by cold weather, or drying out from heat or sunburn. Chamomile flowers contain the volatile oil azuleen that has a very soothing effect on the skin, especially on rashes, skin irritations, and sensitive skin. Comfrey contains alantoin and mucilage, which helps reduce the swelling caused by bruising, and binds tissues together helping to repair the skin when chapped or cracked. It is also great at regenerating skin.

Direction: Use on large areas of dry, flaky, cracked skin or as an intensive night treatment after a bath or shower.

Supplied by: Shirley's Herbal Care, Ireland.

Eczema cream: This cream helps to soothe itching and irritation, relieve discomfort created by the drying out of the skin, even when it cracks at joints such as elbows and knees. It is beneficial to any irritations caused by insect bites and itchy rashes. This is extremely gentle for sensitive skin and can be used as a facial moisturiser for eczema sufferers.

Ingredients: Chickweed (*Stellaria media*) glycerine extraction, chickweed (*Stellaria media*) oil, mallow (*Malva Sylvestris*) infusion, mallow (*Malva Sylvestris*) oil, chamomile (*Matricaria chamomilla*) glycerine extraction, chamomile (*Matricaria chamomilla*) oil, violet (*Viola tricolor*) oil, violet (*Viola tricolor*) infusion, shea butter jojoba wax, beeswax, vegetable emulsifier, *Aloe vera*. Preservative: Grapefruit seed extract, Geranium Essential oil.

Action: Chickweed contain saponins that give its herbal extraction and natural anti-irritant properties. Mallow contains high amounts of mucilage, which gives mallow its soothing and anti-irritant properties and can be well-tolerated by sensitive skin. The oil extraction we make from the petals creates a delicate, pure, white oil, which is gentle and soothing. Chamomile flowers contain the volatile oil azuleen that has a very soothing effect on the skin, especially on rashes, skin irritations, and sensitive skin. Violet has cleansing saponins and anti-inflammatory properties which are known to help relieve skin problems and is especially beneficial to weeping eczema.

Direction: Apply twice daily or more if required. Massage gently into the affected area when break outs occur, and plentifully at night as an intensive treatment. Use sparingly as a facial moisturiser after cleansing and toning, and gently massage into the skin until fully absorbed.

Supplied by: Shirley's Herbal Care, Ireland.

Various Preparations

Flowers

Infusion: The herbal remedy can be taken as a treatment for irritable bowel syndrome, it can be used to boost a poor appetite, and it can also be used in the treatment of persistent indigestion affecting the patient. Patients can drink a cup of the herbal infusion during the night for alleviating the symptoms of insomnia, to beat back anxiety, and to alleviate stress. Strained infusion of about 200–400 ml can be added to a baby's bath water at night to encourage the sleep of the child.

Tincture: This remedy can be taken as an herbal treatment for irritable bowels, and to treat cases of insomnia and as a relief from tension.

Ointment: This herbal remedy can be used as treatment for insect bites, it can be applied to wounds, it can be used in the treatment of itching eczema, and for the treatment of all forms of anal or vulval irritation in patients.

Mouthwash: The herbal remedy is used in the form of an infusion to treat inflammations affecting the mouth and the oral orifice.

Eyewash: Prepare the herbal remedy by dissolving 5–10 drops of herbal tincture in some warm water, and use this herbal remedy for the treatment of persistent conjunctivitis or to alleviate strained eyes.

Inhalation: This form of the herbal remedy can be prepared by adding two teaspoonful of chamomile flowers to a basin of boiling water and the steam can be inhaled to remove accumulated phlegm, the steam inhalation is also effective against hay fever, it can be used as a treatment for asthma, and also for the treatment of bronchitis related symptoms.

Essential Oil

Lotion: This form of the herb can be used in the treatment of skin disorders such as eczema; use about 5 drops of the oil of chamomile added to 50 ml of distilled witch hazel herbal oil—this combination herbal lotion can be applied directly as a topical remedy on the affected parts of the skin.

Inhalation: The steam inhalation can be used for the treatment of bad nasal mucus accumulations, it can also be used in the treatment of cases of asthma, or when carried out under medical supervision—for the treatment of cases of whooping cough, as a general rule, affected individuals can benefit by placing 2–3 drops of the essential oil in a saucer of warm water, leaving this in the room at night—the vapors will infuse around the room and aid in alleviating many symptoms and disorders.

Bath additive: Use 1 pound flowers with 5 quart cold water. Bring to a boil, then steep covered for 10 minutes. Strain and add to bath water. Used as a hair wash, it will brighten the hair.

Rubbing oil: Steep 1 oz fresh or dried flowers in olive oil for 24 hours or more. Strain before using.

Storage of Products

All the products should be stored in well closed containers and away from sunlight.

Dosage

Chamomile is often taken as a tea that can be drunk three to four times daily between meals. Common alternatives are to use tablets, capsules, or tinctures. Herbal tablets, herbal capsules, or the chamomile tincture. Doses for these alternatives can be about 2–3 grams of the capsules or tablets every day or about 4–6 ml of the herbal tincture thrice every day in between the daily meals.

Chamomile can be used as a tea by steeping the crushed flowers (2–3 teaspoonful) in a cup of boiling water for 5 to 10 minutes and then draining. Adding 1 cup of the dried petals in a linen bag to hot bath water will make a relaxing bath. The oils can be used on the skin as treatment for skin conditions such as eczema. The creams and gels made from the oils can be used externally throughout the day. A typical dosage of 10 to 15 grams (about 3 to 5 tablespoonful) daily shows promise.

Apply cream with a 3–10% crude drug chamomile content for psoriasis, eczema, or dry and flaky skin. It can be used as paste by mixing powdered herb with water and apply on the inflamed skin.

Patents

The disclosure herein is directed to organic skin-based products that comprise an excess of 70% organic ingredients. The skin care products described herein are composed of a plurality of organic juices and other organic materials (Jochim and Behnke, 2008).

The invention was directed to a composition for the treatment of the hair roots in order to prevent or at least to reduce the loss of hairs, to restore the growth of the hairs and to prevent or at least to reduce the formation of white hairs with the composition contains an extract of at least one *Chamomilla* plant and/or of at least one *Achillea* plant (Jamel et al., 2010).

A natural gel/ointment for the treatment of hemorrhoids. Bromelain is one primary active ingredient in the gel/ointment. A second active ingredient such as Hamamelis Virginiana (witch hazel) may be combined with the bromelain. Secondary ingredients include alcohol, de-ionized water, butylene glycol, PEG-40 hydrogenated castor oil, hydroxyethylcellulose, *Eucalyptus globulus* oil, *Aloe barbadensis* leaf extract, *Chamomilla recutita* flower extract, *Camellia sinensis* leaf extract, *Avena sativa* kernel flour, Benzophenone-4, disodium EDTA, sodium citrate, sodium hydroxide, citric acid, caprylyl glycol, chlorphenesin and methyliso-thiazolinone. Based on case studies the gel/ointment outperforms other hemorrhoid medications (Thornton, 2011).

A composition and method of use therefore allows for permanent straightening or curl of human hair that is not irritating to the skin, that does not have an obnoxious odor, that is significantly less damaging to hair, that allows for the immediate shampooing of hair thereafter and that provides for re-treatment of the hair without additional damage, primarily using urea and high temperatures to break and reform disulfide bonds as a means to allow for the penetration of low molecular weight proteins into the shaft of the hair (Saute and Saute, 2011).

BIBLIOGRAPHY

1. Aertgeerts P, Albring M, Klaschka F, et al. Comparative testing of Kamillosan cream and steroidal (0.25% hydrocortisone, 0.75% fluocortin butyl ester) and nonsteroidal (5% bufexamac) dermatologic agents in maintenance therapy of eczematous diseases. Z Hautkr 1985;60: 270–277.

2. Fatouma MAL, Nabil M, Prosper E, Adwa AA, Samatar OD, Louis O et al. Anti-microbial and antioxidant activities of essential oil and methanol extract of *Matricaria chamomilla* L. from Djibouti. Journal of Medicinal Plants Research. 2011; 5(9):1512–1517.

3. Fidler P, Loprinzi CL, O'Fallon JR, et al. Prospective evaluation of a chamomile mouthwash for prevention of 5-FU-induced oral mucositis. Cancer 1996; 77:522–525.

4. http://en.wikipedia.org/wiki/Matricaria_chamomilla

5. http://holisticonline.com/Herbal-Med/_Herbs/h44.htm

6. http://www.herbal-supplement-resource.com/german-chamomile.html

7. http://www.herbs2000.com/herbs/herbs_chamomile_ger.htm

8. http://www.herbsandcures.com/239/matricarta-details

9. http://www.motherherbs.com/matricaria-chamomilla.html

10. http://www.phcogrev.com/article.asp?issn=0973-7847;year=2011; volume=5;issue =9; spage=82; epage=95;aulast=Singh

11. http://www.rain-tree.com/chamomile.htm

12. http://www.umm.edu/altmed/articles/german-chamomile-000232.htm

13. Jamel YA, Ahmed Al-dabooni RM, Hamed Al SKM. Composition for the treatment of the hair roots. United States Patent Application 20100124577. Application no: 12/325867. Publication date: 05/20/2010. http://www.freepatentsonline. com/y2010/0124577.html

14. Jarrahi M, Vafaei AA, Taherian AA, Miladi H, Rashidi Pour A. Evaluation of topical *Matricaria chamomilla* extract activity on linear incisional wound healing in albino rats. Nat Prod Res. 2010; 24(8):697–702.

15. Jochim M, Behnke K. Compositions for juice-based skin cleansers. United States Patent Application 20080166313. Application no: 11/933414. Publication date: 07/10/2008. http://www.freepatentsonline. com/y2008/0166313.html

16. Maiche AG, Grohn P, Maki-Hokkonen H. Effect of chamomile cream and almond ointment on acute radiation skin reaction. Acta Oncol 1991;30:395–396.

17. Saied KD, Ali N. Antiulcerogenic effects of *Matricaria Chamomilla* extract in experimental gastric ulcer in mice. *Iran J Med Sci.* 2009; 34(3): 198–203.

18. Saute R, Saute S. Composition and method for hair straightening. United States Patent Application 20110146699. Application no: 12/655209. Publi-cation date: 06/23/2011. http://www. freepatentsonline. com/y2011/0146699. html

19. Thornton JP. Topical medication for the treatment of hemorrhoids and method of use. United States Patent Application 20110243918. Application no: 12/547241. Publication date: 10/06/2011. http://www. freepatentsonline.com/y2011/0243918. html

20. Tolouee M, Alinezhad S, Saberi R, Eslamifar A, Zad SJ, Jaimand K et al., Effect of *Matricaria chamomilla* L. flower essential oil on the growth and ultrastructure of Aspergillus niger van Tieghem. Int J Food Microbiol. 2010; 139(3):127–33.

Nardostachys jatamansi

Plant profile

Kingdom: *Plantae* – Plants
Superdivision: *Spermatophyta* – Seed plants
Class: *Magnoliopsida* – Dicotyledons
Family: Valerianaceae
Species: *Nardostachys jatamansi*

Subkingdom: *Tracheobionta* – Vascular plants
Division: *Magnoliophyta* – Flowering plants
Order: *Dipsacales*
Genus: *Nardostachys*

Vernacular name

San: MiNsi, Jai, Jaili, **Assam:** Jatamansi, Jatamangshi, **Ben:** Jatamamsi, **Eng:** Nardus root, **Guj:** Baalchad, Kalichad, **Hin:** Balchara, **Kan:** Bhootajata, Ganagila maste, **Kashmiri:** Bhutijata, **Mal:** Manchi, Jatamanchi, **Mar:** Jatamansi, **Ori:** Jatamansi, **Punj:** Billilotan, Balchhar, Chharguddi, **Tam:** Jatamanji, **Tel:** Jatamams

Common names: Jatamansi, Sambul lateeb, Sumbul-ut-teeb, balchar

Ayurvedic properties

Rasa: Tikta, Kashiya, **Guna:** Laghu, **Veerya:** Shita (Cold), **Vipaka:** Katu, **Karma:** Medhya, Tridoshanut, Varnya, Nidrijanana, Kushhaghna.

Ayurvedic preparations

Mamsi ghrta, Mamsi taila, Mamsyadi kvatha, Raksoghna ghrta, Sarvausadhi snana, etc.

INTRODUCTION

It is a perennial herb propagated by cutting of the underground parts. The plant is about 10–60 cm in height and with stout and long woody root stocks. The leaves are 15–20 cm in height and with stout and long woody root stocks. The leaves are 15–20 cm long, sessile and oblongovate in shape. The flowers are rosy, slightly pink or blue in dense cymes. The rhizome is thick, covered with brown fibers and fragrant.

Distribution

The herb grows in Alpine Himalayas at an altitude of 3000–5000 metres. It is cultivated from Punjab to Skkim and in Bhutan.

Chemical Constituents

A volatile essential oil 0.5%, resin, sugar, starch, bitter extractive matter and gum are obtained from the rhizome. Roots contain a variety of sesquiterpenes and coumarins. The sedative sesquiterpene valeranone, which is also found in valerian and other plants, is a major component of the root essential oil, at least in some samples. Other terpenoids include spirojatamol, nardostachysin, jatamols A and B, and calarenol. Sesquiterpene ketone like jatamansone isolated from rhizomes, malliene and calarene from oil; a new terpenic coumarin jatamansin and oroselol from roots. A new dietheniod bicyclic ketone nardostachone isolated from roots. Lignan, (+)-92-isovaleroxylariciresinol was also isolated. Three terpenoid compounds were

mild sedative, hence, is beneficial in mental stress, hypertension, insomnia and headache. Jatamansi is one of the medhya resayanas—

Jatamansone   Jatamanins H   Jatamanins I

(+)-9'-isovaleroxylariciresinol

isolated from *Nardostachys jatamansi* DC rhizomes which are nardal, jatamansic acid and nardin.

Medicinal uses

It increases appetite and improves digestion. It is cholegouge, hence useful in hepatitis and enlargement of the liver. It helps relieving the phlegm in cough and asthma. Jatamamsi effectively relieves the symptoms like vertigo, seizures, etc. in the fever due to vitiation of majjagata pitta. As it improves the quantity of urine, it is salutary in urinary problems like dysuria and inflammation of urinary bladder. Jatamansi can be of use as an adjunct in the treatment of sexual debility and impotence. It also exerts a cleansing effect on the uterus, hence is used in menstrual ailments like dysmenorrheal and inflammation of the uterus. Combined with dasamula, it works well in postpartum hysteria. Jatamansi is one of the best herbs used in the treatment of epilepsy. It combines well with brahmi, vaca, jyotismati and yastimadhu. Jatamansi ghrta is a very effective anti-epileptic formulation commonly used. The herb exerts calming and soothing effect on mind because of its sattvika attribute. It works as a tranquillizer and a

brain tonics, useful in enhancing the memory. It is a rejuvenative to majja dhatu.

Cosmetic uses

The paste prepared in cold water is beneficial to reduce the burning sensation. It also imparts fair complexion to the skin and alleviates the pain and swelling. These properties of jatamansi are immensely benevolent in the treatment of *Acne vulgaris*. Topical application of herbal formulations in *Acne vulgaris* was also evaluated. In excessive sweating, jatamansi powder, being fragrant, is invaluable as a deodorant. In the skin ailments, associated with burning and in erysipelas, the decoction of jatamansi is poured frequently on the affected areas. The medicated oil for massage is extremely useful for smooth, silky and healthy hair.

Pharmacological Action

The present study explores the hepatoprotective activity of various extracts of *Ferula asafoetida*, *Momordica charantia* Linn and *Nardostachys jatamansi* against experimental hepatotoxicity. Polyherbal suspensions were formulated using extracts showing significant activity and evaluated for both physicochemical and hepatoprotective activity in comparison with LIV-52 as standard. Petroleum ether (60–80°), chloroform, benzene, ethanol and aqueous extracts of *Ferula asafetida*, *Momordica charantia* Linn and *Nardostachys jatamansi* were evaluated for hepatoprotective activity against carbon tetrachloride-induced liver toxicity in wistar rats. The results concluded that the formulation containing chloroform, petroleum ether and aqueous extracts of *Ferula asafetida*, petroleum ether and ethanol extracts of *Momordica charantia* Linn and petroleum ether and ethanol extracts of *Nardostachys jatamansi* demonstrated significant hepatoprotective activity, that might be due to combined effect of all these extracts (Dandagi et al., 2008).

An aqueous root extract from *Nardostachys jatamansi* was investigated for its antioxidant and anticataleptic effects in the haloperidol-induced catalepsy rat model of the disease by measuring various behavioral and biochemical parameters. A significant ($P <$ 0.01) reduction in the cataleptic scores were observed in the *Nardostachys jatamansi* (250 and 500 mg/kg body weight) administered group. To estimate biochemical parameters: The generation of thiobarbituric acid reactive substances (TBARS); reduced glutathione (GSH) content and glutathione-dependent enzymes; catalase; and superoxide dismutase (SOD), in the brain were assessed. The study concluded that the antioxidant potential has contributed to the reduction in the oxidative stress and catalepsy induced by haloperidol administration (Rasheed et al., 2010).

The present study is designed to evaluate anti-amnesic activity by studying the effect of methanolic extract of *Nardostacys jatamansi* DC (MENJ) rhizome on sleep deprived (SD) amnesic mice. Animals were pretreated with MENJ for 14 days (200 and 400 mg/kg, orally) and Piracetam (200 mg/kg, orally) as standard drug and followed by 5 days sleep deprivation using multiple platform method. Behavioral changes were evaluated against normal control and negative control animals using elevated plus maze, passive shock

avoidance, Y-maze, Morris water-maze and object recognition tests. MENJ at doses of 200 mg/kg and 400mg/kg treated groups showed a significant improvement in learning and cognition parameters in behavioral tests. These findings suggest MENJ exerts a protective effect against loss of memory and cognitive deficits due to sleep deprivation (Rahman and Muralidharan, 2010).

Nardostachys jatamansi DC rhizomes were subjected to extraction, fractionation, and isolation of terpenoid compounds. Three terpenoid compounds were isolated which are nardal, jatamansic acid, and nardin. These compounds were identified based on physical and spectral data (UV, IR, ^{1}H and ^{13}C NMR, 2D NMR, Mass) and comparison with authentic compounds. The crude extract, fractions, and two of the isolated compounds were tested for their hair growth activity. The hair growth studies showed good activities for the extract, fraction, and the isolated compounds (Gottumukkala et al., 2011).

Ethanolic extract of *Nardostachys jatamansi* (valerianaceae) and *Cyperus rotundus* (cyperacae) was evaluated for hair growth on albino rats. The hair growth activity that was worked on chemotherapy induced alopecia model were invest igated by using various parameters like hair density, lymphocyte count and testosterone level along with histopathological study. Hair growth initiation time was markedly reduced to half on treatment with extract as compared to control animal, the time required for complete hair growth was also significantly reduced. Overall results of the study suggest that the both extracts possess significant hair growth activity (Yadav et al., 2011).

Toxicological Studies

Acute toxicity: LD$_{50}$ of jatamansone was found to be 80.3 mg/kg, while in the case of quinidine it is 55 mg/kg. This shows that intravenous toxicity in mice in the case of jatamansone is less than that with quinidine. Large doses of jatamansone can cause vomiting and diarrhea (web article).

The present study was investigated the toxic effects of a phytochemical combination on albino wistar rats. The phytochemicals, Curcumin and nardostachys jatamansi root extract were taken in combination at different concentrations to evaluate acute and sub-chronic toxicity. In the acute toxicity study, the phytochemical combination caused neither mortality nor any acute signs of toxicity in the short term for 48 h and long term throughout the 14 days period of observation. In the subchronic toxicity study, there was no mortality recorded among the animals when varying doses of the phytochemical combination (100, 250 and 1000 mg/kg body weight) was administered orally for 28 days consecutively. The examination of toxic signs and health monitoring showed no abnormalities in the test groups as compared to their respective controls. The results suggest that the phytochemical combination has no significant toxic effects in acute toxicity, while subchronic toxicity studies at lower doses of 100 and 250 mg/kg showed no significant toxicity, but at higher dose of 1000 mg/kg, it caused mild toxic effects (Naveen et al., 2011).

Clinical Investigation

The study was investigated and compared the effects of strength training (shoulder girdle and respiratory muscles) and drug (jatamansi) on reducing hand tremor in archers. Pre and Post Hand Steadiness Score (Arm-Hand-Steadiness Tester on 4 mm hole) and Performance Test Scores (AAPHER Archery Test) were evaluated. 45 National Level Archers within age 16–28 years were assigned. Medicine was consumed for 4 weeks continuously 3 gm TDS and Thera-band strength training was given for 3 times weekly for 4 weeks. It was hypothesized that strength training is more effective in reducing hand tremors than control groups (Laishram et al., 2008).

Market Formulations

Scalp N Hair Herbal Nourishing Shampoo

Scalp n Hair shampoo improves texture and lustre of hair keeping them scalp and hair healthy. Scalp n Hair shampoo prevents brittleness and splitting of hair. It promotes healthy hair growth.

Ingredients

Each 100 ml contains

Ghritkumari (*Aloe barbadensis*) extract: 12 g, Japa (*Hibiscus rosa sinensis*) extract: 12 g, Methika (*Trigonella foneum-graceum*) oil: 2 g, Jatamansi (*Nardostachys jatamansi*) oil: 2 g.

Action: Aloe vera extracts have anti-bacterial and antifungal activities, strengthens the hair, making them shiny, strong and conditioned. Japa provides essential nutrients and proteins to promote hair growth, strengthens hair roots, minimizes hair fall. Methika—provides important proteins to further strengthen hair follicles. Jatamansi-hair conditioner effects—softens hairs, provides lustre and radiances to hairs, enhances hair bounce and volume too.

Direction: Take approx. 2–3 g of the shampoo and massage gently on the hair and scalp till it is builds a rich lather. After 5–10 minutes rinse with water. Apply daily.

Supplied by: Goldenlife Maharashtra, India.

Herbal Hair Oil with Amla, Bhringraj and Jatamansi-revitalizing Stimulating

It prevents hair loss and promotes hair growth. This is suitable for all types of hair.

Ingredients: Almond oil, amla, bhringraj, jatamansi, brahmi, shikakai and triphala extract.

Action: Almond oil nourishes the hair. Triphala and amla retards aging process and prevents hair loss. Shikakai and Brahmi extracts condition the hair and prevent dandruff. Jatamansi prevents brittle hair and split ends. Bhringraj arrests premature graying of hair.

Direction: Take appropriate amount and massage on hair and scalp with fingers so that it reaches the roots of hair. At least twice in a week should be used.

Supplied by: HerbalHealthCareStore.com

Jatamansi Cream

This cream nourishes and conditioning vata and kapha skin. It regulates skin temperature, rejuvenates dull skin, relieves dryness, dehydration.

Ingredients: Jatamansi, Ghee, Kokum butter, Rosewater.

Action

Direction: Use daily following the application of facial oil. Ideally used in the evening as a night cream. Very dry or dehydrated skin types can use twice a day.

Supplied by: Ayoma, San Jose, CA.

Amla and Beal Revitalizing Hair Tonic

Jovees revitalizing hair tonic contains jatamansi, amla, beal and other herbal and botanical extracts that help to prevent hair fall, improves texture of hair and strengthens the hair shaft. This non-greasy tonic penetrates hair shafts and nourishes the scalp while adding body and volume to hair.

Ingredients: Jatamansi ext., amla ext., beal ext.

Direction: Rub a few drops into the scalp with finger tips and leave on.

Supplied by: Jovees Herbals, Coimbatore.

Various Preparations

Four parts of Jatamansi and one part each of *Cinnamomum tamala, Cubeba officinalis,* aniseeds, dry ginger and two parts of sugar can be powdered and used as an effective home remedy for flatulence, colic pains and hysteria. Jatamansi can be combined with saffron and used as a home remedy for scorpion stings and tubercular adenitis. When used as a tonic, it is effective in general weakness and epilepsy.

In milk, Jatamansi powder mixed with honey and turmeric powder, then make a paste form. Apply on the face with rubbing which helps to remove the black heads and make the skin glow.

Relaxing Aromatherapy Body Lotion

To an 8 ounce bottle of unscented body lotion, add the following aromatherapy oils, cap the bottle, and shake it thoroughly: 3 drops lavender (*Lavendula angustifolia*) essential oil, 2 drops spikenard (*Nardostachys jatamansi*) essential oil, 4 drops neroli (*Citrus aurantium bigaradia*) essential oil.

Storage of Products

All the products should be stored in well closed containers and away from sunlight.

Dosage

2–3 g of the drug in powder form.
5–10 g of the drug for decoction.
1–10 g dried root or 3–12 ml/day of a 1:3 @ 45% tincture.

Patents

The invention relates to an antifungal composition for the treatment of human nails containing extracts of walnut hull, pulverized roots of *Nardostachys jatamansi* or vetiveria zizanioides or catharanthus roseus, polyols, fixed oil, non-ionic emulsifiers, thickening agent plasticizer and base. The invention also relates to a process for the preparation of the above synergistic composition (Bindra et al., 2001).

A method for promoting restful, quality sleep in an individual comprising the administration of a composition comprising hops extract, *Eclipta alba* extract powder, and *Nardostachys jatamansi* extract (Heuer et al., 2008).

BIBLIOGRAPHY

1. Bindra RL, Singh AK, Shawl AS, Kumar S. Anti-fungal herbal formulation for treatment of human nails fungus and process thereof. United States Patent 6296838. Application no. 09/535514. Publication date: 10/02/2001. http://www.freepatentsonline.com/6296838. html

2. Dandagi PM, Patil MB, Mastiholimath VS, Gadad AP, Dhumansure RH. Development and evaluation of Hepatoprotective polyherbal formulation containing some indigenous medicinal plants. Indian J Pharm Sci. 2008 Mar–Apr; 70(2): 265–268.

3. Gottumukkala VR, Annamalai T, Muk-hopadhyay T. Phytochemical investigation and hair growth studies on the rhizomes of *Nardostachys jatamansi* DC. Phcog Mag 2011; 7: 146–50.

4. Heuer M, Clement K, Chaudhuri S, Thomas M. Composition for a feeling of calmness. United States Patent 7374785. Application no. 11/696945. Publication date: 05/20/2008. http://www. freepatentsonline.com/7374785. html

5. http://logayurveda.com/plant-profiles/156-nardostachys-jatamansi.html

6. http://mdidea.com/products/proper/proper05804.html

7. http://www.ayurvedictalk.com/jatamansi-musk-root/110/

8. http://www.drugs.com/npp/jatamansi. html

9. http://www.herbalcureindia.com/herbs/nardostachys-jatamansi.htm

10. Laishram D, Kumar R, Sandhu JS. Effects of strength training and jatamansi on reducing hand tremor amongst archers. Archivos Venezolanos de Farmacología y Terapéutica. 2008; 27 (2): 105–109.

11. Naveen P, Kumari NS, Sanjeev G, Gowda KMD, Shetty P. Assessment of acute and subchronic (28 day) oral toxicity of phytochemical combination in rats. Journal of Pharmacy Research. 2011; 4(8): 2696–2698.

12. Rahman H, Muralidharan P. Nardostacys jatamansi DC protects from the loss of memory and cognition deficits in sleep deprived alzheimer's disease (AD) mice model. International Journal of Pharmaceutical Sciences Review and Research. 2010; 5(3): 160–167.

13. Rasheed AS, Venkataraman S, Jayaveera KN, Mohammed Fazil A, Yasodha KJ, Aleem MA et al., Evaluation of toxicological and antioxidant potential of *Nardostachys jatamansi* in reversing haloperidol-induced catalepsy in rats. Int J Gen Med. 2010; 3: 127–136.

14. Yadav SK, Gupta SK, Prabha S. Hair growth activity of *Nardostachys jatamansi* and *Cyperus rotundus* rhizomes extract on chemotherapy induced alopecia. International Journal of Drug Discovery and Herbal Research. 2011; 1(2): 52–54.

Ocimum sanctum

Plant profile

Kingdom: *Plantae* – Plants
Superdivision: *Spermatophyta* – Seed plants
Class: *Magnoliopsida* – Dicotyledons
Order: *Lamiales*
Genus: *Ocimum* L. – Tulsi

Subkingdom: *Tracheobionta* – Vascular plants
Division: *Magnoliophyta* – Flowering plants
Subclass: *Asteridae*
Family: Lamiaceae–Mint family
Species: *Ocimum sanctum* – Tulsi plant

Vernacular name

San: Manjari/Krishna tulsi, Gouri, Bhutaghni, Devadundubhi, **Eng:** Holy Basil, **Hind:** Tulsi, **Ben:** Tulsi
Kan: Tulasi, **Mal:** Trittavu, **Tam:** Thulsi, **Mar:** Tulshi, **Tel:** Thulsi, **Gujarath:** Thulasi

Common name

Basil

Ayurvedic properties

Rasa: Katu (sharp), Tikta (bitter), **Veerya:** Ushna (hot), **Guna:** Lakhu, Rooksa, **Vipaka:** Katu (sharp).

Ayurvedic preparations

Tribhuvana kirti, Caturbhuja rasa, Tulasi svarasa, Tulasi taila, Mukta pancamrta, Jvarasamharaka rasa

INTRODUCTION

Ocimum sanctum is an annual, erect, herbaceous, much-branched, softly hairy plant
with annual purple or crimson flowers. This
herb is belonging to the mint family having
least 150 varieties. The herb is an erect,
erbaceous, much branched, soft hairy plant.
Ocimum sanctum grows to 1–2 ft high. The
toothed leaves are often purple. The flowers
vary in color from white to red, sometimes
with a tinge of purple, appear from June to
September. Ocimum sanctum emits a spicy
scent when bruised. It is believed to purify air
by discharging negative ions. Holy Basil is
also believed to possess spiritual powers.

Distribution

Ocimum sanctum is native to India, Iran and
now cultivated in Egypt, France, Hungary,
Italy, Morocco, USA. Basil is naturally found
wild in the tropical and subtropical regions
of the world. Basil thrives in warm and temperate climates.

Chemical Constituents

Leaves contain an essential oil, which contains
eugenol, eugenal, carvacrol, methylchavicol,
limatrol and caryophylline. Seeds contain oil
composed of fatty acids and sitosterol. Roots
contain sitosterol and three triterpenes
A, B, and C. Leaves also contain a steroid
ursolic acid and n-triacontanol. Eugenol
(70.5%), its methyl ether (4.8%), nerol (6.4%),
caryophyllene (7.5%), terpinen–4 (0.4%),
decylaldehyde (0.2%), selinene (0.4%),
pinene (0.4%), camphene (2.0%) and a-
pinene (3.5%) identified in essential oil
by GC. The oil also consists of eugenol,

Eugenol

Ursolic acid

Rosmarinic acid

Nerol

ursolic acid, rosmarinic acid, thymol, linalool, methyl chavicol, citral and beta-caryophyllene.

Medicinal uses

In ayurveda the *Ocimum sanctum* leaves, flowers and occasionally the whole plant is used medicinally in the treatment of heart and blood diseases, leucoderma, strangury, asthma, bronchitis, lumbago and purulent discharge of the ear. The leaf juice possesses diaphoretic, antiperiodic, stimulant and expectorant properties. It is used to treat infantile cough, cold, bronchitis, diarrhea and dysentery and it is applied to the skin to treat ringworm and other skin diseases. An infusion of the leaves is used as a stomachic for gastric disorders in childrens. The oil extracted from the leaves is reported to possess antibacterial and insecticidal properties, and is effective as a mosquito repellent.

Cosmetic uses

Basil oil has been used over the century as medicated oils and ointments for respiratory infections especially for children. Basil oil is used in flavoring, cosmetics, soap, pharmaceuticals and perfumery. It is used for body scrub. It helps in controlling dandruff. It reduces eczema and psoriasis. With its antibacterial properties it can reduces skin diseases like staph infection. It flushes out the toxins and regenerates skin. Tulsi contains ursolic acid which prevents wrinkles and helps retain the elasticity prevalent in young faces. Tulsi powder can remove the spots from the face when applied on face by mixed with lemon juice. Tulsi paste can also be used for treatment of acne.

Pharmacological Action

The present study was planned to compare the oral and topical *Ocimum sanctum* for wound healing property by excision and incision resutured wound models in albino rats. The study carried out with complete epithelisation time, wound contraction, histopathological study and tensile strength of the wounds. The time taken for 50% wound contraction and complete eipthelization by oral *Ocimum sanctum*, topical *Ocimum sanctum* was significantly (p < 0.001) less compared to oral and topical controls. Further, mean tensile strength of oral and topical *Ocimum sanctum* treated wound was significantly great (p < 0.001) compared to controls. Finally it was revealed that oral and topical *Ocimum sanctum* promoted better granulation tissue, early and complete epithelisation and better tensile strength compared to both controls (Asha et al., 2011).

The study was evaluated the hepato-protective activity of *Ocimum sanctum* and observed its synergistic hepatoprotection with silymarin in albino rats. The alcoholic extract of *Ocimum Sanctum* leaves (OSE) 100 mg/kg and 200 mg/kg BW/day p.o. was administered. The hepatoprotective effect was evaluated by performing an assay of the serum proteins, albumin–globulin ratio, alkaline phosphatase, transaminases and liver histopathology. Finally the results revealed the *Ocimum sanctum* alcoholic leaf extract showed significant hepatoprotective activity and synergism with silymarin (Lahon and Das, 2011).

This study was carried out to observe the antibacterial activity of aqueous extract, chloroform extract, alcohol extract and oil obtained from leaves of *Ocimum sanctum* against the selected bacteria, i.e. *E. coli*, *P. aeruginosa*, *S. typhimurium* and *S. aureus*. The antibacterial activity of Ocimum was evaluated by liquid inhibition test. Chloroform extract were found most effective against *P. aeruginosa*. Extract obtained from *Ocimum sanctum* were observed equally effective against the gram negative and gram positive bacteria. Present investigation reveals that *Ocimum sanctum* may be a better alternative as a preservative in Food Industries (Mishra and Mishra, 2011).

Present study was evaluated the immunomodulatory effect of *Ocimum sanctum* in rat. Aqueous extract of *Ocimum sanctum* were administered orally at doses of 100, 200 mg/

kg/day for 45 days in wistar albino rats and showed increasing antibody production in dose dependent manner. Finally the results concluded that an oral administration of the aqueous extract of O. sanctum showed immunomodulatory effect in rat (Jeba et al., 2011).

This study evaluated a new herbal gel preparation containing extract from leaves of Ocimum sanctum (OS) for its topical anti-inflammatory activity against carrageenan induced edema and anti-nociceptive effect on wistar rats. Gelling agent used in this study was 1% w/w concentration of carbopol- 940. Change in oedema volume of the rat hind paw was measured. From the study it was concluded that combination of Ocimum sanctum (OS) potentiated the anti-inflammatory and anti-nociceptive effect topically (Phanse et al., 2011).

Using haloperidol induced catalepsy in rat and muscle rigidity in mice effects of aq. extract of Ocimum sanctum (OSEaq) were studied. Haloperidol was administered to induce either short term model (0.5 mg/kg i.p) or long term model (4 mg/kg s.c.) to observe changes in catalepsy and muscle rigidity as measured using rota-rod test and chimney test. Out of 100 mg, 200 mg, 300 mg and 600 mg/kg p.o. doses of O. sanctum, 300 mg/kg dose showed maximum shortening of onset and duration of catalepsy as compared with other dose levels in short term and long term model. Pretreatment with OSEaq (300 mg and 450 mg/kg p.o.) significantly reduces initial decrease in activity in rota rod and significantly improves performance of mice in chimney test as compared with control in short-term and long-term models. In conclusion, OSEaq in the dose 300 mg/kg i.p. shows anticataleptic action in rats as well as in the dose 300 and 450 mg/kg i.p. improves the performance of mice in rota-rod and chimney test indicating reduction of muscle rigidity (Joshi et al., 2011).

Toxicological Studies

The study was attempted to investigate the effect of an oral administration of Ocimum sanctum leaf extract on induction of toxicity, mutagenicity, and anti-mutagenicity using rat bone marrow micronucleus test. Cyclophosphamide was used to elucidate the possible mechanism of anti-mutagenicity property of this extract. Ocimum sanctum leaf extract was given daily via oral administration to Sprague-Dawley rats to determine acute toxicity, mutagenicity and anti-mutagenicity against cyclophosphamide. The results showed that Ocimum sanctum leaf extract was not toxic up to the dose of 15 g/kg bw in 5-week-old rats. The mutagenicity, represented as micronucleus induction, was assessed by bone marrow using i.p. 80 mg/kg bw of cyclophosphamide 30 hours prior to bone marrow collection and further showed at a dose level of 5 g/kg bw did not cause mutagenicity in the rats (Khumphant and Lawson, 2002).

The median lethal dose (LD_{50}) of Ocimum sanctum (OS) fixed oil was determined after ip administration in mice. The fixed oil was well tolerated up to 30 ml/kg, while 100% mortality was recorded with a dose of 55 ml/kg. The LD_{50} of oil was 42.5 ml/kg. No untoward effect was found on subacute toxicity study of OS fixed oil at a dose of 3 ml/kg/day, ip for 14 days in rats (Singh et al., 2007). In another study, healthy rats were orally fed with increasing doses (400 mg/kg to 6 g/kg body weight) of alcoholic leaf extract of Ocimum sanctum for 14 days. The results showed the doses up to 4 g/kg body weight did not produce any toxicity and mortality (Shetty et al., 2008).

The acute toxicity of 80% methanolic extract of Ocimum sanctum was determined as per the OECD guideline no. 423 (Acute Toxic Class Method). It was observed that the test extract was not mortal even at 2000 mg/kg dose (Latha et al., 2010).

Clinical Investigation

The present study was planned to evaluate the clinical efficacy and safety of Ophthacare eye drops, on topical application, in acute and chronic conjunctivitis. The study was an open non-comparative clinical trial. Ophthacare eye drops is a polyherbal formulation indicated for acute and chronic conjunctivitis and it contains extracts of *Carum copticum, Terminalia bellirica, Emblica officinalis, Curcuma longa, Ocimum sanctum, Rosa damascena, Cinnamomum camphora* with purified Mel despumatum (honey). Results revealed there was a significant reduction in the mean score for conjunctivitis and significant reduction in the mean score for congestion at the end of the study period. The clinical trial observed a remarkably rapid symptomatic relief and significant reduction in the mean scores for conjunctivitis, congestion, papillae and follicle in the patients with acute and chronic conjunctivitis, without any clinically significant adverse events (Srivastava and Kolhapure, 2004).

Market Formulations

ACLEAR Topical

ACLEAR topical is an effective non-comedogenic topical cream for acne and scars. It is an anti-acne cream formulated with non-comedogenic base for the first time in the field of Ayurveda. This cream begins to reduce inflammation, redness and pain associated with acne within 24–48 hrs of its application. This cream with its non-comedogenic activity prevents the formation of white heads and black heads, which are the most common type of acne. As it inhibits the growth of *P. acnes* (bacteria involved in causing acne), it prevents the acne lesions from further damage. This cream prevents scar formation, smoothens skin and brings back the original glow.

Ingredients: Each 10 gm contains: *Salmalia malabarica* (Shalmali): 1.5 gm, *Sphaeranthus indicus* (mundi): 1.5 gm, *Ocimum sanctum* (tulasi): 1.5 gm, *Cinnamomum camphora* (Karpura): 1.5 gm, *Azadirachta indica* (Nimba):1.5 gm, non-comedogenic base: q.s.

Direction: Clean the affected area with lukewarm water and apply ACLEAR in adequate quantity over the lesions 2–3 times a day.

Supplied by: Ayurveda Elements, Chatswood NSW.

Neem and Tulsi Hand Made Soap

This natural herbal soap helps in prevention of acne, pimple, black acne and skin infections. It also cleanses skin pores and closed them. Hence formation of acne and pimple gets prevented.

Ingredients: Neem, Tulsi.

Action: Neem helps in protecting our skin from microorganisms like bacteria, fungus and protozoa. Hence it acts as a natural anti bacterial, antifungal and antiprotozoal herb. Neem helps in quick healing of acne, pimples eczema and ring worms. It also prevents clogging of pores and further infections. Tulsi helps in skin disorders. Medicinal properties of tulsi help in reducing acne and pimple. It prevents outbreak of acne and pimple.

Supplied by: Ayurhelp.com., Mysore.

Neem and Tulsi Soap

This soap is used for cleansing, aromatherapy purpose.

Ingredients: Saponified oils of coconut, palm and olive; neem leaf, neem oil, tulsi (holy basil) leaf, essential oils of coriander seed, camphor, lemon grass. Also contains shea butter and vitamin E.

Action: Neem's activity against everyday germs and bacteria. Tulsi has been shown to be helpful in helping irritated skin feel better.

Direction: Product contains essential oils, hence it is advised to use the soap externally and keep from splashing into eyes.

Supplied by: Nature's Formulary, USA.

Anti-acne Facial Kit

Anti-acne facial kit is an oil control formula which regulates the excessive oil secretion, thereby detoxifying the skin. It is used for all types of skin.

Ingredients: Aloe vera (ghrat kumari), basil oil (tulsi), clove oil (laung), camphor (kapur), neem, turmeric (haldi), tea tree oil, vitamin E.

Supplied by: Rising Sun Aromas and Spiritual (P) Ltd., New Delhi.

Various Preparations

Tulasi has specified actions on the respiratory system—pranavaha srotasa. It effectively liquefies the phlegm due to its hot and sharp attributes. It gives excellent results in cough due to kapha, allergic bronchitis, asthma and eosinophilia. Combined with honey, the juice works well to control the hiccup. In tubercular cough, tulasi is also beneficial. It is an effective panacea for fever, especially of kapha type, while given with honey and marica fruit powder. In such conditions it effectively controls colds and reduces pain. Tulasi juice works as amapacana, meaning it digests and destroys ama (the toxins). Tea prepared with the leaves of tulasi is a common domestic remedy for cold, cough, milk indigestion, diminished appetite and malaise.

The massage with the leaves juice improves the circulation beneath the skin and augments the sensation in the skin. In the headache due to sinusitis, the instillation of juice in the nose facilitates the secretions of kapha and relieves the headache. The dried powder of the leaves can be inhaled, like a snuff, for the same purpose.

Storage of Products

All the products should be stored in well closed containers and away from sunlight.

Dosage

Tincture: 1–2 ml, two times daily.

Infusion: 1–2 cups daily.

Standardized *Ocimum sanctum* extract 400 mg daily can protect from the radiation.

Patents

The invention provides a novel herbal formulation useful for the treatment of skin disorders and comprising two or more plant extracts selected from *Tagetes erecta*, *Moringa oleifera*, *Ocimum sanctum*, *Tridax procumbens*, *Aloe vera*, and *Gum olibanum* together with conventional additives (Ramakrishna et al., 2002).

The invention provides an analgesic and refreshing herbal composition useful as dentrifrices, said composition comprising 50–60% wt. of betle extract (from piper betle leaves); 40–50% wt. of one or more group I essential oil selected from *Levender officinale*, dementholised oil (ex *Mentha arvensis*), fennel oil and *Ocimum gratissimum*; 3.5–6% wt. of one or more group II essential oils and their isolates selected from *Ocimum sanctum*, Pulegone (ex *Mentha pulegonium*), carvone (ex. dill seed) and menthol (ex *Mentha arvensis*); 1–5% wt. of one or more group III essential oils selected from camphor, turpentine oil, cedarwood oil and safrole oil, along with 0.5–2% wt. of thymol and 0.25–1% wt. of preservative/antioxidant, and a process for preparing the composition (Singh et al., 2003).

The present invention provides a novel nicotine free anticigarette herbal formulation as an antidote to the poisoning effects of tobacco products such as cigarettes, gutka, pan masala and other similar tobacco related products. Formulation(s) comprises sterilized dried plant powder/extracts together with the conventional additives to form the oral dosage forms, which include tablets, capsules, syrup and powders ready for suspension and mouth spray. The anti-tobacco addiction herbal formulation comprises *Sesbania grandiflora*, *Catharanthus roseus*, *Ocimum sanctum*, *Myristica fragrans*, *Elettaria cardamomum*, *Carum copticum*, *Syzygium aromaticum*, *Cinnamomum tamala*, *Acorus calamus*, *Zingibale officinale*, *Piper nigrum*, *Cinnamomum zeylanicum*, *Cuminum cyminum*, *Nigella sativum*, *Cinnamomum camphora*, *Piper longum Ocimum gratissimum* and *Hemidesmus indicus* (Karerat et al., 2009).

BIBLIOGRAPHY

1. Asha B, Nagabhushan A, Shashikala GH. Comparative study of wound healing activity of topical and oral *Ocimum Sanctum* Linn in albino rats. Al Ameen J Med Sci. 2011; 4(4): 309–314.

2. http://ayurveda-foryou.com/ayurveda_herb/tulsi.html

3. http://ayurvedicmedicinalplants.com/plants/381.html

4. http://www.basilleaves.com/

5. http://www.ehow.com/about_5594921_use-tulsi-cosmetics.html

6. http://www.gyanessentialoils.com/holy-basil-oil-ocimum-sanctum-oil.htm

7. http://www.herbalcureindia.com/herbs/ocimum-sanctum.htm

8. http://www.motherherbs.com/ocimum-sanctum-extract.html

9. Jeba RC, Vaidyanathan R, Rameshkumar G. Immunomodulatory activity of aqueous extract of *Ocimum sanctum* in rat. International Journal on Pharmaceutical and Biomedical Research. 2011; 2(1): 33–38.

10. Joshi SV, Bothara SB, Surana SJ. Evaluation of aqueous extract of *Ocimum sanctum* in experimentally induced parkinsonism. J. Chem. Pharm. Res., 2011; 3(1):478–487.

11. Karerat AK, Ifthikar ORM, Varghese J, Vellappillil AV, Palpu P, Rawat AKS, Rao CV, Govindarajan R. Anti-cigarette herbal formulation as an antidote to tobacco. United States Patent 7534454. Application No: 11/024024. Publication date: 05/19/2009. http://www.freepatentsonline.com/7534454. html

12. Khumphant E, Lawson DB. Acute toxicity, mutagenicity and antimutagenicity of ethanol *Ocimum sanctum* leaf extract using rat bone marrow micronucleus assay. Kasetsart Journal: Natural Science. 2002; 36(2): 166–174.

13. Lahon K, Das S. Hepatoprotective activity of *Ocimum sanctum* alcoholic leaf extract against paracetamol-induced liver damage in albino rats. Pharmacognosy Res. 2011; 3(1): 13–18.

14. Latha P, Saravana Kumar A, Mohana Lakshmi S, Harini Chowdary V, Nagarjuna D, Udaya Sri N et al. , *Ocimum sanctum* linn. attenuates haloperidol induced tardive dyskinesia and associated behavioural, biochemical and neuro-chemical changes in rats. International Journal of Phytopharmacology. 2010; 1(2): 74–81.

15. Mishra P, Mishra S. Study of anti-bacterial activity of *Ocimum sanctum* extract against gram positive and gram negative bacteria. American Journal of Food Technology. 2011; 6(4): 336–341.

16. Phanse M, Kulkarni P, Kewatkar S, Lande M, Bhujbal S, Chaudhari P. Evaluation of anti-inflammatory activity of herbal gel formulation. J. Nat. Prod. Plant Resour., 2011, 1 (2): 25–28.

17. Ramakrishna S, Sitaramam BS, Diwan PVR, Raghavan KV. Herbal formulation useful for treatment of skin disorders. United States Patent 6383495. Application No: 09/659124. Publication date: 05/07/2002. http://www.freepatentsonline. com/6383495.html

18. Shetty S, Udupa S, Udupa L. Evaluation of Antioxidant and Wound Healing Effects of Alcoholic and Aqueous Extract of *Ocimum sanctum Linn* in Rats. Evid Based Complement Alternat Med. 2008; 5(1): 95–101.

19. Singh AK, Bindra RL, Gupta R, Shukla YN, Kumar S. Analgesic and refreshing herbal composition and a process for preparing the same. United States Patent 6531115. Application No: 09/752822. Publication date: 03/11/2003. http://www.freepatentsonline.com/6531115. html

20. Singh S, Taneja M, Majumdar DK. Biological activities of *Ocimum sanctum* L. fixed oil-An overview. Indian J Exp Biol.2007; 45: 403–412.

21. Srivastava U, Kolhapure SA. Evaluation of efficacy and safety of Ophthacare eye drops in acute and chronic con-junctivitis. The Antiseptic. 2004; 101(6): 211–217.

Primula veris

Plant profile

Kingdom: *Plantae* – Plants
Superdivision: *Spermatophyta* – Seed plants
Class: *Magnoliopsida* – Dicotyledons
Order: *Primulales*
Genus: *Primula* L. – Primrose

Subkingdom: *Tracheobionta* – Vascular plants
Division: *Magnoliophyta* – Flowering plants
Subclass: *Dilleniidae*
Family: *Primulaceae*–Primrose family
Species: *Primula veris* –Cowslip primrose

Common name

Cowslip primrose, key flower, key of heaven, palsywort, fairy cups, primrose.

INTRODUCTION

It is a herbaceous perennial plant, low growing, to 10–30 cm tall, with a basal rosette of leaves. Leaves are 5–25 cm long and 2–6 cm broad with an irregularly crenate to dentate margin, and a usually short leaf stem. The flowers are 2–4 cm in diameter, borne singly on a slender stem, pale yellow, white, red, or purple, actinomorphic with a superior ovary which later forms a capsule which opens by valves to release the small black seeds. The flowers are hermaphrodite but heterostylous. Blooms are sweet smelling and yellow in colour, marked with orange dots. It is generally one of the first flowers to bloom in spring.

Distribution

This plant is native to Western and Southern Europe (from the Faroe Islands and Norway south to Portugal, and east to Germany, Ukraine, the Crimea and the Balkans), northwest Africa (Algeria), and southwest Asia (Turkey east to Iran). In Northern Belgium, it is common, but mainly occurs in fragmented habitats.

Chemical Constituents

The plant contains saponins, which have an expectorant effect, and salicylates which are the main ingredients of aspirin and have anodyne. The roots and the flowers have somewhat of the odor of anise, due to their containing some volatile oil identical with mannite. Root contains *Phenol glycosides* (0.2 to 2.3%, high values in the spring): Primulaverin (3%, 2-hydroxy-5-methoxy-benzoic acid methyl ester-O-xyloglucoside) changing over during dehydration into the characteristic-smelling 5-methoxy-methyl salicylate, *Triterpene saponins* (5 to 10%): Chief components primulic acid A (chief aglycone protoprimulagenin). Flower contains *flavonoids* (3%) including rutin, kaempferol-3-O-rutinoside, isorhamnetin-3-O-glucoside; isorhamnetin rhamnosyl robinoside, isorhamnetin robinoside, isorhamnetin rutinoside, kaempferol robinoside, limocitrin-3-O-glucoside, quercetin gentiobioside, quercetin-3-O-glucoside, quercetin robinoside.

Primulaverin

Medicinal uses

It is mainly use in treating conditions involving spasms, cramps, paralysis and rheumatic pains. It has anti-inflammatory and febrifuge effects. Flowers and the leaves are anodyne, diaphoretic, diuretic and expectorant. The yellow corolla of the flower is antispasmodic and sedative. They are recommended for treating over-activity and sleeplessness, especially in children. They are potentially valuable in the treatment of asthma and other allergic conditions. Oil from the flowers has an antiecchymotic effect (treats bruising). The root contains 5–10% triterpenoid saponins which are strongly expectorant, stimulating a more liquid mucous and so easing the clearance of phlegm. It has been dried and made into a powder, then used as a sternutatory. The root is also mildly diuretic, antirheumatic and slows the clotting of blood. It is used in the treatment of chronic coughs (especially those associated with chronic bronchitis and catarrhal congestion), flu and other febrile conditions. It is used in the treatment of kidney complaints and catarrh. The cowslip roots are expectorant, and can be used to treat cold and flu-like symptoms. They are widely held to break up mucus so that it can be more easily expelled by the body. It is useful in treating epilepsy, tremors, and may even be developed as a treatment for Parkinson's disease.

Cosmetic uses

Modern herbalists still make a skin cleaning lotion from cowslip. It is said to be useful in treating acne, pimples, and other skin blemishes. Its unique cleansing properties are said to remove dirt, and open the pores of the skin, allowing for a fresher, smoother look. A flower tincture or essential oil works well for insomnia, anxiety, or over-excitement. Ointments with cowslip flowers are used on sunburns and skin blemishes.

Pharmacological Action

Aqueous and ethanolic extracts of 42 plants used in Danish folk medicine for the treatment of epilepsy and convulsions, or for inducing sedation, were tested for affinity to the $GABA_A$–benzodiazepine receptor in the flumazenil-binding assay. Ethanolic extracts of leaves of *Primula elatior* and *Primula veris* and aerial parts of *Tanacetum parthenium* exhibited good, dose-dependent affinity (Jager et al., 2006).

Several flower extracts from *Primula veris* L. has been tested for antibacterial activity and decoction from the flowers has been tested for antimitotic activity. Antibacterial activity was determined by the well diffusion method and *Allium cepa* L. has been used for evaluating cytotoxicity. Decoction of flowers was toxic on root number and root length in *A. cepa* L. and reduced the mitotic index significantly. All of the tested *P. veris* L. extracts showed inhibitory effect against both gram-positive and gram-negative microorganisms at varying degrees. The most effective fraction was found to be the ethanolic (Gamze et al., 2008).

Toxicological Studies

There are only LD values for the saponin fraction from *P. veris* root extract (LD_{50} mouse, i.p. 24.5 mg/kg b.w.) or for primula acid (LD_{50} rat, i.v. 1.2 mg/kg b.w.) available, which have no relevance for the oral administration of Primula flower preparations (Hansel et al., 1994).

The oral toxicity of saponins in mammals is relatively low due to their poor absorption. LD_{50} values are in the range of 50 mg/kg (which is not very low when the figures are correct) and 1000 mg/kg (Hostettman and Marston 1995).

There are no data on genotoxicity, carcinogenicity, reproductive and developmental toxicity was published.

Clinical Investigation

Recent clinical studies with preparation containing solely Primula flower was not reported.

Market Formulations

Neem Oil and Botanical Fading Cream

Clinically proven, botanical blend, neem oil fading cream optimally formulated to brighten your skin. It helps reduce the appearance of dark spots. Neem oil and herbal emollients balance formula. Gentle, effective action with regular use.

Ingredients: Certified organic *Melia azadirachta* (neem) leaf extract, *Aloe barbadensis* gel, glycerin, *Helianthus annuus* (sunflower) oil, *Simmondsia chinensis* (jojoba) oil, *Persea gratissima* (avocado) oil, *Melia azadirachta* (neem) oil, sodium acrylates copolymer, *Malva sylvestris* (mallow) extract, *Menta piperita* (peppermint) leaf extract, paraffinum liquidum, phenoxyethanol, *Primula veris* (cowslip primrose) extract, *Alchemilla vulgaris* (lady's mantle) extract, *Veronica officinalis* (spe-edwell) extract, *Melissa officinalis* (lemon balm) extract, *Achillea millefolium* (yarrow) extract, PPG-1 Trideceth-6, fragrance (Parfum from natural ingredients) methylparaben, isopropylparaben, isobutylparaben, butylparaben.

Supplied by: Herbal Remedies, India.

Soft and Beautiful Skin Hand Cream

Ingredients: Water (aqua/eau), glycerin, petrolatum, mineral oil (paraffinum liquidum, huile minerale), ceteryl alcohol, caprylic/capric acid, triglyceride, stearic acid, cetyl alcohol, cyclopentasiloxane, dimethicone, olive oil peg 7 esters, phenoxyethanol, cyclohexasiloxane, ceteryl glucoside, urea, theobroma cacao seed butter (cocoa), butyrospermum parkii (shea butter), dicetyl phosphate, ceteth 10 phosphate, acrylates/C10 30 alkyl acrylate crosspolymer, triethanolamine, mangifera indica seed butter (mango), methylparaben, fragrance (parfum), disodium EDTA, butylphenyl methylpropional, linalool, hexyl cinnamal, limonene, citronellol, benzyl salicylate, alpha isomethyl ionone, amyl cinnamal, geraniol, benzyl benzoate, *Prunus Amygdalus dulcis* oil (sweet almond oil), polysorbate 20, propylene glycol, hydrogenated sweet almond oil, sorbitol, tocopheryl acetate, panthenol, ascorbic acid, *Sambucus nigra* flower extract, *Primula veris* extract, *Helianthus annuus* extract (sunflower), *Chamomilla recutita* flower extract (matricaria), niacinamide, biotin.

Direction: Apply and massage into hands as needed.

Supplied by: Environmental Working Group, Washington.

Illuminating Cream, Skin Brightener and Spot Fader

CamoCare Organics illuminating cream has the right balance of richness and light refreshment for more beautiful-looking skin. Clinically it is an anti-aging, rejuvenate/beautify. Create the appearance of more even, flawless-looking skin with this advanced formula of eleven brightening botanicals, light diffusing particles and potent antioxidants that also help: Reduce appearance of discolorations and spots, give skin a more luminous, perfected appearance, protect skin against free radicals.

Ingredients: *Aloe barbadensis* (*Aloe vera* gel), Water, SD Alcohol 38-B (organic lavender alcohol), *Malva sylvestris* (mallow) extract, *Mentha piperita* (peppermint) leaf extract, *Primula veris* extract, *Alchemilla vulgaris* extract, *Veronica officinalis* extract, *Melissa officinalis* leaf extract, *Achillea millefolium* extract (gigawhite), *Helianthus annuus* (sunflower) seed oil, cetyl alcohol (from coconut/palm), glyceryl stearate (from sugar/palm), *Butyrospermum parkii* seed (shea) butter, *Punica granatum* (pomegranate) seed extract, *Carthamus tinctorius* (safflower) seed oil, *Lavandula angustifolia* (lavender) extract, *Matricaria recutita* (german chamomile) flower extract, *Helianthus annuus* (sunflower) seed oil, *Glycyrrhiza glabra* (licorice) extract, *Morus alba* (mulberry) extract, glycine max (soyabean) extract, *Curcuma longa* (turmeric) extract, tocopheryl acetate (vitamin E), Tocotrienols, Ascorbic acid (vitamin C), lecithin (from soy), citric acid, cetyl hydroxyethylcellulose (modified plant cellulose), mica (mineral), titanium dioxide (mineral).

Direction: At mornings and evening, apply small amount to face and neck. It may also be used as a spot treatment on discolored skin, age/sun spots and freckles. It can also be used on face, neck, shoulders, chest and back of hands. For the best results, use twice a day for 6–8 weeks.

Storage: Store in a cool, dry place below 25°C (77°F).

Supplied by: Nature's Way Products, USA.

Various Preparations

Cowslip has been used for centuries to make sedative tea. Its leaves are said to be mildly narcotic, and it is used as an herbal remedy for insomnia as well as hyperactivity.

Massage oil from the flower is used to treat nerve pain, or applied to the temples for migraines. Cowslip root decoctions and tinctures can be taken to clear stubborn phlegm, also relieve arthritis and rheumatism as well.

Extract: Percolation with 50 parts water and 50 parts ethanol, then filtration and vacuum drying. The residue is dissolved in 60 parts ethanol and 40 parts water and neutralized with ammonia. It is then cooled for 24 hours and filtered again. It is finally dehydrated to produce a dry extract under low pressure. Liquid extract: Primula extract is dissolved in a mixture of ethanol (30 parts), glycerol 85% (20 parts) and water (20 parts) and filtered when cool.

Tincture: 20 parts root and 100 parts diluted ethanol are processed to a tincture by maceration procedure.

Syrup: 1.5 parts cowslip are dissolved in 20 parts water while being heated. It is then mixed with 10 parts 85% glycerol and 68.5 parts simple syrup.

Tea: 0.2 to 0.5 g finely cut drug are added to cold water and brought to the boil, left to draw for 5 minutes and strained (1 teaspoon corresponds to approximately 3.5 g drug).

Storage of Products

All the products should be stored in well closed containers and away from sunlight.

Dosage

When prepared as a tea or infusion, the patient should drink one cup 3 times per day, roughly corresponding to mealtimes. A 2–4 ml dose of cowslip tincture may also be used three times daily. As an ointment or cream, cowslip may be applied as needed, but usually no more than 2–3 times per day.

As a tea for promoting phlegm, use 1.5–3 teaspoons (2–4 g) of cut and dried flowers daily.

In using the tincture, try 1.5–3 g (about 0.25–0.5 teaspoonfuls) daily.

The average daily dose is 1 g of drug. The single dose is 0.5 g of drug.

Tincture: The daily dose is 7.5 g.

Extract: The single dose is 0.1 to 0.2 g.

Liquid extract: The single dose is 0.5 g.

Patents

Recent patents on *Primula veris* in cosmetic formulations are not available.

BIBLIOGRAPHY

1. Gamze Basbulbul, Ali Ozmen, Halil Biyik H, Ozge Sen. Antimitotic and antibacterial effects of the *Primula veris* L. flower extracts. Caryologia. 2008; 61(1): 88–91.

2. Hänsel, R., Keller, K., Rimpler, H., Schneider, G., editors. Primula. In: Hagers Handbuch der Pharmazeutischen Praxis, 5th ed. Band 6, Drogen P-Z. 269–87. Berlin. Springer-Verlag, 1994.

3. Hostettman, K., Marston, A. Saponins: Cambridge: Cambridge University Press. 1995.

4. http://botanical.com/botanical/mgmh/c/cowsl112.html

5. http://en.wikipedia.org/wiki/Primula_vulgaris

6. http://www.health24.com/natural/Herbs/17-666-676,65502.asp

7. http://www.herbalremedies.com/cowslip-information.html

8. http://www.herbal-supplement-resource.com/cowslip-herb.html

9. http://www.pfaf.org/user/Plant.aspx?LatinName=Primula+veris

10. http://www.purematters.com/herbs-supplements/c/cowslip# indicationsusage

11. Jager AK, Gauguin B, Adsersen A, Gudiksen L. Screening of plants used in Danish folk medicine to treat epilepsy and convulsions. J Ethanopharmacol. 2006; 105(1–2): 294–300.

Prunus dulcis

Plant profile

Kingdom: *Plantae* – Plants
Superdivision: *Spermatophyta* – Seed plants
Class: *Magnoliopsida* – Dicotyledons
Order: *Rosales*
Genus: *Prunus* L. – Plum

Subkingdom: *Tracheobionta* – Vascular plants
Division: *Magnoliophyta* – Flowering plants
Subclass: *Rosidae*
Family: *Rosaceae* – rose family
Species: *Prunus dulcis* D.A. Webb. – Sweet almond

Vernacular name

San: Vatada, Vatama, **Eng:** Almond, Sweet Almond, Indian Almond, **Hin:** Badam, **Ben:** Badaam
Kan: Badami, **Mal:** Badam, Badam kotta, **Tam:** Badaam, **Mar:** Badam, **Gujarath:** Badaam

Common name

Almond

Ayurvedic properties

Rasa: Madhura, Tikta (bitter), **Veerya:** Ushna (hot), **Guna:** Guru, Snigdha.

INTRODUCTION

The almond is a small deciduous tree, growing 4–10 m (13–33 ft) in height, with a trunk of up to 30 cm (12 in) in diameter. The young twigs are green at first, becoming purplish where exposed to sunlight, then grey in their second year. The leaves are 3–5 inches long, with a serrated margin and a 2.5 cm (1 in) petiole. The flowers are white or pale pink, 3–5 cm (1–2 in) diameter with five petals, produced singly or in pairs before the leaves in early spring. Almonds begin bearing an economic crop in the third year after planting. Trees reach full bearing after five to six years after planting. The fruit is mature in the autumn, 7–8 months after flowering. The almond fruit is not a nut, but a drupe 3.5–6 cm (1–2 in) long. The outer covering is called the hull. Inside the hull is a reticulated hard woody shell (like the outside of a peach pit) called the endocarp. Generally, one seed is present, but occasionally there are two.

Distribution

The native region of the almond tree is southwest Asia. Nevertheless, for centuries the almond tree has been cultivated in the warm regions of the Mediterranean countries as well as in California, Southern Australia and South Africa. In India it is cultivated in colder parts of Himalayan peninsula, and in Punjab.

Chemical Constituents

Almonds (bitter as well as sweet) contain about 50% of a fatty oil, which is, though, too expensive to be used for cooking. It is made up of glycerides (80% oleic acid, 15% linoleic acid, 5% palmitic acid). Bitter almonds contain 3 to 5% amygdalin, a so-called *cyanogenic glycoside* composed of mandelic nitrile and gentobiose. Vegetative parts of the almond tree accumulate the analogous prunasin (with glucose as sugar component). On enzymatic hydrolysis of these glycosides by α-glucosidases, the aglycon mandelic nitrile (2-hydroxy-3-phenylacetonitrile) is liberated. A second enzyme (mandelonitrile lyase) converts mandelic nitrile quickly to benzaldehyde (C_6H_5–CHO) and hydrocyanic acid (HCN, also known as prussic acid).

Amygdalin

Medicinal uses

The oil is used in folk remedies for cancer (especially bladder, breast, mouth, spleen, and uterus), carcinomata, condylomata, corns, indurations and tumors. Reported to be alterative, astringent, carminative, cyano-genetic, demulcent, discutient, diuretic, emollient, laxative, lithontryptic, nervine, sedative, stimulant and tonic, almond is a folk remedy for asthma, cold, corns, cough, dyspnea, eruptions, gingivitis, heartburn, itch, lungs, prurigo, skin, sores, spasms, stomatitis, and ulcers. The kernels are valued in diet for peptic ulcers. The seed is demulcent, emollient, laxative, nutritive and pectoral. The leaves are used in the treatment of diabetes. The plant contains the antitumour compound taxifolin.

Cosmetic uses

Sweet almond oil is used as an emollient and emulsifier in cosmetic products. Almond meal is used as a skin cleanser and in medicated soaps. Sweet almond oil is a favorite in aromatherapy as a carrier oil and is valued in natural skin care for its ability to soften and condition the skin. Almond oil is rich in protein, it is great for all skin types. Almond oil is good for all skin types and helps to relieve dryness, itching and inflammation. Finely ground almonds also serve as a facial skin scrub. The gentle abrasive action of the almond, helps to remove dead skin cells and are the perfect natural replacement for facial soap.

For face

- If massaged on face for 10–15 minutes in circular motion, it will improve blood circulation and most certainly will give you a clearer complexion.
- Almond oil when applied on face helps get rid of dark circles.
- It is useful for delaying ageing process. Apply on delicate areas like under eyes and side of eyes, it will help prevent wrinkles crow's feet.
- It can be used for skin rashes, skin irritation and inflammation and chapped lips.
- Apply it on eyebrows along with castor oil and olive oil, it will help grow thicker eyebrows.
- In short use it as a night cream.

For body

- Almond oil is especially helpful in the winter, when skin becomes dry and rough.
- Apply almond oil on face, neck and hands every night before go to sleep. Do not fret about being oily all night, almond oil gets absorbed into the skin within 30 minutes.
- At the end of the day, rub heated almond oil on hands and feet, including the sole, to relax and rejuvenate them. This will help moisturize the tough skin, and give softer, smoother skin.

Pharmacological Action

Seeds of *Prunus dulcis* were traditionally known for its hair growth activity. The study was aimed to investigate the efficacy of various extracts of *Prunus dulcis* as a potential hair growth promoter. Petroleum ether, methanol, chloroform and water extracts of *Prunus dulcis* seeds incorporated in oleaginous ointment base were applied topically on shaved denuded skin of albino rats and screened for hair growth activity. Petroleum ether extract showed consistent and significant increase in the length of hair ($p < 0.001$) and also showed a good percentage of hair follicles in the anagen phase after histological studies. This study concluded that the seed extracts of *Prunus dulcis* exhibits a significant potency in promoting hair growth (Suraj et al., 2009).

The modulatory effects of sweet almond extract on lipid peroxidation, serum lipids, and histology of pancreas in alloxan-induced diabetic wistar rats was studied. Serum lipids and lipid peroxidation levels were determined using appropriate chemical techniques. Total cholesterol (TC) and low density lipoprotein (LDL) activities significantly increased in the diabetic groups of animals treated with 300 mg/kg, 400 mg/kg and 500 mg/kg body weight of sweet almond extract in a dose dependent fashion while high density lipoprotein (HDL) showed a significant ($p < 0.05$) decrease. Triglyceride (TG), very low density lipoprotein (V L D L) and lipid peroxidation (LP), decreased significantly. The anti-lipid per oxidation and serum lipid modulatory activities of sweet almond extract shows ability to protect, reduce pancreatic damage and risk of diabetic complication (Etim et al., 2011).

Toxicological Studies

Pharmacological studies reveal that Sweet almond oil is absorbed slowly through intact skin, whereas it is easily absorbed and digested following oral administration. It is nontoxic when ingested, and products containing up to 25%. Sweet almond seeds do not contain amygdalin and can be eaten safely. Sweet almond oil are practically nonirritating to rabbit skin and only minimally irritating to rabbit eyes. In subchronic studies, sweet almond oil at 100% concentrations was only slightly irritating to rabbit skin (web article).

Clinical Investigation

In clinical studies, undiluted sweet almond oil and products containing up to 25% sweet almond were practically nonirritating and nonsensitizing. Formulations containing up to 2% sweet almond meal were practically nonirritating and nonsensitizing when tested in a repeated insult patch test. On the basis of the available data and clinical experience, it is concluded that sweet almond oil and almond meal are safe for topical application to humans in the present practices of use and concentration (web article).

Market Formulations

Dabur Vatika Olive Enriched Hair Oil

Dabur vatika olive enriched hair oil contains virgin olive oil and nourishing extracts of almonds, cactus and lemon. Vatika's unique formulation ensures deeper oil penetration to give hair and scalp complete nourishment for problem free, beautiful hair.

Ingredients: Mineral oil, RBD palmolein oil, canola oil, virgin olive oil, silicon oil, almond oil, perfume cactus extract in oil, castor oil parsol 1789, vitamin E acetate, lemon oil, tbhq, color C I no.12010, 12740, 61565.

Action: Almond—coats, conditions and softens hair. Cactus—gives hair volume and health. Lemon—regulates sebum flow and helps keep dandruff away.

Direction: Massage Dabur Vatika olive enriched hair oil into the scalp and then cover with a steaming towel for at least half-an-hour. Wash it off and the hair will get a shine like never before. It strengthens and nourishes the hair from within and also gives a very appealing fragrance.

Supplied by: Dabur India, India.

Bajaj Almond Drops

Being a light, non-sticky hair oil, it provides the traditional do-good benefits of nourishment without stickiness. With real almond extracts, Bajaj Almond Drops has 300% more vitamin E than coconut oil which helps nourish the hair roots and makes them strong and healthy.

Ingredients: Mineral oil, vegetable oil, perfume, sweet almond oil, vitamin E.

Direction: Apply almond oil 2–4 hours before shampoo your hair, and hair will be shinier and smoother after washing. Almond oil not only promotes hair growth but also nourishes hair to a high extent. It controls hair fall and makes hair stronger.

Supplied by: Bajaj Corp Ltd, India.

Moisturizing Almond Soap

This soap is used as skin moisturizer and also provides cooling effect to the skin.

Ingredients: Rosa damascena Syn. R.centifolia (satapatri), *Citrus limon* (nimbuka, lemon), *Vetiveria zizanioides* (ushira, khus khus), *Rubia cordifolia* (Indian madder, manjistha), *Cocos nucifera* (coconut oil, narikela), *Prunus amygdalus* (almond oil, vatada), *Prunus armeniaca* (apricot, urumana), *Triticum sativum* (wheatgerm oil, godhuma), *Pongamia glabra* (karanja, pongamia oil).

Action: Enriched with almond extracts that moisturize, soothe and revitalize the skin and rose, which cools and refreshes the skin.

Direction: Wet body and face, apply generously and rinse.

Supplied by: Himalaya Drug Company, India.

Almond Cleansing Lotion

This lotion is applied gently on sensitive skin. It removes make-up and impurities without irritating the skin and prevents cutaneous dryness. It brings the natural self-regulating properties of the skin into balance and minimize risk of allergic reaction as it does not contain essential oils.

Ingredients: CTFA (INCI): Water (aqua), sweet almond (*Prunus dulcis*) oil, glyceryl stearate SE, Blackthorn (*Prunus spinosa*) distillate, witch hazel (*Hamamelis virginiana*) distillate, alcohol, silver sulphate (in highly diluted form).

Action: It cleans the pores without irritating sensitive skin. Free of perfumes and artificial cleansing agents, it naturally helps the skin's self-regulating powers. It contains light almond oil to protect the skin and guard against drying, and a distillation of blackthorn blossoms soothes even the most sensitive skin.

Direction: Apply a small amount on a damp cotton wool pad or with the fingertips to the face, neck and chest. Gently removes with lukewarm water or a damp cotton wool pad. Apply morning and evening.

Supplied by: Weleda Company, India.

Almond Facial Oil

Almond facial oil helps to restore and protect sensitive skin, especially after exposure to harsh weather. It aids in the removal of eye make-up and offers a regenerative, intensive treatment of dry skin at night.

Ingredients: CTFA (INCI): Sweet almond (*Prunus dulcis*) oil, Chamomile (*Chamomile recutita*) extract, Calendula (*Calendula officinalis*) extract.

Action: It is a deeply hydrating, cosmetic oil contains the finest almond oil and extracts of chamomile and Calendula flowers to soften and soothe the skin.

Direction: Pour a few drops into the palm of one hand and then apply gently to cleansed, moist skin. To remove eye make-up apply a few drops to a moist cotton wool pad and gently stroke over the eye area. For night time care on extremely dry skin, gently massage a small amount into the skin using the fingertips. This is particularly effective on sensitive skin areas which are prone to development of wrinkles such as eyelids, neck and chest. Apply morning and evening.

Supplied by: Weleda Company, India.

Almond Facial Masque

It is oil in water emulsion form. It deep cleanses and soothes sensitive skin, leaving it feeling soft and smooth. Its carefully selected ingredients include fine day used to help remove irritants.

Ingredients: CTFA (INCI) water (aqua), carrot (*Daucus carota*) extract, glycerin, sweet almond (*Prunus dulcis*) oil, glyceryl stearate SE, jojoba (*Buxus chinensis*) oil, alcohol, algin, cucumber (*Cucumis sativas*) extract, birch (*Betula alba*) bark extract, witch hazel (*Hamamelis virginiana*) distillate, hectorite.

Action: Extracts of carrot, cucumber, and birch help to promote the natural balance of the skin.

Direction: Apply 1–2 mm layer to face, neck and chest. Avoid the area around the eyes and lips. Allow to work for 15–30 minutes. The mask does not tighten. Gently remove with cosmetic tissues or cotton wool pads, and rinse in lukewarm water. Apply almond moisture cream. Apply once or twice a week to washed/cleansed skin.

Supplied by: Weleda Company, India.

Various Preparations

Sweet almond oil is used topically to moisturize dry skin, soothe chapped lips, and relieve itching due to dryness. Because sweet almond oil is not greasy, it is absorbed quickly. Sweet almond oil generally does not irritate skin and it does not appear to cause sensitization that may lead to allergic reactions. It helps the skin to balance water loss and absorption of moisture. Because the oil does not penetrate the skin overly quickly, it is a good massage medium to use to help spread the oil and essential oil mixture, while still allowing time to do a good massage before it is absorbed by the skin. Not only does almond oil help protect the surface of the skin, but also it has great value, acting as an emollient. It also helps to relieve muscular aches and pains.

Facial Mask

Ingredients: 1 tablespoon chamomile flowers, 1 tablespoon carrots, fully mashed, 1 tablespoon marigold flowers, 1 teaspoon wheat germ oil, 4 ounces water, 1 tablespoon lecithin granules, almond oil, as much as necessary.

Preparation: In a medium-sized container, take 4 ounces of water and allow it to boil. When the water reaches boiling point, take it off from the stove and pour it into the vessel that contains marigold and chamomile flowers in it. Allow the flowers to stay in the water for sometime by covering it firmly with a lid. Meanwhile, take the mashed carrots in a bowl and add wheat germ oil to it. Mix well and then, drain the water from the flowers and add them to the bowl containing carrots. Put in the lecithin granules and blend thoroughly with the blender. Add a little bit of almond to this mixture and mix. To apply this mixture, first wash your face with water and spread the mixture all over your face. Allow the mixture to stay on your face for fifteen minutes and then take it off with tissue papers. Rinse your face with warm water followed by refreshing cold water.

Natural Lip Gloss

Ingredients: 2 teaspoons beeswax, 1 teaspoon honey, 7–8 teaspoons almond oil, 5 drops of any essential oil, vitamin E capsule.

Preparation: If you want to know about homemade make up recipes, you can try out this lip gloss recipe. Take beeswax and oil in a microwave friendly dish and microwave it for 1–2 minutes. Take the dish out of the microwave and set aside. With the help of a pin, pierce the vitamin E capsule and put the contents into the dish followed by honey. Stir the mixture till it is fully set and then transfer it into a small container.

Herbal Cosmetics for Dry Skin

To cleanse

Preparation: Mix 1 tsp almond powder + 1/2 tsp dry milk powder + 1 pinch of sugar and store in a spice jar.

Application: In palm, make a fine paste using 1/4 tsp of the above combination with the required quantity of warm water. Apply this freshly prepared paste all over the face and neck regions and gently massage it into the skin for about one minute. Do not scrub, rinse well with lukewarm water. Do not dry.

To moisturize

Preparation: Melt 1/2 oz cocoa butter in a double boiler, add 4 oz avocado or almond oil and remove it from heat. Using a dropper, add 1 oz orange peel oil, one drop at a time, while stirring. Add 3 or 4 drops each of coconut and rose oil to the mixture when it cools.

Herbal Cosmetics for Sensitive Skin

To cleanse

Preparation: Mix 1 tsp almond oil + 1/2 tsp orange peel + 1/2 tsp dry milk, and store the mixture in a spice jar.

Application: Make a fine paste by mixing 1/4 tsp of the above preparation with required quantity of pure rose water. Apply this fresh paste all over the face and neck region and gently massage it into the skin for about one minute. Do not scrub; rinse well with cool water and do not dry.

To nourish

Preparation: Mix 1 oz almond oil + 10 drops each of rose oil and sandalwood oil. Mix well and store it in a dark glass bottle with a dropper. It is good nourishing oil.

Application: Take 2 to 3 drops of the above nourishing oil and add 4 to 6 drops of water. When the skin is wet, gently massage this mixture all over the face and neck for about one minute.

Storage of Products

All the products should be stored in well closed containers and away from sunlight.

Dosage

Trials of almond dietary supplementation or laetrile in adults have used 50 to 75 g/day of sweet almonds.

Patents

The present invention is based on surprising discovery that a detailed selection and mixture of manufactures extracted from four different kinds of plants (hundreds of officinal plants available) containing aviary egg deutoplasm (or egg yolk or calf) have permitted creating a new mixture with an efficacious no-dandruff effect together exactly as efficacious prevention effect for loss of hair or coat and stimulating the same growth again. The mixture including manufactures of leaves as *Nicotiana tabacum*, from entire plant of *Lavanda officinalis*, or bulb of *Alliums* pp., seed oil of *Prunus amygdalus* or *dulcis* and aviary egg deutoplasm, procedures for its manufacturing, cosmetic products including the above-mentioned mixture, procedures for its manufacturing and its utilization. Although these plants are well-known, their mixtures aer unknown including a selection of manufactures of these plants in addition to aviary egg yolk. The union of these components, proved in the present invention, bringing to create a particular mixture well, as no-dandruff, capable to stimulate a growth-again of the hair or coat and to prevent loss of them, devoid from secondary effects annoying for all users (Soulimani, 2007).

A therapeutic blended oil composition is provided that is useful as a topical application to be absorbed into skin, protect the skin from sun without inducing acne, reduce skin wrinkles lines and dryness. The therapeutic oil composition is created by mixing grape seed oil, soyabean oil, sweet almond oil, sea buckthom seed oil and essential oils. The volume of each constituent added can be modified according to the desired treatment. The composition may be applied topically to absorb into the skin to aid skin tone, tanning, moisturizing, eczema, and bum symptoms. The composition is light oil that does not leave the skin feeling oily and is an effective and safe makeup remover (Venturi and Venturi Jackson, 2007).

The disclosure herein is directed to organic skin-based products that comprise an excess of 70% organic ingredients. The skin care products described herein are composed of a plurality of organic juices and other organic materials (Jochim and Behnke, 2008).

BIBLIOGRAPHY

1. Etim OE, Bassey EI, Ita SO, Udonkang IC, Etim IE, Jackson IL. Modulatory effect of sweet almond extract on serum lipids, lipid peroxidation and histology of pancreas in alloxan-induced diabetic rats. International Research Journal of Pharmacy and Pharmacology. 2011; 1(7): 155–161.

2. hhttp://en.wikipedia.org/wiki/Almond

3. http://101beautysalon.com/skin/homeskincare.html

4. http://ayurvedicmedicinalplants.com/plants/2504.html

5. http://www.ageless.co.za/herb-almondoil. htm

6. http://www.anniesremedy.com/herb_detail74.php

7. http://www.drugs.com/npp/almond-almond-oil.html

8. http://www.herbs2000.com/herbs/herbs_almond.htm

9. http://www.hort.purdue.edu/newcrop/duke_energy/Prunus_dulcis.html

10. http://www.naturalmedicinalherbs. net/herbs/p/prunus-dulcis=almond. php

11. http://www.uni-graz.at/~katzer/engl/Prun_dul.html

12. Jochim M, Behnke K. Compositions for juice-based skin cleansers. United States Patent Application 20080166313. Application No: 11/933414. Publi-cation date: 07/10/2008. http://www. freepatentsonline.com/y2008/0166313. html

13. Soulimani A. Cosmetic mixture for hair. United States Patent Application 20070184123. Application No: 11/578210. Publication date: 08/09/2007. http://www.freepatentsonline.com/y2007/0184123.html

14. Suraj R, Rejitha G, Anbu Jeba Sunilson J, Anandarajagopal K, Promwichit P. *In vivo* hair growth activity of *Prunus dulcis* seeds in rats. Biology and Medicine. 2009; 1 (4): 34–38.

15. Venturi M, Venturi jackson T. Therapeutic blended oil composition and method. United States Patent Application 20070178061. Application No: 11/346033. Publication date: 08/02/2007. http://www.freepatentsonline.com/y2007/0178061.html

16. Web article: Final report on the safety assessment of sweet almond oil and almond meal. International Journal of Toxicology. 1983; 2: 85–99.

Rosmarinus officinalis

Plant profile

Kingdom: *Plantae* – Plants

Superdivision: *Spermatophyta* – Seed plants

Class: *Magnoliopsida* – Dicotyledons

Family: *Lamiaceae* – Mint family

Species: *Rosmarinus officinalis*

Subkingdom: *Tracheobionta* – Vascular plants

Division: *Magnoliophyta* – Flowering plants

Order: *Lamiales*

Genus: *Rosmarinus* L.

Vernacular name

Eng: Rosemary, **Hin:** Rusmari, **Ben:** Rusmari, **Tam:** Rosemary, **Mar:** Rusmari.

Common name

Compass weed, compass plant, dew of the sea.

INTRODUCTION

Rosemary is an evergreen woody shrub with aromatic, needle-like leaves and gray, scaly bark. Rosemary bushes can grow up to 6 ft (1.8 m) tall with a spread of 4–5 ft (1.2–1.5 m). The plants stay smaller in pots. The leaves resemble needles and are about 1 in (2.5 cm) long with a pungent fragrance, somewhat reminiscent of pine. The flowers appear in winter and spring, are pale blue, about 1 in (2.5 cm) long, and arranged in clusters of 2 or 3.

Distribution

Rosemary was originally from the Mediterranean region, where it grows in dry, sandy or rocky soils in a climate characterized by warm summers and mild, dry winters. Recently, rosemary is cultivated in nearly all countries around the Mediterranean Sea, furthermore in England, the US, France, Italy, North Africa, India and Mexico. In India, it is grown in Nilgiris.

Chemical Constituents

The leaves contain about 1 to 2.5% essential oil. Therein, 1,8-cineol (30%), camphor (15 to 25%), borneol (16 to 20%), bornyl acetate (max. 7%), α-pinene (max. 25%) and others contribute to the complex taste. It also contains caffeic acid, ursolic acid, betulinic acid, rosmanol. It contains flavonoids like cirsimarin, diosmin, hesperidin and diterpenes like carnosolic acid, isorosmanol, rosmadial and rosmaridiphenol.

infusion of rosemary leaves refreshes and stimulates. Rosemary oils are known to have antibacterial properties. Rosemary leaves add a fresh, piney scent to sachets and potpourris; to soaps, lotions and perfumes; and to clothes and linens in the drawer. Rosemary is said to deter clothes moths, and an infusion of leaves works as a topical insect repellent. Rosemary flowers are very attractive to honeybees, and a fine honey is produced. Rosemary leaf in a bath stimulates the blood circulation. Use the leaf in a face steam bath. Rosemary oil serves as an extremely effective mouthwash. Current cosmetic uses of rosemary include treating cellulite and

Caffeic acid   Rosmarinic acid   Rosmanol

Medicinal uses

Rosemary was traditionally used as an antiseptic, astringent and a food preservative. It is used in facial cleansers for oily skin, and in compresses in treatment of bruises and sprains. It is a very helpful muscle relaxant, and can relieve muscle cramps and improve circulation in the affected area. It also improves relaxation of muscles of the digestive tract and uterus. It can be used in cases of digestive complaints and menstrual cramps. Some studies suggest Rosemary's beneficial effects on our brain. Carnosic acid, one of the components found in Rosemary, can protect our brain from free radicals, thus being helpful as prevention against strokes and neurodegenerative diseases. Used in aromatherapy, Rosemary oil is said to improve concentration and memory.

Cosmetic uses

Rosemary leaves and flowers contain a volatile oil that increases blood flow just beneath the skin. In the bath water, an

wrinkles, and normalizing excessive oil secretion of the skin.

Pharmacological Action

Several animal tests have demonstrated liver protective, anticonvulsive, antimutagenic and tumor inhibiting effects.

Rosmarinus officinalis L. crude hydro-alcoholic (70%) extract was evaluated for antiulcerogenic activity employing different experimental models. The crude hydroalcoholic extract (CHE) decreased the ulcerative lesion index produced by indomethacin, ethanol and reserpine in rats. No antisecretory activity was observed on pyloric ligation model (Dias et al., 2000).

The study was investigated the effects of some essential oils on *Limantria dispar* (Lepidoptera: Lymantridae, gypsy moth) larvae, one of the most serious pests of cork oak forests. The essential oils were first formulated as oil in water (o/w) emulsions and used in laboratory bioassays to assess their lethal concentration (LC_{50}). Microcapsules containing the most promising oils (*Rosmarinus officinalis* and *Thymus herbabarona*) were then prepared by a phase separation process, followed by freeze-drying. The results showed that the tested oils possess interesting larvicidal effects that make them suitable for application in integrated control strategies (Moretti et al., 2002).

The antioxidant efficacy of rosemary (*Rosmarinus officinalis* L.), sage (*Salvia fruticosa* L.), and sumac (*Rhus coriaria* L.) extracts and combinations at 4% concentrations (wt/vol, extract/oil) were investigated. Methanolic extracts of rosemary, sage, sumac, and their combinations were applied to peanut oil stored at 80°C for 24 h. The antioxidant effect was determined by measuring the peroxide value. The antioxidant effect of all extracts was low compared with that of butylated hydroxytoluene. Rosemary extract (except for 3 and 4 h) exhibited the most antioxidant effect compared with other individual extracts (Ozcan, 2004).

The radioprotective effects of carnosic acid (CA), carnosol (COL), and rosmarinic acid (RO) against chromosomal damage induced by gamma-rays, compared with those of L-ascorbic acid (AA) and the S-containing compound dimethyl sulfoxide (DMSO), were determined by use of the micronucleus test for antimutagenic activity, evaluating the reduction in the frequency of micronuclei (MN) in cytokinesis-blocked cells of human lymphocytes before and after gamma-ray irradiation. The radioprotective effects (antimutagenic) with treatment after gamma-irradiation were lower, and the most effective compounds were CA and COL. RO and AA presented small radioprotective activity, and the sulfur-containing compound DMSO lacked gamma-ray radioprotection capacity. Therefore, CA and COL are the only compounds that showed a significant antimutagenic activity both before and after gamma-irradiation treatments (Del Bano et al., 2006).

The radioprotective effect of *Rosemarinus officinalis* extract (ROE) was studied in mice exposed to 3 Gy gamma radiation. Crypt survival, villus length, apoptotic cells, mitotic figures and goblet cells in intestine were studied at different autopsy intervals, i.e. 12 hrs to 30 days after irradiation. Maximum changes in all the intestinal parameters were observed on day 3 after irradiation. Irradiation of animals resulted an elevation in lipid peroxidation and a reduction in glutathione concentration in the intestine at 1 hr post-irradiation (Jindal et al., 2010).

Toxicological studies

An acute safety study of rosemary extracts was conducted in wistar rats at a single oral gavage dosage of 2,000 mg/kg of body weight. Rosemary extracts were well tolerated; no adverse effects or mortality were observed during the 2-week observation period. No abnormal signs, behavioral changes, body weight changes, or change in food and water consumption occurred. Two weeks after a single oral rosemary extract dose of 2,000 mg/kg of body weight, there were no changes in hematological and serum chemistry values, organ weights, or gross or histological characteristics. Rosemary extracts appear to have low acute toxicity, and the oral lethal doses (LD_{50}) for male and female rats are greater than 2,000 mg/kg of body weight (Anadon et al., 2008).

Clinical Investigation

Rosemary can improve hair growth that was supported by 7 months, randomized double blind study of 86 patients. The results showed rubbing oil (thyme, lavender, rosemary and cedarwood) into the scalp helped with alopecia for 44% of patients versus 15% of control (Hay et al., 1998).

Market Formulations
Anti-dandruff Hair Cream

This cream is effective against dandruff and hair fall.

Ingredients: *Melaleuca leucadendron* (tea tree, kayaputi), *Coriandrum sativum* (dhanyaka), *Rosmarinus officinalis* (rosemary, rusmari).

Action: Contains tea tree oil and rosemary, which provide effective anti-dandruff action.

Direction: Massage gently on scalp in circular motion. Use regularly before and after shampooing for dandruff-free, soft and shiny hair. It is suitable for all hair types and safe to use on artificially colored or permed hair.

Supplied by: Himalaya Drug Company, India.

Anti-dandruff Hair Oil

Anti-dandruff hair oil prevents dandruff by eliminating microbial infections of the scalp. This oil reduces hair fall.

Ingredients: Wrightia tinctoria (sweet Indrajao, hyamaraka), *Rosmarinus officinalis* (rosemary, rusmari), *Melaleuca leucadendron* (tea tree, kayaputi), *Azadirachta indica* (neem, nimba).

Action: Massaging of hair and scalp with a proper nutrient hair oil gives additional nutrition to the scalp and prevents hair loss. Massaging also increases the blood circulation in the scalp and this keeps the hair roots strong.

Direction: Apply oil all over the scalp. Massage the scalp gently with fingers in a circular motion, so that the oil gets absorbed into the scalp, gradually. Leave for an hour or more before washing with an Himalaya Herbals' cleansing shampoo.

Supplied by: Himalaya Drug Company, India.

Various Preparations

Rosemary flowers kept in the linen cupboard drive away moths. The wash kept above a rosemary bush to dry, pulls the aroma in. Place fresh branches of rosemary in a space the air to refresh. Boil during 10 minutes a hand full rosemary leaf to get through a half litter water of an antiseptic solution in the bathroom.

Rosemary Honey Hair Conditioner

This nourishing conditioner blends honey for shine; olive oil for moisture and essential oil of rosemary to stimulate hair growth.

Ingredients: 1/2 cup honey, 1/4 cup warmed olive oil (2T for normal to oily hair), 4 drops of essential oil of rosemary, 1 tsp. xanthum gum (available in health food stores).

Method: Place all the ingredients in a small bowl and mix thoroughly. Pour into a clean plastic bottle with a tight fitting stopper or lid.

Direction: Apply a small amount at a time to slightly dampened hair. Massage scalp and work mixture through hair until completely coated. Cover hair with a warm towel (towel can be heated in a microwave or dryer) or shower cap; leave on to nourish and condition for 30 minutes. Remove towel or shower cap; shampoo lightly and rinse with cool water. Dry as normal and enjoy shinier, softer and healthier hair the natural way.

Rosemary Hair Tonic

This tonic is helpful for activating the scalp, so it is good for people experiencing hair loss.

Ingredients: 1 and 1/2 pint of water, 15–20 g. dried rosemary.

Preparation: Put the water to boil, then put the rosemary. Let it simmering (low fire) for about 10 min. Turn off the heat and leave it to cool off with the lid on. Sieve it. After washing the hair, spill it slowly on your head, being careful so it reaches all points of scalp. Do not rinse off.

Direction: Use at least 2–3 times a week.

Rosemary Hair Conditioner

Ingredients: 4 drops of rosemary essential oil, 1/2 cup of honey, 1/4 cup of olive oil (warm) or 2 tablespoons if hair is normal or dry.

Preparation: Mix all ingredients well and pour into a clean bottle. Dampen the hair and apply the mix gently rubbing the scalp and making sure the mix is spread well all over the scalp and hair. Use a warm towel and wrap the hair up leaving it on for 20–30 minutes (or use a shower cap). Wash the hair with a shampoo and rinse off with cool water.

Storage of Products

All the products should be stored in well closed containers and away from sunlight and moisture.

Dosage

Tincture (1:5): Single-dose: 20–40 drops. Liquid extract: Single dose: 2–4 ml.

Externally: Semi-solid and liquid forms with 6–10% essential oil can be applied directly on the skin.

Bath additive: 50 g drug to 1 lt hot water added to full or hip depth bath. **Washes:** 1% infusion

Patents

A herbal hair treatments and method of making the same for protecting and strengthening hair. The herbal hair treatments and method of making the same include mixing sage, horsetail, kelp, nettle leaf, horsetail, rosemary and mineral water to form a mixture. Heating the mixture to a full boil and straining the mixture to form an herb liquid. Mixing into the herb liquid rosemary oil, vitamin E, castor oil and a soap to form a shampoo (Fluker, 2002).

The present invention relates to a topical composition comprising a malodor-reducing effective amount of a rosemary extract, or active fraction thereof, in combination with a self-tanning effective amount of DHA. The invention also provides a method of reducing the potential for malodor generation of a DHA composition comprising adding to the composition an effective amount of rosemary extract or active fraction thereof (Sokolinsky et al., 2004).

The invention related to a cosmetic composition comprising: (1) rosemary (*Rosmarinus officinalis* L. extract) — 0.001–10%; (2) henna (*Lawsonia inermis* L. extract) —0.001–10%; (3) melanin — 0.000005–5%; and also to compositions containing (1) rosemary (*Rosmarinus officinalis* L. extract) — 0.001–10%; (2) henna (*Lawsonia inermis* L. extract) — 0.001–10%; (3) melanin — 0.000005–5%; and (4) sunflower

(*Helianthus annuus* L. extract) — 0.0001–10%. Combination of the above naturally-occurring components is suitable for preparing finished cosmetic compositions in association with an acceptable cosmetic base. The compositions are particularly useful for dark hair and exhibit sun protection against UV radiation, protection of hair fiber chemical structure, protection and retention of hair color and appropriate gloss (Masiero, 2004).

Addition of rosemary extract to various dentifrice compositions results in toothpaste, mouth rinses, and other compositions that are suitable for treating and preventing a variety of oral disease including gingivitis, plaque build-up, and the like. Rosemary extract, containing major amounts or ursolic acid and camosic acid, is added to dentifrice compositions so that the amount delivered to the oral cavity upon use is effective to provide an antibacterial, antioxidant, and/or anti-inflammatory effect. In various embodiments, the components of rosemary extract are combined with triclosan or other phenolic antibacterial agents to provide enhanced activity (Trivedi et al., 2006).

A composition for the prevention or inhibition of free radical-induced effect on the skin and/or hair comprises at least three antioxidant agents selected from the group consisting of ginkgo extract, emblica extract, dimethylmethoxy chromanol, pine bark extract, and rosemary extract (Barton et al., 2010).

BIBLIOGRAPHY

1. Anadon A, Martinez-Larranga MR, Martinez MA, Ares I, Garcia Risco MR, Senorans FJ, Reglero G. Acute oral safety study of rosemary extracts in rats. J Food Prot. 2008; 71(4): 790–795.

2. Barton SP, Johnson M, Tomlinson PJ. Compositions and methods for the skin and hair. United States Patent Application 20100221202. Application No: 12/681410. Publication date: 09/02/2010. http://www.freepatentsonline.com/y2010/0221202.html

3. Del Bano MJ, Castillo J, Benavente-Garcia O, Lorente J, Martín-Gil R, Acevedo C, Alcaraz M. Radioprotective anti-mutagenic effects of rosemary phenolics against chromosomal damage induced by gamma-rays. J Agric Food Chem. 2006; 54(6): 2064–2068.

4. Dias PC, Foglio MA, Possenti A, de Carvalho JE. Antiulcerogenic activity of crude hydro-alcoholic extract of *Rosmarinus officinalis* L. J Ethanopharmacology. 2000; 69(1): 57–62.

5. Fluker A. Herbal hair treatments and method of making the same. United States Patent 6475476. Application No: 09/718330. Publication date: 11/05/2002. http://www.free patentsonline. com/6475476.html

6. Hey IC, Jamieson M, Ormerod AD. Randomized trial of aromatherapy. Successful treatment for *alopecia areata*. Archives of Dermatology. 1998; 134: 1349–1352.

7. http://health-from-nature.net/Rosemary. html

8. http://www.floridata.com/ref/r/rose_ off. cfm

9. http://www.indianmirror.com/ayurveda/ rosemary.html

10. http://www.natuurlijkerwijs.com/english/ Rosemary.htm

11. http://www.purematters.com/herbs-sup plements/r/rosemary#dosage

12. Jindal A, Agrawal A, Goyal PK. Influence of *Rosemarinus officinalis* extract on radiation-induced intestinal injury in mice. J Environ Pathol Toxicol Oncol. 2010; 29(3):169–79.

13. Masiero S. Cosmetic composition, especially for dark hair. United States Patent Application 20040117921. Appli-cation No: 10/ 714721. Publication date: 06/24/2004. http://www.freepatentsonline.com/y2004/ 0117921.html

14. Moretti MD, Sanna Passino G, Demontis S, Bazzoni E. Essential oil formulations useful as a new tool for insect pest control. AAPS Pharm Sci Tech. 2002; 3(2): E13.

15. Ozcan M. Antioxidant activities of rosemary, sage, and sumac extracts and their combinations on stability of natural peanut oil. Journal of Medicinal Food. 2004; 6(3): 267–270.

16. Sokolinsky M, Ostrovskaya A, Landa PA, Maes DH. Cosmetic compositions containing rosemary extract and dha. United States Patent Application 20040042979. Application No: 10/451565. Publication date: 03/04/ 2004. http://www.freepatentsonline.com/ y2004/0042979.html

17. Trivedi HM, Xu T, Worrell CL, Panaligan K. Oral compositions containing extracts of rosmarinus and related methods. United States Patent Application 20060134025. Application No: 11/256861. Publication date: 06/22/2006. http://www.freepa tentsonline.com/y2006/0134025.html

Santalum album

Plant profile

Kingdom: *Plantae* – Plants

Superdivision: *Spermatophyta* – Seed plants

Subkingdom: *Tracheobionta* – Vascular plants

Division: *Magnoliophyta* – Flowering plants

Class: *Magnoliopsida* – Dicotyledons

Subclass: *Rosidae*

Order: *Santalales*

Family: *Santalaceae* – sandalwood family

Genus: *Santalum* L. – Sandalwood

Species: *Santalum album* L.– Sandalwood

Vernacular name

Ben: Chandan, peetchandan, srikhanda, sufaid-chandan, **Eng:** Cendana, East Indian sandalwood, sandal, sandal tree, sandalwood, white Indian sandalwood, **Mar:** Chandan, **Oriya:** Valgaka, **Hin:** Chandal, chandan, safed-chandan, sandal, **San:** Ananditam, chandana, taliaparnam, Srikhandam, **Tam:** Kulavuri, sandanam, santhanam, srigandam, ulocidam; **Tel:** Bhadrasri, **Kan:** Agarugandha, bavanna, bhadrasri, **Mal:** Chandanam, chandana-mutti

Common name

Sandalwood, Chandan.

Ayurvedic properties

Rasa: Kashaya, Madhura, Tikta, **Veerya**: Seeta, **Guna:** Guru.

Ayurvedic preparations

Chandanabala laxadi taila, Chandrakala rasa.

INTRODUCTION

Santalum album is a small evergreen tree that grows to a height of 20 m; girth of up to 2.4 m, with slender drooping branchlets. Bark is tight, dark brown, reddish, dark grey or nearly black, smooth in young trees, rough with deep vertical cracks in older trees, red inside. Leaves thin, usually opposite, ovate or ovate elliptical, 3–8 × 3–5 cm, glabrous and shining green above, glaucous and slightly paler beneath; tip rounded or pointed; stalk grooved, 5–15 cm long; venation noticeably reticulate. Flowers purplish-brown, small, straw colored, reddish, green or violet, about 4–6 mm long, up to 6 in small terminal or axillary clusters, unscented in axillary or terminal, paniculate cymes. Fruit a globose, fleshy drupe; red, purple to black when ripe, about 1 cm in diameter, with hard ribbed endocarp and crowned with a scar, almost stalkless, smooth, single seeded. The generic name is derived from the Greek 'santalon' meaning 'sandalwood', and the species name from the Latin 'albus' meaning 'white', in allusion to the bark.

Distribution

S. album is indigenous to the tropical belt of the Indian peninsula, eastern Indonesia and Northern Australia. There is still debate as to whether *S. album* is endemic to Australia or was introduced by fishermen or birds from eastern Indonesia centuries ago. The main distribution is in the drier tropical regions of India and the Indonesian islands of Timor and Sumba. The principal sandal tracts are most parts of Karnataka and adjoining districts of Maharashtra, Tamil Nadu and Andhra Pradesh in India. The species is mostly found in dry deciduous and scrub forests in this region.

Chemical Constituents

The main constituent of sandalwood oil is santalol. This primary sesquiterpene alcohol forms more than 90% of the oil and is present as a mixture of two isomers, alfa-santalol and beta-santalol, the former predominating. The characteristic odor and medicinal properties of sandalwood oil are mainly due to the santalols. The other constituents reported in sandalwood oil include: The hydrocarbons santene, nor-tricycloekasantalene and alfa- and beta-santalenes; the alcohols santenol and teresantalol; the aldehydes nor-tricycloekasantalal, and isovaleraldehyde; the ketones l-santenone and santalone.

Alfa-santalol

Medicinal uses

Sandalwood oil is used to cool the body during fevers and heat stroke. It is also used to aid in the passing of kidney and gall stones, and for infections in the urinary tract. It is also use as antidepressant, counters inflammation, antiseptic (urinary and pulmonary), antispasmodic, aphrodisiac, astringent, bactericidal, diuretic, expectorant, insecticidal, sedative, tonic, settles digestion, etc.

Cosmetic uses

Sandalwood oil is an excellent cleansing, astringent addition to massage and facial oils, bath oils, aftershaves, lotions and creams. It relieves itching and inflammation, and as a mild astringent can also profit those with oily skin. *In aromatherapy, it is used to calm and relax (stress), and for meditation.*

Skin care: It is used for skin protection. Its antiseptic property helps to treat rashes, spots, blackheads and other skin eruptions. Its germicidal quality inhibits the growth of bacteria. It has also been found to act as skin moisturizer. Sandalwood paste tones up the skin. Due to its astringent in nature, it is used especially for oily skin. Several other uses like in treatment of acne, dry cracked and chapped skin, after-shave (for barber's rash), eczema, hair and scalp tonic moisturizer, aging, mature skin, etc.

Sandalwood paste has a soothing effect and helps relieve itching in scabies. The red variety of sandalwood protects against harmful rays of the sun and the paste can be applied to ease sunburn. It is widely used in sunscreens and other cosmetics.

Pharmacological Action

The purpose of the investigation was to study the skin cancer chemopreventive effects of α-santalol, a principal component of sandalwood oil in CD-1 and SENCAR mice. Chemopreventive effects of α-santalol were determined during initiation and promotion phase in female CD-1 and SENCAR mice. α-santalol treatment during promotion phase delayed the papilloma development by 2 weeks in both CD-1 and SENCAR strains of mice. α-Santalol treatment during promotion phase significantly ($p < 0.05$) decreased the papilloma incidence and multiplicity when compared with control and treatment during initiation phase during 20 weeks of promotion in both CD-1 and SENCAR strains of mice. Hence it could be an effective chemopreventive agent for skin cancer (Chandradhar et al., 2003).

The objective of the investigation was to study the effects of α-santalol on ultraviolet B (UVB) radiation-induced skin tumor development and UVB-caused increase in epidermal ornithine decarboxylase (ODC) activity in female hairless SKH-1 mice. The study was terminated at 30 weeks. Topical application of α-santalol significantly ($p < 0.05$) decreased tumor incidence and multiplicity in all the three protocols, suggesting its chemopreventive efficacy against UVB radiation-caused tumor initiation, tumor promotion and complete carcinogenesis. In a short-term biochemical study, topical application of α-santalol also significantly ($p < 0.05$) inhibited UVB-induced epidermal ODC activity. Finally it was concluded that α-santalol could be a potential chemopreventive agent against UVB-induced skin tumor development (Chandradhar et al., 2006).

Sandalwood (*Santalum album* L.) is a fragrant wood from which oil is derived for use in food and cosmetics. Sandalwood oil is used in the food industry as a flavor ingredient with a daily consumption of 0.0074 mg/kg. Over 100 constituents have been identified in sandalwood oil with the major constituent being alpha-santalol. Sandalwood oil and its major constituent have low acute oral and dermal toxicity in laboratory animals. Sandalwood oil was not mutagenic in spore Rec assay and was found to have anticarcinogenic, antiviral and bactericidal activity (Burdock and Carabin, 2008).

The purpose of the study was to determine the antibacterial activities of honey, sandal oil and black pepper by taking a selected standard medicine. The bacterial isolates obtained from clinical samples were *Escherichia coli*, *Salmonella* sp., *Staphylococcus aureus* and *Bacillus subtilis*. Results of all sample products showed zone of inhibition against all selected bacterial isolates. Sandal 1 oil showed inhibition zones against all isolates except *E.coli*, whereas sandal 2 was effective against all. The results obtained from black pepper 1 and 2 did not gave effective results both at 50 μ liter: 50 μ liter and 25 μ liter: 75 μ liter concentration, both were found to be effective against *Bacillus subtilis* (Ghori and Ahmed, 2009).

Toxicological Studies

It is considered a non-toxic, non-irritant and non-sensitizing oil. The acute oral toxicity (LD_{50}) of sandalwood oil in rats was reported as 5.58 g/kg of body weight (Bar and Griepentrog, 1967). The acute dermal toxicity (LD_{50}) of sandalwood oil in rabbits was reported as >5 g/kg of body weight (Shelanski, 1971). The acute oral LD_{50} of the major constituent of sandalwood oil, a-santalol, in rats was reported as 3.8 g/kg. The acute dermal LD_{50} of a-santalol in rabbit was reported as > 5 g/kg (Keating, 1972).

Clinical Investigation

Emmons and Marks (1985) investigated the incidence and etiology of cutaneous reactions to cosmetic ingredients, including sandalwood oil. A total of 50 subjects, 19 control and 31 patch test clinic patients (16 with history of adverse cosmetic reactions) were studied by open and patch testing using North American Contact Dermatitis Group (NACDG) fragrance screening series. In both open as well as patch testing, sandalwood oil at a concentration of 2% in

petrolatum did not cause contact urticaria in any of the subjects.

Paulsen and Andersen (2005) applied a series of essential oils to the backs of 318 patients for two days, using Finn Chambers and Scanpor tape; readings were taken on days 3 or 4 and 5–7 according to IRCDRG guidelines. Sandalwood oil was administered at two concentrations (2% and 10%) in petrolatum. One subject exhibited a response to sandalwood oil at 10%.

Results of the pilot study of the four counties randomised controlled trial to evaluate the effectiveness of aromatherapy massage with 1% *Santalum album* (sandalwood) (group A) when compared with massage with sweet almond carrier oil, (group B) or sandalwood oil via an aromastone (group C), in reducing levels of anxiety in palliative care. The primary end points of the research were to report a statistically significant difference in anxiety scores between experimental group (B) and comparison groups (A and C) and to influence the integration of aromatherapy into all aspects of palliative care. The results do seem to support the notion that Sandalwood oil is effective in reducing anxiety (Kyle, 2006).

Market Formulations
Anti-wrinkle Cream

This cream is effective against wrinkles, fine lines and age spots. It also reduces signs of early aging.

Ingredients: *Aloe vera* (barbados aloe, kumari), *Papaver rhoeas* (red poppy, raktaposta), *Vitis vinifera* (grapes, draksha), *Citrus limon* (lemon, nimbaka) and *Solanum lycopersicum* (tomato, raktamaci), *Santalum album* (sandalwood tree, chandana).

Action: Anti-wrinkle cream delays wrinkles and smoothes fine lines. This cream is enriched with alpha hydroxy acids (AHAs), skin nutrients and vitamin E. Its regular use prevents oxidative skin damage, skin laxity and appearance of wrinkles.

Direction: Massage gently over cleansed face twice a day. Use regularly.

Supplied by: Himalaya Drug Company, India.

Baby Powder

A specially formulated cooling talcum powder that helps in excess sweating and relieving itching on the baby's tender skin, besides controlling the body odor.

Ingredients: Zinc calx (*Yashad bhasma*), sandalwood (*Santalum album*), vetiver, khaskhas (*Vetiveria zizanioides*), olive oil (*Olea europaea*).

Action: The baby powder contains natural zinc that has antiseptic properties and accelerates healing, sandalwood extract that cools and olive oil that nourishes the baby's tender skin.

Direction: Sprinkle on the baby's skin liberally after baby's bath, nappy change and before bedtime.

Supplied by: Himalaya Drug Company, India.

Apart from that many other market products are available like sandal soaps, Vicco turmeric cream, bath soaps, moisturizers, etc.

Various Preparations

Moisturizer/Face Cream

Ingredients: 2 tablespoons beeswax, grated, 6 tablespoons jojoba oil, 2 tablespoons rose water, 3 teaspoons vitamin E oil, 4 drops rose essential oil or 8 drops sandalwood essential oil, 3 teaspoons pure aloe vera.

Preparation: Once you are done with cleansing and toning, the next important step is to moisturize the face. To make homemade moisturizer, the first thing one has to do is take jojoba oil in a bowl and blend it with beeswax and vitamin E. Boil water in a shallow pan and place the bowl in it and heat it till the beeswax melts completely. Take the bowl from the pan and keep it for cooling at room temperature. While it is being cooled, take aloe vera and rose water in a food processor

and blend. Once this is done, pour the cooled oil in the processor and blend on a high speed till all the ingredients are completely mixed. Put in some drops of rose or sandalwood essential oils in this mixture and mix well. Transfer this mixture into a glass jar and apply it instead of your regular moisturizer.

Mix chandan with besan and a pinch of turmeric and apply to the skin, leave it for half an hour, let it dry then, wash it thouroughly with warm water. Do it twice a week. It gives skin a soothing effect, makes the skin soft and glow.

Papaya and Sandalwood (Chandan) Face Mask

Papaya and sandalwood face mask is extremely beneficial for people with all skin types.

Ingredients: 1 Papaya, 1 tablespoon of sandalwood powder.

Action: The medicinal properties of sandalwood and the cleansing action of papaya will combine to give you a glowing and healthy skin after you apply this beauty pack with papaya and sandalwood. Papaya helps to remove dead skin, pollution and dirt from the facial skin, leaving it clean, fresh and radiant. Sandalwood is a cooling and soothing agent for the skin. It is suitable for all skin types and also helps to fight against common skin problems such as rashes, inflammation, acne, dryness, dehydration and chapped skin.

Preparation: Mash the papaya finely till it turns to pulp. Mix it with the sandalwood powder till both blends to become a smooth paste. Then apply the paste all over your face and neck. Leave the mask for around 20 minutes. Finally wash it off with water.

Face mask

Ingredients: 1 tsp of sandalwood powder, 1tbsp of honey, ½ tsp of turmeric powder.

Method: Mix well and apply on the face and neck. Leave for 20 minutes. Then wash off with cold water. It leaves the skin tingly fresh.

Storage of Products

All the products should be stored in well closed containers and away from sunlight.

Dosage

The average daily dose is 10 g of the drug and 1–1.5 g of the essential oil. Topically up to 5% sandalwood oil can reduce skin cancer.

Patents

The present invention relates to a formulation of herbal toothpowder or toothpaste for gums and teeth, which comprises powder or paste of *Zanthoxylum armatum* (20–25%), *Zingiber officinale* (25–30%), *Santalum album* (8.25–8.5%), *Spilanthes calva* (2.0–2.5%), *Pistacia lentiseus* (2.0–2.5%), *Quercus infectoria* (8.0–8.5%), *Usnea longissima* (1–4%), as well as roasted alum and common salt (Farooqi et al., 2001).

There is body care composition including botanical oils. The composition includes tea tree oil; lavender oil; myrrh; and/or sandalwood oil. There is a surfactant base such but not limited to shampoo and/or a mild dishwashing detergent. The tea tree oil may be from about 2%, 5%, or 8% to about 15% by volume of the composition. The lavender oil may be from about 0.5%, 1%, 2%, or 4% to about 5% or 6% by volume of the composition. The myrrh may be from about 0.5% or 1% to about 5% by volume of the composition. The sandalwood oil may be from about 0.5% or 1% to about 5% by volume of the composition (Ord, 2007).

BIBLIOGRAPHY

1. Bar FU, Griepentrog F. Die situation in der gesundheitlichen beurteilung der aromatisierungsmittel fur lebensmittel. Medizin Ernaher. 1967; 8: 244.

2. Burdock GA, Carabin IG. Safety asse-ssment of sandalwood oil. Food Chem Toxicol. 2008; 46(2):421–432.

3. Chandradhar D, Valluri HB, Xiangming G, Agarwal R. Chemopreventive effects of a-santalol on ultraviolet B radiation-induced skin tumor development in SKH-1 hairless mice. Carcinogenesis. 2006; 27 (9): 1917–1922.

4. Chandradhar D, Xiangming G, Wendy L. H, Alison L. V, Dawn E. G, Erin M. K, Kelly M. J, Valluri HB, Matthees DP. Chemopreventive effects of α-santalol on skin tumor development in CD-1 and SENCAR mice. Cancer Epidemiol Biomarkers Prev.2003; 12: 151.

5. Emmons WW, Marks Jr., JG. Immediate and delayed reactions to cosmetic ingredients. Contact Dermatitis. 1985; 13: 258–265.

6. Farooqi AHA, Sharma S, Khan A, Kumar R, Kumar S. Formulation useful as a natural herbal tooth powder. United States Patent 6264926. Application no: 09/268334. Publication date: 07/24/2001. http://www. freepatentsonline. com/6264926.html

7. Ghori I, Ahmed SS. Antibacterial activities of honey, sandal oil and black pepper. Pak. J. Bot. 2009; 41(1): 461–466.

8. http://ayurvedicmedicinalplants.com/plants/169.html

9. http://logos_endless_summer. tripod.com/id148.html

10. http://www.beautyandgroomingtips.com/2007/11/sandalwood-chandan-for-beauty.html

11. http://www.flowersofindia.in/catalog/slides/Sandalwood.html

12. http://www.la-medicca.com/raw-herbs-santalum-album.html

13. http://www.newdiet4all.com/beauty-tips/home-made-beauty-tips/papaya-sandalwood-face-mask.html

14. http://www.worldagroforestrycentre. org/sea/products/afdbases/af/asp/Species Info.asp?SpID=1481

15. Keating JW. Report to RIFM, 7 June, 1972.

16. Kyle G. Evaluating the effectiveness of aromatherapy in reducing levels of anxiety in palliative care patients: Results of a pilot

study. Complementary Therapies in Clinical Practice. 2006; 12(2): 148–155.

17. Ord MG. Body care composition. United States Patent Application 20070196317. Application no: 11/307714. Publication date: 08/23/2007. http://www.freepatentsonline.com/y2007/0196317.html

18. Paulsen E, Andersen KE. Colophonium and composite mix as markers of fragrance allergy: Cross-reactivity between fragrance terpenes, colophonium and composite plant extracts. Contact Dermatitis. 2005; 53: 285–291.

19. Shelanski MV. Report to RIFM, 14 November, 1971.

Stevia rebaudiana

Plant profile

Kingdom: *Plantae* – Plants

Superdivision: *Spermatophyta* – Seed plants

Class: *Magnoliopsida* – Dicotyledons

Order: *Asterales*

Genus: *Stevia*

Subkingdom: *Tracheobionta* – Vascular plants

Division: *Magnoliophyta* – Flowering plants

Subclass: *Asteridae*

Family: *Asteraceae* – Aster family

Species: *Stevia rebaudiana.* – Candy leaf

Vernacular name

San: Svaduchada, Bhoomisarkara, Sarkarachada, **Eng:** Stevia, **Ben:** Stevia, **Kan:** Sakare gidda, **Mal:** Stevia, **Tel:** Madhu patri, **Tam:** Seeni tulsi, **Mar:** Madhu parani, **Punjabi:** Madhu pattha.

Common name

Honey-Leaf, Sweet-Leaf, Sweet-Herb.

Ayurvedic properties

Rasa: Madhura, **Veerya:** Seeta, **Guna:** Lakhu

INTRODUCTION

The plant is about 3–4 ft height. Leaves are alternate, herbaceous perennial. The leaves are 5–8 cm long and 1–2.5 cm wide. Flowers are in indeterminate heads, small, white with pale purple throat. The seeds are small and brown in color. Seeds contains little endosperm, dispersed in the wind via a hairy pappus.

Distribution

An herbaceous perennial shrub, *Stevia Rebaudiana* is grown in the highlands of

Paraguay and sections of Argentina and Brazil. It has now commercially cultivated in Mexico, Central America, Japan, China, Malaysia, Israel, Thailand, England, Russia, India and South Korea. In Europe it is reported to be cultivated in Spain, and Belgium. In India, it is cultivated in Gujarat, West Bengal, Himachal Pradesh, Uttaranchal, Punjab, Karnataka and Tamil Nadu.

Chemical Constituents

The leaves contain a complex mixture of natural sweet diterpene glycosides. These are stevioside (4–20% dry weight), steviolbioside (trace), the rebaudiosides A (2–4%), B (trace), C (1–2%), D (trace), E (trace) and dulcoside A (0.4–0.7%). The dry weight composition is given as protein ~6.2%, lipids ~5.6%, total carbohydrates ~52.8%, stevioside~15% and about 42% water-soluble substances. The following non-sweet constituents have been identified: labdane diterpene, triterpenes, sterols, flavonoids, volatile oil constituents, pigments, gums and inorganic matters.

Stevioside: $C_{38}H_{60}O_{18}$

Rebaudioside A: $C_{44}H_{70}O_{23}$

Rebaudioside C: $C_{44}H_{70}O_{22}$

Dulcoside A: $C_{38}H_{60}O_{17}$

Medicinal uses

It is used as a tonic, antifungal and antibacterial agent. It is used for sore throats, colds, flu, allergies, sinus congestion, headaches, fatigue, pain, rheumatism, diabetes, food poisoning, balancing glucose levels in the blood, regulating blood pressure, supportive action to pancreas, spleen, liver and heart. It is also strengthening nerve and immune systems. Stevia is a good source of potassium, a major mineral for healing, muscle function, digestion, brain power, nerve conductivity, fluid balance and the elimination of toxic wastes. Stevia is also very rich in manganese (important to the healthy function of the

Stevioside

Rebaudioside A

glandular system, hormone production, and transmission of impulses between nerves and muscles), and chromium (for metabolism and promoting efficient insulin function). It acts as an antioxidant. Recent studies have shown Stevia has the potential to increase mental alertness, decrease fatigue, improve digestion, regulate blood pressure and ease hypoglycemia.

Cosmetic uses

Aids in the treatment of skin disorders, it has antioxidant properties which acts as moisturizer and also helps to smooth out wrinkles, helps to heal skin blemishes and acne, helps to tighten the skin like a facial mask. The hydrating effect of Stevia is proven 3 times better than glycerin enhances fragrances. The anti-bacterial effect of the product gives it increased shelf life so it is beneficial for both the hair and scalp. Stevia liquid concentrate effectively treats seborrhea, dermatitis and eczema. It helps to protect the skin from environmental, sun damage and visible signs of premature aging. Stevia acts as plaque retardant anti-caries and prevents cavities. Smoother skin, softer to the touch is claimed to result from the frequent application of Stevia poultices and extract. Stevia leaves and extracts placed directly on cuts and wounds give more rapid healing, without scarring. The herbal body soaps are also found to use Stevia to ensure the human skin with a nourishing effect. Shampoos and hair conditioners who contain Stevia render the hair with a soft, silky, and shinning appearance.

Pharmacological Action

The aim of the study was to assess the *in vitro* potential of ethanolic leaf extract of *Stevia rebaudiana* as a natural antioxidant. The DPPH activity of the extract (20, 40, 50, 100 and 200 lg/ml) was increased in a dose dependent manner, which was found in the range of 36.93–68.76% as compared to ascorbic acid 64.26–82.58%. The ethanolic extract was also found to scavenge the superoxide generated by EDTA/NBT system. Measurement of total phenolic content of the ethanolic extract of *S. rebaudiana* was achieved using Folin–Ciocalteau reagent containing 61.50 mg/g of phenolic content, which was found significantly higher when compared to reference standard gallic acid. The results obtained in this study clearly indicate that *S. rebaudiana* has a significant potential to use as a natural antioxidant agent (Shukla et al., 2009).

Anti-inflammatory activity of the EtOAc fraction of *Stevia rebaudiana* (SR) was evaluated. In the inflammatory mediator inhibitory assay from lipopolysaccharide (LPS)-activated macrophages, the EtOAc fraction significantly, and dose dependently, inhibited the enhanced production of nitric oxide (NO) and inducible nitric oxide synthase (iNOS) expression. It was also found that treatment of cells with the EtOAc fraction significantly inhibited LPS-stimulated nuclear factor-κB (NF-κB) reporter gene expression. Such inhibition of NF-κB was closely associated with the inhibition of interleukin-6 (IL-6) and the monocyte chemoattractant protein-1 (MCP-1). Finally, the result concluded that SR has the potential for development as a functional food for the treatment of immune diseases, such as rheumatoid arthritis and lupus (Jeong et al., 2010).

Suspension of ethanolic extract of leaves of *Stevia rebaudiana* Bert in gum acacia were administered orally to evaluate the memory enhancing activity in the aged rats. Memory enhancing effect was evaluated by using Morris water maze and elevated plus maze. Mean escape latency time (ELT) calculated each day during acquisition trial was used as an index of acquisition and mean time spent in target quadrant in search of missing platform provided an index of retrieval. Significant reduction in transfer latency (TL) value of retention indicated improvement of memory. Significant memory enhancing effect of ethanolic extract *Stevia rebaudiana* was observed in aged rats at 200 mg/kg p.o. dose (Juyal et al., 2010).

Anticancer activity of ethanolic extract of *Stevia rebaudiana* leaf has been evaluated in the rat by induced Erlisch's Ascites carcinoma. At doses of 100 and 300 mg/kg/ i.p., the extract inhibited proliferation of cell count significantly and the action was dose dependent manner (Rajesh et al., 2010).

The present study carried out to investigate the antihyperglycemic, oral glucose tolerance test (OGTT) and antihyperlipidemic (total cholesterol and triglycerides) effects of the different fractions (petroleum ether, ethyl acetate and chloroform) of ethanolic extract of *Stevia rebaudiana* leaves. The different fractions of the extract were administered orally as a single dose of 150 mg/kg body weight to alloxan induced hyperglycemic rats and found to reduce blood glucose level significantly ($p < 0.05$). The different fractions resulted in the significant reduction of lipid content which was increased in hyperglycemic rats. The plant fractions also improve the glucose tolerance in the glucose induced rats (Hossain et al., 2011).

The study was investigated the medicative effects of medium-polar (benzene: acetone, 1:1, v/v) extract of leaves from *Stevia rebaudiana* (family Asteraceae) on alloxan-induced diabetic rats. Medium-polar extract was administered orally at daily dose of 200 and 400 mg/kg body wt. basis for 10 days. Medium-polar leaf extract of *S. rebaudiana* (200 and 400 mg/kg) produced a delayed but significant ($p < 0.01$) decrease in the blood glucose level, without producing condition of hypoglycemia after treatment, together with lesser loss in the body weight as compared with standard positive control drug glibenclamide. The results concluded that *Stevia* extract was found to antagonize the necrotic action of alloxan and thus had a re-vitalizing effect on β-cells of pancreas (Misra et al., 2011).

Toxicological Studies

Stevioside, a main glycoside of stevia, was found to be nontoxic in acute toxicity studies in a variety of laboratory animals. Stevia is not mutagenic or genotoxic. One report found that constituents of stevioside and steviol were not mutagenic *in vitro*. Chronic administration of stevia to male rats had no effect in fertility versus controls. Another report concluded that stevioside in high doses did not affect growth or reproduction in hamsters of both sexes. Rebaudioside A showed no toxic effects when given to rats at dosages of up to 2,000 mg/kg/day for 90 days (Nikiforov and Eapen, 2008). Stevioside at a dose as high as 2,500 mg/kg did not do any harm to these animals. Hence stevioside at a dose as high as 2.5 g per kilogram of body weight affects neither the growth nor reproduction in hamsters.

Acute Oral Toxicity

The acute oral toxicity studies of aqueous, ether and methanolic extracts of Stevia were undertaken as per the Organization for Economic Co-operation and Development (OECD) guidelines. Control group received only the vehicle and each treatment group received orally the ether, methanolic and aqueous extracts of the studied plant in the limit test at a rate of 2000 mg/kg body weight was conducted and terminated after four survivals out of four animals. Again a higher dose of 5000 mg/kg of all extracts were given to three groups of rats. Animals were kept under close observation for 4 h after administering the extracts and then they were observed daily for 3 days for any change in general behavior and other physical activities. Study revealed that the administration of graded doses of three crude aqueous, ether, and methanol extracts (up to a dose of 5000 mg/kg) of *S. rebaudiana* did not produce significant changes in behaviors such as alertness, motor activity, breathing, restlessness, diarrhea, convulsions, coma, and appearance of the animals. No death was observed up to the dose of 5 g/kg body weight. The mice were physically active. Results showed that in single dose, the plant extracts had no adverse effect, indicating that the medium lethal dose (LD_{50}) could be greater than 5 g/kg body weight in mice.

Further sub-acute toxicity of aqueous, ether and methanolic extract of *S. rebaudiana* leaves were studied in albino rats of either sex (*n* = 24). Rats were divided randomly into four groups. Group I (*n* = 6) served as control and the other three groups were used as experimental groups. Group II, AE (*n* = 6), group III, EE (*n* = 6), and group IV, ME (*n* = 6) were given 2 g/kg, i.p. of *S. rebaudiana* leaves per day for 4 weeks. No significant difference in the mean concentration was found after 4 week of study (Kujur et al., 2010).

Clinical Investigation

The aim of this study was to formulate and evaluate the noble herbal moisturise gel containing Stevia extract. The cosmetic gel formulation was designed by using aqueous extract of *Stevia rebaudiana* leaves in varied concentrations (2.5% and 5.0%) and evaluated using physiological measurements in comparison with a control placebo gel. The initial physicochemical parameters of formulations SF-I and SF-II, i.e. pH, viscosity, spreadability, extrudability and stability were examined. Furthermore, both formulations were studied for toxicity and skin irritancy on animal model. The study was revealed that both the formulations showed no toxicity and no skin irritation on animals. This formulation of Stevia extract could be suggested as a safe and beneficial moisturiser (Das et al., 2009).

Market Formulations

Stevia is known for skin shining and tightening properties, and has found its way in several commercial skin tightening products or anti-wrinkle products.

Moisturizing Toner

This moisturizing toner is hydrating and softening the skin. This toner is useful for dry and tired skin.

Ingredients: Water, *Rosa damascene* flower extract, alcohol, glycerin, honey, levulinic acid, sodium benzoate, sodium levulinate, *Stevia rebaudiana* leaf extract, sodium hydroxide, tetrasodium glutamate diacetate, citric acid, terpineol, *Helichrysum italicum* extract.

Action: Rose flower extract is used for flavoring agent. Honey and Stevia is used for skin smoothening effect.

Direction: Apply on neck and face with a cotton pad.

Supplied by: Melvita, France.

Stevia Toothpaste–Biodent

Ingredients: Stevia, water, veg. glycerine, green clay, xanthan, chamomile, olive leaf extract, allantoin, spearmint.

Supplied by: Stevia Germany.

Stevia Soap

It is a mild soap, the delicate scent of Stevia. The soap is made for extremely soft to the skin and has a mild peeling effect due to the presence of crushed leaves of Stevia.

Ingredients: crushed leaves of Stevia (2.5%), other soap ingredients but not parabens and other animal fats.

Action: Stevia is used in cosmetics for its cleansing and healing.

Direction: This soap is used at any time of day and can also be used in shaving cream to apply on the skin before shaving.

Supplied by: La Maison du Stevia-Ltd company, Le Havre.

Various Preparations

Stevia concentration can also be applied directly to heal any skin cuts. It can also appear handy in treating the problems of acne and other skin blemishes.

Lip Balm

Ingredients: 5 g (1 teaspoon) olive oil, 15 g Jojoba oil, 15 g sweet almond oil, 20 g beeswax pellets, 5 drops liquid stevia, 20 drops flavor oil.

Method of preparation: Melt beeswax, add other oils and stir until all are melted. If solidified reheat over a low heat. Add the Stevia and flavor oils and pour into pots or tubes. This is a firm balm that works great in the tubes.

Organic face pack: A paste made of Stevia leaf powder, honey and milk can be made and applied as organic face pack. It helps to cure eczema, dermatitis, etc.

Toothpaste

Ingredients: 2 tablespoons coconut oil, 3 tablespoons baking soda, 1/2 small packet of Stevia powder, 20–25 drops of peppermint oil.

Method: Mix all ingredients together in a small bowl, using a fork.

A few drops of water-based Stevia are placed directly on the skin and left there to dry for approximately 1/2 hour to one hour. Rinse off with cold water. It will reduce wrinkle within few days.

Storage of Products

All the products should be stored in well closed containers and away from sunlight.

Dosage

The claimed intake per person per day in Europe from these uses of the fresh and powdered leaves is 2.4 g dry powder equivalent to 400 mg stevioside.

Patents

A green tea based dietary supplement, nutritional beverage or dental hygiene product is revealed with enhanced flavor characteristics. Green tea sources are combined with extracts of the fruit of Lo Han and/or extracts from the leaves of *Stevia rebaudiana* and/or extracts from the leaves of Chinese Blackberry. The resulting product is a pleasant tasting dietary supplement, which is easily absorbed by the body in liquid form and provides substantially therapeutic effects (Selzer and Vasile, 2005).

The invention provides an oral cavity cleaning composition, comprising 35–50 parts by weight of tea leaves extract, 10–25 parts by weight of Stevia leaf extract, 10–25 parts by weight of lemon extract, 10–15 parts by weight of mint leaf extract, 10–20 parts by weight of lonicerae flos extract, 20–30 parts by weight of kuding tea leaf extract, 5–10 parts by weight of xylitol and 25–35 parts by

weight of ethanol. All ingredients in the composition according to the present invention are obtained from edible plants that are very effective in oral cavity care. Furthermore, the composition according to the present invention is nonirritant to the body and can be used for long term (Wang, 2008).

The invention relates to a novel nutraceutical composition containing Stevia extract or its constituents, such as steviol and stevioside, as active ingredient(s). The term "nutraceutical" as used herein denotes usefulness in nutritional, pharmaceutical and veterinary fields of application. The compositions are useful for improvement of cognitive functions, such as learning, memory and alertness, and psychotic stability (Fowler et al., 2011).

The invention provides methods of purifying Rebaudioside D from the *Stevia rebaudiana* Bertoni plant extract along with Rebaudioside A. The methods are useful for producing high purity Rebaudioside D and Rebaudioside A. The invention further provides a low-calorie toothpaste composition containing the purified Rebaudioside D and a process for making the low-calorie tooth paste composition containing the purified Rebaudioside D (Abelyan et al., 2011).

BIBLIOGRAPHY

1. Abelyan V, Markosyan A, Abelyan L. High-purity Rebaudioside D and lowcalorie tooth paste composition containing the same. United States Patent Application 20110091394. Application No: 12/786413, Publication date: 04/21/2011. http://www.freepatentsonline.com/y2011/0091394.html

2. Das K, Dang R, Machale MU. Formulation and evaluation of a novel herbal gel of Stevia extract. 2009; 12(4): 117–122.

3. Fowler A, Goralczyk R, Kilpert C, Maynemechan AO, Mussler B, Wyss A. Novel nutraceutical compositions containing Stevia extract or Stevia extract constituents and uses thereof. United States Patent Application 20110038957. Application No: 12/745421,

Publication date: 02/17/2011. http://www.freepatentsonline.com/y2011/0038957. html

4. Hossain MS, Alam MB, Asadujjaman M, Islam MM, Rahman MA, Islam MA, Islam A. Antihyperglycemic and antihyperlipidemic effects of different fractions of *Stevia rebaudiana* leaves in alloxan induced diabetic rats. Int J Pharm Sci Res. 2011; 2(7): 1722–1729.

5. http://ayurvedicmedicinalplants.com/plants/916.html

6. http://elqalatawy.hubpages.com/hub/Lotion-Bar-Recipes

7. http://www.essentialdayspa.com/forum/viewthread.php?tid=35298& start=50

8. http://www.herbsarespecial.com.au/free-herb-information/stevia.html

9. Jeong Y, Lee HJ, Jin CH, Park YD, Choi DS, Kang MA. Anti-inflammatory activity of *Stevia rebaudiana* in LPS-induced RAW 264.7 cells. J Food Sci Nutri. 2010; 15: 14–18.

10. Juyal DS, Bisht G, Kumar A. Memory Enhancing Effect of Ethanolic Extract of *Stevia rebaudiana* (Bert.). International Journal of Phytomedicine. 2010; 2: 166–171.

11. Kujur RS, Singh V, Ram M, Yadava HN, Singh KK, Kumari S, Roy BK. Anti-diabetic activity and phytochemical screening of crude extract of *Stevia rebaudiana* in alloxan-induced diabetic rats. Pharmacognosy Res. 2010; 2(4): 258–263.

12. Misra H, Soni M, Silawat N, Mehta D, Mehta BK, Jain DC. Antidiabetic activity of medium-polar extract from the leaves of *Stevia rebaudiana* Bert. (Bertoni) on alloxan-induced diabetic rats. J Pharm Bioall Sci. 2011;3: 242–248

13. Nikiforov AI, Eapen AK. A 90-day oral (dietary) toxicity study of rebaudioside A in Sprague-Dawley rats. Int J Toxicol. 2008; 27(1):65–80.

14. Rajesh P, Kannan VR, Durai MT. Effect of *Stevia rebaudiana* Bertoni ethanolic extract on anticancer activity of Erlisch's Ascites carcinoma induced mice. Current Biotica. 2010; 3(4): 549–554.

15. Selzer JA, Vasile CD. All natural flavor enhancers for green tea beverages and dental hygiene product. United States Patent Application 20050152997. Appli-cation No: 10/757876, Publication date: 07/14/2005. http://www.freepatentsonline.com/y2005/0152997.html

16. Shukla S, Mehta M, Bajpai VK, Shukla S. *In vitro* antioxidant activity and total phenolic content of ethanolic leaf extract of *Stevia rebaudiana* Bert. Food and Chemical toxicology. 2009; 47: 2338–2343.

17. Wang C. Oral cavity cleaning composition. United States Patent Application 20080069783. Application No: 11/710518, Publication date: 03/20/2008. http://www. freepatentsonline.com/y2008/0069783. html

Thymus vulgaris

Plant profile

Kingdom: *Plantae* – Plants

Superdivision: *Spermatophyta* – Seed plants

Subkingdom: *Tracheobionta* – Vascular plants

Division: *Magnoliophyta* – Flowering plants

Class: *Magnoliopsida* – Dicotyledons

Order: *Lamiales*

Genus: *Thymus L.* – Thyme

Subclass: *Asteridae*

Family: *Lamiaceae* – mint family

Species: *Thymus vulgaris* L. – Garden thyme

Vernacular name

San: Vanya Yavani, **Eng:** Continental wild thyme, Creeping thyme, **Hind:** Banajbwain, **Punjabi:** Marizha, Masho, Rangsbur, **Tel:** Maruvam, **Mal:** Thottathulasi, **Tam:** Omam

Common names

Black Thyme, Common Thyme, English Thyme, French Thyme, Garden Thyme, German Thyme, Serpyllum, Tomillo and Winter Thyme.

Ayurvedic properties

Rasa: Pungent, astringent, sweet; **Veerya:** Heating; **Vipaka** (post-digestive effect): Pungent

INTRODUCTION

Thyme is a low-growing perennial to around 40 cm tall with fibrous roots and woody stalks. Tiny aromatic leaves are born on erect stems which are about 1/8 inch long and 1/16 inch broad, narrow and elliptical, greenish-grey in color, reflexed at the margins, and set in pairs upon very small foot-stalks. Small lilac flowers are borne in summer and they are quite attractive. The calyx is tubular, striated, closed at the mouth with small hairs and divided into two lips, the uppermost cut into three teeth and the lower into two. The corolla consists of a tube about the length of the calyx, spreading at the top into two lips of a pale purple colour, the upper lip erect or turned back and notched at the end, the under lip longer and divided into three segments. The plant has an agreeable aromatic smell and a warm pungent taste.

Distribution

It is a perennial plant native to Europe and Asia. It is widely distributed in the mountains of Spain and other European countries bordering on the Mediterranean, flourishing also in Asia Minor, Algeria and Tunis. It is cultivated now in most countries with temperate climates including France, Spain, Portugal and Greece, as well as in the western U.S. In India, it is cultivated in the western temperate Himalayas and Nilgiris.

Chemical Constituents

It contains 20–25% phenols: Thymol, carvacrol, cymene, pinene, methone as well as bitters, flavonoids, saponins and tannins. The volatile oil components of thyme are now known to include carvacolo, borneol, geraniol, but most importantly, thymol.

Thymol   Carvacrol   Pinene

Medicinal uses

Thuymus vulgaris possesses antiseptic, antiviral, antifungal, antibiotic, antioxidant, antibacterial, antispasmodic and anti-inflammatory activities. Its oil or infusions are widely recommended for nervous depression, spasm, bruises and sprains, bronchitis, laryngitis, digestive complaints, asthma, and many other ailments. Oil of thyme is employed as a rubefacient and counter-irritant in rheumatism. Thyme reduces water retention, infections of the urinary tract, rheumatism and gout. It helps ringworm, athlete's foot, thrush, scabies and lice. It is useful in promoting perspiration at the commencement of a cold, and in fever and febrile complaints generally.

Cosmetic uses

Externally thyme ointment/oil/tincture is used for easing rheumatic pain (bath) 2 drops thyme oil, with 3 drops eucalyptus oil, 2 drops clary sage oil, hot swellings, sciatica, warts. It is also used as a gargle or mouthwash for mouth and gum infections, sore throat. Thyme has been used to fight tooth infections and toothache. It is used for acne, arthritis, asthma, athlete's foot, blemishes, bronchitis, bruises, burns, candida, colds, crabs, dandruff, dental decay, depression, eye soreness, flu, fungal infection, halitosis, insect bites, insect stings, laryngitis, lice, mastitis, mouth sores, muscle soreness, parasites, plaque, rheumatism, ringworm, scabies, sciatica, sore throat, thrush, tonsillitis and wounds.

Pharmacological Action

Essential oils (EOs) and methanol extracts obtained from aerial parts of *Thymus vulgaris* and *Pimpinella anisum* seeds were evaluated for their single and combined antibacterial activities against nine Gram-positive and Gram-negative pathogenic bacteria: *Staphylococcus aureus, Bacillus cereus, Escherichia coli, Proteus vulgaris, Proteus mirabilis, Salmonella typhi, Salmonella typhimurium, Klebsiella pneumoniae* and *Pseudomonas aeruginosa*. The essential oils and methanol extracts revealed pro-mising antibacterial activities against most pathogens using broth microdilution method. Maximum activity of *Thymus vulgaris* and *Pimpinella anisum* essential oils and methanol extracts (MIC 15.6 and 62.5 mug/ml) were observed against *Staphylococcus aureus, Bacillus cereus* and *Proteus vulgaris*. Com-binations of essential oils and methanol extracts showed an additive action against most tested pathogens especially *Pseudomonas aeruginosa* (Al-Bayati, 2008).

The effects of *Thymus vulgaris* hydro-alcoholic extract on the contractile responses of the isolated guinea pig ileum were investigated. Contraction changes in the terminal ileum of guinea pigs were monitored using a force displacement transducer amplifier connected to a physiograph. *Thymus vulgaris* extract inhibited the contractile responses in a dose-dependent manner and also decreased the amplitude of peristaltic waves. It is concluded that *T. vulgaris* has an anti-spasmodic action on guinea pig ileum by decreasing the amplitudes of the muscle contractions during peristalsis. The EC_{50} was calculated as 1.7 mg/ml. In guinea-pig ileum the extract led to an antispasmodic effect, possibly by affecting the anticholinergic and serotoninergic pathways (Babaei et al., 2008).

A polycarbophil-based gel, containing *Thymus vulgaris* essential oil (1%, w/w), was formulated and its antifungal activity was evaluated *in vitro* against four strains of *Candida* commonly involved in *vulvovaginal candidosis*. Results show that the gel allows the essential oil to maintain its activity at pH values near those observed in the healthy vagina. Also, the lack of significant anti-fungal activity of the polycarbophil-based gel seems to be advantageous when considering its use as a placebo formulation (Neves et al., 2009).

Cytotoxicity of thyme essential oil was investigated on the HNSCC cell line, UMSCC1. The IC_{50} of thyme essential oil extract was 369 μg/ml. Genes involved in the cell cycle, cell death and cancer were involved in the cytotoxic activity of thyme essential oil at the transcriptional level. The three most significantly regulated pathways by thyme essential oil were interferon signaling, N-glycan biosynthesis and extracellular signal-regulated kinase 5 (ERK5) signaling. Finally, the result concluded that thyme essential oil inhibits human HNSCC cell growth (Serkan et al., 2011).

In the research, 600 days old LOHMAN broilers in 5 groups of 30 selected as experimental and control groups; and in all of these groups vaccination process and also determination of the final weight (FCR) have been used, which in experimental group different doses of essence have been received and after vaccination the samples were taken, and study process of histopathological changes were done. Regarding to findings and statistical studies which were done SPSS software, the trachea edema factors, tracheitis, pulmonary edema, bronchitis, have been surveyed. In the results obtained from the 6th day onwards IB vaccine (H120) of all above cases, experimental group reduced by increasing dose (Adel et al., 2012).

Toxicological Studies

The safety limit of thyme EO was determined by recording LC50 value on mice. Mice were selected as test animal for the mammalian toxicity experiments. Requisite amount of thyme essential oil (EO) was mixed properly with tween-80 and distilled water (2:1) to prepare different solutions containing desired dose of oil from 100–300 μl. 0.5 ml of each solution of EO was orally administered separately through a syringe with catherator to each set containing 12 mice. After 72 h, the mortality of the animals was recorded and LC_{50} was calculated in terms of per kg body weight of mice. During safety limit trials on mice, the LC_{50} of thyme oil was recorded as 7142.857 μl/kg body weight where 50% population of test animals were found to be killed on oral administration, indicating a non-mammalian toxic nature of the oil (Kumar et al., 2008).

Clinical Investigation

One of thyme's main constituents, thymol, has antibacterial effects. Thymol is included as one of several ingredients in antiseptic mouthwashes such as Listerine®. Clinical studies have reported efficacy of Listerine® in decreasing plaque formation and gingivitis, although human evidence for thymol alone is limited.

For hair loss, 2–3 drops of an essential oil combination (thyme, lavender, rosemary and cedarwood added to grape seed and jojoba oil) massaged into the scalp every night for

seven months has been studied (Hay et al., 1998). For paronychia (skin infection around a finger or toenail), 1 drop of 1–2% thymol in chloroform to the affected area three times daily, or 1 drop of 4% thymol in chloroform to a chronically affected area three times daily has been used. Diluted thyme oil has been applied as needed in 1–2% ointments for a variety of skin disorders.

Market Formulations

Alpaflor® Nectapure PF

It is a high-quality Alpaflor® specialty with excellent antioxidative and moisturizing properties. It consists of anti-aging formulations, sun and after-sun formulations and day care formulations. Nectapure helps to prevent wrinkles, loss of elasticity and loss of skin vitality.

Ingredients: Glycerin, water, *Buddleja davidii* extract, *Thymus vulgaris* (thyme) extract.

Action: Buddleja davidii and *Thymus vulgaris* have been carefully selected for their capability to fight against pollution being very harmful to the skin.

Supplied by: DSM Nutritional Products Ltd., Switzerland.

Listerine mouthwash: Listerine mouthwash effectively kills oral bacteria, reduces buildup of plaque, and helps prevent gingivitis.

Ingredients: Thymol IP: 0.06%, eucalyptol PCx: 0.09%, benzoic acid IP: 0.15%, Menthol IP: 0.04%, ethanol (95%) IP 26% v/v.

Action: It helps prevent and reduce supragingival plaque accumulation and gingivitis when used as direct in a conscientiously applied program of oral hygiene and regular professional care.

Direction: Rinse around teeth and gums with about 20 ml of undiluted Listerine, gargle, then spit out after 30 seconds. Use after brushing for fresh breath, healthy teeth and gums. Use it regularly to complement your daily oral hygiene routine.

Snoreeze nasal spray: Snoreeze nasal spray is ideal for snoring caused by a cold, allergies or a blocked nose.

Ingredients: Aqua (water), sorbitan stearate (vegetal origin), polysorbate 60 (from vegetable oil), sodium chloride (salt), glycerine (vegetal origin), *Calendula officinalis* (flower extract), menthyl lactate, xanthan gum (from sugar and molasses), cellulose gum (vegetal origin), guar gum (vegetal origin), *Thymus vulgaris* (thyme oil), *Lavandula angustifolia* (lavender oil).

Action: Essential oils of lavender and thyme—to aid easy breathing, menthyl lactate —to moisturise the nasal passages, Xanthan gum is used for lubrication.

Direction: Apply each night (or as required). Blow your nose to clear the nasal passages. Shake bottle. Spray twice in each nostril.

Supplied by: Feelunique.com, UK.

Neem Herbal Skin Conditioning Spray

Ingredients: Purified water, *Aloe Barbadensis* (aloe vera) leaf juice, glycerin, polysorbate-80, *Helianthus annuus* (sunflower) seed oil, *Azadica indica* (neem) oil, *Hydrastis canadensis* (golden seal) extract, *Berberis vulgaris* (barberry) extract, *Anthemis nobilis* (chamomile) extract, *Thymus vulgaris* (thyme) oil, *Juniperus mexicana* (cedarwood) oil, phenoxyethanol ethyl hexyl glycerin, *Citrus aurantium* (sweet orange) oil, *Cympobogon flexuosus* (lemon-grass) oil, *Lavandula hybrida* (lavandin) oil, *Cymbopogon nardus* (citronella) oil.

Action: Neem aura can soothe the effects of irritated skin.

Direction: Use this spray before and after exposure to irritating elements such as wind and sun.

Supplied by: VitaSprings, USA.

Gardeners dream cream: This cream is used to relief of skin conditions and weird. It is also used for rashes, dry cracked hands; knees; elbows and cuticles, acne and cold sores.

Ingredients: Aqua (deionized water), *Butrospermum parkii* (shea butter), *Helianthus annuus* (sunflower) seed oil, cetearyl alcohol, *Vitis vinifera* (grape) seed oil, caprylic/capric triglyceride, *Cera alba* (beeswax), *Arnica montana* (arnica) extract, *Lavandula angustifolia* (lavender) oil, isostearic acid, tocopheryl acetate (vitamin E), *Citrus grandis* (grape fruit) seed extract, *Papaver somniferum* (poppy) seed oil, *Mentha Piperita* (peppermint) oil, *Rosmarinus officinalis* (rosemary) leaf oil, *Eucalyptus globulus* (eucalyptus) oil, *Cinnamomum zeylanicum* (cinnamon) leaf oil, *Retinyl palmitate* (vitamin A), *Michelia champaca* flower (chamaca) oil, *Origanum vulgare* (wild margoram) extract, *Thymus vulgaris* (thyme) extract, *Cinnamomum zeylanicum* (cinnamon) extract, *Hydrastis canadensis* (golden seal) root extract, coconut derived.

Action: Lavender: Excellent skin cell regenerator; antiseptic; helps heal infections; effective aid in healing burns; minor cuts and wounds. Peppermint: Helps relieve inflammation, indigestion, sinus and lung congestion, itching, insomnia, and shock. It can also help with lack of focus. Rosemary: Stimulates circulation; relieves congestion and sore muscles; improves memory and confidence; helps to balance the mind and body. Cinnamon: Helps with infections, indigestion, and congestion. Considered to be an aphrodisiac and is useful for fatigue and depression. Eucalyptus: Helps clear congestion, infections, pimples and insect bites; increases energy. Arnica: A natural herbal extract effective for relief of joint pains, inflammation and bruises. Thyme: It is used for cooling effect.

Direction: Apply liberally to affected area as often as necessary. Store in a cool, dry place.

Supplied by: Feel Good Natural Health Store, Canada.

Various Preparations

It is used in gargle and mouthwash for dental decay, laryngitis, mouth sores, plaque formation, sore throat, thrush, tonsillitis, and bad breath. Thyme poultice is used for wounds, mastitis, insect bites and stings. It is used to wash for fungal infections such as athlete's foot and ringworm, and use against parasites such as crabs, lice and scabies. Essential oil is added to soaps and anti-depressant inhalations. Added to massage oils for sore muscles, rheumatism and sciatica, and applied directly to warts.

Antifungal soap for ringworm: This soap's penetrative therapeutic effect treats and cures ringworm. The soap can be used daily while having a shower, after which one needs to apply the ringworm treatment cream to get rid of it.

Ingredients: 300 gm white soap, 5 tbsp of jojoba oil, 10 ml of tea tree oil, 1.5 ml of *thymus vulgaris*, 500 ml rose water, cheese grater, Some dried marigold petals, greaseproof paper.

Preparation: Take the white soap bar and grate it using the cheese grater. Transfer these soap shavings to a bowl and add rosewater into it. Place a pan filled halfway with water on the burner and let the water boil. Once the water has begun boiling, place the bowl of soap and rosewater into the water. Secondary heating will enable the soap to melt uniformly. Once the soap melts, remove from the burner and add the jojoba oil and stir well. Also add the tea tree oil and *thymus vulgaris* and sprinkle in the dried marigold petals. Mix this soap solution well and put it into a shallow dish, suitable to prepare a soap loaf. Leave the soap solution to cool and set for the next 10–14 days. Once they have set, slice the soap loaf into small cubes and wrap them in greaseproof paper. Store them in a cool dry place.

Facial Tonner

Ingredients: 1 teaspoon dried rosemary, 1 teaspoon dried thyme, 1 teaspoon finely chopped fresh mint leaves, 1 cup white wine.

Preparation: Put all ingredients in a pan and let it gently simmer for 10 minutes. Leave it to steep and cool down for at least one hour. Strain it and throw away the herbs.

Storage of Products

All the products should be stored in well closed containers and away from sunlight.

Dosage

The usual therapeutic dosage of thyme fluid extract is 3–12 ml per day. The usual therapeutic dosage of thyme tincture is 6–18 ml per day.

Patents

A natural preservative composition obtained from plant materials provides antimicrobial activity for use as an antifungal agent. The antifungal agent is effective in inhibiting the growth of *Epidermophyton floccosum*, *Trichophyton mentagrophytes* and *Microsporum canis*. The antimicrobial agent has a MIC as low as 0.03 µl/ml capable of inhibiting and/or killing these organisms. The antimicrobial agent includes selective mixtures of *Origanum vulgare* L., *Thymus vulgaris* L., *Cinnamomum zeylanicum* Nees, *Rosmarinus officinalis* L., *Lavandula officinalis* L., *Mentha piperita* L., *Citrus Limon* L., *Hydrastis canadensis* L. and *Olea europaea* L (D'amelio Sr and Mirhom, 2006).

A hair coloring composition contains an oxidative hair dye and at least one anti-oxidant agent selected from the group consisting of *Rosmarinus officinalis*, *Origanum vulgare*, *Camellia sinensis*, *Camellia oleifera*, *Salvia officinalis*, *Apium graveolen*, *Thymus vulgaris*, *Rosa canina* and *Coriandrum sativum* in combination with a suitable diluent or carrier. The hair coloring compositions are gentler on the skin and hair in that they reduce the effects or the oxidising agent on the hair and skin (Smith and Long, 2007).

The invention discloses a chemical formula for the composition of a liquid topical analgesic. It is produced from a unique blend of herbal essential oils and plant- and/or vegetable-derived oils. Although the precise proportions and method of mixture are held in confidence, it comprises at least olive oil, castor oil, grapeseed oil, almond oil, apricot kernel oil, vitamin E, various herbal essential oils, and a stabilizer. Such essential oils may include eucalyptus (*Eucalyptus globulus*, among others), lavender (*Lavandula angustifolia*), tea tree (*Melaleuca alternifloria*), ginger (*Zingiber officinale*), lemongrass (*Cymbopogon citratus*, among others), red thyme (*Thymus vulgaris*), sweet birch, (*Betula lenta*) and ylang ylang (*Cananga odorata*). Alternate and additional essential oils or plant- and/or vegetable-derived oils are anticipated to be used as well. It will reduce pain from sore muscles, bruised ligaments and tendons, lower back pain, arthritis and other similar ailments associated with common aches and pains (Anderson, 2008).

BIBLIOGRAPHY

1. Adel F, Farhad D, Mehrdad N. The effect of *Thymus vulgaris* volatile oils in controlling of the complications resulted from spray of H120 vaccine in broiler chicks. Journal of Animal and Veterinary Advances. 2012; 11: 357–360.

2. Al-Bayati FA. Synergistic antibacterial activity between *Thymus vulgaris* and *Pimpinella anisum*. J Ethnopharmacol. 2008;116(3): 403–406.

3. Anderson LA. Herbal healing oil. United States Patent 7368135. Application No. 11/491581. Publication date: 05/06/2008. http://www.freepatentsonline.com/7368135. html

4. Babaei M, Abarghoei ME, Ansari R, Vafaei AA, Taherian AA, Akhavan MM, Toussy G, Mousavi S. Antispasmodic effect of hydroalcoholic extract of *Thymus vulgaris* on guinea pig ileum. Nat Prod Res. 2008; 22(13): 1143–1150.

5. D'amelio Sr., FS, Mirhom YW. Antifungal composition, its fungicidal effect on pathogenic dermatophytes, and process for inhibiting growth of fungi. United States Patent Application 20060073218. Application No. 11/211102. Publication date: 04/06/2006. http://www.freepatentsonline.com/y2006/0073218. html

6. Hay IC, Jamieson M, Ormerod AD. Randomized trial of aromatherapy. Successful treatment for *alopecia areata*. Arch Dermatol. 1998;134(11):1349–1352.

7. http://www.aminaherbs.com/product.php?id_product=275

8. http://www.globalherbalsupplies.com/herb_information/thyme.htm

9. http://www.healthline.com/natstandardcontent/thyme-thymol#2

10. http://www.herbsorganic.co.za/pages/working%20on/thyme/template%20thyme.htm

11. http://www.indianmirror.com/ayurveda/thyme.html

12. http://www.indianspices.com/html/s0625 thyme.htm

13. http://www.motherherbs.com/thymusvulgaris.html

14. Kumar A, Shukla R, Singh P, Prasad CS, Dubey NK. Assessment of *Thymus vulgaris* L. essential oil as a safe botanical preservative against post harvest fungal infestation of food commodities. Innovative Food Science and Emerging Technologies. 2008; 9: 575–580.

15. Neves JD, Pinto U, Amaral MH, Bahia MF. Antifungal activity of a gel containing *Thymus vulgaris* essential oil against *Candida* species commonly involved in *vulvovaginal candidosis*. Pharmaceutical Biology. 2009; 47(2): 151–153.

16. Serkan S, Eichhorn T, Peter K.P, Thomas E. Cytotoxicity of *Thymus vulgaris* essential oil towards human oral cavity squamous Cell Carcinoma. Anticancer Research. 2011; 31(1): 81–87.

17. Smith LR, Long SP. Hair coloring compositions. United States Patent 7172632. Application No. 10/486577. Publication date: 02/06/2007. http://www.freepatentsonline.com/7172632. html

Trigonella foenum-graecum

Plant profile

Kingdom: *Plantae* – Plants

Subkingdom: *Tracheobionta* – Vascular plants

Superdivision: *Spermatophyta* – Seed plants

Division: *Magnoliophyta* – Flowering plants

Class: *Magnoliopsida* – Dicotyledons

Subclass: *Rosidae*

Order: *Fabales*

Family: *Fabaceae* – Pea family

Genus: *Trigonella* L. – Fenugreek

Species: *Trigonella Foenum graecum.* – Fruit fenugreek

Vernacular name

San: Methika, **Eng:** Fenugreek, **Hin:** Methi, **Ben:** Methi, **Ori:** Methi sag, **Kan:**Menthina soppu, **Mal:** Uluva ila, **Guj:** Methi, **Tam:** Meti, Vendayam, Vetani, Vendaya kirai, **Mar:** Methi, **Tel:** Menthkoora

Common name
Fenugreek

Ayurvedic properties

Rasa (Taste): Katu (Pungent), **Guna (Characteristics):** Laghu (Light), Snigdha (Unctuous), **Veerya (Potency):** Ushna (Hot), **Vipaka (Post-digestion effect):** Katu (Pungent).

Ayurvedic action

Arochakahar: Methi alleviates sensation of tastelessness from mouth, **Deeptikar:** Methi ignites the digestive fire in the stomach. **Vata pranashini:** Methi is one of the best vata pacifying herb in Ayurveda, **Jwarnashini:** Methi is useful in chronic fever.

INTRODUCTION

An erect annual to 50 cm. which may be branched, the leaves are trifoliate and the leaflets oblong-lanceolate, to 5 cm. Its yellowish flowers 12–18 mm long (1–2) are in the leaf axils. The fruits are almost straight and flattened with a pronounced beak. They are 50–110 mm long excluding the beak of 10–35 mm. Chromosome number 2n = 16. The seeds are brownish, about 1/8 inch long, oblong, rhomboidal, with a deep furrow dividing them into two unequal lobes; they are contained, ten to twenty together, in long, narrow, sickle-like pods.

Distribution

The plant is native of southeastern Europe and Western Asia, fenugreek grows today in many parts of the world, including India, Northern Africa, and the United States.

Chemical Constituents

Fenugreek leaves contain moisture 86.1 percent, protein 4.4 percent, fat 0.9 percent, minerals 1.5 percent, fiber 1.1 percent and carbohydrates 6.0 percent per 100 grams of edible portion. The mineral and vitamin contents are calcium, iron, phosphorous, carotene, thiamine, riboflavin, niacin and vitamin C.

Fenugreek seeds contain moisture 13.7 percent, protein 26.2 percent, fat 5.8 percent, minerals 3.0 percent, fiber 7.2 percent and carbohydrates 44.1 percent per 100 grams. The mineral and vitamin contents are calcium, phosphorous, carotene, thiamine, niacin and riboflavin. Several alkaloids have been found in fenugreek seeds.

The seeds contain alkaloid trigonelline and choline, essential oil and saponin. Trigonelline has highly toxic action on neuromuscular preparations. The seeds also contain fixed and volatile oil, mucilage, bitter extractive and a yellow coloring substance. Air-dried seeds contain a little amount of trigonelline and nicotinic acid.

Trigonelline   Choline

Medicinal uses

In ayurveda it is used medicinally for the treatment of wounds, abscesses, arthritis, bronchitis, and digestive disorders. In traditional Chinese medicine it is also used for kidney problems and conditions affecting the male reproductive tract. Fenugreek was, and remains, a food and a spice commonly eaten in many parts of the world. Recent researches have proved it beneficial for atherosclerosis, constipation, diabetes, high cholesterol and hyper-triglyceridemia (high triglycerides). The seeds of fenugreek contain the most potent medicinal effects of the plant.

It helps to improve immunity and protects heart, brain and other vital organs of body through its medicinal properties.

Cosmetic uses

Hair oil made of methi seeds helps to prevents premature graying of hair and makes them strong and free from dandruff. Apply a paste of methi leaves to the palms and soles of the feet, alleviates burning sensation.

Fenugreek leaves as a beauty aid: Fresh fenugreek leaves paste and coconut milk applied over the scalp is believed to prevent hairloss, promote hair growth, preserve its natural color, delay graying of hair and make it silky soft. A paste of methi leaves mixed with vinegar is good for treating dandruff. A pimple and blackhead prone skin when treated with the paste of fenugreek leaves and turmeric, improves the skin tone. Fenugreek leaves also has anti-aging properties. Make a paste of fenugreek leaves and add boiled milk to it. Applying this to your face delays the appearance of fine lines and face wrinkles. It not only improves the complexion but also makes one looks younger.

Fenugreek leaves and acne treatment: Fenugreek leaves paste when applied on acne affected area every night and washed away the following morning with warm water prevents fresh break, out of acne.

Pharmacological Action

Effects of fenugreek (*Trigonella foenugraecum* Linn) on serum lipid profile in hyper-cholesteremic type 2 diabetic patients were studied. Administration of fenugreek seed powder of 25 gm orally twice daily for 3 weeks and 6 weeks produces significant ($p < 0.001$) reduction of serum total cholesterol, triacylglyceride and LDL-cholesterol in hypercholesteremic group but the change of serum HDL-cholesterol was not significant. On the other hand, changes of lipid profile in hypercholesteremic type 2 diabetic patients without fenugreek were not significant ($p < 0.001$). The study was suggested that fenugreek seed powder would be considered as effective agent for lipid lowering purposes (Moosa et al., 2006).

The study was evaluated the anti-arthritic and vascular protective effects of fenugreek, *Boswellia serrata* and *Acacia catechu* alone and in combinations. Arthritis was induced by CFA (6 mg/ml, i.d.) and parameters measured include paw volume, body weight, RA factor, arthritic index, E.S.R., WBC counts. Vascular and endothelial dysfunction study was done by performing CRC of peroxynitrite and acetylcholine respectively using rat thoracic aorta strip pre-contracted with phenylephrine. The results of arthritis model—combination of three drugs exhibited 48.75 % and combination of two drugs such as methi and boswellia exhibited 44.58 %, acacia and boswellia exhibited 36.25 % and methi and acacia exhibited 29.16 % reduction in paw edema after 21 days respectively. Also combination of three drugs showed significant reduction in arthritic index, RA factor, ESR, WBC counts and normalization of body weight in arthritic rats as compare to diseased control or other two drugs combination. In vascular study combination of three drugs showed significant difference in relaxation (p < 0.001) vs. disease control as compare to other two drug combination (p < 0.01) or alone drug (p < 0.05) in Ach CRC. Combination of three drugs showed significant anti-arthritic and vascular protective effect (Vyas Amit et al., 2010).

Crude extracts of *Trigonella foenum-graecum* (Leguminoceae) were evaluated for *in vitro* anthelmintic activity on the Indian adult earthworms *Pheritima posthuma*. The seed extracts of *Trigonella foenum-graecum* had shown a dose dependant inhibition of spontaneous motility (paralysis) of earthworms. It has been observed that alcoholic extract (60 mg/ml) has shown anthelmintic activity, which was compared with albendazole as reference drug. Therefore the seeds could be categorized under anthelmintic herbal drugs and could become a potent key ingredient of such herbal formulation (Khadse and Kakde, 2010).

The current study was undertaken to evaluate anti-inflammatory activity of alcoholic extract of seeds of *Trigonella foenum-graceum* (fenugreek) on wister strain rat. Seeds were collected, air dried, reduced to no: 36 powder and extracted with petroleum ether at 60°–80°C followed by extracted with 50% aqueous methyl alcohol. The acute anti-inflammatory activity of *Trigonella foenum-graceum* (100mg and 200mg/kg b.w) was measured plethysmographically using carrageenan as inflammatory agent. The study indicated that alcoholic extract of *Trigonella foneum-graceum* exhibited anti-inflammatory effect on carrageenan induced paw oedema in rats at dosage of both at 100 mg and 200 mg/kg b.w. Finally it is conclud that the alcoholic and petroleum ether extracts of fenugreek seeds can be used safely as anti-inflammatory agent (Datta and Shanbagh, 2010).

The aim of the study was to screen the medicinal and antibacterial activities of methanol and acetone extracts of fenugreek (*Trigonella foenum* L.) and coriander (*Coriandrum sativum* L.). Crude extract of the spices with methanol and acetone were screened for antibacterial activities against four Gram-negative pathogenic bacteria—*Pseudomonas* spp., *Escherichia coli*, *Shigella dysentiriae* and *Salmonella typhi*. The *in vitro* antibacterial activity was performed by agar well diffusion method. Methanol extract of fenugreek and coriander revealed an elevated antimicrobial activity against *Pseudomonas* spp., whereas acetone extract of spices exhibited highest activity against *Escherichia coli*. Acetone extract of fenugreek and Coriander showed no activity against *Salmonella typhi*. The results obtained in the present study suggest that the methanol extract of *Trigonella foenum* Linn and *Coriandrum sativum* Linn revealed a significant scope to develop a novel broad spectrum of anti-bacterial herbal formulation (Dash et al., 2011).

Toxicological Studies

Toxicological evaluation of 60 diabetic patients who took powdered fenugreek seeds at a dose of 25 g per day for 24 weeks disclosed no clinical hepatic or renal toxicity and no hematological abnormalities (Sharma et al., 1996).

In an animal study, the acute oral LD_{50} was found to be > 5 g/kg in rats, and the acute dermal LD_{50} was found to be > 2 g/ kg in rabbits. In another animal study, fenugreek powder failed to induce any signs of toxicity or mortality in mice and rats who received acute and subchronic regimens. Moreover, there were no significant hematological, hepatic, or histopathological changes in weanling rats fed fenugreek seeds for 90 days (Rao et al., 1996).

In this investigation, toxic effects of debitterized fenugreek (DFG) powder have been assessed following acute and subchronic regimens in mice and rats. In the acute study, DFG powder intragastrically administered to albino mice (CFT-Swiss, *Mus musculus*) and albino rats (CFT-Wistar, *Rattus norvegicus*) of both sexes failed to induce any signs of toxicity or mortality up to a maximum practical dosage of 2 and 5 g/kg body weight, respectively. Further, no significant alterations either in relative organ weights or their histology were discernible at terminal autopsy. In the 90-day subchronic study, DFG fed to weanling rats of both sexes at dietary doses of 0, 1, 5 and 10% in a pure diet had no effect either on the daily food intake or growth. Terminal autopsy revealed no alterations in relative organ weights of various vital organs, or their histoarchitecture (Muralidhara et al., 1999).

Clinical Investigation

The objective of the study involved preparation of herbal hair oil using amla, hibiscus, brahmi, methi and its evaluation for increase in hair growth activity. Each drug was tested for their hair growth activity in a concentration range for 1–10% separately. Based on these results mixture of crude drugs fruits of *Emblica officinalis*, flowers of *Hibiscus rosa sinensis*, leaves of *Bacopa monnieri* and seeds of *Trigonella foenum-graecum* were prepared in varying concentration in the form

of herbal hair oil by three different oils preparation techniques (direct boiling, paste and cloth method) and were tested for hair growth activity. The formulation containing 7.5% of each drug used for the study and showed excellent hair growth activity with standard (2% minoxidil ethanolic solution) by an enlargement of follicular size and prolongation of the anagen phase. Excellent results of hair growth were seen in formulation prepared by boiling method of oil preparation technique (Banerjee et al., 2009).

Market Formulations

Herbal shampoo: Herbal shampoo is recommended for brittle, dull hair and split ends. It prevents hair disorders by cleansing microbial infections of the scalp. It provides relaxing and soothing effects to the scalp skin and gives complete herbal hair care.

Ingredients: Amalaki extract—*Emblica officinalis*, Bhringraja extract—*Eclipta alba*, Methi—*Trigonella foenum-graecum*, *Azadirachta indica* extract, *Psoralea corylifolia* seed extract and shampoo base q.s.

Action: Amla or Indian gooseberry helps hair to retain its natural colour and texture and makes hair strong and healthy. Bhringraja extract revitalizing hair follicles by stimulating blood circulation which is the prime objective of hair rejuvenating formulations. Fenugreek is used in hair care formulations for its excellent properties ranging from itching control, anti-inflammatory and anti-fungal properties. It helps to control dandruff and premature hair fall and seasonal irritations of scalp. Neem extract is used for its antibacterial action. Psoralea seed extract is used to nourishes the hair follicles and remove the scalp.

Direction: Wet hair, massage shampoo gently on the hair. Rinse and repeat, if necessary. In case of contact with the eyes, just wash with normal water.

Supplied by: Amazing Herbal Remedies, New Delhi, India.

Herbal baby shampoo: Natural baby shampoo is a mild, extremely gentle on little scalps as well as your baby's skin. It gently cleanses the hair and scalp without stripping baby's natural protective oils. It provides relaxing and soothing effects to the scalp skin and gives complete herbal hair care.

Ingredients: Amalaki extract—*Emblica officinalis*, bhringraja extract—*Eclipta alba*, Methi—*Trigonella foenum-graecum, Azadirachta indica* extract, *Psoralea corylifolia* seed extract, cetramide and shampoo base q.s.

Action: Amla or Indian gooseberry is the richest source of natural Vitamin C. It helps hair to retain its natural color and texture and makes hair strong and healthy. It is well known as softening treatment for babies. Bhangra has excellent properties for restoring hair, its natural health through revitalizing hair follicles by stimulating blood circulation which is the prime objective of hair rejuvenating formulations. Methi is used in hair care formulations for its excellent properties ranging from itching control, anti-inflammatory and anti-fungal properties. It helps to control dandruff and premature hair fall and seasonal irritations of scalp.

Directions: Wet hair, massage shampoo gently on the hair. Rinse and repeat, if necessary. In case of contact with the eyes, just wash with normal water.

Supplied by: Amazing Herbal Remedies, New Delhi, India.

Parachute Advanced Ayurvedic Hair Oil

It controls hair fall. The oil nourishes hair roots and scalp, effectively controlling hair fall, get long, thick, beautiful hair. It also promotes hair growth, prevents premature greying, reduces dandruff, keeping your hair healthy.

Ingredients: Amla (Amalaki)—12.50 g, ghritakumari (Kumar)—6.25 g, mehendi (madayantika) 2050 g, nagarmotha (mustak) —2.50 g, methi (methika)—1.25 g, sungandhit dravya - 0.85 g, coconut oil/excipients—q.s.

Direction: Massage regular for best results.

Storage: Store in a cool and dry place, protected from heat.

Supplied by: Mall.coimbatore.com, Coimbatore.

Revitalizing hair oil: This hair oil is non-sticky, which nourishes and promotes hair growth.

Ingredients: Emblica officinalis, Sapindus trifoliatus, Azadirachta indica, Trigonella foenum-graecum, Eclipta prostrata, Cicer arientinum, Aegle marmelos.

Action: Emblica officinalis is used to promote the hair growth. *Sapindus trifoliatus* fruits possess the tonicity and astringency. *Azadirachta indica* is used for its antifungal and antibacterial action. *Trigonella foenum-graecum* seeds prevent hair fall and promote hair growth. *Eclipta prostrate* juice of the leaf applied on the scalp for promoting hair growth. *Cicer arientinum* seeds are used to prevent the dandruff. *Aegle marmelos* fruits are used for its astringent property.

Direction: Apply the oil over the scalp. Then gently massage the scalp in a circular motion with finger. Leave for an hour.

Supplied by: Himalaya Drug Company, India.

Various Preparations

A poultice made of the leaves can be applied with benefits in external and internal swellings, sprains and burns.

Hair washed with fenugreek (methi) seed paste prevents dandruff, falling hair, baldness and dandruff keeping the hair long, healthy and black. Just soak the fenugreek seeds overnight in water to soften the seeds and grind in the morning to make paste. Before hair wash, apply this paste on scalp and hair and leave it for half an hour. Wash off with shampoo later.

Soak fenugreek (methi) seeds in yogurt overnight and apply the curd on the scalp for half an hour before washing in the morning. It reduces dandruff.

Methi-shikakai shampoo: 1 kg of shikakai, 250 g of methi, handful of orange or lemon peels. Make a fine powder of all these ingredients. Before washing the hair, mix this powder with half cup of water and keep it for at least 2 hours.

Prevents hair loss: Fenugreek has very potent seeds, which help treat balding, thinning of hair and hair fall. To create the hair tonic, boil the fenugreek seeds, soak them in coconut oil, overnight. The next morning, massage your head with this potion for five minutes, concentrating on the balding areas. Repeat it everyday and for a few months.

Antidote for skin problems: Fenugreek seeds prove to be an excellent beauty product. They help prevent wrinkles, blackheads, pimples, dryness and rashes. Apply a paste of fresh fenugreek leaves, mixed with water, over the face and keep it for twenty minutes. Then wash it with warm water. Wash the face with water, boiled with fenugreek seeds. It can also be applied on the inflamed body parts and as a cosmetic product.

Storage of Products

All the products should be stored in well closed containers and away from sunlight.

Dosage

Powder: 1–3 g.

Patents

A novel solvent free process of obtaining an insoluble fiber rich fraction from *Trigonella foenum-graceum* seeds is disclosed. The multifunctional fiber rich fraction (FRF) and highly purified FRF are useful as excipients for pharmaceutical dosage forms for various routes of administration. These excipients can be used as binder, disintegrant, filler, dispersing agent, coating agent, film forming agent, thickener and the like, for preparation of variety of dosage forms. FRF and highly purified FRF can also be used in a controlled release, targeted release and other specialized

delivery systems, as well as in food and cosmetics formulation (Pilgaonkar et al., 2005).

BIBLIOGRAPHY

1. Banerjee PS, Sharma M, Nema RK. Preparation, evaluation and hair growth stimulating activity of herbal hair oil. Journal of Chemical and Pharmaceutical Research. 2009; 1(1): 261–267.

2. Dash BK, Sultana S, Sultana N. Anti-bacterial activities of methanol and acetone extracts of fenugreek (*Trigonella foenum*) and coriander (*Coriandrum sativum*). Life Sciences and Medicine Research. 2011; LSMR-27: 1–8.

3. Datta D, Shanbagh T. A study of anti-inflammatory activity of alcoholic extract of seeds of *Trigonella foenum graceum* (fenugreek) on wister strain rat. IJPRD. 2010; 2(9): 81–85.

4. http://blogs.medindia.net/diet-nutrition/fenugreek-leaves-for-health-and-beauty/

5. http://www.ayurvedicdietsolutions. com/Methi.php

6. http://www.fao.org/AG/AGP/agpc/doc/Gbase/DATA/pf000412.htm

7. http://www.neeroga.com/methi.aspx

8. Khadse CD, Kakde RB. *In vitro* anth-elmintic activity of fenugreek seeds extract against *Pheritima posthuma*. Int. J. Res. Pharm. Sci. 2010; 1(3): 267–269.

9. Moosa ASM, Rashid MU, Asadi AZS, Ara NA, Uddin MM, Ferdaus A. Hypolipidemic effects of fenugreek seed powder. Bangladesh J Pharmacol. 2006; 1: 64–67.

10. Muralidhara, Narasimhamurthy K, Viswanatha S, Ramesh BS. Acute and subchronic toxicity assessment of debitterized fenugreek in the mouse and rats. Food Chem Toxicol. 1999; 37(8):831–838.

11. Pilgaonkar PS, Rustomjee MT, Gandhi AS, Bhumra, VS. Fiber rich fraction of *Trigonella foenum graceum* seeds and its use as a pharmaceutical excipient. United States Patent Application 20050084549. Application no: 10/945938, Publication date: 04/21/2005. http://www. freepatentsonline.com/y2005/0084549. html

12. Rao PU, Sesikeran B, Rao PS, et al. Short term nutritional and safety evaluation of fenugreek. Nutr Res. 1996;16:1495–1505.

13. Sharma RD, Sarkar A, Hazra DK, et al. Toxicological evaluation of fenugreek seeds: a long term feeding experiment in diabetic patients. Phytother Res. 1996;10:519–520.

14. Vyas Amit S, Patel Nailesh G, Panchal Aashish H, Patel Rameshwar K, Patel Madhabhai M. Anti-arthritic and vascular protective effects of fenugreek, *Boswellia serrata* and *Acacia catechu* alone and in combinations. An international journal of pharmaceutical sciences. 2010; 1(2): 95–111.

Vetiveria Zizanioides

Plant profile

Kingdom: *Plantae* – Plants

Subkingdom: *Tracheobionta* – Vascular plants

Superdivision: *Spermatophyta* – Seed plants

Division: *Magnoliophyta* – Flowering plants

Class: *Liliopsida* – Monocotyledons

Subclass: *Commelinidae*

Order: *Cyperales*

Family: *Poaceae* – Grass family

Genus: *Vetiveria Bory* – Vetiver grass

Species: *Vetiveria zizanioides* L. Nash

Vernacular names

Hindi, Bengali: Khas, Khas-Khas, Khus-Khus, Khus, **Gujarati:** Valo, **English:** Vetiver, Khas-Khas, **San:** Ushira, Reshira, Sugandhimula, **Marathi:** Vala, **Tel:** Kuruveeru, Vettiveellu, Vettiveerum, **Tam:** Vattiver, **Kannada:** Vattiveeru, Laamancha, Kaddu, Karidappasajje Hullu, **Malyalam:** Ramaccham, Vettiveru

Ayurvedic name

Ushira

Ayurvedic properties

Rasa: Tikta, Madhura, **Guna:** Lakhu, Rooksha, **Veerya:** Seeta, **Vipaka:** Katu

INTRODUCTION

Vetiveria zizanioides is commonly known as *Khas Khas* or *Khus* grass in India. It belongs to the family Poaceae. It is a perennial densely tufted grass, found throughout the plains and lower hills of India, particularly on the riverbanks and in rich sandy loam soil. The grass is stout and over 2 m tall, with stout spongy aromatic roots. The leaves are narrow, erect, keeled, glabrous and its margins are scabrid. The inflorescence is a panicle of numerous slender racemes (up to 15–45 cm long) in whorls on a central axis. The spikelets (4–6 mm long) are grey-green or purplish in color and in pairs. One is sessile and the other is pedicelled, 2-flowered; the lower floret is reduced to a lemma, upper bisexual in sessile.

Distribution

Vetiveria grass grows throughout the plains of India ascending up to an altitude of 1200 m. It grows in a wide variety of ecological habitats covering all bio-geographic provinces of India and propagated through seeds. With an annual rainfall of 1000–2000 mm and temperature ranging from 22 to 43°C is favorable for suitable growth of this grass. It is distributed throughout Africa, India, Burma, Sri Lanka, Southeast Asia. In India this grass grows widely in Haryana, Uttar Pradesh, Rajasthan, Gujarat, Bihar, Orissa, Madhya Pradesh and throughout South India. It is systematically cultivated in the North Indian states of Rajasthan, Uttar Pradesh, Punjab and in the South Indian states of Kerala, Tamil Nadu, Karnataka and Andhra Pradesh.

Chemical Constituents

The main constituents are alpha- and beta-vetivone (Champagnat et al., 2006), zizanal, zizaene, khusimol, eudesmol, prezizaene and epizizizanal, and were isolated from the oil (Weyerstahl et al., 2000, Sellier and cazaussus 1991, Martinez et al., 2004). Apart from that roots contain other essential oils like caryophyllene oxide (5.78%), Juniper camphor (3.49%), 2-methylenecholestan-3-ol (7.13%), beta-vatirenene (4.75%), etc.

Medicinal uses

Roots are stimulant, tonic, cooling, stomachic, diuretic, antispasmodic and emmenagogue, and used in fevers, inflammations and irritability of stomach. Essence of the root is used to check vomiting in cholera. Smoke of grass is inhaled to relieve headache (Ghani 2003). Apart from this, it used as insect repellent and soil erosion management tool. Traditionally root paste is used for headaches and leaf paste for rheumatism and sprains. Even tribal people in the subcontinent use different parts of the grass for many of their ailments such as mouth ulcer, fever, boil, epilepsy, burn, snakebite, scorpion sting, rheumatism, etc (Singh and Maheshwari 1983). A decoction of the leaves is recommended as a diaphoretic. The oil is reported to be used as a carminative in flatulence, colic and obstinate vomiting. It is regarded as a stimulant, refrigerant and antibacterial and when applied externally, it

Beta-vetivone   Khusimol   Eudesmol

removes excess heat from the body and gives a cooling effect.

Cosmetic uses

Generally root parts are used for the cosmetic preparations because an essential oil procured from roots are having earthy, sweet, heavy, erotic fragrance. Hence essential oil is used as a fixative in modern perfumery, aromatherapy; treats mature, dry or irritated skin, acne, wounds. It also strengthens the connective tissues. Further it is used in formulation in soaps, lotions, creams, etc. It blends well with the oils of sandalwood, patchouli and rose.

Pharmacological Action

Antibacterial activity of hexane extract of *Vetiveria zizanioides* against wild-type and drug-resistant strains of *Mycobacterium smegmatis* and *Escherichia coli* using disk diffusion and micro-broth dilution methods. The results showed that *Vetiveria zizanioides* has better inhibitory activity against drug-resistant strains of *M. smegmatis* and *E. coli* (Luqman et al., 2005).

Crude methanolic extracts of the six varieties of *V. zizanioides* roots showed antifungal activity against *Trichophyton mentagrophytes* at 1% W/V. Some varieties showed antibacterial activity against *Staphylococcus aureus* ATCC 25923, *Escherichia coli* ATCC 25922 and *Pseudomonas aeruginosa* ATCC 278533 at 10% W/V using the agar well diffusion method (Putiyanan et al., 2005).

Antioxidant capacities of vetiver (*Vetiveria zizanioides*) oil were evaluated by many researchers using different assays. Two *in vitro* assays like the DPPH• free radical scavenging assay and the Fe^{2+}-metal chelating assay were used to establish an antioxidant activity of vetiver oil extracted from root of *Vetiveria zizanioides*. Results showed that the vetiver oil (VO) possessed a strong free radical scavenging activity when compared to standard antioxidants such as butylated hydroxytoluene (BHT) and α-tocopherol (Hyun-Jin Kim et al., 2005).

The research work revealed the *Vetiveria zizanioides* root extract used for kidney and other urinary problems (Lans, 2006).

The significant antifungal activity was exhibited by essential oil extracted from *Vetiveria zizanioides* and it showed dose-dependent inhibitory effect of mycelial growth against the soil-borne pathogens, viz. *Rhizoctonia bataticola* and *Sclerotium rolfsii* by poisoned food technique (Sharma et al., 2009).

The generation of free radicals O_2, H_2O_2, OH and NO were effectively scavenged by the ethanolic extract of *V. zizanioides*. In all these methods, the extract showed strong antioxidant activity in a dose dependent manner and it indicates that *V. zizanioides* scavenges free radicals, ameliorating damage imposed by oxidative stress in different disease conditions and serve as a potential source of natural antioxidant (Subhadradevi et al., 2010).

Insecticidal activity of the root extract of *Vetiveria zizanioides* (L.) in petroleum ether, ethyl acetate, acetone and methanol against four strains of the red flour beetle, *Tribolium castaneum* (herbst) was studied. Experimental extracts were applied on larvae and adult beetles in film residue methods and mortality was recorded after 24 h. In larval bio-assay the highest toxicity was recorded for petroleum ether extract, whereas in adults petroleum ether extract accessible highest toxicity and hence it was concluded that vetiver oil can directly be used for the pest control (Sujatha, 2010).

Antimicrobial activity of root and shoot of an aromatic plant *Vetiveria zizanioides* (recently reclassified as *Chrysopogon zizanioides* L. Roberty), on two pathogenic bacteria *E.coli* (MTCC 443) and *Staphylococcus aureus* (MTCC 737) and two potent pathogenic fungi, *Candida albicans* and *Cryptococcus neoformans* were studied. The root extract showed larger zone of inhibition than leaf extract. But in leaf extract the results showed that the maximum zone of inhibition was found against *E.coli* and *C. neoformans*, respectively. The results of MIC of root and leaf fractions against pathogens were 10 mg/ml and IC_{50} of these fractions varied between 5 and 7.5 mg/ml. The results showed that the extracts of vetiver are pharmacologically important that could be applied for human ailments (Jayashree et al., 2011).

Toxicity Studies

An experimental approach was applied to vetiver oil to elicit toxic responses over an exposure period of 90 days. Acute and subacute toxicity studies were carried out in male and female rats. Acute toxicity determination indicated that vetiver oil has LD_{50} values of 2985.38 mg/kg. The drug is practically non-toxic at oral doses. To carry out the subacute toxicity studies, the oil was administered to rats for 45 and 90 days, and the blood samples drawn from animals were subjected to full biochemical and hematological investigations. It was found that prolonged exposure of rats to vetiver oil exerted a mild hematotoxic effect and may alter lipid metabolism (Tripathi et al., 2006).

Clinical Investigations

A prospective, open, non-comparative, phase III clinical trial was conducted to evaluate the clinical efficacy and safety (short- and long-term) of **"Anti-Dandruff Shampoo"** in the management of dandruff for 35 patients. The shampoo contains polyherbal formulation of the extracts of *Rosmarinus officinalis, Vetiveria zizanioides, Nigella sativa, Santalum album, Ficus bengalensis, Citrus lemon* and oil of *Melaleuca leucodendron*. There was a remarkable decrease in the scaling and itching of scalp after a week's treatment. Subsequently, there was remarkable improvement in the healing of the dandruff lesions from the second week onwards, and after fourth week, there was complete control of dandruff in all the patients (Ravichandran, et al, 2004).

The double-blind, randomized and controlled clinical trial was investigated to evaluate the antimicrobial efficacy and safety (short- and long-term) of PureHands as a **hand sanitizer** for 16 healthcare workers. Alcohol-based polyherbal hand sanitizer was used that contains extracts of *Coriandrum sativum, Citrus limon, Azadirachta indica, Vetiveria zizanioides, Coleus vettiveroides* in ethyl alcohol (w/w 60%). Results revealed that there was a significant reduction in the total microbial load between PureHands group as compared to the placebo group and was due to the synergistic action of its ingredients. The results concluded finally that PureHands is clinically effective and safe for usage as a hand sanitizer, to reduce the risk of nosocomial infections (Kavathekar et al, 2004).

The study was planned to evaluate the efficacy and safety (short- and long-term) of **"Nourishing Baby Oil"** which contains the oil of *Olea europaea*, and the oil extracts of *Sida cordifolia, Withania somnifera, Vetiveria zizanioides* and *Aloe vera*, in total 30 infants, between 1 and 12 months of age. Initially, the baby oil was applied to the back of the elbow in a small quantity, and was kept for 30 minutes. The applied area was observed for any kind of immediate hypersensitivity reaction. The signs and symptoms of immediate skin irritation and delayed hypersentivity reactions were evaluated as per the standard guidelines by the scoring scale. There was no immediate or delayed type of hypersensitivity reaction and the overall compliance to the "Nourishing Baby Oil" was excellent. In this formulation, *Vetiveria zizanioides* acts as moisturizing, soothing agent due to presence of valencene, octadecenamide, hexamethyltetracosahexaene, benzendicarboxylic acid, di-isooctyl ester and terpenoids (Chatterjee et al., 2005).

A prospective clinical trial was conducted for 20 infants, who were suffering from hyperhidrosis, miliaria rubra, and bad body odor, to evaluate the clinical efficacy and safety (short- and long-term) of **"baby powder"**. The "baby powder" is a polyherbal formulation and it contains the oils of *Santalum album, Vetiveria zizanioides* and *Olea europaea*, and the powder of *Yashada* bhasma. The results showed there was a significant improvement in hyperhidrosis, miliaria rubra, and bad body odor in all the babies. There were no clinically significant adverse reactions. These results might have been due to the synergistic activities of the ingredients of the baby powder. Therefore, it may be concluded that the "baby powder" is clinically effective and safe in the management of infantile hyperhidrosis, miliaria rubra, and bad body odor (Chatterjee et al., 2005).

Market Formulations

Dermafex soap: Dermafex soap is a unique soap that effectively manages pH of the skin and checks bacterial and fungal infections, also helps to prevent hand transmitted diseases.

Ingredients

Chakramard (*Cassia tora*) 0.05 %, krishna marich (*Piper nigrum*) 0.05%, dhamasa (*Fagonia arabica*) 0.10%, haldi (*Curcuma longa*), 0.10%, khas (*Vetiveria zizanioides*), 0.10%, kritmala (*Cassia fistula*) 0.10%, majitha (*Rubia cordifolia*) 0.10%, neem (*Azadirachta indica*) 0.10%, tulsi (*Ocimum sanctum*) 0.10%, vasaka (*Adhatoda vasica*) 0.10%, sadafal oil 0.30%, dalchini oil 0.10%, laving oil 0.10%, karanj oil 0.25%, bruhad marichyadi oil 0.25%, edible oil q.s. to 1.00%.

Action

It controls bacterial and fungal infection, strikes at pathogens, relieves itching effective remedy in acne, pimples, pruritus, eczema and skin eruptions. In this formulation the root oil of *Vetiveria zizanioides* acts as antifungal and antibacterial effect and helps to other ingredients to prevent the same. Further it gives soothing effect.

Supplied by: Ban Labs Ltd. Gondal Road, South Rajkot, Gujarat - 360 005, India.

Storage: Store in carton pack.

Dosage: Dermafex soap should be applied as an alternate to general soap during bath or as directed by the physician.

Melaleuca geranium moisturizing soap: It has the essential oils of two melaleucas (*Melaleuca alternifolia* and *Melaleuca ericifolia*), along with geranium (*Pelargonium graveolens*) and vetiver (*Vetiveria zizanioides*) essential oils. This soap is made with pure botanical oils and absorbed into the skin naturally, creating healthy skin. Essential oils increase the soothing, moisturizing benefits.

Ingredients: Saponified *Elaeis guineensis* (palm) oil, Saponified *Cocos nucifera* (coconut) oil, Saponified *Olea Europaea* (olive) fruit oil, vegetable gylcerin, *Lycium barbarum* seed oil, *Avena dativa* (oat) kernel meal, *Simmondsia chinensis* (jojoba) seed oil, *Melaleuca alternifolia* (tea tree) leaf oil, *Pelargoium graveolens* flower oil, *Melaleuca ericifolia* leaf oil, *Aloe barbadensis* leaf juice, *Vetiveria zizanioides* root oil, and *Rosmarinus officinalis* (rosemary) leaf extract.

Action: Vetiveria zizanioides is a potent emollient. In this formulation the root oil of the grass acts as moisturizing, soothing agent due to presence of valencene, octadecenamide, hexamethyl-tetracosahexaene, benzendicarboxylic acid, di-isooctyl ester and terpenoids. Overall it gives synergistic activities with the other ingredients.

Nourishing Baby Oil

Nourishing baby oil is a pure, mild and gentle baby oil. Its daily massage has been shown to benefit a baby's overall growth and development.

Ingredients: Country mallow (*Sida cordifolia*), winter cherry (*Withania somnifera*), Vetiver, khas khas (*Vetiveria zizanioides*), barbados aloe (*Aloe vera*), olive oil (*Olea europaea*).

Indications: Dry skin and undernourishment.

Action: This is enriched with olive oil and vitamins that nourish winter cherry, which improves the skin tone and *Aloe vera* that moisturizes the skin.

Direction: Spread the oil all over the body and gently massage.

Supplied by: Himalaya Drug Company, India.

Purifying Neem Foaming Face Wash

Purifying neem foaming face wash is for regular use for individuals with oily, and pimple prone skin.

Ingredients: Azadirachta indica (neem, nimba), *Curcuma longa* (turmeric, haridra), *Vetiveria zizanioides* (vetiver, ushira).

Action: Purifying neem foaming face wash is a soap-free herbal formulation that gently removes impurities from skin and prevents pimples. Enriched with neem, turmeric and vetiver, neem is well known for its anti-bacterial, antiviral and germicidal activities, helps to control acne, pimples and their recurrences. Turmeric is also known for its antiseptic activity and soothes skin. Vetiver provides coolness and refrigerant activity. It has antibacterial activities.

Direction: Moisten face, apply purifying neem foaming face wash and gently work up lather with a circular motion. Wash off and pat dry. Follow-up with Himalaya's face moisturizing lotion.

Supplied by: Himalaya Drug Company, India.

Patents

The invention relates to an antifungal composition for the treatment of human nails containing extracts of walnut hull, pulverized roots of *Nardostachys jatamansi* or *Vetiveria zizanioides* or *Catharanthus roseus*, polyols, fixed oil, non-ionic emulsifiers, thickening agent plasticizer and base. The invention also relates to a process for the preparation of the above synergistic composition (Bindra et al., 2001).

The present invention includes methods for the treatment and/or prevention of hair loss and methods for the regeneration or restoration of hair growth comprising a step of identifying an individual suffering from or susceptible to hair loss or hair thinning or in need of hair regeneration, and a step of administering an extract of the root of a vetiver

grass. Preferably, the extract is an aqueous extract and is administered topically. Also preferably, the vetiver grass is a subspecies of *Vetiveria zizanioides* and is most preferably *Vetiveria zizanioides* (L.) nash. The present invention also provides a composition, preferably in the form of a lotion, gel, cream, or other suspension, and a distinct chemical compound or class of chemical compounds therein, effective in restoring hair growth, preventing hair loss, and/or reversing the effects of hair thinning. The composition may include an effective amount of a hair loss preventative or hair growth promoting composition isolatable as an extract of the roots of a vetiver grass, together with a pharmaceutically-acceptable topical carrier other than water (Porras and Jochamowitz, 2001).

BIBLIOGRAPHY

1. Bhuiyan Nazrul Islam, Jasim Uddin Chowdhury and Jaripa Begum. Essential oil in roots of *Vetiveria zizanioides* (l.) Nash *ex small* from Bangladesh. Bangladesh j. Bot. 2008, 37(2): 213–215 (December).

2. Bindra RL, Singh, AK, Shawl A S, Kumar S. Anti-fungal herbal formulation for treatment of human nails fungus and process thereof. United States Patent 6296838, Filing Date: 03/24/2000, Publication Date: 10/02/2001.

3. Champagnat, Pascal, Figueredo, Gilles, Chalchat, Jean-Claude, Carnat, Andre Paul and Bessiere, Jean-Marie. A study on the composition of commercial *Vetiveria zizanioides* oils from different geographical origins. J. Essent. Oil Res. 2006, 18(4): 416–422.

4. Chatterjee S, Pramanick N, Chattopadhyay S, Munian K, Kolhapure S.A. Evaluation of the efficacy and safety of "Nourishing Baby Oil" in infantile xerosis. The Antiseptic 2005, 102(4): 179–182.

5. Chatterjee S, Pramanick N, Chat-topadhyay S, Munian K, Kolhapure S.A. Evaluation of the efficacy and safety of "Baby Powder" in

infantile hyperhidrosis, miliaria rubra and bad body odor. The Antiseptic 2005, 102(3): 126–128.

6. Ghani, A. Medicinal plants of Bangladesh with chemical constituents and uses. 2nd ed. Asiatic Society of Bangladesh, Dhaka, Bangladesh, 2003, pp. 425–426.

7. Hyun-Jin Kim, Feng Chen, Xi Wang, Hau Yin Chung and Zhengyu Jin. Evaluation of antioxidant activity of Vetiver (*Vetiveria zizanioides* L.) oil and identification of its antioxidant constituents. *J. Agric. Food Chem.*, 2005, 53 (20): 7691–7695.

8. Jayashree S, Rathinamala J, Lakshmana-perumalsamy P. Antimicrobial activity of *Vetiveria zizanoides* against some pathogenic bacteria and fungi. International Journal of Phytomedicines and Related Industries. 2011; 3(2): 151–156.

9. Kavathekar M, Bharadwaj R, Kolhapure S.A. Evaluation of clinical efficacy and safety of PureHands in hand hygiene. Medicine Update. 2004, 12(3): 49–55.

10. Lans C. A. Ethnomedicines used in Trinidad and Tobago for urinary problems and diabetes mellitus. *Journal of Ethnobiology and Ethnomedicine.* 2006, 2; 45.

11. Luqman S, Srivastava S, Mahendra P. Darokar, and Khanuja S.P.S. Detection of antibacterial activity in spent roots of two genotypes of aromatic grass *Vetiveria zizanioides*. Pharma-ceutical Biology. 2005, 43(8); 732–736.

12. Martinez, J. P., T.V. Rosa, C. Menut, A. Leydet, P. Brat, D. Pallet and Meireles M.A.A. Valorization of Brazilian vetiver (*Vetiveria zizanioides* (L.) Nash ex Small) oil. J. Agr. Food Chem. 2004, 52: 6578–6584.

13. Porras SE, Jochamowitz FG. Hair restorer containing vetiver grass extract. United States Patent 6193976. Application No. 09/318665. Publication date: 02/27/2001. http://www.freepatentsonline.com/6193976.html

14. Putiyanan S, Nantachit K, Bunchoo M, Khantava B, Khamwan C. Phar-macognostic

indentification and anti-microbial activity of *vetiveria zizanioides* (l.) Nash root. Chiang Mai Med Bull. 2005, 44(3):85–90.

15. Ravichandran G, Bharadwaj VS, Kolhapure SA. Evaluation of the clinical efficacy and safety of "Anti-Dandruff Shampoo" in the treatment of dandruff. The Antiseptic 2004, 201(1): 5–8.

16. Sellier, N and Cazaussus A. Structure deter-mination of sesquiterpenes in Chinese vetiver oil by gas chromatographytendem mass spectrometry. J. Chromatogr. 1991, 557: 454–458.

17. Sharma PK, Raina AP, Dureja P. Evaluation of the antifungal and phytotoxic effects of various essential oils against *Sclerotium rolfsii* (Sacc) and *Rhizoctonia bataticola* (Taub). Archives of Phyto-pathology and Plant Protection. 2009, 42(1): 65–72.

18. Singh, K.K. and J.K Maheshwar. Traditional phytotherapy amongst the tribals of Varanasi district U. P. J. Econ. Tax. Bot. 1983, 4: 829–838.

19. Subhadradevi V, Asokkumar K, Uma-maheswari M, Sivashanmugam A.T and Sankaranand R. *In vitro* antioxidant activity of *Vetiveria zizanioides* root extract. Tanzania Journal of Health Research. 2010, 12(02): 1–8.

20. Sujatha S. Essential oil and its insecticidal activity of medicinal aromatic plant *Vetiveria zizanioides* (L.) against the red flour beetle *Tribolium castaneum* (herbst). Asian Journal of Agricultural Sciences. 2010, 2(3): 84–88.

21. Tripathi R, Darpan K, Tripathi A, Rasal V.P, Khan S.A. Acute and subacute toxicity studies on vetiver oil in rats. FABAD J. Pharm. Sci., 2006, 31: 71–77.

22. *Vetiveria zizanioides*, http://www.la-medicca.com/raw-herbs-vetiveria-zizanioides.html

23. Weyerstahl, P., H. Marschall, U. Splittgerber and Wolf D. Analysis of the polar fraction of Haitian vetiver oil. Flavour Fragrance J. 2000; 15: 153–173.

Zingiber officinale

Plant profile

Kingdom: *Plantae* – Plants

Superdivision: *Spermatophyta* – Seed plants

Class: *Liliopsida* – Monocotyledons

Order: *Zingiberales*

Genus: *Zingiber Mill.* – Ginger

Subkingdom: *Tracheobionta* – Vascular plants

Division: *Magnoliophyta* – Flowering plants

Subclass: *Zingiberidae*

Family: *Zingiberaceae* – Ginger family

Species: *Zingiber officinale* – Ginger

Vernacular names

Hindi: Aadrak, **Bengali:** Ada, **Gujarati:** Adrak, **English:** Ginger, **San:** Ardraka, Shundi, Shrngavera, Mahoushadhi, **Marathi:** Alay, **Tel:** allamu, sonthi, **Tam:** Ellam, Inji, **Kannad**a: Shunti, **Malyalam:** Inji

Ayurvedic properties

Rasa: Tikta, Madhura, **Guna:** Lakhu, Rooksha, **Veerya:** Seeta, **Vipaka:** Katu

INTRODUCTION

The creeping herb has thick, branching rhizomes and sturdy, upright stems, with pointed lance-like leaves that are 15 to 30 cm in length. It produces yellow-green flowers, with a deep purple lip and yellow marking. The fruits resemble fleshy capsules. Ginger grows well in fertile, moist and well-drained soil.

Distribution

This is a perennial herb, native of Southern Asia, Jamaica, Nigeria and the West Indies.

The plant is cultivated in Florida, California and Hawaii, China, India and tropical regions. Today ginger is cultivated in all the tropical and subtropical Asian regions and Brazil. Fifty percent of worldwide ginger production is in India. The best quality ginger comes from Jamaica. In India, it has been cultivated in almost all the states. Some reports suggest that the climate conditions of Orissa, West Bengal, North Eastern states and Kerala are more suitable for the growth of *Z. officinale* in India.

Chemical Constituents

The essential oil (1 to 3% of the fresh rhizome) contains mostly sesquiterpenes, e.g. (–)-zingiberene (up to 70%), (+)-ar-curcumene β-sesquiphellandrene, bisabolene and farnesene. Monoterpenoids occur in traces (β-phelladrene, cineol, citral). The major pungent compounds in ginger, from studies of the lipophilic rhizome extracts, have yielded potentially active gingerols, which can be converted to shogaols, zingerone and paradol. The pungent gingeroles degrade to the milder shoagoles during storage; high gingerole content and good pungency thus indicate freshness and quality. The compound 6-gingerol appears to be responsible for its characteristic taste. Zingerone and shogaols are found in small amount in fresh ginger and in larger amount in dried or extracted products.

other painful disorders. Chewing a fresh ginger piece helps reduce these ailments. The herb is an effective remedy for cold and cough. Ginger juice is taken with honey thrice or four times a day for quick relief. Ginger tea is also effective in cough. Fresh ginger juice should be mixed with 1 teaspoon of honey and equal amounts of mint and lime juices. This mixture is effective in treatment of dyspepsia due to indigestion, piles, jaundice and biliousness. Ginger juice should be mixed with honey and fenugreek decoction. It is an excellent remedy to proliferate sweating and reduce fever. It also acts as expectorant in asthma, bronchitis, tuberculosis of lungs and whooping cough. The herb is an excellent painkiller. It helps in cure of headache, toothache and earache. Ginger is pounded and boiled in water and sugar is added. This

Zingerone

Zingiberene

Bisabolene

Paradol

Medicinal uses

Ginger is also a blood thinner, which helps reduce angina episodes by lowering cholesterol. The increase in blood flow helps relieve abdominal cramps and open the pelvis to bring on menstruation. Ginger can be used safely to treat such a wide range of health problems, from simple nausea to arthritis. The roots of the herb have gastrointestinal benefits and carminative and anti-inflammatory properties. Ginger is effective in certain stomach disorders such as colic, spasms, vomiting, dyspepsia, flatulence and

decoction should be taken thrice daily to get relief from painful menstruation. It helps cure impotency, spermatorrhea, involuntary seminal discharge and premature ejaculation.

Cosmetic uses

Ginger, when included in soaps, warms the skin. A stimulant and anti-irritant, its warming, soothing properties are very beneficial to the skin. Sometimes used as a fragrance. Ginger is an excellent ingredient for bath oils and other cosmetics. Commercially, ginger is also used as a fragrance in cosmetics and other products such as air fresheners. It can be diluted with massage oil and applied to skin for warming effect.

Zingiber officinale root oil stimulates detoxification and micro-circulation, purifies blood and lymphatic fluid (expedites the healing of acne), provides anti-bacterial and anti-fungal actions. It is useful as a hair grown inhibitor and a preparation for external use. The water-soluble ginger root extract of the present invention causes less skin irritation because of being substantially free of gingerols.

Ginger on the skin can increase skin's radiance and decrease inflammation that may contribute to conditions such as psoriasis, acne and rosacea. Ginger is a perfect anti-oxidant and it also has amazing anti-inflammation properties.

Pharmacological Action

A study using 144 one-day-old Arbor Acres broilers was conducted to assess the effects of dried ginger root (*Zingiber officinale*) that was processed to particle sizes of 300, 149, 74, 37, and 8.4 µm on growth performance, antioxidant status, and serum metabolites of broiler chickens. Supplementation of ginger at the level of 5 g/kg improved antioxidant status of broilers and the efficacy was enhanced as the particle size was reduced from 300 to 37 µm (Zhang et al., 2009).

Larvicidal activities of the compounds isolated form the roots of Z. *officinale* were performed against the larvae of *Anisakis simplex*. A. *simplex* is a parasitic nematode, which present in fish and other marine mammals. Human get infected by consuming infected raw sea food and diseases is called anisakiasis. The study reveled that the compounds ([10]-shogaol, [6]-shogaol, [10]-gingerol and [6]-gingerol) isolated from the roots of Z. *officinale* kill or reduce spontaneous movement in A. *simplex* larvae. Among all, [10]-gingerol resulted into 100% lethality against the larvae of A. *simplex* (Lin et al., 2010).

Hydroalcoholic extracts of *Curcuma longa*, *Zingiber officinale* and combination of *Curcuma*

longa and *Zingiber officinale* rhizome extracts (1:1) were evaluated for their anthelmintic activity using *Pheretima posthuma* model (Indian earthworm). Three concentrations (10, 20 and 50 mg/ml) of each extracts were used for this study which involved the determination of time of paralysis (vermifuge) and time of death (vermicidal activity) of the worms. Extracts obtained from both rhizomes not only paralyzed but also killed the earthworms. Maximum vermicidal activity was shown in *Zingiber officinale* extract at the concentration of 50 mg/ml. On the basis of the observations, it was concluded that both *Curcuma longa* and *Zingiber officinale* rhizomes extracts bearing a potential anthelmintic property (Singh et al., 2011).

Ginger rhizomes successive extracts (petroleum ether, chloroform and ethanol) were examined against liver fibrosis induced by carbon tetrachloride in rats. The evaluation was done through measuring antioxidant parameters; glutathione (GSH), total superoxide dismutase (SOD) and malondialdehyde (MDA). Succinate and lactate dehydrogenases (SDH and LDH), glucose-6-phosphatase (G-6-Pase), acid phosphatase (AP), 5'-nucleotidase (5'NT), aspartate and alanine aminotransferases (AST and ALT) as well as alkaline phosphatase (ALP), gamma glutamyl transferase (GGT), total bilirubin were estimated. Treatments with the selected extracts significantly increased GSH, SOD, SDH, LDH, G-6-Pase, AP and 5'NT. However, MDA, AST, ALT ALP, GGT and total bilirubin were significantly decreased. Results concluded that the extracts of ginger, particularly the ethanol one resulted in an attractive candidate for the treatment of liver fibrosis induced by CCl_4 (Motawi et al., 2011).

The analgesic activity of the Z. *officinale* oil was evaluated by the acetic acid-induced writhing in mice and hot-plate test in mice. The study reported the significant analgesic effect against chemically and thermally-induced nociceptive pain stimuli in mice (Yong et al., 2011).

The present study was designed to compare the anti-hyperglycemic activity of juice of *Z. officinale* with standard glibenclamide and metformin in alloxan induced diabetic rats. 30 albino rats were randomly divided into 5 equal groups. Fresh juice of *Z. officinale* produced a time dependent decrease in blood glucose level significantly compared to both glibenclamide and metformin. The study shows hypoglycemic action in spite of insulin depletion by alloxan and additional non-insulin related hypoglycemic action could be inferred (Asha et al., 2011).

Toxicological Studies

The toxicity of ginger is generally considered to be negligible. Oral LD_{50} values in various animals of ginger oil exceed 5 gm/kg. In vitro microbial assays have shown both mutagenicity and antimutagenicity for compounds isolated from ginger. The adverse reaction profile of ginger is benign, consonant with its use as a common spice and food (Chen et al., 2007).

Widely differing doses of ethanolic extract of ginger (*Zingiber officinale*) were administered to the animals in acute toxicity test. About 10 adult mice of both sexes were used and 2 mice per group of 5 groups given various doses of the crude extract dissolved in 10% tween 20 solution. This was injected at concentrations of 10, 100, 1000, 2000 and 3000 mg/kg. Injection was carried out intraperitoneally. The acute toxicity tests showed that when ginger extract was administered to the mice at dosages of between 1000 and 2000 mg/kg body weight there were no mortality recorded, however, there was 80% mortality at 3000 mg/kg body weight and above. Thus, the administered extract, dosage used throughout the experiment were non-lethal (Ezeonu et al., 2011).

Clinical Studies

The clinical study was conducted to evaluate the efficacy and safety of Polyherbal formulations (Rumalaya gel) in the management of pain and inflammation associated with the musculoskeletal inflammatory disorders like rheumatoid arthritis, osteoarthritis sprains, spondylosis, post-traumatic stiffness and peri shoulder arthritis. The study was an open, prospective, non-comparative, phase-III clinical trial. A total of 50 patients were selected and advised to apply a small quantity of gel topically to the affected area with gentle massage twice daily. The study observed significant reduction in the mean symptom scores for joint tenderness, joint swelling, joint mobility. The beneficial action may be due to synergistic action of all the ingredients. Finally concluded that rumalaya gel is a safe and an effective in the management of pain and inflammation associated with the musculoskeletal inflammatory disorders like rheumatoid arthritis, osteoarthritis sprains, spondylosis, post-traumatic stiffness and peri-shoulder arthritis (Sharma and Kolhapure, 2005).

Market Formulations

Herbal Rumalaya gel: It is a powerful multi-action topical application, which has potent analgesic and anti-inflammatory effects that offer quick relief from pain. It exerts counter-irritant and rubefacient actions, and ensures better absorption through skin. It restores mobility of the joints. It offers deeper penetration, faster action and quicker relief to guarantee mobility unlimited.

Ingredients: Five-leaved chaste tree (*Vitex negundo*), Himalayan cedar (*Cedrus deodara*), *Zingiber officinale, Boswellia Serrata, Mentha arvensis, Cinnamomum zeylanicum, Pinus roxburghii* and *Gaultheria fragrantissima*.

Action: Pain and inflammation due to musculoskeletal disorders such as sprains, strains, tendonitis, capsulitis and bursitis. It also relief low back pain, neck and shoulder pains, hip and knee joint pains, and muscular pain, muscle stiffness, etc.

Direction: Apply twice to thrice daily in the affected areas with gentle rubbing. Simultaneous use of Herbal Rumalaya tablet ensures complete relief faster.

Supplied by: Himalaya Drug Company, India.

Smoothing Body Refiner

This refiner is an anti-cellulite treatment for all skins, particularly those hiding from the world. Stimulating organic mountain ash, phytoactive ginger and ginkgo increase localised circulation and detoxify, whilst green coffee and bioactive organic fig extract tone for visibly smoother, firmer skin. Bio-active peptides reduce the signs of aging whilst nourishing cupuacu butter keeps skin smooth and supple.

Ingredients

Water (aqua), glycerin (plant sources), ethyl macadamiate (macadamia), alcohol (organic wheat), hexyldecanol (coconut and palm kernel), *Hexyldecyl laurate* (coconut and palm kernel), *Butyrospermum parkii* (shea butter), dipalmitoyl hydroxyproline (amino acid and palmitic acid), sodium hyaluronate (lactic acid and wheat), *Ginkgo biloba* leaf extract, *Zingiber officinale* (ginger) root extract, hydro-xypropyl starch phosphate (corn), xanthan gum (fermented sugar), arginine (plant sources), sclerotium gum (fermented sugar), lactose (milk), *Oryza sativa* (rice) hull powder, natural fragrance (parfum), caramel, *Ficus carica* (fig) bud extract, *Pyrus sorbus* bud extract (mountain ash), *Simmondsia chinensis* (jojoba) seed oil, *Theobroma grandiflorum* seed butter (cupuacu), whey protein (lactis pro-teinum), limonene (essential oil), galactaric acid (apple pectin), *Coffea arabica* (coffee) leaf/seed extract, bifida ferment lysate (milk), milk protein (*Lactis proteinum*), *Helianthus annuus* (sunflower) seed oil, *Rosmarinus officinalis* (rosemary) leaf extract, phenoxyethanol, ethylhexylglycerin, dehydroacetic acid, linalool (essential oil), citral (essential oil), benzyl benzoate (essential oil).

Direction: Apply morning and night to problem areas using firm strokes towards the heart. For best results use after body brushing.

Supplied by: The Hut.com Ltd, UK.

Candied Ginger Body Lotion

This lightweight body moisturiser has the fresh and zingy scent of ginger. It hydrates normal to dry skin leaving it feeling softer and smoother.

Ingredients

Aqua (solvent/diluent), glycerin (humectant), cetearyl alcohol (emulsifier), ethylhexyl palmitate (skin conditioning agent), *Buty-rospermum parkii* (skin-conditioning agent/emollient), glycine soja oil (emollient/skin conditioner), *Prunus amygdalus* dulcis oil (skin-conditioning agent), parfum (fragrance), glyceryl stearate (emulsifier), PEG-100 stearate (surfactant), dimethicone (skin conditioning agent), sesamum indicum seed oil (skin-conditioning agent), *Orbignya oleifera* seed oil (emollient), phenethyl alcohol (fragrance ingredient), caprylyl glycol (skin conditioning agent), *Bertholletia excelsa* seed oil (emollient), carbomer (stabiliser/viscosity modifier), xanthan gum (viscosity modifier), limonene (fragrance ingredient), propylene glycol (humectant), hexyl cinnamal (fragrance ingredient), sodium hydroxide (pH adjuster), disodium EDTA (chelating agent), benzyl salicylate (fragrance ingredient), tocopherol (antioxidant), cinnamal (fragrance ingredient), eugenol (fragrance ingredient), citral (fragrance ingredient), linalool (fragrance ingredient), citronellol (fragrance ingredient), *Zingiber officinale* root oil (natural additive), *Zingiber officinale* root extract (natural additive), citric acid (pH adjuster), caramel (colour), CI 15985 (color), CI 14700 (color).

Direction: Massage into skin and allow to sink in before dressing.

Supplied by: The Body Shop International plc., UK.

Lemongrass and Ginger Botanical Body Soap

It is moisturizing and non-irritating, with dreamy lather and fabulous essential oil fragrances that are exceptionally good for the skin.

Ingredients: Saponified oils of *Cocos nucifera* oil, *Olea Europaea* oil, *Elaeis guineensis fruit* oil, *Elaeis guineensis kernel oil, Ricinus communis* oil, *Cera alba, Butyrospermum parkii butter*, and essential oils of *Cymbopogon schoenanthus, Zingiber officinale*.

Action: Organic lemongrass and ginger is used to invigorate and stimulate body and mind.

Direction: Daily should be used during bath.

Supplied by: W.S. Badger Company, Inc. Gilsum, NH.

Various Preparations

Ginger root, when mixed with oils, can help relieve dry or cracked skin. Grate a 2-inch piece of fresh ginger root. Mix the juice with 2 teaspoons each of vitamin E oil, sesame oil and apricot kernel oil, plus 1/2 cup of cocoa butter. Heat slowly in a pan to melt the cocoa butter and mix the oils into it. After cooling apply the lotion liberally to any areas of dry skin.

Ginger reduces dandruff, as it has antiseptic properties and improves blood flow in the scalp. Juice or finely grate enough ginger to yield 1 tablespoon of juice. Mix the juice with 1 teaspoon of lemon juice and 1 teaspoon of sesame oil. Apply the mixture to the scalp three times each week before shampooing the hair.

The invigorating and rejuvenating properties of ginger root make it an ideal exfoliator. Add 1/2 tablespoon of ground or freshly grated ginger root and 1/2 tablespoon of ground cinnamon to 1 cup of sea salt and 1 cup of almond or olive oil and mix them together. This oil can be used as a facial or body exfoliator to remove dead skin and give a glowing complexion.

Homemade Facial Scrub for Oily Skin

Mix 1 teaspoon of each: sea salt, grated ginger, cinnamon and nutmeg. Add some water to receive smooth paste. Apply to face for 20 minutes. Rinse off.

Homemade Ginger Scrub for Body

Mix 2 glasses of sea salt, 2 tablespoons of finely chopped orange peel, 3 drops of grated ginger. Apply to body, massage in circular motions and rinse off.

Rejuvenating Body Scrub

Mix 1 cup of coffee, 1/2 cup of avocado oil, 1 teaspoon of grated ginger and a few drops of cinnamon leaf oil. Apply to body, massage in circular motions and rinse off. Use it 2–3 times a week.

Storage of Products

All the products should be stored in well closed containers and away from sunlight.

Dosage

For most purposes a typical dose of ginger is 1–4 g daily, taken in divided doses. To prevent motion sickness, it is best to begin treatment 1–2 days before the scheduled trip.

Standardized dose: 75–2000 mg in divided doses with food.

For arthritis pain: 250 mg 4 times daily.

Patents

The invention relates to a cosmetic whose active constituents are specified plant extracts and which has a very good skin-smoothing activity. According to the invention, the skin-smoothing of porous or uneven skin having macroscopically visible elevations and depressions during cosmetic processes occurs without irritating side effects. The cosmetic contains plant extracts from *Bambusa vulgaris, Nymphaea alba, Poterium officinale, Zingiber officinalis, Cinnamomum cassia, Nasturtium officinale* R.Br., *Nelumbo nucifera* Gaertn and contains a powder selected from the group consisting of talcum powder, bamboo powder, kaolin, zinc oxide and mixtures thereof. The inventive cosmetic also contains cosmetic adjuvants, excipients, additional active substances and mixtures thereof (Golz-berner and Zastrow, 2008).

The invention provides a novel herbal composition for treatment of arthritis and inflammation. The herbal composition comprises a therapeutically effective combination of extracts obtained from the plants *Terminalia chebula, Pluchea lanceolata, Desmodium gangeticum, Vitex negunto* and *Zingiber officinale*, optionally along with pharmaceutically acceptable additives. The invention further comprises methods of making the herbal composition and methods of use for the treatment of arthritis and inflammation (Palpu et al., 2009).

The herbal ointment includes about 45 weight percent of herb-infused oil, water or alcohol, or any desired combination thereof, about 30 weight percent purified water, about 10 weight percent emulsifier wax, about 5 weight percent menthol, about 3 weight percent dimethyl isosorbide, about 2 weight percent glycerin, about 2 weight percent hydrogenated methyl abietate, about 0.5 weight percent *lonicera caprifolium* and *lonicera japanica* extract, about 0.5 weight percent tocopherol, about 0.35 weight percent vanillyl butyl ether, about 0.35 weight percent xanthan gum, about 0.3 weight percent citric acid 50% aqueous solution, and about 0.25 weight percent menthyl lactate, the herb-infused oil, water or alcohol including a 1:5 dilution of herbal extracts in a solvent, wherein the herbal extracts are of German Chamomile (*Matricaria recutita*), Valerian (*Valeriana officinalis*), Ginger (*Zingiber officinale* Roscoe), Peppermint (*Mentha×piperita*), Feverfew (*Tanacetum parthenium*), and Lemon Balm (*Melissa officinalis*) (Urschel et al,. 2011).

BIBLIOGRAPHY

1. Asha B, Krishnamurthy KH, Siddappa D. Evaluation of anti-hyperglycaemic activity of *Zingiber officinale* (ginger) in albino rats. J. Chem. Pharm. Res., 2011; 3(1): 452–456.

2. Chen JC, Li-Jiau H, Shih-Lu W, Sheng-Chu K, Tin-Yun H, Chien-Yun H. Ginger and its bioactive component inhibit enterotoxigenic *Escherichia coli* heatlabile enterotoxin-induced diarrhoea in mice. J. Agric. F. Chem. 2007; 55: 8390–8397.

3. Ezeonu CS, Egbuna PAC, Ezeanyika LUS, Nkwonta CG, Idoko ND. Anti-hepatotoxicity studies of crude extract of *Zingiber officinale* on CCl4 induced toxicity and comparison of the extract's fraction D hepatoprotective capacity. Research Journal of Medical Sciences. 2011; 5(2): 102–107.

4. Golz-berner K, Zastrow L. Skin-smo-othing cosmetic based on plant extracts. United States Patent 7338671. Application No. 10/523082. Publication date: 03/04/2008. http://www.freepatentsonline. com/7338671. html

5. http://ayurveda.ygoy.com/2010/08/16/what-are-the-medicinal-uses-of-zingiber-officinale/

6. http://ayurvedicmedicinalplants.com/plants/1188.html

7. http://mdidea.com/products/new/new02105.html

8. http://www.anniesremedy.com/herb_detail27.php

9. http://www.ehow.com/info_8119005_homemade-beauty-uses-ginger-root.html

10. http://www.floridata.com/ref/z/zing_off.cfm

11. http://www.iloveindia.com/indian-herbs/ginger.html

12. http://www.naturalcosmeticsupplies.com/dried-ginger.html

13. http://www.umm.edu/altmed/articles/ginger-000246.htm

14. http://www.uni-graz.at/~katzer/engl/Zing_off.html

15. Lin RJ, Chen CY, Lee JD, Lu CM, Chung LY, Yen CM, Larvicidal constituents of *Zingiber officinale* (ginger) against *Anisakis simplex*. Planta Med. 2010; 76(16): 1852–1858.

16. Motawi TK, Hamed MA, Shabana MH, Hashem RM, Naser AFA. *Zingiber officinale* acts as a nutraceutical agent against liver fibrosis. Nutrition and Metabolism. 2011; 8: 40–51.

17. Palpu P, Venkateswara RC, Raghavan G, Kumar OS, Singh RA, Dayanad RG, Shanta M. Anti-arthritic herbal composition and method thereof. United States Patent 7618663. Application No. 12/019373. Publication date:

11/17/2009. http://www.freepatentsonline.com/7618663.html

18. Sharma A, Kolhapure SA. Evaluation of the efficacy and safety of Rumalaya gel in the management of acute and chronic inflammatory musculoskeletal disorders: An open, prospective, non-comparative, phase-III clinical trial. Medicine update. 2005; 12(10): 39–45.

19. Singh R, Mehta A, Mehta P, Shukla K. Anthelmintic activity of rhizome extracts of *Curcuma longa* and *Zingiber officinale* (zingiberaceae). International Journal of Pharmacy and Pharmaceutical Sciences. 2011; 3(2): 236–237.

20. Urschel MJ, Urschel TL, Moore KD. Herbal Ointment for Musculoskeletal and Joint-Related Conditions. United States Patent Application 20110212193. Application No. 13/034728. Publi-cation date: 09/01/2011. http://www. freepatentsonline.com/y2011/0212193. html

21. Yong-liang J, Jun-ming Z, Lin-hui Z, Bao-shan S, Meng-jing B, Fen-fen L, Jian S, Hui-jun S, Yu-qing Z, Qiang-min X, Analgesic and anti-inflammatory effects of ginger oil, Chinese herbal medicines. 2011; 3(2): 150–155.

22. Zhang GF, Yang ZB, Wang Y, Yang WR, Jiang SZ, Gai GS. Effects of ginger root (*Zingiber officinale*) processed to different particle sizes on growth performance, anti-oxidant status, and serum metabolites of broiler chickens. Poultry Science. 2009; 88; 2159–2166.

Index

Reader's Note

Reader's Note

Reader's Note

Reader's Note